GRADUATE TEXTS IN PHYSICS

Graduate Texts in Physics publishes core learning/teaching material for graduate- and advanced-level undergraduate courses on topics of current and emerging fields within physics, both pure and applied. These textbooks serve students at the MS- or PhD-level and their instructors as comprehensive sources of principles, definitions, derivations, experiments and applications (as relevant) for their mastery and teaching, respectively. International in scope and relevance, the textbooks correspond to course syllabi sufficiently to serve as required reading. Their didactic style, comprehensiveness and coverage of fundamental material also make them suitable as introductions or references for scientists entering, or requiring timely knowledge of, a research field.

Series Editors

Professor Richard Needs
Cavendish Laboratory
JJ Thomson Avenue
Cambridge CB3 0HE, UK
E-mail: rn11@cam.ac.uk

Professor William T. Rhodes
Florida Atlantic University
Imaging Technology Center
Department of Electrical Engineering
777 Glades Road SE, Room 456
Boca Raton, FL 33431, USA
E-mail: wrhodes@fau.edu

Professor H. Eugene Stanley
Boston University
Center for Polymer Studies
Department of Physics
590 Commonwealth Avenue, Room 204B
Boston, MA 02215, USA
E-mail: hes@bu.edu

The Ecole Polytechnique, one of France's top academic institutions, has a long-standing tradition of producing exceptional scientific textbooks for its students. The original lecture notes, the *Cours de l'Ecole Polytechnique*, which were written by Cauchy and Jordan in the nineteenth century, are considered to be landmarks in the development of mathematics.

The present series of textbooks is remarkable in that the texts incorporate the most recent scientific advances in courses designed to provide undergraduate students with the foundations of a scientific discipline. An outstanding level of quality is achieved in each of the seven scientific fields taught at the *Ecole*: pure and applied mathematics, mechanics, physics, chemistry, biology, and economics. The uniform level of excellence is the result of the unique selection of academic staff there which includes, in addition to the best research in its own renowned laboratories, a large number of world-famous scientists, appointed as part-time professors or associate professors, who work in the most advanced research centers France has in each field.

Another distinctive characteristic of these courses is their overall consistency; each course makes appropriate use of relevant concepts introduced in the other textbooks. This is because each student at the Ecole Polytechnique has to acquire basic knowledge in the seven scientific fields taught there, so a substantial link between departments is necessary. The distribution of these courses used to be restricted to the 900 students at the Ecole. Some years ago we were very successful in making these courses available to a larger French-reading audience. We now build on this success by making these textbooks also available in English.

Henri Alloul

Introduction to the Physics of Electrons in Solids

Translated by Stephen Lyle

Prof. Dr. Henri Alloul
Université Paris-Sud XI
CNRS
Laboratoire de Physique des Solides
91405 Orsay CX
Bâtiment 510
France
alloul@lps.u-psud.fr

Original French edition published by Édition de l'École Polytechnique, Palaiseau, France, 2007

ISSN 1868-4513   e-ISSN 1868-4521
ISBN 978-3-642-13564-4   e-ISBN 978-3-642-13565-1
DOI 10.1007/978-3-642-13565-1
Springer Heidelberg Dordrecht London New York

Library of Congress Control Number: 2010932620

© Springer-Verlag Berlin Heidelberg 2011
This work is subject to copyright. All rights are reserved, whether the whole or part of the material is concerned, specifically the rights of translation, reprinting, reuse of illustrations, recitation, broadcasting, reproduction on microfilm or in any other way, and storage in data banks. Duplication of this publication or parts thereof is permitted only under the provisions of the German Copyright Law of September 9, 1965, in its current version, and permission for use must always be obtained from Springer. Violations are liable to prosecution under the German Copyright Law.
The use of general descriptive names, registered names, trademarks, etc. in this publication does not imply, even in the absence of a specific statement, that such names are exempt from the relevant protective laws and regulations and therefore free for general use.

Cover design: WMXDesign GmbH, Heidelberg

Printed on acid-free paper

Springer is part of Springer Science+Business Media (www.springer.com)

Preface

The physics of condensed matter, that is liquid, solid, or amorphous materials, occupies something like 50% of physicists working in fundamental research, and the industrial opportunities in the engineering sciences are very important. The best known applications are in hi-tech, especially communications and information processing (semiconductors, magnetic data storage), but also medical imaging (superconductivity and magnetism). However, this discipline has also led to spectacular progress in the use of more commonplace materials employed in metallurgy (steels, special alloys, composites) or the construction industry (concretes).

By its very nature, condensed matter physics has a broad interface with other scientific disciplines relating to chemistry, biology, and mechanics. And this multidisciplinary aspect is growing steadily, through conceptual and instrumental developments which allow us to tackle ever more complex systems.

Before the advent of modern physics, it would have been difficult to understand the physics of electrons in solids, and this is the key feature of condensed matter physics. Since solids are collections of atoms, it is clear that the quantisation of atomic electron states must play a major role in their properties. The basic concepts of quantum mechanics and statistical physics are thus absolutely essential if we are to understand the macroscopic behaviour of electrons in solids.

Although a whole range of different atoms can be built up from protons, neutrons, and electrons, there is nevertheless a limit to what can be produced. On the other hand, atoms can be associated in infinitely many combinations to make up a solid. But from this apparent lawlessness, a range of original generic types of behaviour emerge, whose properties would be difficult to imagine on the basis of the individual atoms making up the solid. How is it that the electrons circulate freely in some solids (metals), and induce significant magnetic forces in others (magnets)? The great difficulty in predicting behaviour on an a priori basis is a characteristic of this discipline, whence the predominant role of observation.

This is therefore a useful point to specify the spirit of our own teaching, and hence of this textbook. In many countries, teaching traditions have always given pride of place to a formal, and essentially deductive, presentation of the physics, i.e., starting from formal hypotheses and leading up to observable consequences. This deductive approach leaves a purely a posteriori verificational role to observation, and hides the thinking that has gone into building up the models in the first place. Here we shall adopt the opposite approach, which begins with the fact that in science in general, and in solid state physics in particular, the qualitative understanding of a phenomenon is an important step which precedes the formulation of any theoretical development. We thus urge the reader to carry out a careful examination of the deeper significance of experimental observations,[1] in order to understand the need for specific models and carry out realistic approximations.

In this way we can also present the main physical effects without necessarily developing the whole theoretical formalism. This approach is indeed unavoidable, since it is impossible today to explain the properties of all solids within a single theoretical framework. Quite the contrary, in fact, since the main themes discussed in the book, viz., metals, superconductivity, and magnetism, are currently understood through radically different approximations. For this reason, many issues which now occupy researchers can be located precisely at the interface between these themes. The best example is undoubtedly high-temperature superconductivity, discovered in 1986, which involves all three phenomena at the same time!

Another aim of this book is to demonstrate the high level of interplay between fundamental scientific research and the development of modern technologies. Indeed, these physical phenomena underlie many developments that are set to revolutionise technology in the twenty-first century. We have thus decided not to restrict here to the fundamental physical concepts, but to go ahead and introduce those ideas that are essential for describing applications. Note that the nanotechnologies which are so much in the news these days seek to exploit the properties of very small objects, with length scales in the nanometer range. However, it would be hopeless to try to understand the characteristics of nanomaterials without a firm grasp of the properties of larger physical systems and the methods used to study them.

From the earliest times, humans have exploited the properties of materials they found in their natural surroundings, and many of these were solids. Of course, they did not have to wait for the arrival of modern science in order to find uses for them. The natural approach was always to take advantage of some observed behaviour, e.g., the exposed edge of a broken flint for cutting purposes, the ductility of metals for forging tools, or the good conduction of metals for making cooking pots, and so on. The recognition of these properties, even if they may today appeal to elaborate mathematical formulations, in no way requires an understanding of the fundamental underlying reasons. The mechanics of materials is almost always governed by macroscopic constitutive laws based on observation, whose microscopic origins are far from being fully understood. This in no way prevents their use. Magnetism is a case in point. The natural occurrence of rocks able to attract iron has been known since ancient times. The striking magnetic behaviour of iron has been ingeniously exploited in the compass to help explorers find their way. But while observation suffices for simple applications, it was only little by little that the chemical nature of the constitutive elements, the regular structure of crystals, and other microscopic features could eventually be identified.

---

[1] To help things along, a number of questions have been interspersed throughout the book. The answers to these questions can be found at the end of the chapter in which they have been raised.

Preface   vii

The levitation of a magnet by a superconductor provides one of the most amazing manifestations of superconductivity, a source of wonder to all that witness it for the first time, and especially to students who have just themselves synthesised the superconducting ceramic $YBa_2Cu_3O_7$. How can three insulating oxides Y_2O_3, CuO, and BaO react in the solid phase to generate a metallic material which becomes superconducting in liquid nitrogen? This illustrates the fact that, in complex systems, properties are very hard to predict from a straightforward understanding of the separate constituents. The richness of condensed matter physics lies in the experimental revelation of spectacular phenomena arising from complex systems, which leads us to seek rational explanations

Recent Trends in Condensed Matter Physics

The nineteenth and twentieth centuries witnessed a major change in condensed matter physics as developments in metallurgy, then in chemistry, produced more and more artificial materials. For its part, physics tends to investigate natural materials under artificial conditions, in order to characterise certain properties. Physicists try to understand why some materials conduct electricity (Ohm's law!), by studying simple elementary metals. Experimentation can discover novel constitutive laws. Superconductivity is a case in point. It was demonstrated for the first time in 1911 when investigating the properties of metals at very low temperatures. But the theory needed to understand the microscopic origins of the behaviour was not yet available.

The decisive step here was the advent of modern physics in the form of statistical physics and then quantum mechanics, which led to considerable progress. It was found that many types of behaviour are *macroscopic manifestations of microscopic quantum phenomena*. Examples are the quantum origins of magnetism and the relation between the optical properties of solids and the organisation of their electronic energy levels into energy bands, among many others. The process of discovery picked up momentum after the 1950s, with the large-scale expansion of scientific endeavour and the development of cross-disciplinary research: chemical substances provided a host of new types of behaviour for physicists. The rate of fundamental discoveries has held up over the past thirty years, with the quantum Hall effect, scanning tunneling microscopy, high-T_c superconductivity, fullerenes and graphene, experimental observations that led in each case to Nobel prizes for those involved.

In time, these new types of behaviour are put to use in applications. But here scientists no longer limit themselves to understanding, nor engineers to application. The two roles are moving closer and closer together. The discovery of semiconductors was the direct result of a hybrid approach, with applications (the diode and the transistor) following very quickly after the discovery of the new type of behaviour (the junction between two differently doped semiconductors). Likewise gas lasers, and subsequently solid-state lasers, were purely artificial constructions which started life as laboratory curiosities based on a sound understanding of the basic science. The time scales between the discovery of new phenomena and concrete applications are shrinking fast, but still remain rather long (20 years for lasers, and 50 for superconductors). Economic factors are clearly the key, but experience shows that *a deeply novel type of behaviour always leads in the end to some viable application.*

This discussion might suggest that solid-state physics is a purely utilitarian science, but this would be quite wrong. In order to understand a novel observation, sophisticated theoretical models are often required, and these problems are naturally more difficult to solve than those generated by simple systems that prove easier to formulate. The difficulty here comes from trying to understand the macroscopic properties of ever more complex systems. This still involves the use of sometimes simplistic models, but which may be very hard to solve formally. More and more sophisticated instruments are developed to probe the structure and properties of materials on the microscopic scale. One has to deal with problems in which the basic state of the physical system is far from being intuitive, such as superconductivity or the Kondo effect. But novel theoretical ideas and methods are also devised. The theory of the renormalisation group has revolutionised the study of phase transitions, and the notion of disorder has become a subject of theoretical physics. Minimisation methods can determine the energy minimum of a disordered magnetic system, such as a spin glass, but can be broadly generalised to study neural networks or optimisation of the travelling salesman problem.

The style and choice of material in the following lectures on electrons in solids attempts to reflect this kind of approach, which characterises the discipline and which is not generally familiar to students. To achieve this, we have put the emphasis on two themes, namely magnetism and superconductivity, which are manifestations on the macroscopic scale of the quantum properties of solids, and which lead to many current or potential hi-tech applications. Although these same themes have up to now been the subject of detailed independent presentations, it seems judicious here, for historical reasons, to present them in tandem.

Magnetism and Superconductivity

How do magnetism and superconductivity fit in to the general context of condensed matter physics? The two disciplines are clearly distinct, in particular through their history, and have different consequences for technology. Magnetic materials have been known since ancient times, while superconductors were only discovered at the beginning of the twentieth century. The former have well known applications resulting from their behaviour, i.e., strong mutual attraction, and interaction with fields produced by electric currents. The latter also display novel types of behaviour, i.e., zero electrical resistance and diamagnetism, but which are only manifested *under currently unfavourable economic conditions*, viz., low temperatures. Everything would tend to distinguish the industrial impact of these two sciences.

But if one considers their most basic manifestations, the analogy between the magnetic fields produced by currents or by magnets has been common knowledge since the nineteenth century. Even before the end of the 1930s, quantum mechanics had shown that the electronic moments of atoms, like the persistent currents in superconductors, are quantum phenomena. It was soon understood that they result indirectly from electrostatic interactions between electrons. The two scientific disciplines witnessed several novel developments, rewarded by Nobel prizes: Bloch–Purcell–Pound, Mössbauer, Néel, Mott–Anderson–Van Vleck in one, and Bardeen–Cooper–Schrieffer, Josephson–Giaver in the other. Common scientific applications were developed, such as superconducting solenoids used in nuclear magnetic resonance (NMR) in chemistry, physics, and biology, but also instrumentation based on superconductors for work on magnetism (SQUIDs). But there was still no common scientific textbook discussing both disciplines, and lecture courses remained quite separate, reserved for a few hundred specialists. Magnetism is interesting for scientists because it can be used to study more general problems through specific high-performance experimental tools (neutrons, NMR, Mössbauer), while remaining susceptible to simplified and well tested theoretical formulation. In this way, many problems of statistical thermodynamics, such as phase transitions, low-dimensional systems, disorder, spin glasses, and so on, found elegant solutions.

At the beginning of the 1980s, industry had little time for superconductivity, but the applications of magnetism were many and varied. The electrical industrial revolution was made possible thanks to the magnetic materials used in electrical generators and motors. The magnetic recording industry alone represents an annual turnover that can be measured in multiples of 100 million US dollars. But superconductivity has come into the limelight through its association with NMR in *magnetic resonanance imaging* (MRI), bringing together the two disciplines in the first application to affect the general public. But these results were not enough to lead to generalised applications. (At the time, P.G. de Gennes, an original contributor to both fields and future Nobel prizewinner, claimed that these were sciences of the past!)

However, the discovery of high-temperature superconductors in 1986, with the attribution of the Nobel prize to Bednorz and Müller, gave hope for applications in the twenty-first century. Until then, theoreticians working on superconductivity had thought that the critical temperature T_c of superconductivity would not be able to go above 30 K. And yet here was an example of superconductivity with $T_c \approx 150$ K, in systems expected rather to exhibit novel magnetic properties. Understanding this kind of superconductivity has proved to be a tremendous challenge for science. These new superconductors incidentally allow us now to exhibit and popularize the basic superconducting phenomena in simple demonstrations as that depicted in the image of page vii, or using toy levitating trains.

Regarding magnetism, new high magnetisation ferromagnetic materials have been discovered. Magnetic data storage gained little benefit from scientific developments in the 1970s, since increases in data storage density were achieved rather through improved mechanics and accuracy of read heads. But more recently, a new read method has been devised, based on a magnetic phenomenon known as giant magnetoresistance, which has allowed a considerable step forward in this domain. This too was rewarded by the Nobel prize, attributed to A. Fert and P. Grünberg in 2007. This has initiated the novel field known as *spintronics*, which exploits the electron spin in electronic devices, and has also given hope of new applications. It involves new materials and is currently the subject of considerable interest in research centers.

Magnetism and superconductivity thus raise novel problems of a fundamental nature, but also in relation to materials and their applications, and they are now commonly encountered together in the most recent developments. The same materials can be made magnetic or superconducting with the help of minimal chemical adjustments. The instruments used to investigate them are similar, and often even the same. This coexistence can only be strengthened by the advent of nanotechnology, and the trend has reached a point at which it seems essential to bring the two subjects together into a single lecture course for students of science, whether they intend to go into research or engineering. The basic level of knowledge required by the engineer is clearly going to increase in the decades to come!

A Brief Guide to the Course

Why are electrons free to move in some solids, namely metals, but localised in others, namely insulators? To tackle this problem, we shall focus mainly on crystalline solids. The existence of a three-dimensional periodicity in these solids allows one to establish simple rules for the electron energy levels. We shall determine these basic rules using the *independent electron approximation*, i.e., considering only the Coulomb potential of the ions and accounting for the Coulomb repulsion of the electrons through only an average value. In this way, we find that the electron energy levels are distributed over allowed energy bands separated by forbidden bands, or bandgaps. This is shown in Chap. 1 for the simple case of a linear chain of atoms. To deal with 3D solids, some notions of crystallography are introduced in Chap. 2. We can then calculate the energy band structure using some simple approximations. Several experimental methods for determining the characteristics of the band structure are described in Chap. 3. The data displayed in page xiii exemplifies the most efficient technique, constantly improved since the 1990's, with the development of new synchrotrons. This energy level structure is the key to understanding the difference between metal and insulator, and introducing the notion of semiconductor. The physical parameters constraining the conductivity of metals are discussed in Chap. 4. These first chapters of the book already allow some simple applications of methods developed in quantum mechanics and statistical physics. They provide the basis for a subsequent understanding of semiconductors. The specificities of the band structure and transport properties of graphene, for which A. Geim and K. Novoselov have been awarded the 2010 Nobel prize, are highlighted as well.

The results of the first four chapters must be reconsidered *if interactions between the electrons can no longer be neglected*. But this is exactly what happens in superconductors and magnetic solids. Understanding the consequences of these interactions between electrons for the physical properties of the resulting solids is a central problem in our field. It goes well beyond the framework of an undergraduate course, which is the level aimed at here, but we feel it important to provide some kind of overview in the case of simple magnetic and superconducting materials.

Students have already acquired some understanding of phenomena related to magnetism. We have thus chosen to begin with superconductivity as the less familiar of the two disciplines. This should allow a gradual assimilation of the ideas, for we shall then refer back to them on many occasions in the rest of the course. A full discussion of the superconductivity of simple metals, which is now well understood, must be left to the doctoral level. For the present purposes, we shall stick to a limited description based on a historical and experimental presentation. This deliberate choice is used to place the student in the same conditions as those which led to our present understanding of this phenomenon. This is all the more important in that we shall then be able to illustrate an approach that is not altogether natural for students coming from a school education system. In any case, in the specific case of superconductivity, it would be optimistic to carry out a full theoretical presentation without first putting across the subtleties of the physical behaviour of superconductors.

Chapter 5 is thus devoted to the existence of persistent currents, the exclusion of magnetic induction in a superconductor, and the resulting quantisation of the magnetic flux. Chapter 6 studies the thermodynamic properties of superconductors, and the notions of coherence length and mixed phase that follow. Finally, in Chap. 7, we describe the experiments that help us to understand the microscopic origins of superconductivity, and analyse the Cooper calculation which lays the foundation for a full understanding.

In Chap. 8, we briefly discuss the origins of atomic magnetism and the exchange interactions responsible for ferromagnetism and antiferromagnetism. The idea of magnetic anisotropy which explains the existence of magnetic domains and domain walls in ferromagnetic materials is essential for understanding applications of magnetism (Chap. 9). In Chap. 10, we present experimental methods used to characterise the magnetic properties of materials. These reveal a methodological link between magnetism and superconductivity. In both cases, the need to understand physical properties and meet technological requirements has encouraged the development of methods for observing magnetism on microscopic, and even nanoscopic scales. Magnetic resonance detection is introduced in Chap. 11, and leads to a discussion of nuclear magnetic resonance and the importance of hyperfine methods for studying magnetism and superconductivity on the atomic scale. Finally, in Chap. 12, we discuss the elementary excitations (spin waves) in a ferromagnetic material, and also the basic theory of thermodynamic properties close to a phase transition (Landau and Ginzburg–Landau theories).

Preface

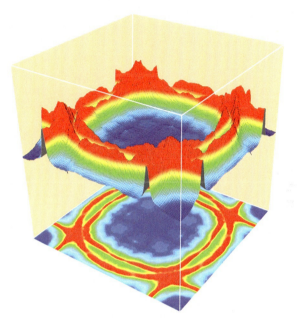

Experimental observation of the electronic states of Sr_2RuO_4 obtained by Angular Resolved Photoemision Spectroscopy (ARPES), a technique that takes benefit from the intense monochromatic light beams produced by synchrotrons. This technique has been intensively developed since the 1990s in particular for the study of high temperature superconductors.

Sr_2RuO_4 is a layered compound whose metallic behaviour is due to RuO_2 planes interleaved with insulating SrO planes. The Ru atoms form a square lattice with oxygen inserted between them. The single crystal sample has been cleaved and exhibits a flat RuO_2 plane. Irradiation by a monochromatic light of sufficient energy emits photoelectrons (this is the photoelectric effect for which A. Einstein has been awarded a Nobel prize). The study of the photoelectron intensity as a function of the emission angle allows one to map out the energy distribution of the electronic levels as a function of their in plane wave vector (k_x, k_y). The projected curves represent the Fermi surface (see question 3.6 in Chapter 3). Image courtesy of A. Damascelli, from experimental results by Damascelli, A., Lu, D.H., Shen, K.M., Armitage, N.P., Ronning, F., Feng, D.L., Kim, C., Shen, Z.-X., Kimura, T., Tokura, Y., Mao, Z.Q., Maeno, Y., Phys. Rev. Lett. **85**, 5194 (2000)

A Guided Choice of Problems

Over the years this lecture course on the physics of electrons in solids has been taught at the Ecole polytechnique (Palaiseau, France), its organisers have produced a significant number of written tests. We have often been led to propose themes focussing on experimental observations, which encourage students to reflect on their physical meaning.

In Chapter 13, we have included those problems which illustrate novel physical effects that are not discussed in detail in the lectures. Indeed, it seemed important to complement the basic ideas discussed in the book with a presentation of the more important aspects of solid-state physics which could not be taught during the year in the limited time available for lectures and supervisions. The aim here is not therefore a mere test of understanding acquired by studying the main course. Students who have independently acquired an understanding of the basic ideas of solid-state physics will be able to use the subjects here to glimpse some of the very active ongoing research themes, illustrated by modern experimental methods.

Direct manifestations of electronic band structure in the optical response of solids are examined in Problem 2: *Reflectance of Aluminium* and Problem 5: *Optical Response of Monovalent Metals*. The total energy of the electronic band states can affect the crystal structure of certain alloys (see the Hume–Rothery rules of Problem 4: *Electronic Energy and Stability of Alloys*) or induce a Peierls transition, which corresponds to a doubling of the unit cell and a metal–insulator transition in 1D compounds (Problem 6: *One-Dimensional TTF-TCNQ Compounds*).

We have illustrated the specific band structures of certain metals that are important for their superconducting properties, such as the high-T_c cuprate $YBa_2Cu_3O_7$ (Problem 3: *Band Structure of $YBa_2Cu_3O_7$*), the so-called A15 compounds, such as V_3Si in which the vanadium atoms are arranged in chains (Problem 14: *Electronic Structure and Superconductivity of V_3Si*), or the new superconducting compound MgB_2, whose highly 2D electronic structure is similar to that of graphene and graphite (Problem 16: *Magnesium Diboride: A New Superconductor?*).

Since semiconductor technology is often the subject of specific lecture courses, as is the case at the Ecole polytechnique, the discussion here simply reviews the relevance of the band structure for their electronic transport properties. The classic cyclotron resonance experiments are used to determine the effective masses of electrons and holes (Problem 8: *Cyclotron Resonance*). The notion of transition from a Mott insulator to a metal is introduced in such systems by examining the structure of donor electron levels introduced by substituting P in Si. This transition results from competition between Coulomb repulsion at an atomic site, which favours electron localisation, and the tendency of electrons to delocalise due to hopping integrals between neighbouring sites. This competition can explain the insulator–metal transition observed in Si when it is strongly doped with phosphorus (Problem 7: *Insulator–Metal Transition*).

The notion of atomic vibrations and their quantisation in terms of phonons is barely touched upon in the main course chapters. Further considerations can be found in Problem 9: *Phonons in Solids*, where their effects on the heat capacity of solids and the resistivity of pure metals are exemplified. Their consequences for X-ray diffraction diagrams are also studied in Problem 1: *Debye–Waller Factor*.

Many problems here are devoted to a deeper investigation of the novel properties associated with superconductivity. For example, one can study the effect of flux quantisation on the thermodynamic properties of superconductors (Little–Parks experiment discussed in Problem 10: *Thermodynamics of a Thin Superconducting Cylinder*) and the direct or alternating Josephson effect in zero field (Problem 11: *Direct and Alternating Josephson Effects in Zero Magnetic Field*). The existence of mixed states in a Josephson junction (Problem 12: *Josephson Junction in a Magnetic Field*), and the irreversibility of magnetisation curves induced by interactions between vortices and the surface of a type II superconductor (Problem 13: *Magnetisation of a Type II Superconductor*) provide a better understanding of the physics of vortices. The observation of these mixed states by scanning tunnelling microscopy (Problem 15: *Superconductivity of $NbSe_2$*) provides a window on the excited states of superconductors and the fine structure of the superconducting gap in compounds like $NbSe_2$ or MgB_2 (Problem 16: *Magnesium Diboride: A New Superconductor?*).

The aspects of magnetism dealt with here are the antiferromagnetism of undoped cuprates (Problem 17: *Electronic Properties of La_2CuO_4*) and the magnetic properties of antiferromagnetic materials in the molecular field approximation (Problem 18: *Properties of an Antiferromagnetic Solid*). Finally, the importance of the magnetism of thin films is illustrated by the use of magneto-optical methods to show how they decompose into domains (Problem 19: *Magnetism of Thin Films and Magneto-Optic Applications*), and an investigation of surface effects for the magnetic anisotropy of thin films (Problem 20: *Magnetism of a Thin Film*).

It is hoped that the reader's curiosity will be stimulated by the problem approach to these exciting physical phenomena and the considerable reflection required to explain the experimental observations discussed here. This problem set should allow the reader to glimpse the rich rewards of the scientific method, which relentlessly confronts observation with theoretical models.

Orsay, October 2010 *Henri Alloul*

Contents

1	**The Quantum Mechanics of Solids**			1
	1.1	General Hamiltonian for a Solid		3
		1.1.1	The Need for Approximations	4
		1.1.2	Wave Functions and Pauli Exclusion Principle	7
	1.2	Energy Bands		8
		1.2.1	Tight-Binding Approximation	8
		1.2.2	Linear Atomic Chain	11
		1.2.3	Energy Bands Associated with Tightly Bound Atomic Levels	20
	1.3	Summary		21
2	**Crystalline Solids: Diffraction**			23
	2.1	Crystal Structures		25
		2.1.1	Crystal Lattice	25
		2.1.2	Two-Dimensional Crystals	27
		2.1.3	Three-Dimensional Crystals	29
		2.1.4	Beyond the Perfect Crystal	33
	2.2	Diffraction		35
		2.2.1	General Principles	35
		2.2.2	Reciprocal Lattice	38
	2.3	Determination of Crystal Structures		39
		2.3.1	On the Bragg Diffraction Condition	39
		2.3.2	Diffraction Intensity and Basis of the Primitive Cell	41
		2.3.3	X-Ray Diffraction	42
	2.4	Summary		44
	2.5	Answers to Questions		45
3	**Electronic Structure of Solids: Metals and Insulators**			51
	3.1	Electrons in a Periodic Potential		53
		3.1.1	Bloch's Theorem in Three Dimensions	53
		3.1.2	General Properties of Bloch States	55

xvii

		3.1.3	Electrons in a Weak Periodic Potential	56
		3.1.4	First Brillouin Zone	59
	3.2	Band Structure of Solids	60
		3.2.1	Linear Chain	60
		3.2.2	Band Structure of Two-Dimensional Solids	64
		3.2.3	Three-Dimensional Band Structures	68
	3.3	Metals, Insulators, and Semiconductors	72
		3.3.1	Statistical Physics of Fermions	72
		3.3.2	Real Solids: Conductor or Insulator?	74
		3.3.3	Introduction to the Semiconductor Case	77
		3.3.4	The Special Case of Graphene	78
		3.3.5	Thermodynamic Properties of Metals	80
	3.4	Experimental Determination of Band Structures	84
		3.4.1	Optical Absorption in Insulators	84
		3.4.2	Tunneling Effect and Scanning Tunneling Microscopy	85
		3.4.3	Angle-Resolved Photoemission Spectroscopy	88
	3.5	Summary	90
	3.6	Answers to Questions	92
4	**Electron Transport in Solids**			99
	4.1	Drude Model for Transport in an Electron Gas: Relaxation Time and Collisions		101
		4.1.1	Electrical Conductivity	102
		4.1.2	Thermal Conductivity and the Wiedemann–Franz Law	105
	4.2	Electron Transport in a Fermion Gas		107
		4.2.1	Electrical Conductivity	108
		4.2.2	Thermal Conductivity and the Wiedemann–Franz Law	110
	4.3	Electrons in a Lattice: Dynamics of Bloch Electrons		111
		4.3.1	Group Velocity	111
		4.3.2	Acceleration in Reciprocal Space and Real Space	113
		4.3.3	Electronic Conductivity in a Crystal	115
	4.4	Origin of Collisions		116
		4.4.1	Experimental Observation	116
		4.4.2	Scattering by Lattice Vibrations	118
		4.4.3	Collisions with Impurities and Defects	119
	4.5	Electrons, Holes, and Dopants in Semiconductors		121
	4.6	Electrons, Holes, and Transport in Graphene		126
	4.7	Summary		132
	4.8	Answers to Questions		133
5	**Introduction to Superconductivity**			145
	5.1	Conditions for Superconductivity		147
		5.1.1	Persistent Currents	147
		5.1.2	Critical Field and Critical Current	150

	5.2	Difference Between a Perfect Conductor and a Superconductor: The Meissner Effect	152
		5.2.1 Effect of a Magnetic Field on a Perfect Conductor	152
		5.2.2 Meissner Effect: Exclusion of the Magnetic Field	154
		5.2.3 London Equations and Penetration Depth	155
	5.3	A Macroscopic Quantum Effect	158
		5.3.1 Macroscopic Wave Function and Current	158
		5.3.2 Quantisation of Magnetic Flux	159
		5.3.3 Measuring the Flux Quantum	160
		5.3.4 Josephson Effect	162
		5.3.5 SQUID	165
	5.4	Summary	167
	5.5	Answers to Questions	168
6	**Thermodynamics of Superconductors**		**175**
	6.1	Thermodynamics of Bulk Superconductors	177
		6.1.1 Free Energy of the Superconducting State	178
		6.1.2 Entropy and Specific Heat	179
	6.2	Thin Films and Coherence Length	181
		6.2.1 Thermodynamics of Thin Films	181
		6.2.2 Coherence Length ξ	184
		6.2.3 Mixed State: A Simple Model	185
	6.3	Two Types of Superconductivity	188
		6.3.1 Type I Superconductors	189
		6.3.2 Simple London Model for a Vortex	189
		6.3.3 Type II Superconductivity	191
		6.3.4 Applications of Type II Superconductors	194
	6.4	Summary	195
	6.5	Answers to Questions	196
7	**Microscopic Origins of Superconductivity**		**201**
	7.1	Conventional Metal Superconductors	203
		7.1.1 Superconducting Metals and Alloys	203
		7.1.2 Attractive Interaction Due to Phonons	205
		7.1.3 The Band Gap in the Superconducting State	208
	7.2	Cooper Instability	210
		7.2.1 Electrons in the Conduction Band of a Semiconductor	212
		7.2.2 Extra Electrons in a Metal	213
	7.3	BCS Theory: Experimental Evidence	214
		7.3.1 Ground State and Band Gap	215
		7.3.2 Excited States in the BCS Theory	216
		7.3.3 Electron Tunneling Determination of the Gap	217
		7.3.4 Critical Temperature and BCS Theory	220
		7.3.5 Coherence Length	221

	7.4	High-T_c Superconductors 223
		7.4.1 Cuprates 223
		7.4.2 Other Families of Superconductors 225
	7.5	Summary 226
	7.6	Answers to Questions 227

8 Magnetism of Insulators 231
    8.1 Magnetic Behaviour of Solids................................. 234
    8.2 Magnetism of Atoms 236
       8.2.1 Hydrogen Atom 236
       8.2.2 Multielectron Atoms with Filled Shells 237
       8.2.3 An Atom with a Partially Filled Shell: Carbon 237
       8.2.4 Atoms with Partially Filled Shells: Hund Rules 240
    8.3 Paramagnetism of an Ensemble of Isolated Ions.................. 243
    8.4 Ordered Magnetic States 244
       8.4.1 Interatomic Exchange Interaction....................... 245
       8.4.2 Heisenberg Model 247
       8.4.3 Ferromagnetism: Molecular Field Approximation 249
    8.5 Antiferromagnetism and Ferrimagnetism 252
    8.6 From Insulator Magnetism to Metallic Magnetism 255
       8.6.1 Mott–Hubbard Insulator 256
       8.6.2 Mott Transition and Doped Mott–Hubbard Insulators 258
       8.6.3 Magnetism and Superconductivity 259
       8.6.4 Metallicity and Magnetism in a Band Approach 261
    8.7 Summary 265
    8.8 Answers to Questions 266

9 Magnetic Anisotropy, Domains, and Walls 269
    9.1 Magnetic Anisotropy 272
       9.1.1 Magnetocrystalline Anisotropy 272
       9.1.2 Effect of Anisotropy: Irreversibility 273
    9.2 Dipole Interactions, Demagnetising Fields, and Domains 277
       9.2.1 Demagnetising Fields 277
       9.2.2 Demagnetisation Energy and Magnetic Domains.......... 282
    9.3 Bloch Walls... 285
    9.4 Magnetic Hysteresis Cycles 288
       9.4.1 Bitter Method for Observing Domains 288
       9.4.2 Magnetisation Process and Hysteresis 289
    9.5 Summary 291
    9.6 Answers to Questions 292

10 Measurements in Magnetism: From the Macroscopic to the Microscopic Scale 295
    10.1 Macroscopic Magnetic Measurements......................... 297
       10.1.1 Susceptibility or Magnetisation 298
       10.1.2 Examples of Measurements on Superconductors 301

	10.1.3	Hysteresimeters: Toroidal Geometry 303
10.2	Magnetic Surface Measurements and Magnetic Imaging 304	
	10.2.1	Microscale Measurements and Surface Imaging 305
	10.2.2	Towards Submicron Scales 308
10.3	Summary 312	
10.4	Answers to Questions 312	

11 Spin Dynamics and Magnetic Resonance 317
- 11.1 Dynamics in Magnetism: General Considerations 319
  - 11.1.1 Linear Response and Dissipation 320
  - 11.1.2 Pulse Response and Frequency Response 321
- 11.2 Dynamics in Ferromagnets 322
  - 11.2.1 Low Frequency Losses in Ferromagnets 322
  - 11.2.2 Ferromagnetic Resonance 324
- 11.3 Resonance in the Paramagnetic Regime 331
  - 11.3.1 Resonance for a Thermodynamic Ensemble of Spins 331
  - 11.3.2 Nuclear Magnetic Resonance (NMR) 333
  - 11.3.3 Spin Echoes: Transverse Relaxation 336
  - 11.3.4 Applications 339
- 11.4 Summary 341
- 11.5 Answers to Questions 342

12 The Thermodynamics of Ferromagnets 345
- 12.1 Excited States and Low Temperature Properties.................. 347
  - 12.1.1 Magnons 347
  - 12.1.2 Low Temperature Thermodynamic Properties 349
  - 12.1.3 Experimental Detection of Magnons 351
- 12.2 The Magnetic Phase Transition................................ 352
  - 12.2.1 Mean Field Theory................................... 352
  - 12.2.2 Landau and Ginzburg–Landau Theories 354
  - 12.2.3 Critical Behaviour 357
- 12.3 Summary... 359
- 12.4 Answers to Questions 360

13 Problem Set 363
Problem 1	Debye–Waller Factor 367
Problem 2	Reflectance of Aluminium.............................. 373
Problem 3	Band Structure of $YBa_2Cu_3O_7$ 377
	3.1 Isolated Copper–Oxygen Chain 378
	3.2 Isolated Copper–Oxygen Plane..................... 379
	3.3 Chain and Plane 380
	3.4 Realistic Models of $YBa_2Cu_3O_7$ 381
Problem 4	Electronic Energy and Stability of Alloys 391
Problem 5	Optical Response of Monovalent Metals 399
Problem 6	One-Dimensional TTF-TCNQ Compounds 405
	6.1 Isolated Chains 405

	6.2 Experimental Observations 406	
	6.3 Dimerised Chain 407	
	6.4 Peierls Transition 409	
Problem 7	Insulator–Metal Transition 419	
	7.1 Tight-Binding Method for Hydrogen-Like Orbitals ... 419	
	7.2 Interactions Between Electrons 420	
	7.3 Alkali Elements and Hydrogen 423	
	7.4 Insulator–Metal Transition in Si–P 423	
Problem 8	Cyclotron Resonance 441	
	8.1 Real and Reciprocal Space Paths of an Electron State . 441	
	8.2 A Semiconductor: Silicon 442	
	8.3 Metals 444	
Problem 9	Phonons in Solids 453	
	9.1 Einstein Model 453	
	9.2 Debye Model 453	
	9.3 Experimental Detection of Phonons 455	
	9.4 Thermodynamic Properties 456	
	9.5 Resistivity 457	
Problem 10	Thermodynamics of a Thin Superconducting Cylinder:	
	Little–Parks Experiment 469	
Problem 11	Direct and Alternating Josephson Effects	
	in Zero Magnetic Field 481	
	11.1 Model Josephson Junction 481	
	11.2 Realistic Josephson Junction 481	
	11.3 Josephson Junction in a Microwave Field 483	
Problem 12	Josephson Junction in a Magnetic Field 491	
	12.1 Current Distribution 492	
	12.2 Screening of the Magnetic Field 493	
	12.3 Josephson Plasma Resonance 495	
Problem 13	Magnetisation of a Type II Superconductor 511	
Problem 14	Electronic Structure and Superconductivity of V_3Si 523	
Problem 15	Superconductivity of $NbSe_2$ 531	
Problem 16	Magnesium Diboride: A New Superconductor? 541	
	16.1 Atomic and Electronic Structure of MgB_2 541	
	16.2 Superconductivity of MgB_2 543	
Problem 17	Electronic Properties of La_2CuO_4 557	
Problem 18	Properties of an Antiferromagnetic Solid 563	
	18.1 Preliminaries: The Ferromagnetic Case 563	
	18.2 Antiferromagnetic Transition 564	
	18.3 Susceptibility in the Antiferromagnetic State 565	
Problem 19	Magnetism of Thin Films and Magneto-Optic Applications . 573	
Problem 20	Magnetism of a Thin Film 579	
	20.1 Uniform Magnetisation 580	
	20.2 Non-uniform Situations 581	
	20.3 Detailed Investigation of Non-uniform Situations 583	

Appendix A	**Physical Constants**	593
Appendix B	**Some Useful Functions and Relations**	595
Appendix C	**Standard Notation**	599
Appendix D	**Specific Notation**	601

References 607

Index 613

Acknowledgements

This is a good point to acknowledge that the very spirit and style of these lectures results from a long collaboration with my colleague Heinz Schulz. While teaching a range of different topics to small groups of students, we gradually refined this approach, which looks for a good balance between experimental observations, theoretical developments, and applications. Heinz Schulz actually wrote some parts of the original manuscript for this book, but unfortunately his untimely death brought this joint project to a premature end. Since then, the initial structure of the book has been radically changed as a result of further interaction with students, and a subsequent harmonisation of the overall layout. I have nevertheless retained some trace of his altogether original contributions, especially in Chaps. 1, 8, and 12, in homage to this sadly departed colleague. To a lesser extent, I have also retained some elements of the relevant contribution by C. Hermann, who helped us to set up this lecture series in the first year. I extend my warmest thanks to her.

I would also like to thank Florence Albenque, Silke Biermann, Pierre Ledoussal, and Gilles Montambaux for their critical readings of and comments on certain parts of this manuscript. The French version of this book was prepared in the publishing department of the Ecole Polytechnique at Palaiseau in France. Over the years it has been a great pleasure to work with Mmes M. Digot, M. Maguer, V. Pellouin, and D. Toustou. I thank them for their efficient, patient, and friendly assistance.

The translation of the French manuscript has been ensured by Stephen Lyle. It has been a great pleasure for me to have this opportunity to learn many subtleties of the English language through numerous e-mail exchanges. I am pleased to thank him warmly here for his extremely efficient and friendly collaboration.

Orsay, October 2010 *Henri Alloul*

Heinz Jürgen Schulz (1954–1998)

H. Schulz was a brilliant theoretician in the field of solid state physics, who always paid careful attention to the dialogue between theory and experiment. His scientific contributions are distinguished by the use of elegant and powerful methods of statistical physics. His main interest was in low-dimensional electron systems, in highly competitive areas, hotly debated by the international scientific community, such as one-dimensional organic conductors, high-temperature superconductors, and strongly correlated fermions. Heinz was highly esteemed, especially by the students at the Ecole Polytechnique, for his broad scientific interests and his depth of understanding, but also his ability to listen and his discrete kindness. He was taken from us in November 1998 by a sudden illness which carried him away in less than six months.

Chapter 1
The Quantum Mechanics of Solids

Contents

1.1	General Hamiltonian for a Solid	3
	1.1.1 The Need for Approximations	4
	1.1.2 Wave Functions and Pauli Exclusion Principle	7
1.2	Energy Bands	8
	1.2.1 Tight-Binding Approximation	8
	1.2.2 Linear Atomic Chain	11
	1.2.3 Energy Bands Associated with Tightly Bound Atomic Levels	20
1.3	Summary	21

Image obtained by scanning tunneling microscopy of the surface of a single-crystal TTF-TCNQ (scale 150 Å×150 Å). The vertical chains formed by stacking up organic molecules are only very weakly coupled to one another and hence constitute molecular wires with an electrical conductivity about 10^3 times greater in the vertical direction than in the horizontal directions. See Problem 6: *One-Dimensional TTF-TCNQ Compounds* for more details. Image courtesy of Wang, Z.Z. (LPN, Marcoussis)

In this chapter, we discuss the quantum description of a solid. As we shall see in the next section, the problem is much too complex to be treated in its most general form. We must therefore introduce approximations motivated by physical considerations. The first will be to neglect the dynamics of the atomic nuclei, treating them as fixed in space, and in fact arranged in a regular crystal lattice. The second approximation will be to treat the Coulomb interaction between the electrons only through an average. We shall see that, using these approximations, the basic ideas can be formulated and important physical phenomena can be explained, in particular, the difference between metals and insulators. Our aim here will be not so much a detailed quantitative understanding of the properties of a given material as a qualitative approach to the phenomena that can be used to classify the various types of behaviour. With a more detailed knowledge of a great many solids, these basic ideas can be developed to produce quantitative results in many cases, provided that we have recourse to a sufficient capacity for numerical computation. However, even a brief outline of the arguments and methods involved in this kind of approach goes well beyond the scope of the book.

In later chapters, we shall consider physical phenomena involving highly novel behaviour that cannot be explained within the framework of these approximations. We shall see in particular that the atomic motions must be taken into account in order to explain the resistivity of metals, or the appearance of superconductivity in certain metals. Likewise, magnetism arises in materials as a direct result of the strong local Coulomb interactions between the electrons.

1.1 General Hamiltonian for a Solid

Let us construct the general Hamiltonian for a solid, and hence observe the impossibility of solving the corresponding Schrödinger equation. We thus discover the need to make approximations.

An arbitrary solid can be treated as made up of a very large number N_n of atomic nuclei and N_e of electrons of mass m_0. To simplify the notation, we assume that the nuclei are all identical, with mass M and charge Ze. Since the solid as a whole is electrically neutral, it follows that $N_e = ZN_n$. In principle, the arguments in this section can be generalised to the case where there are several species of atomic nuclei without great difficulty. The Hamiltonian for this system is thus a sum of five terms: the kinetic energies of the nuclei and the electrons, and the Coulomb interactions between the nuclei, between electrons and nuclei, and between the electrons. It therefore has the form:

$$H = \frac{1}{2M}\sum_{i=1}^{N_n} \mathbf{P}_i^2 + \frac{1}{2m_0}\sum_{j=1}^{N_e} \mathbf{p}_j^2 + \frac{Z^2}{2}\sum_{i,j=1, i\neq j}^{N_n} V_c(\mathbf{R}_i - \mathbf{R}_j)$$
$$-Z\sum_{i=1}^{N_n}\sum_{j=1}^{N_e} V_c(\mathbf{r}_j - \mathbf{R}_i) + \frac{1}{2}\sum_{i,j=1, i\neq j}^{N_e} V_c(\mathbf{r}_i - \mathbf{r}_j). \qquad (1.1)$$

Here the $\mathbf{R}_i$ are the positions of the nuclei, the $\mathbf{r}_j$ are the positions of the electrons, and the momentum operators are given by $\mathbf{P} = -i\hbar \nabla_\mathbf{R}$ and $\mathbf{p} = -i\hbar \nabla_\mathbf{r}$ for the nuclei and electrons, respectively. Finally, the Coulomb potential has the well known form

$$V_c(\mathbf{r}) = \frac{e^2}{4\pi\varepsilon_0 |\mathbf{r}|} \ . \tag{1.2}$$

1.1.1 The Need for Approximations

If we could solve the Schrödinger equation with the Hamiltonian (1.1), we could in principle predict all the properties of the system, e.g., its structure, including the arrangement of all the atomic nuclei, its thermodynamic behaviour, and its electrical or thermal transport properties. However, such a full solution is clearly impossible. In a macroscopic solid, there are something like 10^{23} nuclei, and hence a similar number of electrons. Due to the Coulomb interactions, the Hamiltonian cannot be separated into distinct terms only affecting a limited number of independent particles. We must therefore solve an equation in something of the order of 10^{23} variables, which is just not feasible. As a comparison, recall that the problem can be solved for the hydrogen atom because the corresponding Schrödinger equation expressed in polar coordinates (r, θ, ϕ) separates into two differential equations. Even better, they each have analytic solutions, so it is possible in that case to obtain analytic expressions for the wave function [2, Chap. 11]. No such miracle is to be expected here. Indeed, there is not even an analytic solution for the simplest problem after the hydrogen atom, namely the helium atom (with one nucleus and two electrons).

To make any progress here, we are thus compelled to make approximations, motivated if possible by physical considerations. The first thing to note is that the nuclei are generally 10^4–10^5 times more massive than the electrons, except in the case of the lightest elements. For this reason, they will have much slower dynamics than the electrons, and their kinetic energy can thus be neglected to a first approximation. Experience confirms this idea. Indeed, crystallographic studies show that, in most solids, the atoms make up a static, i.e., time-independent, arrangement, with at most very small oscillations about their equilibrium positions.

> In monatomic solids, the only exception to this rule is helium. Owing to its relatively small atomic mass and the weakness of the interatomic forces, the zero point quantum motion of the atoms is large enough to prevent any solidification at atmospheric pressure, and helium remains liquid even right down to the lowest temperatures. However, a moderate pressure can induce solidification, but there remain significant effects due to zero point motion even in this solid state. Quite remarkable differences between the properties of the isotopes ^{3}He (a fermion of spin $1/2$) and ^{4}He (a boson of spin 0) are then observed.
>
> In molecular solids, certain motions persist and play a role in the physical properties. This happens in particular in C_{60} compounds, for example. This molecule has the shape of a football (see Chap. 2) and can rotate or distort, stretching into the shape of a rugby ball, for example.

1.1 General Hamiltonian for a Solid

In this chapter, we shall therefore omit the kinetic energy of the nuclei and assume that they are held fixed at points $\mathbf{R}_i$, determined experimentally. In this case, the Coulomb interaction between the nuclei gives a constant contribution to the energy and can thus be dropped. The Hamiltonian (1.1) then simplifies to

$$H_e = \frac{1}{2m_0}\sum_{j=1}^{N_e}\mathbf{p}_j^2 - Z\sum_{i=1}^{N_n}\sum_{j=1}^{N_e}V_c(\mathbf{r}_j - \mathbf{R}_i) + \frac{1}{2}\sum_{i,j=1, i\neq j}^{N_e}V_c(\mathbf{r}_i - \mathbf{r}_j), \quad (1.3)$$

and the only quantum operators remaining are the electron position and momentum operators $\mathbf{r}_j$ and $\mathbf{p}_j$, respectively (hence the subscript e on the Hamiltonian), since the positions of the nuclei are now fixed parameters. In the following, we shall be concerned only with crystalline solids, so the $\mathbf{R}_j$ will be arranged in a regular way on the crystal lattice.

But the problems raised by the Hamiltonian (1.3) are almost as intractable as those raised by the original problem. There are of the order of 10^{23} electron variables, and owing to the Coulomb interaction between the electrons, H_e still cannot be separated into a sum of terms involving only a small number of electrons.

We could clearly separate the electron variables by simply neglecting the Coulomb interaction term between the electrons. Under this assumption, consider the situation in which an extra electron is added to the globally neutral solid containing N_e electrons. By neglecting the interaction term between the electrons and keeping the term in $-ZV_c$ in (1.3), this extra electron will be subjected to the potential of the enormous total positive charge $ZN_n e$, a situation which obviously bears no resemblance to the actual physical situation.

We can only separate the electron variables and (partially) solve this problem by making a more careful approximation, wherein the electron–electron interactions are treated by means of an average. A given electron is no longer subjected to a potential depending on the instantaneous positions of all the other electrons, but rather to a potential corresponding to the average distribution of the electrons. Such an approximation raises no major difficulties if the electrons are not too close together, but it becomes critical if they may approach each other. It is well justified when the kinetic energies of the electrons are greater than their average potential energy, but very often this criterion can only be accepted with hindsight, when the results of simulations or experiment provide some a posteriori justification.

> It is interesting to note here that exact solutions of the N-body problem of the kind posed by the Hamiltonian (1.3) with $N = N_e$ do in fact exist for *one-dimensional systems*. Restriction to one dimension may seem to make the discussion somewhat academic. However, it should be borne in mind that there are in fact many experimental systems that come rather close to this situation: quasi-1D conductors or magnetic compounds, i.e., with highly anistropic structures, 1D semiconducting nanostructures called *quantum wires*, *carbon nanotubes*, and many others. The exact solutions to the N-body problem in the 1D model often provide valuable information about the physics of these materials. Note, however, that they involve difficulties of their own. The underlying idea of the method for obtaining almost all the exact solutions currently known

was developed by Hans Bethe in 1931, along with the first applications. The first results that could be directly compared with certain experimental observations were obtained in the 1960s, and this subject is still one of the main themes of theoretical physics today.

For our present purposes, averaging the electron–electron interactions means replacing (1.3) by the separable Hamiltonian

$$H_{\mathrm{e}} = \frac{1}{2m_0}\sum_{j=1}^{N_{\mathrm{e}}}\mathbf{p}_j^2 + \sum_{i=1}^{N_{\mathrm{n}}}\sum_{j=1}^{N_{\mathrm{e}}} V_{\mathrm{at}}(\mathbf{r}_j - \mathbf{R}_i)$$

$$= \sum_{j=1}^{N_{\mathrm{e}}} \left[\frac{\mathbf{p}_j^2}{2m_0} + \sum_{i=1}^{N_{\mathrm{n}}} V_{\mathrm{at}}(\mathbf{r}_j - \mathbf{R}_i) \right] = \sum_{j=1}^{N_{\mathrm{e}}} H_j, \quad (1.4)$$

where we have replaced the Coulomb electron–nucleus and electron–electron interaction potentials affecting a given electron by the single term V_{at}, or *atomic potential*, so that it becomes possible to separate H_{e} into a sum of N_{e} independent terms. In this case, an extra electron added to the solid will 'feel' a total charge of zero, since the positive nuclear charges and negative electron charges now balance one another.

What can be said about the form of V_{at}? If the extra electron is indeed to feel a globally zero charge, this potential must correspond, via Poisson's equation, to a zero total charge. Using Fourier transforms, it can be shown that any potential falling off faster than $1/r$ at large distances satisfies this constraint. To be more precise, note that V_{at} represents the joint effect of the nuclei and the average distribution of the Z electrons associated with it. This gives rise to an exponential decrease of V_{at} at large distances. On the other hand, when an electron is located very close to the nucleus, it will essentially feel the full nuclear charge. At short distances we will thus find $V_{\mathrm{at}}(\mathbf{r}) = -Ze^2/4\pi\varepsilon_0 r$. The full form for V_{at} is thus strongly dependent on the relevant element, *but also on the solid in which it finds itself*. We shall not attempt here to discuss the exact form, since it will have little importance in the following. The only relevant point is that it is attractive at short range and thus allows bound states. We shall also assume that it is invariant under rotations, i.e., $V_{\mathrm{at}}(\mathbf{r}) = V_{\mathrm{at}}(|\mathbf{r}|)$.

To illustrate how the precise form of V_{at} might be determined in a specific example, consider the case of a single atom in the context of the simplest approximation, where we assume that a given electron interacts with the other electrons only through their average density. We thus have

$$V_{\mathrm{at}}(\mathbf{r}) = \frac{e^2}{4\pi\varepsilon_0}\left[-\frac{Z}{r} + \int d^3 r'\, \frac{\rho(\mathbf{r}')}{|\mathbf{r}-\mathbf{r}'|}\right], \quad (1.5)$$

where the first term represents the interaction with the nucleus and the second the interaction with the average electron density $\rho(\mathbf{r}')$ at the point $\mathbf{r}'$. We must now solve the Schrödinger equation with the potential (1.5). This clearly brings out the complexity of the problem. The density ρ is given by

$$\rho(\mathbf{r}) = \sum_k |\psi_k(\mathbf{r})|^2, \quad (1.6)$$

1.1 General Hamiltonian for a Solid

where the wave functions ψ_k are themselves determined by the Schrödinger equation, and the sum is over the occupied states of the atom. We thus do not know the form of the potential a priori. It must be determined in some self-consistent way. For example, we can use an iterative method. We begin with a first approximation for V_{at}, by calculating the ψ_k^0 when there is no electron–electron interaction, then solve the Schrödinger equation with ρ given by (1.6) for the wave functions ψ_k^0. This leads to another set of eigenfunctions ψ_k^1, and these can be used with (1.5) and (1.6) to obtain a new Schrödinger equation. This process continues until it converges on a solution to the required accuracy. A procedure of this kind clearly involves a considerable degree of numerical computation, but one that is perfectly accessible to today's computers, at least in the case of a single isolated atom. These calculations nevertheless become much more time consuming if we have to treat large solids.

The approximation (1.5) is called the *Hartree approximation*. It does not take into account the electron spins, and does not satisfy the Pauli principle (see Sect. 1.1.2). The analogous approximation taking into account the Pauli principle, known as the *Hartree–Fock approximation*, requires even greater computation time. In solid state physics today, the most efficient method, and hence the one most commonly used, is based on the *density functional theory* (DFT). All such methods involve large amounts of computation time, and we shall not try to describe them here.

1.1.2 Wave Functions and Pauli Exclusion Principle

In the second line of (1.4), we obtained a separable form of H_e, written as a sum of identical terms, each one of which refers only to the coordinates and momentum of one electron. The Hamiltonian (1.4) can thus be treated quite simply if we can determine the eigenstates $\psi_k(\mathbf{r})$ and energy eigenvalues E_k of the one-electron Schrödinger equation corresponding to the general term

$$H_1 = \frac{\mathbf{p}^2}{2m_0} + \sum_{l=1}^{N_n} V_{\text{at}}(\mathbf{r} - \mathbf{R}_l) . \tag{1.7}$$

A general solution of the Schrödinger equation (1.4) for the N_e electrons has the form (see, for example, any textbook on statistical physics; this is the form of the wave function used in the Hartree approximation)

$$\Psi(\mathbf{r}_1, \mathbf{r}_2, \ldots, \mathbf{r}_{N_e}) = \psi_{k_1}(\mathbf{r}_1)\psi_{k_2}(\mathbf{r}_2)\ldots\psi_{k_{N_e}}(\mathbf{r}_{N_e}) , \tag{1.8}$$

with energy eigenvalue

$$E = \sum_{i=1}^{N_e} E_{k_i} . \tag{1.9}$$

However, the function in (1.8) does not satisfy the Pauli exclusion principle. According to this, since electrons are fermions, their wave functions must be antisymmetric when the positions of any two electrons are exchanged. Although this is not the case in (1.8), the problem can be corrected by replacing (1.8) by the Slater determinant (the wave function used in the Hartree–Fock approximation [2]):

$$\Psi(\mathbf{r}_1,\mathbf{r}_2,\ldots,\mathbf{r}_{N_e}) = \frac{1}{\sqrt{N_e!}} \begin{vmatrix} \psi_{k_1}(\mathbf{r}_1) & \psi_{k_1}(\mathbf{r}_2) & . & . & . \\ \psi_{k_2}(\mathbf{r}_1) & . & . & . & . \\ . & . & . & . & \psi_{k_{N_e-1}}(\mathbf{r}_{N_e}) \\ \psi_{k_{N_e}}(\mathbf{r}_1) & . & \psi_{k_{N_e}}(\mathbf{r}_{N_e-1}) & \psi_{k_{N_e}}(\mathbf{r}_{N_e}) \end{vmatrix} . \quad (1.10)$$

This function corresponds to the same energy eigenvalue (1.9) as the not yet antisymmetrised function in (1.8). Note that the determinant structure of this function is what ensures the antisymmetry of the wave functions and hence satisfaction of the Pauli exclusion principle. Indeed, if two of the quantum numbers k_j were the same, the corresponding rows of the determinant would be identical, hence (1.10) would not be an admissible wave function.

1.2 Energy Bands

The next step in the solution of this problem, to determine the electron states of the macroscopic solid, is therefore to find the one-particle wave functions $\psi_k(\mathbf{r})$. Of course, it would be impossible to solve the Schrödinger equation for the Hamiltonian (1.7) without knowing the exact form of the potential V_{at}. However, what interests us here is to understand what happens to the electron states of the atoms when they are put together to form a solid. The situation will be radically different depending on whether we consider the states corresponding to electrons that are strongly bound to the nucleus or electrons that are only weakly bound. These two cases correspond respectively to strong and weak potentials V_{at}, respectively. In this book, we shall consider these two simple limits, dealing with the first in the remainder of this chapter and the second in Chap. 3. When the electrons are strongly bound, we speak of the tight-binding approximation. Note that this refers to the binding of the electrons to the nucleus, not to chemical bonds between the atoms.

1.2.1 Tight-Binding Approximation

In this context it is useful to consider the solid as a chemist might view it, as an ensemble of rather isolated atoms, interacting only weakly together. A quantum treatment begins by solving the Schrödinger equation for a single atom:

1.2 Energy Bands

$$\left[\frac{\mathbf{p}^2}{2m_0} + V_{\text{at}}(\mathbf{r})\right] \chi_n(\mathbf{r}) = E_n \chi_n(\mathbf{r}). \tag{1.11}$$

Consider the case where V_{at} corresponds to the potential of a neutral 'atom' whose lowest eigenstates with energies $E_1, E_2, \ldots, E_n, \ldots, E_v$ are occupied by the Z electrons (see Fig. 1.1a). To fix ideas, we consider only those electron states of the solid resulting from the last occupied electron state of the neutral atom (the valence state), with energy E_v and wave function χ_v (see Fig. 1.1b).

Let $\chi_v(\mathbf{r} - \mathbf{R}_n)$ be the eigenfunctions of degenerate energy E_v corresponding to the atomic states centered on the various sites $\mathbf{R}_n$. The tight-binding hypothesis assumes that the eigenfunctions $\psi_k(\mathbf{r})$ of the system of N_n atomic potentials centered on $\mathbf{R}_n$ have energy eigenvalues E_k that differ only slightly from E_v, but are well removed from the energies E_{v-1} and E_e of the closest atomic states of the neutral atom.

In this case, it seems reasonable to assume (and it can be checked after the event) that the functions $\psi_k(\mathbf{r})$ should be expressible as linear combinations of the atomic orbitals $\chi_v(\mathbf{r} - \mathbf{R}_n)$ (or LCAOs):

$$\psi_k(\mathbf{r}) = \frac{1}{\sqrt{N_n}} \sum_{l=1}^{N_n} a_{k,l} \chi_v(\mathbf{r} - \mathbf{R}_l),$$

corresponding to the quantum states

$$|\psi_k\rangle = \frac{1}{\sqrt{N_n}} \sum_{l=1}^{N_n} a_{k,l} |\mathbf{R}_l\rangle. \tag{1.12}$$

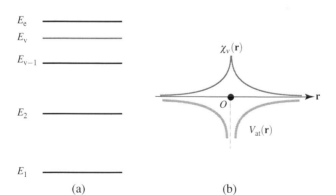

Fig. 1.1 (a) The eigenstates of the Hamiltonian (1.11) with energies $E_1, E_2, \ldots, E_v$ are occupied by the Z electrons of the atom. The level E_e is the energy of the first excited state. (b) Attractive atomic potential V_{at} centered at $\mathbf{r} = 0$, and schematic representation of the wave function $\chi_v(\mathbf{r})$ corresponding to the energy E_v

Here, the k are the quantum numbers of the eigenstates of the solid, while the $a_{k,l}$ are the coefficients of the linear combination, which remain to be determined from the Schrödinger equation. In (1.12), we have introduced the ket $|\mathbf{R}_l\rangle$ associated with the wave function $\chi_v(\mathbf{r} - \mathbf{R}_l)$.

In the rest of this book, the exact form of $\chi_v(\mathbf{r})$ will be of little importance. Our aim will be only to describe the method used to determine eigenfunctions and eigenvalues for simple situations to begin with, then for more realistic cases. We will study the case of a non-degenerate state, for which χ_v will be real and even, i.e., $\chi_v(\mathbf{r}) = \chi_v(-\mathbf{r})$.

Schrödinger Equation in the LCAO Approximation

We now seek the eigenfunctions and eigenvalues of the Hamiltonian (1.7) in the context of the tight-binding approximation. To do this, we first transform the Schrödinger equation, inserting the form (1.12) of the wave function in $H_1|\psi_k\rangle = E_k|\psi_k\rangle$ to obtain

$$\sum_{l=1}^{N_n} a_{k,l} H_1 |\mathbf{R}_l\rangle = \sum_{l=1}^{N_n} a_{k,l} E_k |\mathbf{R}_l\rangle . \qquad (1.13)$$

We now separate the potential into two parts, one due to the atom at site l and the other due to all the other sites. Combining (1.7) and (1.11), this leads to

$$H_1 |\mathbf{R}_l\rangle = E_v |\mathbf{R}_l\rangle + V_l(\mathbf{r}) |\mathbf{R}_l\rangle , \qquad (1.14)$$

where we have defined

$$V_l(\mathbf{r}) = \sum_{m=1, m \neq l}^{N_n} V_{\text{at}}(\mathbf{r} - \mathbf{R}_m) . \qquad (1.15)$$

Using (1.14), we may rewrite (1.13) in the form

$$\sum_{l=1}^{N_n} a_{k,l} \Big[E_v |\mathbf{R}_l\rangle + V_l(\mathbf{r}) |\mathbf{R}_l\rangle \Big] = \sum_{l=1}^{N_n} a_{k,l} E_k |\mathbf{R}_l\rangle . \qquad (1.16)$$

In order to simplify the calculation, we begin by assuming that the $|\mathbf{R}_l\rangle$ form an orthonormal set, i.e.,

$$\langle \mathbf{R}_n | \mathbf{R}_l \rangle = \delta_{n,l} , \qquad (1.17)$$

where δ is the Kronecker symbol ($\delta_{l,m} = 1$ if $l = m$, $\delta_{l,m} = 0$ otherwise). Multiplying (1.16) on the left by $\langle R_n |$ and introducing the transfer (or *hopping*) integrals

$$-t_{n,l} = \langle \mathbf{R}_n | V_l(\mathbf{r}) | \mathbf{R}_l \rangle , \qquad (1.18)$$

1.2 Energy Bands

we obtain

$$-\sum_{l=1}^{N_n} t_{n,l} a_{k,l} = (E_k - E_v) a_{k,n} . \quad (1.19)$$

Note that the choice of sign in (1.18) gives $t_{n,l} > 0$ in most cases. The N_n equations (1.19) (the value of n can vary from 1 to N_n) constitute a homogeneous linear system in the N_n variables $a_{k,n}$. Such a system has a solution only if $E_k - E_v$ is an eigenvalue of the matrix with entries $-t_{n,l}$. We have thus transformed the initial Schrödinger equation into a matrix eigenvalue problem.

1.2.2 Linear Atomic Chain

The above argument applies in practice to any type of solid. However, many properties of electron states in a solid can already be understood through the simplest possible example, namely a linear chain of atoms. This example may seem rather academic, but the generalisation to more realistic cases will be greatly simplified by a prior discussion of this case, which will allow us to simplify the notation. The 1D situation is clearly also of some practical interest for understanding the properties of the quasi-1D conductors mentioned earlier.

In the linear chain, the atomic positions are specified by the vectors $\mathbf{R}_m = R_m \mathbf{x}$, where $\mathbf{x}$ is a fixed unit vector in the x direction and the R_m are scalars given by

$$R_m = ma . \quad (1.20)$$

The distance a between nearest neighbour atoms (see Fig. 1.2) is usually called the *lattice constant*, for reasons to be discussed in more detail in Chap. 2.

In the special case of the linear chain, it is easy to give a graphical representation of the total potential in (1.7), which is the sum of the atomic potentials centered on the sites of the chain (see Fig. 1.3a). The potential $V_l(\mathbf{r})$ deduced by simply suppressing the term $V_{at}(\mathbf{r} - \mathbf{R}_l)$ is shown in Fig. 1.3b. The wave functions centered on n and l, and the potential $V_l(\mathbf{r})$ used to calculate the transfer integral $t_{n,l}$ in (1.18), are shown in Fig. 1.3c.

a. Periodic Boundary Conditions and Bloch's Theorem

A general analysis of (1.19) is neither straightforward nor useful without knowing the coefficients $t_{n,l}$. Even if all these coefficients are known, the solutions remain

Fig. 1.2 Linear chain of atoms (*dots*) with lattice constant a

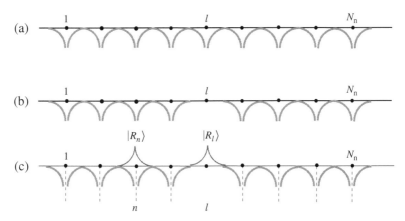

Fig. 1.3 (a) Periodic potential of (1.7). (b) Potential $V_l(r)$. (c) Elements used to calculate $t_{n,l}$ with the help of (1.18)

complicated for a finite chain of N_n sites, due to effects arising from the endpoints of the chain. We shall therefore slightly modify the geometry of our chain by forming a ring, as shown in Fig. 1.4a, in such a way that the site N_n ends up at distance a from site 1. In general, one would expect this different geometry to lead to different properties as compared with the finite chain. However, in the *thermodynamic limit* where N_n is very large, which is practically the only one we shall be concerned with, the angle between consecutive bonds tend to zero and the two systems become equivalent. The chain is then invariant under translation through whole multiples of the constant a.

This approach is equivalent to imposing *periodic boundary conditions* on the finite chain (see Fig. 1.4b). These Born–von Kármán conditions [3, Chap. 7] are used to make the system periodic by extending the wave functions in such a way that $\psi_k(x+N_n a) \equiv \psi_k(x)$, or

$$a_{k,n+N_n} = a_{k,n}. \tag{1.21}$$

The translation invariance of the chain under shifts through integer multiples of a, and the symmetry properties of V_{at} and χ_v chosen above imply that the coefficients $t_{n,l}$ depend only on the distance between the sites n and l:

$$t_{n,l} = t_{|n-l|}. \tag{1.22}$$

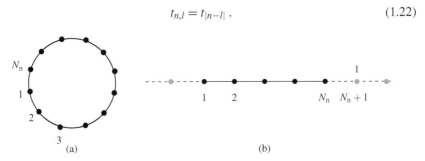

Fig. 1.4 (a) Ring of N_n atoms. (b) Periodic boundary conditions for the chain

1.2 Energy Bands

Formally, these coefficients are only nonzero for sites that are no further apart than $(N_n - 1)a$, but we shall see hereafter that they decrease quickly with distance, so that only a few terms will be retained anyway. In the case of the ring, or the chain with periodic boundary conditions, we can rewrite (1.19) by singling out the site n and collecting together terms associated with equidistant neighbours:

$$-t_0 a_{k,n} - \sum_{m=1}^{N_n/2-1} t_m(a_{k,n+m} + a_{k,n-m}) = (E_k - E_v)a_{k,n},$$

which can be written, after replacing the dummy index m by l

$$-\sum_{l=1}^{N_n/2-1} t_l(a_{k,n+l} + a_{k,n-l}) = (E_k - E_v + t_0)a_{k,n}. \tag{1.23}$$

Before trying to solve (1.23), note that (for the ring, or for the chain with periodic boundary conditions) electrons separated by a distance a feel the same crystal potential, so the Hamiltonian $\hat{H}_1$ in (1.7) is invariant under translation by a. For this kind of periodic potential, Bloch's theorem (see the box below) shows that the eigenfunctions of $\hat{H}_1$, known as Bloch functions, have the general form

$$\psi_k(x) = e^{ikx} u_k(x), \quad \text{with } u_k(x+a) = u_k(x), \tag{1.24}$$

whence the function $u_k(x)$ is itself periodic.

If the form (1.12) for $\psi_k(x)$ is to satisfy (1.24), we therefore require

$$u_k(x) = \frac{1}{\sqrt{N_n}} \sum_{l=1}^{N_n} a_{k,l} e^{-ikx} \chi_v(x - R_l)$$

to be periodic, i.e.,

$$u_k(x+a) = \frac{1}{\sqrt{N_n}} \sum_{l=1}^{N_n} a_{k,l} e^{-ika} e^{-ikx} \chi_v(x - R_{l-1}) \equiv u_k(x),$$

whence

$$a_{k,l} e^{-ika} = a_{k,l-1},$$

which implies

$$a_{k,l} = a_{k,0} e^{ikla}.$$

The periodic boundary condition (1.21) then implies that $e^{ikN_n a} = 1$, which leads to the following quantisation condition for k:

$$\boxed{k = \frac{2\pi m}{N_n a}}, \quad m \text{ integer}. \tag{1.25}$$

Further, normalisation of the wave function ψ_k leads to $a_{k,0} = 1$, and the $a_{k,n}$ can thus be expressed in the general form

$$a_{k,n} = e^{ikR_n} = e^{ikna} . \tag{1.26}$$

Bloch's Theorem

The potential felt by the electron is periodic with $V(x+a) = V(x)$. Let $\hat{T}$ be the elementary translation operator through a distance a, such that

$$\hat{T}f(x) = f(x+a) ,$$

for any function $f(x)$. This operator commutes with the Hamiltonian $\hat{H}_1$, so there is a common basis of eigenstates. The eigenstates of $\hat{H}_1$ can thus be sought amongst those of $\hat{T}$. Since $\hat{T}$ is a unitary operator, the eigenvalue associated with any eigenstate of $\hat{T}$ has unit modulus. The eigenstates of $\hat{T}$ can thus be characterised by an index k which determines their eigenvalue

$$\hat{T}\psi_k = e^{ika}\psi_k .$$

For a state ψ_k to be an eigenstate of $\hat{H}_1$, it must satisfy the periodic boundary conditions. We must therefore have

$$\psi_k(x + N_n a) = (\hat{T})^{N_n} \psi_k(x) = e^{ikN_n a} \psi_k(x) = \psi_k(x) .$$

This leads to the quantisation condition (1.26) for k. Setting $u_k(x) = e^{-ikx}\psi_k(x)$, we find that

$$u_k(x+a) = e^{-ik(x+a)} \psi_k(x+a) = e^{-ikx}\psi_k(x) = u_k(x) .$$

Hence the function $\psi_k(x)$ does indeed have the form (1.24) of a Bloch function.

b. Solutions Under the Nearest-Neighbour Approximation

Equation (1.23) will receive a general solution below. Owing to the fast exponential decline of the wave functions χ_ν with distance, it generally suffices to consider only the nearest-neighbour hopping integrals (see Fig. 1.5):

$$t_n = 0 \quad \text{if } |n| > 1 . \tag{1.27}$$

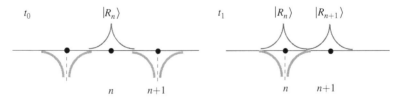

Fig. 1.5 Potentials and wave functions used to calculate the hopping integrals t_0 and t_1 in the nearest-neighbour approximation

1.2 Energy Bands

Equation (1.23) then reduces to

$$-t_1(a_{k,n+1} + a_{k,n-1}) = (E_k - E_v + t_0)a_{k,n} \,. \tag{1.28}$$

The condition (1.26) on the $a_{k,n}$ can be used to determine the energy eigenvalue associated with the states $|\psi_k\rangle$ from (1.28):

$$\boxed{E_k = E_v - t_0 - 2t_1 \cos ka} \,. \tag{1.29}$$

The energy eigenvalues (1.29) are shown graphically in Fig. 1.6 as a function of the quantisation index k. There are several points to note about this result which will also apply in more complex situations, in particular, in the case of dimensions higher than one:

1. The energy levels (1.29) are shifted by t_0 from the discrete atomic level E_v, and also broadened into an *allowed energy band*, with lower bound at $E_v - t_0 - 2t_1$ and maximum at $E_v - t_0 + 2t_1$, so that the total width of this band is $4t_1$. The function E_k of k describes the structure of the energy band (hence the term 'band structure'). The LCAO approximation is only valid if all the states in the band lie below the first excited state E_e of the atomic potential, i.e., we must have $2t_1 \ll E_e - E_v$.
2. The state of energy E_k is given by

$$\boxed{|\psi_k\rangle = \frac{1}{\sqrt{N_n}} \sum_{l=1}^{N_n} e^{ikla} |R_l\rangle} \,. \tag{1.30}$$

We now note that, if we replace k by $k + 2\pi/a$, we obtain the same wave function, i.e., the states $|\psi_k\rangle$ and $|\psi_{k+2\pi/a}\rangle$ are in fact the same, and k is only defined

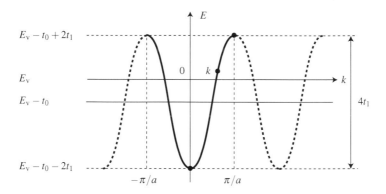

Fig. 1.6 Tight-binding approximation. Energy band arising from the atomic valence level for the linear atomic chain. To represent all the energy states, it suffices to restrict to the first Brillouin zone $(-\pi/a, \pi/a)$

modulo $2\pi/a$. In practice, to represent this periodic function, it is convenient to choose k in the interval (for N_n even)

$$\boxed{-\frac{\pi}{a} < k \leq \frac{\pi}{a}}. \tag{1.31}$$

This interval is known as the *first Brillouin zone*. Clearly, other choices are also possible. The motivation for this particular choice will be spelt out in more detail in Chap. 3. A graphical representation of the band structure in the first Brillouin zone is shown in Fig. 1.6.

3. The form (1.26) for the coefficients $a_{k,l}$ looks like a plane wave, and is reminiscent of the wave function of a free particle, subject to periodic boundary conditions. This is illustrated in Fig. 1.7, which shows the general form of the real part of ψ_k, and the corresponding limiting cases at $k = 0$ and $k = \pi/a$. The first case corresponds to a situation in chemistry where the orbitals are bonding, while the second corresponds to antibonding orbitals. We may therefore wonder whether, as in the case of free particles, there is a direct relation ($p = \hbar k$) between the wave vector k and the momentum p. The same question is raised for any Bloch function. We shall see later that this relation is no longer valid, but that some features of the behaviour of plane waves in vacuum will survive. For this reason, the quantity $\hbar k$ is called the *crystal momentum* or *quasi-momentum*.

4. We should ask whether the states $|\psi_k\rangle$ form a complete set. Owing to the orthonormalisation relation (1.17), the $|\psi_k\rangle$ form an orthonormal set:

$$\langle \psi_{k'} | \psi_k \rangle = \delta_{k',k} ,$$

where δ is taken modulo $2\pi/a$. Furthermore, we have one orbital per atom, so the Hilbert space has dimension N_n. Finally, the quantisation rule (1.25) shows that, in the interval (1.31), there are also N_n states, so we find that the $|\psi_k\rangle$ do indeed constitute a complete orthonormal basis for the Hilbert space.

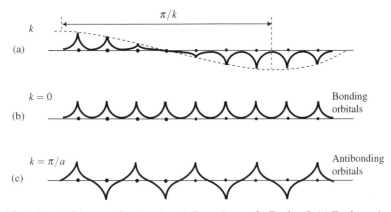

Fig. 1.7 Real part of the wave function ψ_k. (**a**) General case. (**b**) For $k = 0$. (**c**) For $k = \pi/a$

1.2 Energy Bands

c. Density of States

When calculating thermodynamic properties, transport properties such as electrical and thermal conductivity, and many other physical quantities, one often has to evaluate sums of the form

$$G = \sum_{BZ1} g(E_k) = \frac{N_n a}{2\pi} \int_{BZ1} g(E_k) dk, \quad (1.32)$$

where g is an arbitrary function, and the sum is generally over the whole of the first Brillouin zone. To obtain the second equality, we used the fact that, in the thermodynamic limit, the allowed values of k, with spacing $2\pi/N_n a$, almost form a continuum, which justifies replacing the sum by an integral over k. The important point in (1.32) is that the integrand only depends on k through the energy E_k. It is thus convenient to be able to rewrite (1.32) in the form

$$G = N_n \int dE \, g(E) D(E), \quad (1.33)$$

where the *density of states* $D(E)$ is such that $N_n D(E) dE$ represents the number of states with energies between E and $E + dE$ for the solid comprising N_n atoms. While it is already convenient for the 1D case, this ploy will become even more useful in the 2D and 3D situations.

Note that there are various k_i values for which $E(k_i) = E$. For each k_i value the interval of energy dE corresponds to an interval of k values such that $dE = (dE/dk_i)dk_i$. The number of corresponding states close to k_i is therefore

$$\frac{N_n a}{2\pi} |dk_i| = \frac{N_n a}{2\pi} \frac{dE}{|dE/dk_i|}, \quad (1.34)$$

so the total density of states of energy E per atom is

$$D(E) = \frac{a}{2\pi} \sum_{k_i} \frac{1}{|E'_{k_i}|}. \quad (1.35)$$

Let us now consider the linear chain with E_k given by (1.29). If $|E - E_v + t_0| > 2|t_1|$, there is no eigenstate and the density of states is zero:

$$D(E) = 0 \quad \text{if } |E - E_v + t_0| > 2|t_1|. \quad (1.36)$$

If this is not the case, there are two symmetric solutions for k_i such that

$$\cos k_i a = -\frac{E - E_v + t_0}{2t_1}.$$

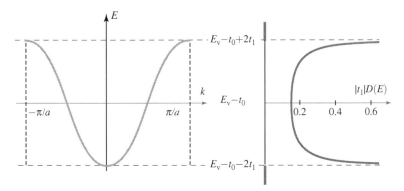

Fig. 1.8 Dispersion relation (*left*) and density of states (*right*) of the linear atomic chain in the LCAO approximation

As $|E'_{k_i}| = 2a|t_1|\sin k_i a = 2a|t_1|\sqrt{1-\cos^2 k_i a}$, it follows from (1.35) that

$$D(E) = \frac{1}{2\pi|t_1|}\frac{1}{\sqrt{1-(E-E_v+t_0)^2/4t_1^2}} \quad \text{if } |E-E_v+t_0| < 2|t_1|. \quad (1.37)$$

The density of states is plotted versus the energy in Fig. 1.8. Note the square root divergences at the band edges. These divergences arise for the same reason as those in the free electron gas model where $D(E) \propto E^{d/2-1}$, with d the space dimension (see Chap. 3).

Generalisations

In the last section, we made two approximations. The first was the nearest-neighbour approximation (1.27), in which we neglect all hopping integrals linking the given site to those beyond its immediate neighbours. The second assumed orthogonality of the orbitals at different sites, expressed by (1.17). If instead we retain only the orthogonality of the orbitals but drop the nearest-neighbour approximation, using the full equation (1.23) rather than (1.28), it is easy to see that the wave functions are unchanged, but that their eigenvalues become

$$E_k = E_v - t_0 - 2\sum_{l=1}^{N_n/2-1} t_l \cos kla - t_{N_n/2}e^{ikN_n a/2}. \quad (1.38)$$

Here we have considered the case where N_n is even. Regarding the summation range, we ought to make the following comment. Owing to the periodicity of the ring arrangement, we have $t_{l+N_n} = t_l$, so we may write $t_{N_n-1} = t_1$, for example. Furthermore, the maximal distance between any two sites is $N_n a/2$, but this distance appears only once in the sum, whereas all the other distances appear twice, once when starting from any point and moving to the right, and once when moving to the left. In the sum over l, the upper bound is thus correct. Note that the quantisation condition for the k does indeed lead to a real value for the last term. When N_n is odd, the upper bound of the sum is replaced by $(N_n - 1)/2$ and the last term in (1.38) does not arise. Note finally that, at large distances, the hopping integrals t_l decrease exponentially with distance, and the sum (1.38) converges rapidly.

1.2 Energy Bands

In the thermodynamic limit $N_n \to \infty$, the sum in (1.38) can be replaced by

$$E_k = E_v - t_0 - 2\sum_{l=1}^{\infty} t_l \cos kla \ . \tag{1.39}$$

In the usual case where the hopping integrals decrease quickly with the distance l, the general form of E_k does not change from what was obtained with the nearest-neighbour approximation, and in particular retains a global maximum and a global minimum, without intervening local extrema. This gives rise to a similar density of states curve to the one shown in Fig. 1.8, with just two singularities at the band edges. However, we can envisage some (rather exceptional) cases in which the t_l are such that E_k has other, local extrema. In this case, the density of states would be changed more radically by the existence of further singularities, also involving square root divergences, in the middle of the band, and more precisely, at the locations of the local extrema.

The situation becomes somewhat more involved if we drop the orthonormalisation condition (1.17). It is replaced by the general expression

$$\langle R_n | R_l \rangle = \alpha_{l-n} = \alpha_{n-l} \ , \tag{1.40}$$

where the fact that the coefficients α_{l-n} (called *overlap integrals*) depend only on the distance between the two sites results from the same symmetry properties as for the t_{l-n}. We still consider normalised wave functions, which corresponds to $\alpha_0 = 1$.

To calculate eigenfunctions and eigenvalues, we return to (1.16), which has changed slightly to become

$$\sum_{l=1}^{N_n} a_{k,l} V_l(\mathbf{r}) |R_l\rangle = (E_k - E_v) \sum_{l=1}^{N_n} a_{k,l} |R_l\rangle \ . \tag{1.41}$$

We multiply on the left by $\langle R_n |$ to obtain, after making the change of variable $l \to l+n$,

$$-\sum_{l=1}^{N_n} a_{k,l+n} t_l = (E_k - E_v) \sum_{l=1}^{N_n} a_{k,l+n} \alpha_l \ . \tag{1.42}$$

Thanks to the periodic boundary conditions on the $a_{k,l}$ and the periodicity properties of the t_l and α_l, there is no need to change the summation limits. The coefficients $a_{k,l}$ have the same form as before, viz., $a_{k,l} = \exp(ikla)$, and after cancelling a factor of $\exp(ikna)$ on each side of (1.42), we finally obtain the eigenvalues:

$$E_k = E_v - \frac{t_0 + 2\sum_{l=1}^{\infty} t_l \cos kla}{1 + 2\sum_{l=1}^{\infty} \alpha_l \cos kla} \ , \tag{1.43}$$

where we have already taken the thermodynamic limit.

Note that the result (1.43) is a periodic function of k of period $2\pi/a$, as in the result (1.39) for the orthogonal case. Expanding in a Fourier series, (1.43) can thus be written in the same form as (1.39), modifying only the coefficients by $t_l \to t'_l$. The wave functions take the same form as in the simpler case of (1.30). A similar calculation to the one leading to (1.43) gives the orthogonality relation

$$\langle \psi_{k'} | \psi_k \rangle = \delta_{k,k'} \left(1 + 2\sum_{l=1}^{\infty} \alpha_l \cos kla \right) \ . \tag{1.44}$$

The $|\psi_k\rangle$ are thus orthogonal but not normalised (although they could easily be normalised, of course). Note that the norm of any wave function must be positive, which guarantees that the denominator in (1.43) is always positive.

1.2.3 Energy Bands Associated with Tightly Bound Atomic Levels

So far we have been considering the case of the atomic valence level E_v of Fig. 1.1a. Clearly, the same approach can be used for the deeper levels $E_1, E_2, \ldots, E_n$, and will lead to an analogous situation, i.e., each of these atomic levels will lead to an energy band with width determined by an integral like t_1. Note that this hopping integral is given by (see Fig. 1.5)

$$t_1 = \int \chi_m^*(\mathbf{r} - \mathbf{R}_0) V_{\text{at}}(\mathbf{r} - \mathbf{R}_0) \chi_m(\mathbf{r} - \mathbf{R}_1) \mathrm{d}^3 r, \tag{1.45}$$

where we have taken here into account the realistic three dimensional atomic wave functions $\chi_m^*(\mathbf{r} - \mathbf{R}_0)$ and $\chi_m(\mathbf{r} - \mathbf{R}_1)$ for the given level, centered on two neighbouring atoms in the chain. Now the atomic orbital wave functions have the property that their spatial extents are narrower for more tightly bound electron levels. This can be seen in Fig. 1.9a, which shows the functions $r\chi_m(r)$ corresponding to the various states $1s, 2s, 2p$, etc., of sodium, on two sites separated by a distance corresponding to the interatomic distance in metallic sodium.

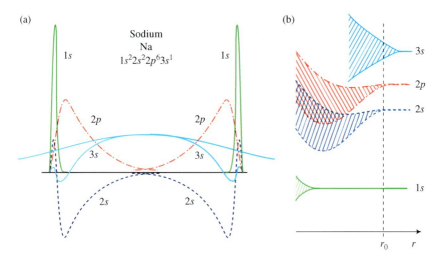

Fig. 1.9 (a) The product $|\mathbf{r} - \mathbf{R}_i| \chi(|\mathbf{r} - \mathbf{R}_i|)$ is shown for the wave functions of two nearest-neighbour atoms in sodium (Na). This choice of representations magnifies the degree of overlap between the orbitals. It increases continuously for the states $1s, 2s, 2p$, and $3s$ (Adapted from [4], p. 177). (b) Energy bands arising from the atomic levels $1s, 2s, 2p$, and $3s$ for Na as a function of the interatomic distance r

The figure shows clearly that, in the relation (1.45), the weight of the product $\chi_m^*(r - \mathbf{R}_0)\chi_m(r - \mathbf{R}_1)$ is such that the values of t_1 are smaller for deeper energy levels. It thus follows that the tight-binding approximation leads to energy bands in the solid that get broader as the energy of the relevant atomic level gets higher.

This is depicted schematically in Fig. 1.9b, which shows the energy bands resulting from the atomic levels $1s, 2s, 2p$, and $3s$ for a hypothetical solid made up of sodium atoms arranged with a spacing of r. As expected from the behaviour of t_1 as a function of r, the bands broaden when the distance is reduced. The deep levels give rise to extremely narrow bands (a few micro electronvolts or meV), while the valence levels give rise to broad bands (in the electronvolt range). In the case of natural Na, with spacing r_0 in Fig. 1.9b, the width of the band arising from the $3s$ level is of the order of the distance between the $2p$ and $3s$ levels, and the tight-binding approximation is no longer applicable. We shall see in Chap. 3 how the structure of this band can then be determined using a different approximation, wherein it is assumed that these electron levels correspond to nearly free electrons.

1.3 Summary

Two subjects have been discussed in this chapter. To begin with, we described the quantum mechanics of a solid comprising a number of particles of the order of the Avogadro number. It is easy to write down a Hamiltonian which takes into account the kinetic energies of the atomic nuclei and the electrons, together with their mutual Coulomb interactions. However, we soon realised that this problem would prove insoluble in its full generality. A more practical and realistic approach here is to begin by making two approximations:

1. We assumed that the *atomic nuclei were static*, which amounts to saying that the nuclei sit at fixed positions in space. We also restricted to the case of crystalline solids, characterised by a regular and periodic arrangement of the nuclei. This is valid for a great many solids, but nevertheless rules out the interesting but complex category of disordered solids.
2. The problem resulting from this approximation, which concerns only the dynamics of the electrons, is still too difficult to handle. We would have to account for the interactions between some 10^{23} electrons, which would be strictly impossible. We thus further restricted to *an average description of the Coulomb interactions between the electrons*, which is often quite adequate. In this case, the problem involving a large number of electrons reduces to a large number of problems involving just one electron, which can be dealt with by well known methods of quantum mechanics and statistical physics.

We then studied the problem of a single electron in a crystalline solid, where it feels a periodic potential. Instead of trying to investigate the general situation, for which there are in any case no explicit solutions, we considered the so-called *tight-binding approximation*, in which the electrons are tightly bound to the atoms. The wave functions of an electron are then written as a linear combination of atomic orbitals (LCAO). Having introduced this approximation for an arbitrary solid, we focused on the specific case of a linear chain of atoms, for which we were able to carry out the calculation exactly. The main result is that the influence of neighbouring atoms in the solid leads to a broadening of the discrete atomic levels into a band of allowed energies, characterised by a function $E(\mathbf{k})$ which gives the energy of a quantum state in terms of its crystal momentum $\mathbf{k}$. For the linear chain of atoms, the function $E(\mathbf{k})$ is given by

$$E_k = E_v - t_0 - 2t_1 \cos ka,$$

where t_0 and t_1 are hopping integrals, characterising the possibility of transfer of an electron from one atom to the neighbouring site. The spectrum arising from a given atomic level is thus bounded but quasi-continuous. All the eigenstates can be obtained for values of k lying in the interval $(-\pi/a, \pi/a)$, which is called the *first Brillouin zone*.

Chapter 2
Crystalline Solids: Diffraction

Contents

2.1 Crystal Structures 25
    2.1.1 Crystal Lattice 25
    2.1.2 Two-Dimensional Crystals 27
    2.1.3 Three-Dimensional Crystals.. 29
    2.1.4 Beyond the Perfect Crystal 33
2.2 Diffraction 35
    2.2.1 General Principles 35
    2.2.2 Reciprocal Lattice 38
2.3 Determination of Crystal Structures 39
    2.3.1 On the Bragg Diffraction Condition 39
    2.3.2 Diffraction Intensity and Basis of the Primitive Cell 41
    2.3.3 X-Ray Diffraction 42
2.4 Summary 44
2.5 Answers to Questions 45

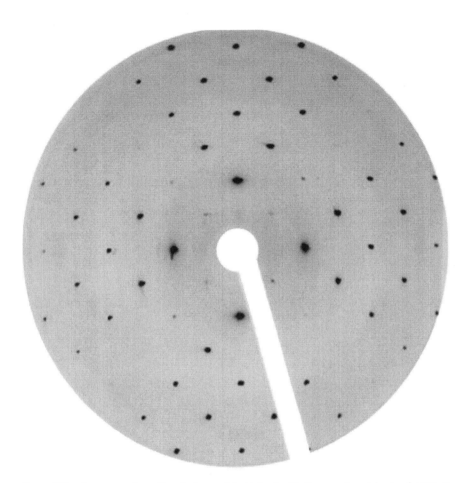

X-ray diffraction pattern for a C_{60} single crystal obtained with the experimental setup known as a precession chamber. This directly visualises the Bragg spots corresponding to a plane of the reciprocal lattice of the crystal (to solve Question 2.6 one has to take into account that the actual diameter of the film in 12 cm). Image courtesy of Launois, P., Moret, R.: Laboratoire de Physique des Solides. Orsay, France

2.1 Crystal Structures

Our main concern in this book is to describe the electronic properties of crystalline solids. The existence of translation symmetries associated with such ordered crystal structures leads to specific features in the electronic structure and to a specific representation of the energy states in wave vector space. In Sect. 2.1, we explain how a crystal structure can be described formally by defining a crystal lattice of nonmaterial points together with a repeated material motif, known as the basis. This periodic structure of matter causes diffraction of electromagnetic waves, or equivalently, of quantum particles. We shall see in Sect. 2.2 that these diffraction phenomena lead to the notion of reciprocal lattice in wave vector space, and this will be important later for characterising the electronic states of these solids. In Sect. 2.3, we outline the experimental methods used to determine the crystal structures of solids using diffraction methods.

2.1 Crystal Structures

It is usual to consider a crystal as a natural object with regular external geometric features, as found for example in rock salt, diamonds, and so on. By the end of the nineteenth century, the systematic study of the external shapes of such natural crystals led scientists to conclude that this regularity of the outer faces must be due to structural regularities on the microscopic scale. The molecules or atoms had to be assembled in a periodic manner to make a crystal. In this chapter, we shall see how to specify the arrangement of a crystal structure. This structure can be ascertained experimentally either by direct observation, or by light diffraction (X-ray crystallography). These experimental methods show that many solids actually have a crystal structure, even when their outer surfaces do not give this impression. They are in fact polycrystalline, i.e., made up of a host of small crystals, which may themselves be of micrometric dimensions, but which are nevertheless of macroscopic size. We begin in Sect. 2.1.1 by describing the notions of crystal lattice and unit cell. In Sect. 2.1.2, these ideas will be exemplified in two dimensions, i.e., for planar crystal structures, where the geometric representations are simpler. We then illustrate some simple 3D systems in Sect. 2.1.3.

2.1.1 Crystal Lattice

A crystal is an arrangement of atoms or molecules that is invariant under translations in three space directions constituting a triad $(\mathbf{a}_1, \mathbf{a}_2, \mathbf{a}_3)$. Many human constructions have ordered structures exhibiting such characteristics, especially in two dimensions: wall paper and floor tiling often have periodic structures that can be considered as 2D crystals. In these cases, the primitive material basis is repeated through two translations. A molecular example, observed by a modern microscopic method, is shown in Fig. 2.1. These are alkane molecules with chemical formula

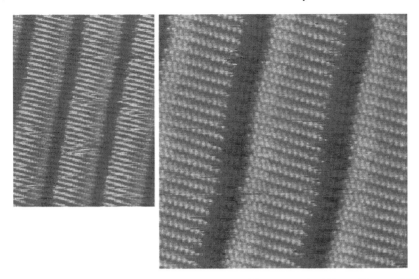

Fig. 2.1 A monolayer of alkane $C_{33}H_{68}$ deposited on graphite arranges itself into a 2D crystal. The right-hand figure was obtained with twice the resolution to reveal details of the atomic structure. The distance between molecules is 4.5 Å, and their length is 45 Å. Image courtesy of Cousty, J.: SPCSI/CEA. Saclay, France

$C_{33}H_{68}$, deposited on a graphite surface. Note that, if we choose an origin O in space, the crystal can be reconstructed in two steps:

- We produce a lattice of points obtained from O by all the translations

$$\mathbf{R_l} = l_1\mathbf{a}_1 + l_2\mathbf{a}_2 + l_3\mathbf{a}_3 , \qquad (2.1)$$

where l_1, l_2, l_3 are integers. This set of points (the *lattice points* or *nodes*) constitutes the *crystal lattice* or *Bravais lattice*.
- In the second step, we arrange the material basis relative to these nodes in such a way as to completely tile the space.

The Bravais lattice is thus a set of non-material points, whereas the atoms and molecules make up the material basis, which we shall call the 'basis' from now on, when no confusion is possible. For elementary solids, containing only one atomic species, the atoms may coincide with the lattice nodes, since the basis then often comprises a single atom. But as soon as the solid contains several atomic species, such a situation is no longer possible.

Note that the lattice of points provides a way of defining a *primitive unit cell* for the material basis, namely, as the smallest volume that can tile the space by applying the translations $\mathbf{R_l}$. If n_p is the density of lattice points, the volume of the primitive cell is $v = 1/n_p$. The primitive cell and the triad $(\mathbf{a}_1, \mathbf{a}_2, \mathbf{a}_3)$ are not unambiguously defined, as can be seen from Fig. 2.2. The triad $(\mathbf{a}_1, \mathbf{a}_2, \mathbf{a}_3)$ is often taken to be the set of vectors that best reveals the symmetries of the lattice, e.g., $(\mathbf{a}_1, \mathbf{a}_2)$ rather than $(\mathbf{b}_1, \mathbf{b}_2)$ for the square and rectangular planar lattices in Fig. 2.2. A primitive cell

2.1 Crystal Structures

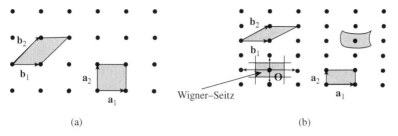

Fig. 2.2 (**a**) *Square* lattice. (**b**) *Rectangular* lattice. These can be specified by the two vectors $(\mathbf{a}_1, \mathbf{a}_2)$ or the two vectors $(\mathbf{b}_1, \mathbf{b}_2)$. The vectors $(\mathbf{a}_1, \mathbf{a}_2)$ better reveal the symmetries of the lattice. Four primitive cells are shown for the *rectangular* lattice, those specified by $(\mathbf{a}_1, \mathbf{a}_2)$ and $(\mathbf{b}_1, \mathbf{b}_2)$, an arbitrary cell, and the Wigner–Seitz cell containing the lattice point O. The latter is obtained by constructing the orthogonally bisecting planes of the four vectors $(O\mathbf{a}_1, O\mathbf{a}_2, -O\mathbf{a}_1, -O\mathbf{a}_2)$

tiling the space may be chosen with an arbitrary shape as displayed in Fig. 2.2(b), but we often opt for the rhombohedral primitive cell specified by $(\mathbf{a}_1, \mathbf{a}_2, \mathbf{a}_3)$.

Of particular importance is a primitive unit cell known as the *Wigner–Seitz cell*. This is constructed in such a way that *every point of the cell is closer to one lattice point* (for example, O) *than to any other lattice point* (see Fig. 2.2). It is bounded by the orthogonally bisecting planes of the vectors $\mathbf{R}_l$ with origin the chosen node. For an elementary solid, this volume constructed with one atom at the center represents in some sense the region of influence of this atom.[1]

Note that the lattice points can be grouped together in parallel planes in infinitely many different ways. These are called *lattice planes*. In two dimensions, they constitute parallel rows (see Fig. 2.3). The lattice planes group together points that can be obtained from one another by two of the translations $\mathbf{R}_l$.

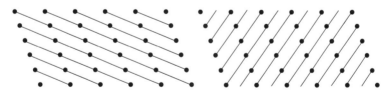

Fig. 2.3 Two families of lattice rows in the same Bravais lattice (oblique lattice in a 2D space)

2.1.2 Two-Dimensional Crystals

In the above, we illustrated the idea of a crystal by 2D representations. Since real crystals are three-dimensional, these 2D representations may appear rather

---

[1] We shall see that, in the reciprocal lattice to be defined hereafter, the corresponding unit cell will specify the first Brillouin zone.

academic. However, important cases of 2D physics are becoming more and more common today:

- The surface of any 3D crystal is obviously a 2D structure. One might think that this would be that of the lattice plane corresponding to the infinite crystal. However, in many cases, the translational symmetry breaking associated with the existence of a surface leads to a significant modification of the surface structure. We refer to this as surface reconstruction.
- Many 3D crystals occur in a layered form, with widely spaced molecular or atomic layers. An example is graphite, or the high-T_c cuprate superconductors. The structural and electronic properties of these materials are strongly affected by the 2D nature of the material.
- Finally, novel fabrication methods devised in nanotechnology are now used to deposit monomolecular layers (see Fig. 2.1), or even monatomic layers, on the surfaces of crystalline substrates. In 2004, it became possible to peel off graphite sheets and hence study isolated layers of graphene, which is an almost ideal 2D crystal form of carbon with highly original electronic properties.

Quite generally, a given (Bravais) crystal lattice is characterised by the symmetry operations that preserve its structure, and which include, apart from the lattice translations, axes of symmetry under rotation through some angle θ (or n-fold symmetry axes, where $\theta = 2\pi/n$), and planes of symmetry (*mirror planes*). In two dimensions, there are only five types of Bravais lattice. These are, in order of increasing symmetry:

- The oblique lattice (Fig. 2.3), which has the minimal 2D symmetry, i.e., only one two-fold symmetry axis perpendicular to the plane.
- The rectangular lattice (Fig. 2.2b), which also has mirror planes parallel to the shortest translation axes of the lattice.
- The centered rectangular lattice (see Fig. 2.4a), with mirror planes distinct from the shortest translation axes of the lattice.
- The hexagonal lattice (Fig. 2.4b), with a 3-fold (and hence a 6-fold) symmetry axis.
- The square lattice (Fig. 2.2a), with a 4-fold symmetry axis.

In the centered rectangular lattice of Fig. 2.4a, the primitive cell constructed from $(\mathbf{a}_1, \mathbf{a}_2)$ does not help us to visualise the lattice symmetries as clearly as the *conventional unit cell* $(\mathbf{a}, \mathbf{b})$, which is not a primitive unit cell, since it has double the

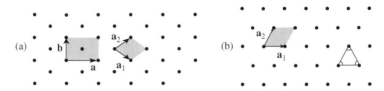

Fig. 2.4 Plane lattices. (**a**) Centered *rectangular*, (**b**) *hexagonal*. In the first case, we observe the conventional centered *rectangular* cell specified by $(\mathbf{a}, \mathbf{b})$ and the primitive cell $(\mathbf{a}_1, \mathbf{a}_2)$. In the plane *hexagonal* lattice, all the primitive *triangles* are equilateral

2.1 Crystal Structures

area. In fact the latter contains two lattice points, viz., the point at the origin and the center of the rectangle. These two points are indeed equivalent, as required by the notion of a Bravais lattice, since each one is the center of the rectangle formed by its four nearest neighbours. The conventional cell must be considered as a cell with one basis (the two lattice points), and the crystal will be obtained by introducing twice the material basis of the primitive cell around these points.

> **Question 2.1.** 1. Determine the Bravais lattice and the primitive cell for the alkane crystal observed by scanning tunneling microscopy in Fig. 2.1.
> 2. Determine the Wigner–Seitz cells associated with the centered rectangular and hexagonal lattices of Fig. 2.4.

The 2D structures attracting most attention since 2005 are the graphene honeycomb structure of Fig. 2.5a and the so-called Kagomé structure. Some compounds like hebersmithite $ZnCu_3(OH)_6Cl_2$ are built up from alternating planes of $ZnCl_2$ with a simple 2D structure and Kagomé planes of $Cu_3(OH)_6$. The Cu^{2+} ions are arranged as in Fig. 2.5b, while the $(OH)^-$ serve to bind the Cu^{2+}. The latter carry total electron spin 1/2, making this a very interesting material in quantum magnetism.

> **Question 2.2.** Determine the Bravais lattice and a primitive cell for each of the two structures of Fig. 2.5.

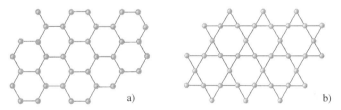

Fig. 2.5 (a) Crystal structure of C atoms in graphene. (b) Atomic arragement known as 'Kagomé' (after a typical pattern of straw baskets made in Japan)

2.1.3 Three-Dimensional Crystals

In three dimensions, the simplest lattice to visualise is the cubic lattice. The three primitive translation vectors $(\mathbf{a}_1, \mathbf{a}_2, \mathbf{a}_3)$ form an orthogonal triad, and each of them has length equal to the side a of the cube constituting a primitive cell (see Fig. 2.6a). Chemical elements crystallise scarcely into such a simple cubic lattice, but it is encountered in many polyatomic crystals (we shall discuss the example of CsCl below).

However, many chemical elements crystallise into body-centered cubic (bcc) crystal lattices (see Table 2.1). This Bravais lattice shown in Fig. 2.6b is the 3D

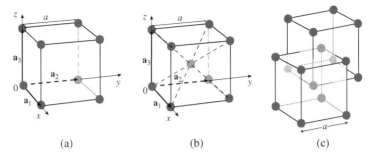

Fig. 2.6 (a) Simple cubic lattice. (b) Body-centered cubic lattice. (c) Equivalence of nodes in the body-centered cubic lattice

Table 2.1 Lattice constants a of the conventional cell for several elementary solids which form a body-centered cubic structure with one atom per primitive unit cell

Element	a [Å]	Element	a [Å]	Element	a [Å]
Ba	5.02	Li	3.49	Ta	3.31
Cr	2.88	Mo	3.15	Tl	3.88
Cs	6.05	Na	4.23	V	3.02
Fe	2.87	Nb	3.30	W	3.16
K	5.23	Rb	5.59		

analog of the 2D centered rectangular lattice. In the bcc lattice, the conventional cell contains two lattice points, one at the origin and the other at the center of the cube. These two lattice nodes are indeed equivalent as each is the center of a cube formed by its eight nearest neighbours (see Fig. 2.6c).

The most common lattice for elementary solids is the face-centered cubic lattice (fcc). This is what is usually obtained when we try to stack hard spherical balls. It is a common structure for many metals (see Table 2.2). The face-centered

Table 2.2 Lattice constants a of the conventional cell in several elementary solids which form face-centered cubic structures

Element	a [Å]	Element	a [Å]	Element	a [Å]
Ar	5.26	Ir	3.84	Pt	3.92
Ag	4.09	Kr	5.72	Pu	4.64
Al	4.05	La	5.30	Rh	3.80
Au	4.08	Ne	4.43	Sc	4.54
Ca	5.58	Ni	3.52	Sr	6.08
Ce	5.16	Pb	4.95	Th	5.08
Co	3.55	Pd	3.89	Xe	6.20
Cu	3.61	Pr	5.16	Yb	5.49

2.1 Crystal Structures

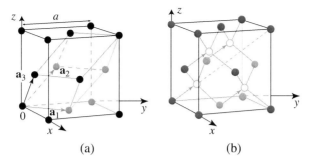

Fig. 2.7 (**a**) Face-centered cubic lattice. The primitive cell is the rhombohedron specified by the vectors $\mathbf{a}_1 = (a/2)(\mathbf{x}+\mathbf{y})$, $\mathbf{a}_2 = (a/2)(\mathbf{y}+\mathbf{z})$, and $\mathbf{a}_3 = (a/2)(\mathbf{x}+\mathbf{z})$. The conventional cell contains four lattice nodes. (**b**) Cubic crystal structure of diamond. The atoms are all chemically identical in diamond, Si, and Ge. In the case of InP or GaAs, the two species differ, and are represented by *empty spheres* and *full spheres*

cubic lattice also has a conventional cubic unit cell containing a four-node basis comprising one vertex of the cube and the three centers of the adjacent faces (see Fig. 2.7a). Note that the primitive translations of the Bravais lattice are the three vectors joining the cube vertex to the centers of the adjacent faces, so a rhombohedral primitive cell can be constructed (see Fig. 2.7a). Among the systems crystallising in this structure are diamond and many semiconductors such as Si, Ge, GaAs, and InP. All atoms within the conventional cell are shown in Fig. 2.7b, where the full and empty spheres correspond to the two atomic species in the case of binary compounds, but are identical in the case of Si or Ge.

> **Question 2.3.** Check that the crystals in Fig. 2.7b do indeed correspond to a face-centered cubic Bravais lattice, and determine the atomic basis.

When we consider polyatomic crystals, the primitive cell necessarily contains several atoms. A simple illustration is given in Fig. 2.8 for the alkali halides CsCl and NaCl. Although Cs is at the center of a Cl cube, the associated Bravais lattice is the

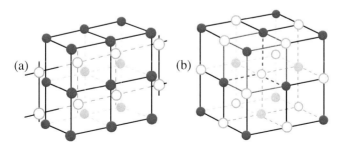

Fig. 2.8 (**a**) Crystal of CsCl, with simple cubic primitive cell. The basis comprises one atom of Cl at the vertex of the cube and one atom of Cs at the center of the cube. (**b**) Crystal of NaCl with fcc primitive cell comprising one atom of Cl and one atom of Na per primitive cell of Fig. 2.7a

simple cubic lattice of side a, with a basis comprising one atom of Cl at the vertex of the cube and one atom of Cs at the center of the cube.

Many mixed oxides of transition metals crystallise into a cubic structure called the *perovskite structure*, with primitive cell ABO_3, in which A and B are cations with different size and valence. The small cation, generally A^{2+}, is surrounded by an octahedron of oxygen atoms, while the large cation B^{n+} (in general $n = 3$ or 4) is surrounded by 12 oxygen atoms, as can be seen in Fig. 2.9a. These oxides can exhibit a wide range of physical properties, from ferromagnetism in the manganites $LaMnO_3$ and cobaltites $LaCoO_3$ to antiferromagnetism in iron-based perovskites like $LaFeO_3$. Below 120°C, slight structural distortions with respect to the ideal structure of Fig. 2.9a induce ferroelectric properties in the case of $BaTiO_3$.

Families of metal oxides with a highly 2D structure can sometimes be obtained by combining planes with perovskite structure with square MO planes, where M is a third metal cation. A classic example of such hybrid structures is provided by the high-T_c cuprate superconductors, whose discoverers, Müller and Bednorz, were rewarded by the Nobel prize. An example of such a structure is found in $HgBa_2CuO_5$. This is shown in Fig. 2.9b. It is generated by intercalating sheets of Ba_2CuO_4, which has the perovskite structure, with planes of HgO, or can alternatively be considered simply as alternating planes of $CuO_2/BaO/HgO/BaO/CuO_2$, and so on. The primitive cell of this structure is a right-angled parallelepiped whose sides a and b are equal (tetragonal structure). In other cuprates with $a \neq b$, the structure is said to be orthorhombic.

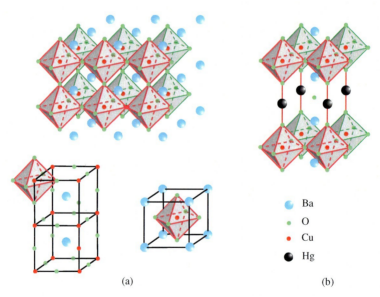

Fig. 2.9 (a) ABO_3 cubic perovskite structure. Below the 3D representation designed to show the octahedra surrounding the A cations are two primitive cells centered respectively on the A and B cations. (b) Quadratic structure of a high-T_c cuprate superconductor made up of planes of HgO inserted between sheets of perovskite Ba_2CuO_4. This structure confers 2D physical properties on the material

2.1 Crystal Structures

Naturally, there are many other 3D crystal systems with even fewer symmetries than the simple lattices considered above. There is no question here of undertaking an exhaustive study: there are 14 Bravais lattices in three dimensions!

2.1.4 Beyond the Perfect Crystal

Not all natural or artificial solids are crystalline. In many cases, there is no long range order in the atomic arrangement. In particular, when a liquid is suddenly cooled down below its solidification temperature, we can obtain a solid state which simply freezes in the arrangement of atoms as it occurred in the liquid state. A glass is obtained in this way by quenching the liquid, whereupon the atoms arrange themselves in a way that suffers only one constraint, namely that atoms are not allowed to interpenetrate. If the atoms are thought of as hard spheres, the resulting glass structure looks like what would be obtained by putting beads in a container and shaking them up. This glassy state is generally metastable, in the sense that the system can have a lower free energy when the atoms are arranged into an fcc or hcp crystal structure, which correspond to the closest packing of the beads. Crystallisation can then be obtained by heat treatment, which amounts to shaking the box of beads in our analogy.

But many other situations can be observed, with varying degrees of order. Consider for example what happens for some *alloys* of two metals. A structure close to a crystal structure can often be seen in these materials. The atoms distribute themselves randomly at the lattice points of a perfect crystal structure. This is called a *solid solution*. This happens for example for the alloys $Au_{1-x}Cu_x$, which can be made with an arbitrary concentration x of Cu. The Cu and Au atoms distribute themselves randomly over the fcc lattice sites of pure Au, and the lattice spacing varies slightly depending on the Cu concentration. This structure is not strictly speaking a perfect crystal structure, although it can be treated as such in many respects.

Question 2.4. For some values of x, a heat treatment allows the atoms to arrange themselves into a perfect crystal structure on the lattice. In the case of the $Au_{1-x}Cu_x$ alloys, indicate one or more values x_0 of x for which a perfect arrangement could in principle be obtained. What would then be the associated primitive cell and Bravais lattice? In your opinion, if the $Au_{1-x}Cu_x$ alloy forms a perfect crystal arrangement for $x = x_0$, is it likely from a physical point of view that the same arrangement will be obtained for $x = 1 - x_0$?

Imperfect crystal arrangements are observed in many other cases, in particular for complex molecular structures. For example, the real crystals of cuprate superconductors shown in Fig. 2.9a are such that the HgO plane is highly deficient in oxygen. The chemical formula is then $HgBa_2CuO_{4+\delta}$, and the oxygen vacancies are important in determining the physical properties.

A novel illustration of disorder in crystals is shown below in the case of the fullerene C_{60}, a molecule discovered in 1985, which has a football shape. Its face-centered cubic structure, with large empty spaces between the C_{60} molecules,

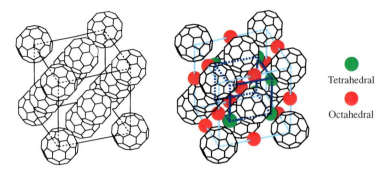

Fig. 2.10 *Left*: C_{60} crystal. *Right*: Rb_3C_{60} crystal. These lattices are face-centered cubic. In Rb_3C_{60} the rhombohedral primitive cell contains a C_{60} molecule and three Rb atoms

is shown in Fig. 2.10. It is easy to insert cations between the C_{60} molecules and thereby create compounds of the form A_nC_{60}. The compound Rb_3C_{60} has attracted considerable attention as it happens to be a metal that becomes a superconductor below 27 K. Its fcc structure contains 3 rubidium atoms and one molecule of C_{60} per primitive cell. The rubidium atoms have two different types of position: one, located in the middle of an edge of the cube, has an octahedral C_{60} environment, while the other two have a tetrahedral C_{60} environment (one vertex of the cube and three face centers). Note that the C_{60} molecule has symmetries that are not compatible with the face-centered cubic lattice. Indeed, there is no way of orientating the C_{60} molecule so that it can map onto itself under all the symmetries of the lattice. There are not really any 3D crystal structures whose primitive cells are given by those shown in Fig. 2.10. These structures can nevertheless be considered as crystalline, but with orientational disorder of the C_{60} molecules.

Crystals and Molecular Motions

At high temperatures, the C_{60} molecules are not immobile, but have rotational motions. These rapid rotational movements are such that, on average, the C_{60} molecules behave like spheres, and one can consider that the symmetry of the C_{60} molecule is no longer relevant. The average structure is as shown in Fig. 2.10, treating the C_{60} as simple spheres. Since the molecules are not fixed, we do not strictly have a crystal. Such systems in which molecular motions occur are called *plastic crystals*.

For pure C_{60}, a phase transition takes place at 260 K from the face-centered cubic high temperature structure of the plastic crystal to a body-centered cubic plastic crystal, as the rotational motions of the C_{60} molecules occur in a correlated manner about particular axes relative to the crystal axes.

At low temperatures, these rotational motions freeze, but the relative orientations of the C_{60} molecules in low temperature phases are not yet perfectly understood. Although one can speak of an average face-centered cubic structure in Rb_3C_{60}, the state of relative disorder or order of the C_{60} molecules has not yet been completely characterised.

2.2 Diffraction

In Sect. 2.2.1, we describe the general principles governing the diffraction of waves by a periodic pattern, and show that, for a crystal, the directions in which diffraction can occur are of course associated with the crystal structure. These diffraction conditions are used in Sect. 2.2.2 to define a lattice of points in the wave vector space, which is known as the *reciprocal lattice* of the crystal lattice.

2.2.1 General Principles

This effect was originally demonstrated by von Laue and the Braggs (father and son) in 1912–1913. The electromagnetic waves (photons) are generally X rays, with wavelength given in angstrom units (Å) and corresponding photon energy ε given in keV. The latter quantities are related by

$$\lambda_{\text{Å}} = 12.4/\varepsilon_{\text{keV}} . \tag{2.2}$$

The X rays used typically have energies in the range $10 < \varepsilon < 50$ keV. In specific cases, the radiation may also be in the form of neutrons or electrons. The wavelength is then given by the de Broglie relation $\lambda = h/p$.

To understand how diffraction works, consider first the scattering of an arbitrary plane (electromagnetic or matter) wave by some obstacle, usually an atom located at the origin (see Fig. 2.11). The amplitude of the incident wave will then have the form

$$a_{\text{in}}(\mathbf{r},t) = a_0 e^{i(\mathbf{k}_0 \cdot \mathbf{r} - \omega t)} , \tag{2.3}$$

where $\omega = c|\mathbf{k}_0|$ for an electromagnetic wave, with c the speed of light, and $\omega = \hbar \mathbf{k}_0^2/2m$ for a matter wave, with m the mass of the incident particles. Here, a_0 is a vector-valued amplitude in the case of electric or magnetic fields and a complex-valued amplitude in the case of quantum matter waves. This amplitude depends only on the intensity of the incident radiation. In general, the scattered wave will have the form

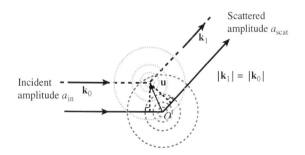

Fig. 2.11 Scattering of an incident plane wave with wave vector $\mathbf{k}_0$. Waves scattered in the direction $\mathbf{k}_1$ reveal the phase difference between beams scattered by two objects at 0 and $\mathbf{u}$

$$a_{\text{scat}}(\mathbf{r},t) = a_0 \sum_{\mathbf{k}} \alpha_{\mathbf{k}_0,\mathbf{k}} e^{i(\mathbf{k}\cdot\mathbf{r}-\omega t)}, \tag{2.4}$$

where conservation of energy implies that $|\mathbf{k}| = |\mathbf{k}_0|$. The determination of the coefficients $\alpha_{\mathbf{k}_0,\mathbf{k}}$ is a (rather complex) problem of quantum mechanics which depends on the type of wave and the quantum properties of the scattering object. However, the exact expression for these coefficients is not needed to understand the underlying principle of the diffraction methods described below.

Let us now ask what happens when the object is displaced through $\mathbf{u}$ from the origin. Clearly, the scattered amplitude must be taken from the point $\mathbf{u}$ rather than from the origin. We must therefore replace $\mathbf{u}$ by $\mathbf{r} - \mathbf{u}$ in the exponential of the expression in (2.4). But in addition, the incident wave will have a phase offset of $\mathbf{k}_0 \cdot \mathbf{u}$ (see Fig. 2.11), and we therefore obtain

$$a_{\text{scat}}(\mathbf{r},t) = a_0 e^{i\mathbf{k}_0\cdot\mathbf{u}} \sum_{\mathbf{k}} \alpha_{\mathbf{k}_0,\mathbf{k}} e^{i[\mathbf{k}\cdot(\mathbf{r}-\mathbf{u})-\omega t]}. \tag{2.5}$$

Finally, we use a detector that only detects waves scattered in a specific direction $\mathbf{k}_1$. The amplitude in this specific direction will then be

$$a_{\text{scat}}(\mathbf{k}_1;\mathbf{r},t) = a_0 \alpha_{\mathbf{k}_0,\mathbf{k}_1} e^{i(\mathbf{k}_1\cdot\mathbf{r}-\omega t)} e^{-i\mathbf{K}\cdot\mathbf{u}}, \tag{2.6}$$

with

$$\mathbf{K} = \mathbf{k}_1 - \mathbf{k}_0. \tag{2.7}$$

There is therefore a phase difference between the waves scattered in the direction $\mathbf{k}_1$ by the objects located at $\mathbf{O}$ and $\mathbf{u}$. In the general case of a crystal where the identical scatterers are the material bases of the primitive cell repeated at the different lattice points, these scattered amplitudes will generally undergo different phase shifts $\exp(i\mathbf{K}\cdot\mathbf{u})$, leading to a low value of the total scattered intensity (sum of the squared amplitudes). However, as the Braggs and von Laue observed, if all the phase factors are the same in certain directions $\mathbf{k}_1$, the resulting *phase coherence* between the amplitudes scattered in these directions will lead to a high diffracted intensity.

a. The Bragg Formulation

We consider the crystal lattice as an ensemble of lattice planes as shown in Fig. 2.12, with O and u two lattice points belonging to one of the lattice planes, and we take the scatterers to be the material bases of the primitive cell centered at O and u. If the wave vector $\mathbf{k}_0$ of the incident wave makes an angle of incidence θ with the lattice plane, it is easy to see that the phase difference $\mathbf{K}\cdot\mathbf{u}$ vanishes if $\mathbf{k}_1$ corresponds to a reflection of $\mathbf{k}_0$ in the lattice plane. So in this direction, there will be phase coherence between the intensities scattered by all the bases associated with this

2.2 Diffraction

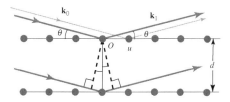

Fig. 2.12 Reflection of incident radiation on lattice planes. The Bragg condition obtains when the path difference is a multiple of the wavelength

particular lattice plane. If in addition we wish to observe phase coherence between the amplitudes scattered by the different parallel lattice planes, examination of Fig. 2.12 shows that the condition that must be satisfied by θ is

$$2d\sin\theta = n\lambda, \quad (2.8)$$

where d is the distance between the lattice planes and n is an integer. This diffraction condition, known as the *Bragg condition*, is given here in a form that corresponds to the representation of the crystal lattice in lattice planes.

b. General von Laue Formulation

We now consider the scattering by a very large number of atomic bases of the primitive cell arranged in positions $\mathbf{R}_n$. The total scattered wave will simply be the superposition of many terms of the form (2.6):

$$a_{\text{scat}}^T(\mathbf{k}_1;\mathbf{r},t) = a_0 e^{i(\mathbf{k}_1\cdot\mathbf{r}-\omega t)}\mathscr{A}(\mathbf{K}), \quad (2.9)$$

with

$$\boxed{\mathscr{A}(\mathbf{K}) = \alpha_{\mathbf{k}_0,\mathbf{k}_1}\sum_n e^{-i\mathbf{K}\cdot\mathbf{R}_n}}. \quad (2.10)$$

In this equation, $\alpha_{\mathbf{k}_0,\mathbf{k}_1}$ characterises the radiation and the arrangement of atoms in the primitive cell, while the sum over n is only associated with the spatial positions of the Bravais lattice points. If $\mathbf{K}$ is chosen such that

$$\boxed{e^{i\mathbf{K}\cdot\mathbf{R}_n} = 1, \quad \text{for all translation vectors } \mathbf{R}_n \text{ in the crystal lattice}}, \quad (2.11)$$

we obtain $\mathscr{A}(\mathbf{K}) = N\alpha_{\mathbf{k}_0,\mathbf{k}_1}$, where N is the number of primitive cells in the crystal. Under these conditions, the diffracted intensity in the direction $\mathbf{k}_1$ is

$$I_{\text{diff}} \propto |\mathscr{A}(\mathbf{K})|^2 = N^2|\alpha_{\mathbf{k}_0,\mathbf{k}_1}|^2. \quad (2.12)$$

Equations (2.11) and (2.7) for $\mathbf{k}_1$ and $\mathbf{k}_0$ provide another expression of the *Bragg condition*.

2.2.2 Reciprocal Lattice

Here we interpret (2.11) in the special case of a linear chain. In one dimension, we have seen that the Bravais lattice is specified by $\mathbf{R_m} = m\mathbf{a}_1 = ma\mathbf{x}$, where a is the period and $\mathbf{x}$ a (dimensionless) unit vector along the Ox axis. The relation (2.11) implies that $\mathbf{K}$ is an integer multiple of $\mathbf{a}^* = (2\pi/a)\mathbf{x}$.

In three dimensions, if $\mathbf{K}$ is specified relative to a frame $(\mathbf{a}_1^*, \mathbf{a}_2^*, \mathbf{a}_3^*)$ by

$$\mathbf{K} = x_1 \mathbf{a}_1^* + x_2 \mathbf{a}_2^* + x_3 \mathbf{a}_3^*, \tag{2.13}$$

then (2.11) takes the form

$$(x_1 \mathbf{a}_1^* + x_2 \mathbf{a}_2^* + x_3 \mathbf{a}_3^*)(l_1 \mathbf{a}_1 + l_2 \mathbf{a}_2 + l_3 \mathbf{a}_3) = 2\pi n_\mathbf{l}, \tag{2.14}$$

with $n_\mathbf{l}$ an integer, and this for all $\mathbf{R_l}$, i.e., for all integer values of (l_1, l_2, l_3). We thus see that, by choosing the basis $(\mathbf{a}_1^*, \mathbf{a}_2^*, \mathbf{a}_3^*)$ such that

$$\boxed{\mathbf{a}_i \cdot \mathbf{a}_j^* = 2\pi \delta_{ij}}, \tag{2.15}$$

the relation (2.14) reduces to

$$x_1 l_1 + x_2 l_2 + x_3 l_3 = n_\mathbf{l} \text{ integer}, \tag{2.16}$$

and this for all (l_1, l_2, l_3), implying that x_1, x_2, and x_3 are also integers. The components of $\mathbf{K}$ relative to the frame $(\mathbf{a}_1^*, \mathbf{a}_2^*, \mathbf{a}_3^*)$ are called the *Miller indices*, and are usually denoted by (h, k, l).

As a consequence, the vectors $\mathbf{K}$ satisfying (2.11) generate a Bravais lattice in the $\mathbf{k}$ space. This is the *reciprocal lattice* associated with the Bravais lattice in position space, called hereafter *real* or *direct*. The reference frame $(\mathbf{a}_1^*, \mathbf{a}_2^*, \mathbf{a}_3^*)$ of the reciprocal space is defined in terms of the real space frame $(\mathbf{a}_1, \mathbf{a}_2, \mathbf{a}_3)$ by (2.15). It is easy to check that the $\mathbf{a}_i^*$ are given by

$$\boxed{\mathbf{a}_1^* = 2\pi \frac{\mathbf{a}_2 \wedge \mathbf{a}_3}{\mathbf{a}_1 \cdot (\mathbf{a}_2 \wedge \mathbf{a}_3)}}, \tag{2.17}$$

and cyclic permutations. Here the denominator is precisely the volume of the primitive cell of the direct lattice.

Note that, according to (2.11), the direct and reciprocal lattices play symmetric roles. In particular, the reciprocal lattice of the reciprocal lattice is just the direct lattice. However, the real crystal is a lattice of atoms or molecules, or more generally, a lattice of what we have called material bases, whereas the reciprocal lattice is a *lattice of points that are independent of the bases of the real crystal*.

For example, the reciprocal lattice of a simple cubic lattice with lattice constant a is a simple cubic lattice with lattice constant $2\pi/a$. The reciprocal lattice of an fcc lattice with lattice constant a is a body-centered cubic lattice of lattice constant $4\pi/a$, e.g., Al, Si, GaAs. Conversely, the reciprocal lattice of a body-centered cubic lattice is an fcc lattice, e.g., Fe.

Fig. 2.13 Determination of the Bragg plane

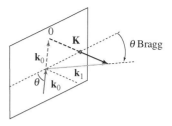

The notion of reciprocal lattice can be used to relate the Bragg and von Laue representations of the diffraction conditions. Indeed, since $|\mathbf{k}_0| = |\mathbf{k}_1|$, the diffraction condition $\mathbf{k}_1 - \mathbf{k}_0 = \mathbf{K}$ implies that

$$2\mathbf{k}_1 \cdot \mathbf{K} = K^2 . \tag{2.18}$$

As illustrated in Fig. 2.13, this means that $\mathbf{k}_0$ and $\mathbf{k}_1$ are obtained from one another by a reflection in the orthogonally bisecting plane of the vector $\mathbf{K}$. The direction of diffraction $\mathbf{k}_1$ therefore corresponds to a Bragg diffraction on the lattice planes of the crystal parallel to the orthogonally bisecting plane of the vector $\mathbf{K}$. There is thus a one–one correspondence between the lattice planes of the crystal and the orthogonally bisecting planes of the reciprocal lattice vectors, which are known conventionally as *Bragg planes*.

2.3 Determination of Crystal Structures

In Sect. 2.3.1, we show to begin with that the directions satisfying the Bragg condition are well defined experimentally, but that they are not always easy to detect. Once they have been found, the crystal lattice can be determined. In Sect. 2.3.2, we describe how the basis of the primitive cell can be ascertained by determining the intensities of the different Bragg diffraction spots. This will be exemplified for the case of X-ray diffraction in Sect. 2.3.3.

2.3.1 On the Bragg Diffraction Condition

We have seen that the diffracted intensity is high when the factor

$$A_0(\mathbf{K}) = \sum_n e^{-i\mathbf{K} \cdot \mathbf{R}_n}$$

is large, and this occurs when the Bragg condition holds:

$$\boxed{\mathbf{k}_1 - \mathbf{k}_0 = \mathbf{K}, \quad \mathbf{K} \in \text{reciprocal lattice}} . \tag{2.19}$$

Consider now the scattered intensity if the Bragg condition is not satisfied. To simplify the notation, we discuss the case of a simple cubic structure with lattice constant a and length La in each of the three space directions. The sum in (2.10) then becomes the product of three geometric sums, and direct calculation gives

$$|\mathscr{A}(\mathbf{K})|^2 = |\alpha_{\mathbf{k}_0, \mathbf{k}_1}|^2 f(K_x a) f(K_y a) f(K_z a), \quad \text{with} \quad f(x) = \frac{\sin^2 \frac{xL}{2}}{\sin^2 \frac{x}{2}}, \quad (2.20)$$

where $\mathbf{K}$ is not a reciprocal lattice vector here. Figure 2.14 shows $f(x)$ for two values of L. We observe that, apart from the Bragg diffraction points $2n\pi$, where $\mathbf{K}$ is a reciprocal lattice vector, this function remains small everywhere. *If the Bragg condition is not satisfied, the scattered intensity will therefore remain very low.* However, the maxima at points $x = 2n\pi$ become sharper and sharper as L increases, which supports the result (2.12). We thus conclude that the diffraction conditions are very precisely specified, and this will only be limited experimentally by the size of the diffracting crystal and the wavelength dispersion of the incident radiation.

> **Question 2.5.** For the alloy $Cu_{1-x}Au_x$ of Question 2.4, how does the diffraction pattern change when the Cu and Au atoms arrange themselves for the concentration x_0?

To determine the reciprocal lattice, and hence the Bravais lattice, it remains only to determine all the space directions in which Bragg diffraction occurs. Experimentally, it is not totally obvious how to determine the directions in which Bragg diffraction will occur. This is illustrated in Fig. 2.15a, which shows the *Ewald construction*, used to determine the diffraction directions for an incident wave $\mathbf{k}_0$ and a given position of the crystal. The points of the reciprocal lattice corresponding to this position of the crystal are shown in the left-hand figure, taking the origin Γ of the reciprocal lattice to be the point corresponding to the end of the vector $\mathbf{k}_0$. The diffraction directions are obtained by determining the points of the reciprocal lattice on a sphere, known as the Ewald sphere, with radius $|\mathbf{k}_0|$ and centered at the

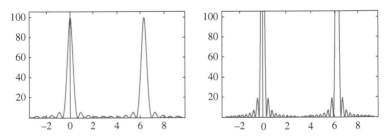

Fig. 2.14 The function $f(x)$ for x between $-\pi$ and 3π. *Left*: $L = 10$. *Right*: $L = 20$. Note that, on the right-hand graph, the peak is at the ordinate value 400

2.3 Determination of Crystal Structures

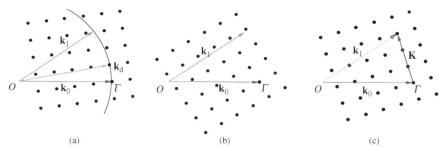

Fig. 2.15 Geometry of Bragg diffraction in reciprocal space. For an incident vector $\mathbf{k}_0$, diffraction directions such as $\mathbf{k}_d$ are obtained using the Ewald construction as shown in (**a**). The reciprocal lattice with origin Γ is depicted. For an incident vector $\mathbf{k}_0$ and a detector in the direction $\mathbf{k}_1$, no Bragg diffraction will generally be detected (**b**), except for certain specific orientations of the crystal, such that the associated reciprocal lattice is oriented as shown in (**c**), for example

origin O of $\mathbf{k}_0$. Diffraction directions such as $\mathbf{k}_d$ which join O to these points of the reciprocal lattice are few and far between.

If we have only one detector set in the direction $\mathbf{k}_1$, for example, there will generally be little diffraction in this direction. If the crystal is then rotated in space, according to (2.17), the reciprocal space will also rotate about its origin Γ. As can be seen from Fig. 2.15, Bragg diffraction will only be detected in the given direction $\mathbf{k}_1$ for very specific orientations of the crystal. Given that we are working in 3D space, it is clear that, for a randomly chosen orientation of the crystal relative to the incident radiation, we will only very rarely observe a Bragg diffraction peak (or spot).

There are several experimental methods to get round this difficulty, e.g., using powders rather than single crystals, but we shall not discuss the issue further here. Knowing the orientations at which diffraction occurs, we may then characterise the reciprocal lattice, and hence also the crystal symmetries and the size of the primitive cell in the crystal lattice itself.

2.3.2 Diffraction Intensity and Basis of the Primitive Cell

So far we have not considered the structure of the diffracting object in any detail. Suppose now that we observe the diffraction by a crystal whose primitive cell comprises N_a atoms, possibly of different chemical nature, at positions $\mathbf{r}_l$ inside the unit cell. In this case, a coefficient $\alpha^{(l)}_{\mathbf{k}_0,\mathbf{k}_1}$ is associated with each atom. The sum in (2.9) over all sites l leads to a diffracted amplitude in the direction $\mathbf{k}_1$ given by

$$\mathscr{A}(\mathbf{K}) = \sum_{l=1}^{N_a} \alpha^{(l)}_{\mathbf{k}_0,\mathbf{k}_1} \sum_n e^{-i\mathbf{K}\cdot(\mathbf{R}_n+\mathbf{r}_l)} = \left[\sum_{l=1}^{N_a} \alpha^{(l)}_{\mathbf{k}_0,\mathbf{k}_1} e^{-i\mathbf{K}\cdot\mathbf{r}_l}\right]\left(\sum_n e^{-i\mathbf{K}\cdot\mathbf{R}_n}\right). \quad (2.21)$$

There are thus two multiplicative factors in the expression for the diffracted intensity. One of these,

$$\mathscr{A}_0(\mathbf{K}) = \sum_n e^{-i\mathbf{K}\cdot\mathbf{R}_n},$$

depends only on the Bravais lattice of the crystal and is only nonzero when the Bragg condition (2.19) is satisfied. The second,

$$S(\mathbf{K}) = \sum_{l=1}^{N_a} \alpha_{\mathbf{k}_0,\mathbf{k}_1}^{(l)} e^{-i\mathbf{K}\cdot\mathbf{r}_l}, \qquad (2.22)$$

called the *structure factor*, depends only on the basis of the primitive cell and the nature of the diffracted radiation through the coefficients $\alpha_{\mathbf{k}_0,\mathbf{k}_1}^{(l)}$. If we measure the intensities of many Bragg diffraction spots, we can then deduce the structure of the primitive cell, provided we have a good understanding of the physical processes involved in the scattering of radiation by the atoms, since it is these processes that determine the values of the coefficients $\alpha_{\mathbf{k}_0,\mathbf{k}_1}^{(l)}$. Note that these coefficients will be different for X-ray diffraction, the X rays being scattered by electrons, and neutron diffraction, since neutrons are scattered by the atomic nuclei.

2.3.3 X-Ray Diffraction

In the specific case of X-ray diffraction, the coefficients $\alpha_{\mathbf{k}_0,\mathbf{k}_1}$ can be expressed very simply by considering the underlying physical process for the scattering of X rays by atoms. Indeed, at the high frequencies associated with X rays, the electric field of the electromagnetic wave couples predominantly with the electrons in the atom. Each electron vibrates in phase with the incident wave at its frequency ω. The electron charge thus has an oscillatory motion with angular frequency ω, and consequently emits an electromagnetic wave at this same angular frequency, in phase with the incident wave.

Consider an atom centered at the origin O and with electronic density distribution $\rho(\mathbf{u})$, where $\mathbf{u}$ indicates the position of an electron relative to O. The amplitude of the wave scattered in the direction $\mathbf{k}_1$ by the electronic density at O is proportional to the electronic density at this point:

$$a_0 \rho(\mathbf{0}) e^{i(\mathbf{k}_1\cdot\mathbf{r}-\omega t)}. \qquad (2.23)$$

Consider now an arbitrary point at position $\mathbf{u}$. In an analogous way to Fig. 2.11, the amplitude scattered in the direction $\mathbf{k}_1$ can thus be written

$$a_0 \rho(\mathbf{u}) e^{i[\mathbf{k}_1\cdot(\mathbf{r}-\mathbf{u})-\omega t]} e^{i\mathbf{k}_0\cdot\mathbf{u}}. \qquad (2.24)$$

The total wave diffracted in the direction $\mathbf{k}_1$ by all the electrons in the atom is therefore

2.3 Determination of Crystal Structures

$$a_{\text{diff}}(\mathbf{k_1}, \mathbf{r}) = a_0 e^{i(\mathbf{k_1} \cdot \mathbf{r} - \omega t)} \int \rho(\mathbf{u}) e^{-i(\mathbf{k_1} - \mathbf{k_0}) \cdot \mathbf{u}} d^3\mathbf{u}, \quad (2.25)$$

where $|\mathbf{k_1}| = |\mathbf{k_0}|$. This expression, valid for any molecular ensemble of electronic density $\rho(\mathbf{u})$, shows that *the diffraction amplitude of the X rays is directly related to the Fourier transform of the electronic density*. The diffraction pattern produced by an arbitrary object thus contains information about the structure of the diffracting object.

Going back to the case of a single atom and comparing (2.25) and (2.6), we define the *atomic form factor*

$$\boxed{\alpha_{\mathbf{k_0},\mathbf{k_1}} = \int \rho(\mathbf{u}) e^{-i(\mathbf{k_1}-\mathbf{k_0}) \cdot \mathbf{u}} d^3\mathbf{u} = f(\mathbf{k_1} - \mathbf{k_0})}. \quad (2.26)$$

For a crystal whose primitive cell contains N_a atoms at position $\mathbf{r}_l$, the diffraction amplitude is the Fourier transform of the total electronic density, which is the periodic reproduction of the electronic density of the primitive cell. The latter is given by (2.21), where the structure factor is simply

$$\boxed{S(\mathbf{K}) = \sum_{l=1}^{N_a} f_l(\mathbf{K}) e^{-i \mathbf{K} \cdot \mathbf{r}_l}}. \quad (2.27)$$

It can be determined once we know the form factors and the positions of the different atoms in the primitive cell. We thus find that the experimental determination of the intensities of the Bragg diffraction peaks or spots will be extremely useful for ascertaining the arrangement of atoms in the primitive cell of the Bravais lattice.

Let us consider a specific example to illustrate how the basis of the primitive cell affects the intensity of the Bragg peaks. Consider X-ray diffraction by a crystal of C_{60} whose fcc primitive cell is shown in Fig. 2.10. A suitable experimental setup records on a photographic film all the diffraction spots corresponding to the vectors $\mathbf{K}$ in one plane of reciprocal space (see the image on p. 24). The pattern observed can thus be used to directly visualise a plane of the reciprocal lattice and its symmetries. In the image on p. 24, the spots are such that the vectors ℓ_K separating them correspond to the vectors $\mathbf{K}$ in one plane of the reciprocal lattice up to a scale factor given by $|\mathbf{K}| = 2\pi |\ell_K|/\lambda L$. Here λ is the wavelength of the X rays and L is a length depending on the experimental setup. (Here $\lambda = 1.542$ Å and $L = 0.06$ m.)

Question 2.6. 1. Identify the body-centered cubic reciprocal lattice plane visualised in the image of p. 24. Deduce the dimensions of the primitive cell for C_{60}.
2. Some of the diffraction spots are very faint. Which vectors of the reciprocal lattice do these correspond to? To understand this, we take into account the fact that the X rays are scattered by electrons. We may consider the C_{60} molecule as a uniform charge distribution $p(r)$ over the surface of a hollow sphere of radius R_0, viz., $p(r) = A\delta(r - R_0)$. Calculate the structure factor for the C_{60} crystal. Deduce the radius R_0 of C_{60}.

This aspect of X-ray diffraction was very important in determining the spatial structure of complex molecules, such as biological molecules, which are commonly conserved in solution. By crystallising N molecules, one then benefits from the fact that, in a crystal, the Bragg diffraction spots have intensities that increase as N^2, whereas the intensity only increases as N when the molecules do not have a crystalline arrangement. This method was used to determine the structure of DNA and many other biologically important molecules.

Note that disorder or atomic and molecular motions modify the intensities of the Bragg diffraction spots. The effect of lattice vibrations is discussed in Problem 1: *Debye–Waller factor*.

2.4 Summary

A crystal lattice can be described as a combination of two entities: the Bravais lattice, which is a periodic arrangement of lattice points in space, and hence an abstract construction, and a material basis which is the actual physical entity associated with each node of the Bravais lattice. These are given by

$$R_\ell = \ell_1 \mathbf{a}_1 + \ell_2 \mathbf{a}_2 + \ell_3 \mathbf{a}_3 ,$$

where $(\mathbf{a}_1, \mathbf{a}_2, \mathbf{a}_3)$ is a vector basis specifying a primitive cell of the lattice. In the simplest cases, the material basis is a single atom, but it may be a much more complex physical entity, such as an arrangement of atoms, one or more molecules, and so on.

The structures of arbitrary molecular entities can be determined using diffraction methods. An incident wave of wave vector $\mathbf{k}_0$ is elastically scattered by the various objects making up the molecular entity. Interference between the scattered waves leads to a diffraction pattern. In the case of a crystal, the diffracted intensity is only significant in specific directions $\mathbf{k}_1$ satisfying

$$\mathbf{k}_1 - \mathbf{k}_0 = \mathbf{K} ,$$

where $\mathbf{K}$ is a vector belonging to another lattice called the reciprocal lattice for the actual lattice in real space. This relation is the Bragg diffraction condition.

The vectors $\mathbf{K}$ of the reciprocal lattice are defined by the condition

$$e^{i\mathbf{K}\cdot\mathbf{R}_\ell} = 1 ,$$

for any $\mathbf{R}_\ell$ belonging to the Bravais lattice of the crystal. The reference triad $(\mathbf{a}_1^*, \mathbf{a}_2^*, \mathbf{a}_3^*)$ of the reciprocal lattice is given by

$$\mathbf{a}_1^* = 2\pi \frac{\mathbf{a}_2 \wedge \mathbf{a}_3}{(\mathbf{a}_1 \cdot \mathbf{a}_2 \wedge \mathbf{a}_3)} ,$$

2.5 Answers to Questions

and cyclic permutations. The vectors $\mathbf{k}_1$ satisfying the Bragg diffraction condition can be used to determine the reciprocal lattice. The amplitude of the diffraction is related to the basis of the primitive cell of the crystal. It is given by

$$A(\mathbf{K}) = S(\mathbf{K}) \sum_n e^{-i\mathbf{K}\cdot\mathbf{R}_n} ,$$

where $S(\mathbf{K})$ is the structure factor of the basis of the crystal.

For X rays, scattering is due to electrons in the atomic orbitals, and each atom l at $\mathbf{r}_l$ is characterised by its atomic structure factor $f_l(\mathbf{k}_1 - \mathbf{k}_0)$. The structure factor of the primitive cell is then

$$S(\mathbf{K}) = \sum_{l=1}^{N_a} f_l(\mathbf{K}) e^{-i\mathbf{K}\cdot\mathbf{r}_l} .$$

2.5 Answers to Questions

Question 2.1

1. The alkane molecules are arranged in parallel rows. Note that two consecutive rows are staggered in a quincuncial arrangement. This is checked by looking at Fig. 2.1 at grazing incidence with respect to the axis of the molecules. The lattice is therefore centered rectangular. Figure 2.16 shows the centered rectangular conventional unit cell $(\mathbf{a}, \mathbf{b})$ and the primitive unit cell $(\mathbf{a}_1, \mathbf{a}_2)$ containing one alkane molecule. This is shown on part of the image and also on a molecular model.
2. It is an irregular polyhedron for the centered rectangular lattice and a hexagon for the hexagonal lattice (see Fig. 2.17).

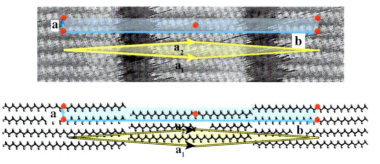

Fig. 2.16 Centered rectangular cell and primitive unit cell for the 2D alkane crystal of Fig. 2.1. These are shown on an enlarged portion of the image of Fig. 2.1 (*upper*) and on a molecular model (*lower*)

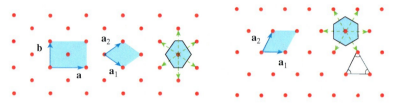

Fig. 2.17 First Brillouin zones of the centered rectangular and hexagonal 2D lattices: construction displayed *on the right* for each reciprocal lattice

Question 2.2

The honeycomb structure of graphene corresponds to a hexagonal lattice in which one in three sites have been removed. The primitive cell ($\mathbf{a}_1, \mathbf{a}_2$) is hexagonal with three times the area of the initial hexagonal cell, i.e., $3a^2\sqrt{3}/2$. It contains two carbon atoms placed at the nodes of two hexagonal sublattices of types A and B, which differ in the opposite orientations of their nearest neighbours. Each atom A has three nearest neighbours B and vice versa.

The Kagomé plane structure corresponds to a hexagonal lattice in which one site in four has been removed (one site out of two and in one line out of two). Its Bravais lattice is hexagonal with twice the lattice constant of the initial structure. The primitive cell contains 3 atoms of Cu^{2+} and 6 $(OH)^-$ (see Fig. 2.18).

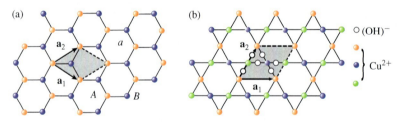

Fig. 2.18 Unit cell and bases of the Bravais lattices for the graphene sheet and for the Kagomé 2D planes of the hebersmithite crystal

Question 2.3

The crystal in Fig. 2.7b corresponds to an fcc lattice with full spheres at the lattice points and empty spheres obtained from the full ones by translation through $(a/4, a/4, a/4)$. There are thus two atoms per polyhedral primitive cell in Fig. 2.7a, shown by two empty spheres in Fig. 2.19, one at the corner of the cube and the other at $(a/4, a/4, a/4)$. There are 8 atoms per fcc unit cell: the vertex, the three centers of adjacent faces, and the 4 empty spheres in Fig. 2.7b.

Fig. 2.19 Primitive unit cell of the diamond structure. The atoms located on the two empty spheres represent the basis for this crystal

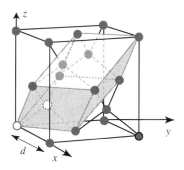

Question 2.4

When the atoms $Au_{1-x}Cu_x$ are randomly distributed over the sites of the fcc Bravais lattice, the crystal structure is not perfect. The atoms must be placed at the sites of the fcc lattice in a periodic manner to obtain a perfect arrangement. With two different atoms, viz., Cu and Au, the primitive cell of the perfect crystal will have at least one two-atom basis and hence a Bravais lattice with unit cell at least doubled in volume.

It should be fairly clear that it is not possible to associate the sites of the fcc structure two by two to define a new Bravais lattice. However, if we take 4 atoms with 3 the same, one of the atoms can be placed at the vertex of the fcc conventional unit cell and the three others at the centers of the adjacent faces (empty in Fig. 2.20a). We thereby construct a crystal with simple cubic Bravais lattice of side a and a basis of 4 atoms per cell.

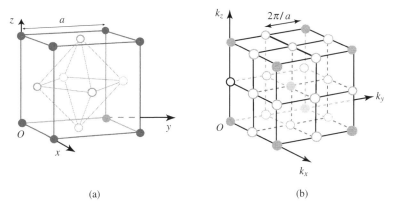

Fig. 2.20 (**a**) Simple cubic crystal structure of Cu_3Au. *Full spheres* represent Au and *empty spheres* Cu. (**b**) The reciprocal lattice of the fcc structure is the bcc lattice shown by *full spheres*. *Empty spheres* represent the extra reciprocal space nodes which appear for the ordered cubic structure of Cu_3Au

There are therefore two simple possibilities, those corresponding to Au_3Cu or Cu_3Au, i.e., $x = 0.25$ or $x = 0.75$. Naturally, we can imagine ordered solutions for lower concentrations of one of the metals, but there is no particular physical reason why they should ever occur.

Note that, in the ordered Cu_3Au structure, the gold atoms have no Au nearest neighbour. This structure is found to be stable on thermodynamical grounds, which indicates that Au atoms strongly repel one another. However, the Au_3Cu structure is not obtained experimentally, probably because the repulsion between copper atoms is not strong enough. The ordered AuCu alloy exists, but crystallises into a cubic lattice which bears no relation to the fcc structure of the pure metals.

Question 2.5

The ordered arrangement of the Au and Cu atoms can be detected by a diffraction method. The fcc structure of the disordered solid solution has a primitive cell of volume $a^3/4$ and a reciprocal lattice with a body-centered cubic conventional cell of side $4\pi/a$. The ordered lattice of Cu_3Au shown in Fig. 2.20a has a cubic primitive cell of volume a^3, and its reciprocal lattice is cubic with side $2\pi/a$. There are therefore many diffraction spots for the ordered alloy. In Fig. 2.20b, the lattice points of the body-centered cubic reciprocal lattice are represented by full spheres, while the extra nodes of the simple cubic lattice are represented by empty spheres. The presence of order can thus be revealed by the appearance of diffraction spots associated with these new vectors in reciprocal space.

Question 2.6

1. The C_{60} lattice is face-centered cubic with conventional unit cell of side a. Its reciprocal lattice is body-centered cubic with conventional cell of side $4\pi/a$ (see Fig. 2.20b). Spots are therefore expected for

$$\mathbf{K} = h\mathbf{a}^* + k\mathbf{b}^* + \ell\mathbf{c}^*, \qquad (2.28)$$

with $|\mathbf{a}^*| = |\mathbf{b}^*| = |\mathbf{c}^*| = 2\pi/a$ and h, k, ℓ of the same parity. In the observed plane of the reciprocal lattice, the spots form a square lattice. The only planes of the body-centered cubic reciprocal lattice with this property are the planes passing through one of the faces of the cube, i.e., $(\mathbf{a}^*, \mathbf{b}^*)$, or $(\mathbf{b}^*, \mathbf{c}^*)$, or $(\mathbf{c}^*, \mathbf{a}^*)$. Assuming that this is the plane $(\mathbf{a}^*, \mathbf{b}^*)$, the $\mathbf{a}^*$ and $\mathbf{b}^*$ axes are shown in Fig. 2.21. The spots correspond to the vectors $\mathbf{K} = (h, k, 0)$ with h and k even. If ℓ_{K_0} is the spacing of the observed square lattice which corresponds to $K_0 = 4\pi/a$, it follows that

$$\frac{4\pi}{a} = \frac{2\pi}{\lambda}\frac{\ell_{K_0}}{L}, \quad \text{and then } a = 2\lambda\frac{L}{\ell_{K_0}}. \qquad (2.29)$$

2.5 Answers to Questions

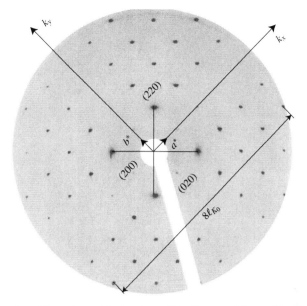

Fig. 2.21 Indexation of the reciprocal lattice plane of the C_{60} crystal detected by X-ray diffraction (image on p. 24). See answer to Question 2.6

By measuring over various rows corresponding to 8 times the lattice spacing, and knowing that the film diameter is 12 cm, we obtain $8\ell_{K_0} = 10.45 \pm 0.05$ cm in Fig. 2.21, so finally,

$$a = 14.17 \pm 0.07 \text{ Å}. \qquad (2.30)$$

2. Note that the spots $(h,0,0)$ and $(0,k,0)$ are rather faint. In fact their intensity is found to be at least 10^3 times lower than the intensity of the spots (220). Set $\rho(r) = A\delta(r-R)$. The total number of electrons is 360 per C_{60}. Hence,

$$360e = \int \rho(r)\mathrm{d}^3 r = \int_0^\infty 4\pi r^2 \delta(r-R) A \mathrm{d}r = A 4\pi R^2,$$

which yields

$$A = \frac{360e}{4\pi R^2}.$$

Then

$$S(\mathbf{K}) = \int \rho(r) e^{i\mathbf{K}\cdot\mathbf{r}} \mathrm{d}^3 r$$

$$= \frac{360e}{4\pi R^2} \int_0^\infty r^2 \delta(r-R) \mathrm{d}r \int_0^{2\pi} \mathrm{d}\phi \int_0^\pi e^{iKr\cos\theta} \sin\theta \mathrm{d}\theta$$

$$= \frac{360e}{2R^2} \int_0^\infty 2\frac{\sin Kr}{Kr} r^2 \delta(r-R) dr$$

$$= 360e \frac{\sin KR}{KR},$$

and the spots have intensity

$$I \propto \frac{\sin^2 KR}{(KR)^2}. \tag{2.31}$$

The intensity vanishes for $KR = n\pi$ with n integer. The fact that the spots corresponding to $K = n(4\pi/a)$ are almost extinguished thus shows that $4\pi R/a \simeq \pi$, and then $R \sim a/4$ or $R \simeq 3.54$ Å. With a relative intensity of 10^{-3} for the spot (200) compared with the spot (220), the error obtained for R is found to be

$$R = 3.54 \pm 0.04 \text{ Å}. \tag{2.32}$$

In fact, using (2.31) and taking into account the measured intensities for all the spots, the error in R is much lower, namely $R = 3.52 \pm 0.01$ Å.

Chapter 3
Electronic Structure of Solids: Metals and Insulators

Contents

- 3.1 Electrons in a Periodic Potential 53
  - 3.1.1 Bloch's Theorem in Three Dimensions 53
  - 3.1.2 General Properties of Bloch States 55
  - 3.1.3 Electrons in a Weak Periodic Potential 56
  - 3.1.4 First Brillouin Zone 59
- 3.2 Band Structure of Solids 60
  - 3.2.1 Linear Chain 60
  - 3.2.2 Band Structure of Two-Dimensional Solids 64
  - 3.2.3 Three-Dimensional Band Structures 68
- 3.3 Metals, Insulators, and Semiconductors 72
  - 3.3.1 Statistical Physics of Fermions 72
  - 3.3.2 Real Solids: Conductor or Insulator? 74
  - 3.3.3 Introduction to the Semiconductor Case 77
  - 3.3.4 The Special Case of Graphene 78
  - 3.3.5 Thermodynamic Properties of Metals 80
- 3.4 Experimental Determination of Band Structures 84
  - 3.4.1 Optical Absorption in Insulators 84
  - 3.4.2 Tunneling Effect and Scanning Tunneling Microscopy 85
  - 3.4.3 Angle-Resolved Photoemission Spectroscopy 88
- 3.5 Summary 90
- 3.6 Answers to Questions 92

H. Alloul, *Introduction to the Physics of Electrons in Solids*, Graduate Texts in Physics,
DOI 10.1007/978-3-642-13565-1_3, © Springer-Verlag Berlin Heidelberg 2011

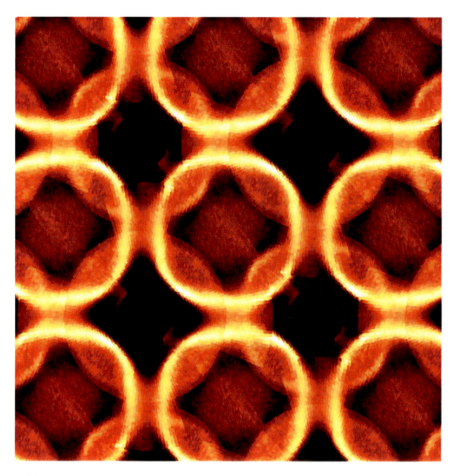

As illustrated on the cover, the Fermi surface of a metallic material can be represented in reciprocal space using angle-resolved photoemission experiments. In this image, the Fermi surface of the CuO_2 plane of the cuprate superconductor $BiSr_2CaCu_2O_8$ is reconstructed experimentally in the reciprocal space plane k_x, k_y. Question 3.4 examines this experimental result in more detail. Image courtesy of S. Borisenko, constructed from experimental results in Kordyuk, A.A., Borisenko, S.V., Golden, M.S., Legner, S., Nenkov, K.A., Knupfer, M., Fink, J., Berger, H., Forró, L., Follath, R.: Phys. Rev. B **66**, 014502 (2002)

3.1 Electrons in a Periodic Potential

In this chapter, we introduce the generalisation needed to discuss the electronic properties of solids in more detail. We restrict the discussion to features that can be described using the approximations presented in Chap. 1. Considering fixed atomic positions, and averaging electron–electron interactions, the task boils down to solving a one-electron problem. We also restrict to the case of crystalline solids, having introduced the basic ideas in Chap. 2.

In Sect. 3.1, we discuss the general form of the Bloch theorem and its implications for the band structure of periodic solids. While the tight-binding approximation provided a first handle on the notion of energy band, we shall see that this notion can also be introduced in the framework of quite the opposite approximation, in which the periodic potential is weak. This will reveal the importance of the reciprocal lattice and lead to a general definition of the Brillouin zones outlined in our discussion of the linear atomic chain in Chap. 1. We also establish a link between the limiting cases of weak and strong atomic potentials. We can then establish certain general rules about the band structures for the chain, but also for two- and three-dimensional solids (see Sect. 3.2). In Sect. 3.3, we use the band model to discuss some physical properties of solids and in particular their thermodynamic properties. We find that this model implies only two possible states for a solid, viz., either conducting (metallic) and paramagnetic, or insulating and non-magnetic. In addition, without carrying out any detailed calculation, we can make several quantitative predictions that can be compared with experiment. A discussion of these points will bring out the strengths and weaknesses of the band model. Finally, in Sect. 3.4, we describe some experimental methods for determining the electron density of states or the band structure of solids.

3.1 Electrons in a Periodic Potential

We saw in Chap. 1 that, under the tight-binding approximation, the atomic electron states broaden to give rise to energy bands. We shall see here that this organisation of the structure of the electronic states into energy bands is in fact a general property of the electronic energy spectrum in the presence of a periodic potential. We begin in Sect. 3.1.1 by generalising the Bloch theorem to three dimensions. We shall then see how the energy band structure can be obtained in a simple way from that of the free electrons in the weak limit of the periodic potential (see Sect. 3.1.3). The notion of Brillouin zone can then be extended in a very general way (see Sect. 3.1.4).

3.1.1 Bloch's Theorem in Three Dimensions

In an infinite crystal, or more precisely, in a crystal subject to periodic boundary conditions, electrons located at positions separated by a direct lattice translation $\mathbf{R}_l$ [of the form (2.1)] will feel the same crystal potential, which means that the one-electron Hamiltonian H_1 will be invariant under the lattice translations associated with the operators $\hat{T}$ defined by $\hat{T}_{R_l}f(\mathbf{r}) = f(\mathbf{r}+\mathbf{R}_l)$, for any function $f(\mathbf{r})$. It is straightforward to generalise Bloch's theorem to a 3D space.[1]

The solutions of the Schrödinger equation

$$\left[\frac{p^2}{2m} + V(\mathbf{r})\right]\psi(\mathbf{r}) = E\psi(\mathbf{r}), \qquad (3.1)$$

where the crystal potential $V(\mathbf{r})$ has the periodicity of the crystal, are called *Bloch functions*. They have the form

$$\psi_{n,\mathbf{k}}(\mathbf{r}) = e^{i\mathbf{k}\cdot\mathbf{r}} u_{n,\mathbf{k}}(\mathbf{r}), \qquad (3.2)$$

where the function $u_{n,\mathbf{k}}(\mathbf{r})$ is periodic for the translations of the direct lattice:

$$u_{n,\mathbf{k}}(\mathbf{r}+\mathbf{R}_l) = u_{n,\mathbf{k}}(\mathbf{r}). \qquad (3.3)$$

A state is thus specified by four quantum numbers, namely the three components k_x, k_y, k_z of the vector $\mathbf{k}$ and an integer n, because (3.1) can have several solutions for a given $\mathbf{k}$. One consequence of (3.3) is that

$$\psi_{n,\mathbf{k}}(\mathbf{r}+\mathbf{R}_l) = e^{i\mathbf{k}\cdot\mathbf{R}_l} \psi_{n,\mathbf{k}}(\mathbf{r}). \qquad (3.4)$$

This wave function is such that the resulting physical properties at points $\mathbf{r}$ and $\mathbf{r}+\mathbf{R}_l$ are the same, because the wave function differs only by a phase factor that is independent of $\mathbf{r}$. In particular, the electronic density $|\psi_{n,\mathbf{k}}(\mathbf{r})|^2$ is periodic in space.

Furthermore, the periodic boundary conditions for a rhombohedral solid containing many primitive cells (see Fig. 3.1a), with edges $N_1\mathbf{a}_1, N_2\mathbf{a}_2, N_3\mathbf{a}_3$, can be written in the form

$$\psi(\mathbf{r}+N_i\mathbf{a}_i) = \psi(\mathbf{r}), \quad i=1,2,3. \qquad (3.5)$$

Comparing (3.5) with the form of the Bloch functions implies that the values of $\mathbf{k}$ must be quantised, with $\mathbf{k}\cdot(N_i\mathbf{a}_i) = 2\pi n_i$, where the n_i are integers, which leads to

$$\mathbf{k} = \sum_{i=1,2,3} \frac{n_i}{N_i}\mathbf{a}_i^*.$$

For a macroscopic solid, these values of $\mathbf{k}$ will be extremely close together on the scale of the primitive unit cell of the reciprocal lattice (see Fig. 3.1b), since the N_i are large. Note that, in the primitive cell $(\mathbf{a}_1^*, \mathbf{a}_2^*, \mathbf{a}_3^*)$ of the reciprocal lattice, *there will be $N_1N_2N_3$ quantised values of $\mathbf{k}$, exactly corresponding to the number of primitive cells in the crystal.*

---

[1] The generalisation to several dimensions is quite simple. We introduce three primitive translation operators $\hat{T}_{\mathbf{a}_i}$, for $i = 1, 2, 3$, which are easily shown to commute with one another and with the Hamiltonian. We can then apply the argument in the box in Sect. 1.2.2a almost word for word, to arrive at (3.2) and (3.3).

3.1 Electrons in a Periodic Potential

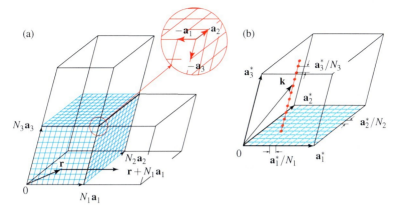

Fig. 3.1 (a) Periodic boundary conditions applied to a solid containing $N_1N_2N_3$ unit cells. (b) Quantised values of **k** in the primitive unit cell $(\mathbf{a}_1^*, \mathbf{a}_2^*, \mathbf{a}_3^*)$ of the reciprocal lattice

The vectors **k** play a key role when describing the properties of crystalline solids. They belong to the reciprocal space attached to the crystal, generated by the basis vectors $(\mathbf{a}_1^*, \mathbf{a}_2^*, \mathbf{a}_3^*)$ of the reciprocal lattice.

3.1.2 General Properties of Bloch States

Consider the Bloch function with wave vector **k**, such that $\mathbf{k} = \mathbf{k}_0 + \mathbf{K}$, where **K** is a vector of the reciprocal lattice. We shall show that this Bloch function $\psi_{n,\mathbf{k}}(\mathbf{r})$ is also a Bloch function for the wave vector $\mathbf{k}_0$:

$$\psi_{n,\mathbf{k}}(\mathbf{r}) = e^{i\mathbf{k}\cdot\mathbf{r}} u_{n,\mathbf{k}}(\mathbf{r}) = e^{i\mathbf{k}_0\cdot\mathbf{r}} e^{i\mathbf{K}\cdot\mathbf{r}} u_{n,\mathbf{k}}(\mathbf{r}) . \tag{3.6}$$

The function $\exp(i\mathbf{K}\cdot\mathbf{r})u_{n,\mathbf{k}}(\mathbf{r})$ is periodic as expressed by (3.3). Indeed, using the definition (2.11) of the reciprocal lattice, it follows that

$$e^{i\mathbf{K}\cdot(\mathbf{r}+\mathbf{R}_l)} u_{n,\mathbf{k}}(\mathbf{r}+\mathbf{R}_l) = e^{i\mathbf{K}\cdot\mathbf{r}} u_{n,\mathbf{k}}(\mathbf{r}) . \tag{3.7}$$

The function (3.6) is thus a solution of the Schrödinger equation with the same energy for both **k** and $\mathbf{k}_0 = \mathbf{k} - \mathbf{K}$. The reciprocal space is therefore much bigger than it need be to specify the Bloch functions, because it contains all the points **k** and $\mathbf{k} - \mathbf{K}$, with **K** an arbitrary vector in the reciprocal lattice, for which the quantum states are identical. Since the quantum states are only defined up to a vector of the reciprocal lattice, they can *all* be specified *by restricting to a particular primitive unit cell of the reciprocal lattice*.

The form (3.2) and (3.3) for the wave functions has one very important consequence for any physical measurement involving external radiation (photons, neutrons, electrons), generally represented by a plane wave $\exp(i\mathbf{q}\cdot\mathbf{r})$. The wave function of an electron in a Bloch state $\psi_{n,\mathbf{k}}(\mathbf{r})$, where **k** is a vector belonging to the primitive cell, can be written as a superposition of plane waves with the general form $\exp[i(\mathbf{k}+\mathbf{K})\cdot\mathbf{r}]$, where $\mathbf{K}$ is a vector of the reciprocal lattice. In an interaction between incident radiation $\mathbf{q}_{in}$ and a Bloch state $\mathbf{k}_{in}$, scattered to $\mathbf{q}_{out}$ and $\mathbf{k}_{out}$, the transition probability can only be nonzero for partial waves conserving the total momentum:

$$\mathbf{q}_{in} + \mathbf{k}_{in} + \mathbf{K}_{in} = \mathbf{q}_{out} + \mathbf{k}_{out} + \mathbf{K}_{out},$$

that is

$$\mathbf{q}_{in} + \mathbf{k}_{in} = \mathbf{q}_{out} + \mathbf{k}_{out} + \Delta\mathbf{K}. \tag{3.8}$$

This means that, in a crystal, *the quasi-momentum (or crystal momentum)* $\mathbf{k}$ *plays the same role of a conserved quantity as the true momentum* $\mathbf{p}$ *in vacuum*. The only precaution required here concerns the existence of the term $\Delta\mathbf{K}$ in (3.8), which is a vector of the reciprocal lattice: if this term vanishes, we speak of normal scattering, whereas for $\Delta\mathbf{K} \neq 0$, we speak of umklapp scattering. For reasons of energy conservation, umklapp effects can often be neglected. It should be noted that, in a crystal, the average of the true momentum $\mathbf{p}$ can never be treated as a conserved quantity. In fact, $\mathbf{p}$ is not conserved, due to the presence of the potential in the Hamiltonian.

3.1.3 Electrons in a Weak Periodic Potential

Band structure calculations based on the tight-binding approximation, as discussed in Chap. 1, apply in particular to the case where there is a wide separation between atomic levels compared with the hopping integrals. In other cases, for example, the case of valence electrons of certain metals, which may have a high kinetic energy, the periodic potential can be treated as weak. For these *nearly free electrons*, the crystal potential may even be treated as a perturbation. We demonstrate below that the wave function of the electron is then close to a plane wave, and that the energy varies to a first approximation as it would for free electrons. However, even though the perturbation due to the periodic potential may be weak, it may become important for specific values related to the crystal periodicity. This idea will lead to a more general definition of the first Brillouin zone.

For a free electron gas of volume Ω, the Hamiltonian

$$H_0 = p^2/2m_0$$

has plane wave eigenfunctions

$$\psi_\mathbf{k}(\mathbf{r}) = \Omega^{-1/2} e^{i\mathbf{k}\cdot\mathbf{r}}, \tag{3.9}$$

with associated energy

$$\varepsilon_\mathbf{k} = \hbar^2 k^2/2m_0. \tag{3.10}$$

In one dimension, the band structure thus reduces to a simple parabola. In a crystal, the one-electron Hamiltonian also includes the potential due to the ions given in (1.7):

3.1 Electrons in a Periodic Potential

$$V(\mathbf{r}) = \sum_{l=1}^{N_n} V_{at}(\mathbf{r} - \mathbf{R}_l) . \tag{3.11}$$

We assume here that the potential is very weak, and can thus be treated as a perturbation in comparison with H_0. We know that, if we consider the non-degenerate states of the unperturbed Hamiltonian, the effect of the perturbation is to alter the energy levels and wave functions in the following way [2, Chap. 9]:

$$E_{\mathbf{k}} = \varepsilon_{\mathbf{k}} + \langle \mathbf{k}|V|\mathbf{k}\rangle + \sum_{\mathbf{k}' \neq \mathbf{k}} \frac{|\langle \mathbf{k}|V|\mathbf{k}'\rangle|^2}{\varepsilon_{\mathbf{k}} - \varepsilon_{\mathbf{k}'}} + \cdots , \tag{3.12}$$

$$|\tilde{\mathbf{k}}\rangle = |\mathbf{k}\rangle + \sum_{\mathbf{k} \neq \mathbf{k}'} \frac{\langle \mathbf{k}|V|\mathbf{k}'\rangle}{\varepsilon_{\mathbf{k}} - \varepsilon_{\mathbf{k}'}} |\mathbf{k}'\rangle + \cdots . \tag{3.13}$$

These results are obtained by treating the perturbation to first and second order and assuming that $|\langle \mathbf{k}|V|\mathbf{k}'\rangle| \ll |\varepsilon_{\mathbf{k}} - \varepsilon_{\mathbf{k}'}|$, i.e., that the perturbation only couples states with widely separated energies.

To first order in perturbation theory, the energy shift is

$$\langle \mathbf{k}|V|\mathbf{k}\rangle = \frac{1}{\Omega} \int e^{-i\mathbf{k}\cdot\mathbf{r}} V(\mathbf{r}) e^{i\mathbf{k}\cdot\mathbf{r}} d^3\mathbf{r} = \frac{1}{\Omega} \int V(\mathbf{r}) d^3\mathbf{r} . \tag{3.14}$$

The energies of the plane waves $|\mathbf{k}\rangle$, which are particular Bloch functions with $u_{\mathbf{k}}(\mathbf{r}) \equiv 1$, are simply shifted rigidly by an energy that is independent of the state under consideration, this energy being the average of the periodic potential. This effect can be neglected, since it does not change the parabolic shape of the variation of $E_{\mathbf{k}}$ as a function of $|\mathbf{k}|$.

Let us consider the case of a linear chain, where it will be easier to represent $E_{\mathbf{k}}$ graphically. The existence of the periodic potential means that periodic energy eigenvalues must be obtained in reciprocal space. To first order in perturbation theory, by continuity with the case where the potential is identically zero, the electron states of the free electron parabola must be translated by all lattice vectors $p(2\pi/a)\mathbf{x}$ (with p an integer) of the reciprocal lattice. This generates curves $E_{\mathbf{k}}$ that are periodic in reciprocal space. We thus obtain the *extended zone band structure* in $\mathbf{k}$ space (see Fig. 3.2a). We can also focus on the primitive unit cell $[-\pi/a, \pi/a]$ of reciprocal space which we called the first Brillouin zone, thereby obtaining the *restricted zone band structure*, in which the parabola appears to be folded up (see Fig. 3.2b), to generate all the values of the energy for $k \in [-\pi/a, \pi/a]$.

According to (3.12), the potential only affects the energy of a state $|\mathbf{k}\rangle$ to second order if it couples it to other states $|\mathbf{k}'\rangle$ with similar energies by a nonzero matrix element $\langle \mathbf{k}|V|\mathbf{k}'\rangle$. To determine this matrix element, we use the fact that the potential $V(\mathbf{r})$ is periodic. In one dimension, $V(x)$ is thus expanded in a Fourier series containing all the harmonics of $(2\pi/a)x$:

$$V(x) = \sum_{p=-\infty}^{+\infty} V_{p(2\pi/a)} e^{ip(2\pi/a)x} . \tag{3.15}$$

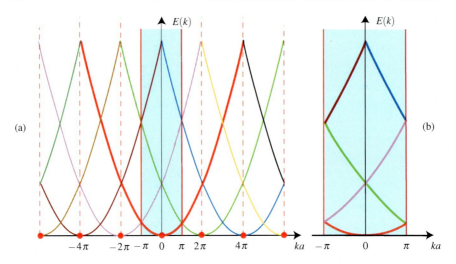

Fig. 3.2 In the presence of a very weak periodic potential, the energies of the free electrons are not modified. However, in the quasi-momentum space, (**a**) the 1D parabolic band is translated in such a way as to reveal the periodicity associated with the potential (extended zone band structure). (**b**) If we restrict to the first Brillouin zone, we obtain an $E(k)$ diagram in which the parabola is folded up (restricted zone band structure)

The terms in this expansion are nothing other than the products $\mathbf{K} \cdot \mathbf{x}$, where $\mathbf{K}$ is a vector of the reciprocal lattice. This Fourier series generalises for 3D periodic functions, and the potential is then described by a Fourier series involving all the vectors $\mathbf{K}$ of the reciprocal lattice:

$$V(\mathbf{r}) = \sum_{\mathbf{K}} V_{\mathbf{K}} e^{i\mathbf{K} \cdot \mathbf{r}}, \qquad (3.16)$$

with $V_{-\mathbf{K}} = V_{\mathbf{K}}^*$, since $V(\mathbf{r})$ is real. We thus have

$$\langle \mathbf{k} | V | \mathbf{k}' \rangle = \frac{1}{\Omega} \sum_{\mathbf{K}} V_{\mathbf{K}} \int e^{i(\mathbf{k}' - \mathbf{k} + \mathbf{K}) \mathbf{r}} d^3 \mathbf{r} = \sum_{\mathbf{K}} V_{\mathbf{K}} \delta(\mathbf{k}' - \mathbf{k} + \mathbf{K}). \qquad (3.17)$$

This matrix element is only nonzero for $\mathbf{k}' = \mathbf{k} - \mathbf{K}$. We may thus write (3.12) and (3.13) in the form

$$E_{\mathbf{k}} = \varepsilon_{\mathbf{k}} + \sum_{\mathbf{K} \neq 0} \frac{|V_{\mathbf{K}}|^2}{\varepsilon_{\mathbf{k}} - \varepsilon_{\mathbf{k}-\mathbf{K}}}, \qquad (3.18)$$

$$|\tilde{\mathbf{k}}\rangle = |\mathbf{k}\rangle + \sum_{\mathbf{K} \neq 0} \frac{V_{\mathbf{K}}}{\varepsilon_{\mathbf{k}} - \varepsilon_{\mathbf{k}-\mathbf{K}}} |\mathbf{k} - \mathbf{K}\rangle. \qquad (3.19)$$

As indicated above, these expansions are only valid if $|V_{\mathbf{K}}| \ll |\varepsilon_{\mathbf{k}} - \varepsilon_{\mathbf{k}-\mathbf{K}}|$. In particular, if $\varepsilon_{\mathbf{k}} \sim \varepsilon_{\mathbf{k}-\mathbf{K}}$, the states are quasi-degenerate and the above perturbation

3.1 Electrons in a Periodic Potential   59

expansion is no longer applicable. Note that the condition $\varepsilon_{\mathbf{k}} = \varepsilon_{\mathbf{k}-\mathbf{K}}$ on the states $\mathbf{k}$, viz., $\mathbf{k}^2 = (\mathbf{k}-\mathbf{K})^2$, is obtained for

$$2\mathbf{k} \cdot \mathbf{K} = \mathbf{K}^2, \tag{3.20}$$

which is satisfied if $\mathbf{k}$ belongs to the orthogonally bisecting plane of $\mathbf{K}$, i.e., if it belongs to a *Bragg plane*. In this case, even a very weak periodic potential will have a significant effect on the electron energy levels. This is not surprising, because we know from the discussion in Chap. 2 that a plane wave with a vector $\mathbf{k}$ belonging to a Bragg plane will be diffracted. We shall see more precisely in Sect. 3.2 how this situation arises by carrying out a detailed calculation in the case of the linear chain. In one dimension, the Bragg condition is satisfied when $k = m\pi/a$, and it is clear from Fig. 3.2a that, in these cases, the levels corresponding to two translated branches of the initial free electron parabola are indeed degenerate.

3.1.4 First Brillouin Zone

The Bragg planes are thus of singular importance for understanding the electronic structure of a solid. These planes, which orthogonally bisect the reciprocal lattice vectors taken from the origin, bound a particular primitive cell of the reciprocal lattice, namely the Wigner–Seitz cell constructed around the origin, as defined in Chap. 2. It is a primitive unit cell of the reciprocal lattice which specifies the *first Brillouin zone*. Recall that it contains all points of reciprocal space that are closer to the origin $\mathbf{k} = 0$ than to any other reciprocal lattice node. It contains all points $\mathbf{k}$ such that the line segment $O\mathbf{k}$ does not intersect any orthogonally bisecting plane of the vectors of the reciprocal lattice leaving O. The volume of the first Brillouin zone is the same as that of any primitive unit cell of the reciprocal lattice, i.e., $(2\pi)^3/v$, where v is the volume of the primitive cell in the direct lattice. In one dimension, it corresponds to the interval $[-\pi/a, \pi/a]$.

> **Question 3.1.** Find the expression for the vectors of the reciprocal lattice in two dimensions. Determine the reciprocal lattices for the hexagonal, centered rectangular, and oblique lattices, together with their first Brillouin zones.

The first Brillouin zone of the cubic lattice, e.g., CsCl (see Fig. 2.8a), is a cube of side $2\pi/a$ with center at the origin $\mathbf{k} = 0$ (see Fig. 3.3a). The first Brillouin zone for a face-centered cubic Bravais lattice, e.g., Si, Ge, GaAs, is shown in Fig. 3.3b. If a is the edge of the cube in the actual crystal and Γ the origin of the reciprocal lattice, the points of symmetry of the reciprocal lattice shown in the figure are respectively $(\pi/a, \pi/a, \pi/a)$ for L and $(2\pi/a, 0, 0)$ for X. The coordinates of equivalent points under the symmetries of the cube are easily deduced.

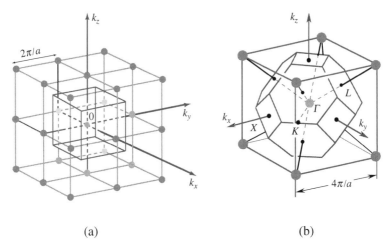

Fig. 3.3 (**a**) Reciprocal lattice of the cubic lattice with side a. The first Brillouin zone, a cube of side $2\pi/a$, is illustrated. (**b**) The body-centered cubic lattice is the reciprocal lattice of the face-centered cubic lattice. The first Brillouin zone is shown. This is the region bounded by the orthogonally bisecting planes of the vectors joining the centre Γ of the cube to its vertices, and by the faces of the cube, which are the orthogonally bisecting planes of the vectors joining Γ to the centers of the neighbouring cubes

3.2 Band Structure of Solids

3.2.1 Linear Chain

To obtain a better understanding of the energy level distribution of the Bloch states for a linear chain, in Sect. 3.2.1a, we continue the perturbation calculation for **k** satisfying the Bragg condition, which is a Bragg point for the 1D case. (In 2D, this would be a Bragg line, and in 3D, a Bragg plane, as we have already referred to it.) We will then be able to show in Sect. 3.2.1b that a range of forbidden energies appears in the distribution of energy levels. Finally, in Sect. 3.2.1c, we shall see how to make a smooth transition from a situation in which the atomic potential is weak to one in which it is strong.

a. Perturbation Calculation for a Bragg Point

Consider the degenerate eigenstates corresponding to the Bragg points $k = \pm\pi/a$, with energy $\varepsilon_0 = \pi^2\hbar^2/2m_0a^2$ and wave functions

$$\psi_{\pi/a} = L^{-1/2}e^{i\pi x/a}, \qquad \psi_{-\pi/a} = L^{-1/2}e^{-i\pi x/a},$$

where L is the length of the chain. To find out how these eigenstates are modified, we keep only those terms corresponding to the Fourier series expansion of the potential, viz.,

3.2 Band Structure of Solids

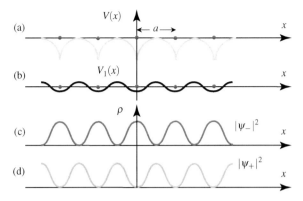

Fig. 3.4 (a) Position dependence of the electrostatic potential energy of an electron subject to the Coulomb field of the ions in a linear chain of atoms. (b) Potential limited to its Fourier coefficients obtained for $k = 2\pi/a$ and $k = -2\pi/a$, viz., $V_1(x)$ given in (3.21). (c) and (d) Probability of finding the electron in different positions for the states ψ_- and ψ_+

$$V_1(x) = V_{2\pi/a}\left[\exp(i2\pi x/a) + \exp(-i2\pi x/a)\right] = 2V_{2\pi/a}\cos(2\pi x/a), \quad (3.21)$$

where $V_{2\pi/a}$ is negative because the potential has minima at the positions of the ions (see Fig. 3.4b). In the basis of states $\psi_{\pi/a}$ and $\psi_{-\pi/a}$, the Hamiltonian takes the matrix form

$$H_1 = \begin{pmatrix} \varepsilon_0 & V_{2\pi/a} \\ V_{2\pi/a} & \varepsilon_0 \end{pmatrix}. \quad (3.22)$$

This has eigenstates

$$\psi_- = \frac{1}{\sqrt{2}}\left(\psi_{\pi/a} + \psi_{-\pi/a}\right) = \sqrt{\frac{2}{L}}\cos\frac{\pi}{a}x, \quad \text{with } E_- = \varepsilon_0 - |V_{2\pi/a}|, \quad (3.23)$$

$$\psi_+ = \frac{1}{\sqrt{2}}\left(\psi_{\pi/a} - \psi_{-\pi/a}\right) = i\sqrt{\frac{2}{L}}\sin\frac{\pi}{a}x, \quad \text{with } E_+ = \varepsilon_0 + |V_{2\pi/a}|. \quad (3.24)$$

The potential $V_1(x)$ thus removes the degeneracy of the levels ε_0, and the wave functions shown in Fig. 3.4 correspond to stationary waves. The probability of finding the electron near an ion is increased for the state ψ_- with lowest energy.

b. Band Structure of Nearly Free Electrons

The above perturbation calculations show that the only free electron states to be modified significantly by the weak periodic potential are those close to a Bragg plane, i.e., with the property that $\mathbf{k} \sim p\pi/a$ for the linear chain. The perturbing potential only couples states for which the vectors $\mathbf{k}$ differ from a vector of the reciprocal lattice, i.e., from $p2\pi/a$ for the linear chain. In Fig. 3.5a, double-headed arrows indicate states coupled by the term $V_1(x)$ in the potential. For $k \simeq \pi/a$, (3.18) gives the change in E_k:

$$E_k = \varepsilon_k + \frac{|V_{2\pi/a}|^2}{\varepsilon_k - \varepsilon_{(k+2\pi/a)}} + \frac{|V_{2\pi/a}|^2}{\varepsilon_k - \varepsilon_{(k-2\pi/a)}}. \quad (3.25)$$

Here the first correction term can be neglected as its energy denominator corresponds to free electron states far apart in energy. The second correction term is negative for $k \lesssim \pi/a$ and positive for $k \gtrsim \pi/a$, which results in a repulsion between the states in the first two bands relative to the free electron parabola, as shown in Fig. 3.5. For $k \to \pi/a$, E_k tends to the values E_- and E_+ determined in the case where the states are degenerate.

In the energy interval

$$\varepsilon_0 - |V_{2\pi/a}| < E < \varepsilon_0 + |V_{2\pi/a}|, \quad (3.26)$$

there is no stationary state of the energy. The presence of the periodic potential $V_1(x)$ forces open a *band gap* near $k = \pm\pi/a$. Figure 3.5 shows how the degeneracy is removed in the unfolded and folded band structures.

Note that by applying the same argument to the other degenerate cases corresponding to the states $k = \pm p\pi/a$, and taking into account the order p Fourier components of the potential

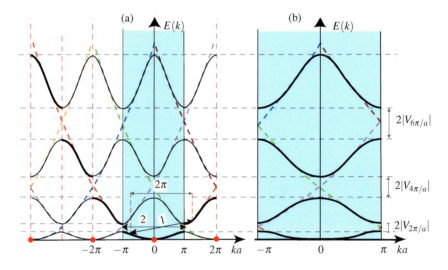

Fig. 3.5 Energies of the electron states for a linear chain in the approximation where the electrons are assumed to be nearly free: (**a**) extended zone and (**b**) restricted zone band structure. The free electron parabolas in Fig. 3.2 are shown as *dashed curves*, while the *continuous curves* take into account the effect of the crystal potential as a perturbation. In (**a**), the states coupled by the Fourier component $V_1(x)$ of the potential are joined by a *double-headed arrow*: the first (1) couples states that are non-degenerate in the absence of the periodic potential, while the second (2) couples states that are degenerate in the absence of $V_1(x)$. Band gaps appear. Their half-widths are the Fourier components $\ell = 1, 2, 3$ of the periodic potential

3.2 Band Structure of Solids

$$V_p(x) = 2V_{p2\pi/a} \cos \frac{2\pi}{a} px, \qquad p \text{ integer}, \qquad (3.27)$$

we obtain band gaps for the energies

$$p^2 \varepsilon_0 - |V_{p2\pi/a}| < E < p^2 \varepsilon_0 + |V_{p2\pi/a}|. \qquad (3.28)$$

This calculation, carried out only in the 1D case for didactic reasons, shows that the periodicity modifies the parabolic dispersion relation of the free electron, causing band gaps to appear at the points which correspond in the reciprocal space to the midpoints of vectors in the reciprocal lattice.

These marked singularities in the band structure, obtained for states corresponding to the vectors **k** close to a Bragg plane, arise because the corresponding plane wave states suffer a Bragg diffraction on the periodic potential of the ions, even if this potential is very weak. The eigenstates for the wave vectors (3.23) or (3.24) are no longer plane waves but stationary waves, obtained by diffraction.

c. General One-Dimensional Case

It thus remains to understand how the above approach based on the idea of nearly free electrons can be related to the tight-binding approximation discussed in Chap. 1. Naturally, the energy eigenvalues for the model in the last section can be obtained numerically to any required accuracy, whatever the amplitude of the periodic potential. The results of such a calculation, in which only the term $V_1(x)$ has been retained, are shown in Fig. 3.6.

We observe that, for a weak potential, the results are indeed those obtained using the nearly-free electron approximation. However, for a stronger potential, the shape of the lower band approaches the cosine shape obtained in the tight-binding approximation. For even stronger potentials, this trend becomes more pronounced and extends to higher bands, provided we take into account the Fourier components $V_p(x)$ of the potential. *We thus move smoothly from the case of nearly free electrons, corresponding to a weak atomic potential, to the LCAO approximation, which corresponds to a strong atomic potential.*

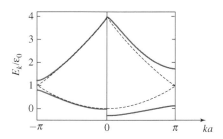

Fig. 3.6 Folded band structure according to the model (3.21) in the first Brillouin zone for $V_{2\pi/a} = 0.2\varepsilon_0$ (*left*) and for $V_{2\pi/a} = 0.8\varepsilon_0$ (*right*). The free electron parabola is shown by a *dashed curve*. E_k is even

3.2.2 Band Structure of Two-Dimensional Solids

We now extend the tight-binding and nearly-free electron methods (Sects. 3.2.2a and b, respectively) to the case of more realistic solids than the linear atomic chain. The notion of reciprocal space becomes essential here for representing the structure of the energy states in two dimensions. The fact that the quantum numbers characterising the Bloch state have two components apart from the band index leads to specific spatial representations. We also show in Sect. 3.2.2c that the existence of several directions in **k** space can lead to band overlap, something that did not arise in the 1D case.

a. Tight-Binding Approximation

We consider a square plane lattice with one atom per primitive cell. Its reciprocal lattice is also square with side $2\pi/a$. The first Brillouin zone is in this case simply a square of side $2\pi/a$ centered at O. It is easy to reproduce the LCAO type of calculation in this case, taking into account the fact that the Bloch wave functions are specified in the 2D **k** space. The wave function is therefore

$$|\psi_\mathbf{k}\rangle = \frac{1}{\sqrt{N_n}} \sum_\mathbf{l} a_{\mathbf{k},\mathbf{l}} |\mathbf{R}_\mathbf{l}\rangle . \tag{3.29}$$

If $\psi_\mathbf{k}$ is required to be a Bloch function (3.2), the coefficients $a_{\mathbf{k},n}$ can be written in the form

$$a_{\mathbf{k},\mathbf{n}} = e^{i\mathbf{k}\cdot\mathbf{R}_\mathbf{n}} . \tag{3.30}$$

The general eigenvalue equation for the $E_\mathbf{k}$ obtained in (1.19) is clearly still valid, i.e.,

$$-\sum_\mathbf{l} t_{\mathbf{n},\mathbf{l}} a_{\mathbf{k},\mathbf{l}} = (E_\mathbf{k} - E_\mathrm{v}) a_{\mathbf{k},\mathbf{n}} , \tag{3.31}$$

with the usual general expression for the hopping integrals:

$$-t_{\mathbf{n},\mathbf{l}} = \langle \mathbf{R}_\mathbf{n} | V_\mathrm{at}(\mathbf{r}) | \mathbf{R}_\mathbf{l} \rangle . \tag{3.32}$$

1. As in the 1D case, we neglect all the hopping integrals beyond nearest neighbours, i.e., for which $|\mathbf{R}_\mathbf{n} - \mathbf{R}_\mathbf{l}| > a$.
2. We take into account the symmetry of the crystal. In particular, the periodic boundary conditions are such that the crystal is perfectly invariant under translation. The $t_{\mathbf{n},\mathbf{l}}$ then only depend on the vector $\mathbf{R}_\mathbf{n} - \mathbf{R}_\mathbf{l}$.
3. In addition, the potential V_at, and often the wave functions too, are invariant under rotation, something we shall assume here.[2] The $t_{\mathbf{n},\mathbf{l}}$ then depend only on

---

[2] Note that for orbitals with nonzero angular momentum, one cannot assume that the wave function is invariant under rotations. From the point of view of computation time, the changes this brings to our analysis can be significant, but the results are not fundamentally different.

3.2 Band Structure of Solids

the length of the vector $\mathbf{R_n} - \mathbf{R_l}$. In our example, there is then only one nonzero hopping integral, apart from t_0, which we denote by t_1.

Equation (3.31) now becomes

$$-t_1 \sum_{\mathbf{d}} a_{\mathbf{k},\mathbf{n}+\mathbf{d}} = (E_{\mathbf{k}} - E_v + t_0)a_{\mathbf{k},\mathbf{n}}, \quad (3.33)$$

where the sum over $\mathbf{d}$ concerns only the nearest neighbours of the site $\mathbf{n}$:

$$\mathbf{d} = \pm a\mathbf{x}, \ \pm a\mathbf{y}. \quad (3.34)$$

Putting the expression (3.30) for the $a_{\mathbf{k},\mathbf{n}}$ into (3.33), we immediately obtain the eigenvalues as a generalisation of the 1D case:

$$\boxed{E_{\mathbf{k}} = E_v - t_0 - 2t_1(\cos k_x a + \cos k_y a)}. \quad (3.35)$$

The surface representing this function $E(k_x, k_y)$ is shown in Fig. 3.7. The constant energy curves, found by intersecting this surface by horizontal planes, are projected onto the base plane. Their general shape in the first Brillouin zone is shown in Fig. 3.7. For small values of $|\mathbf{k}|$, the cosine functions can be expanded to second order in a Taylor series, and the function $E(k_x, k_y)$ approximated by a simple paraboloid. The constant energy curves are then circles. Likewise, the constant energy curves close to the energy maximum of the band are circles centered on the points $(\pm \pi/a, \pm \pi a)$. Note that the band determined like this using the tight-binding approximation has width $8t_1$. Such 2D systems are encountered in high-T_c cuprate superconductors.[3]

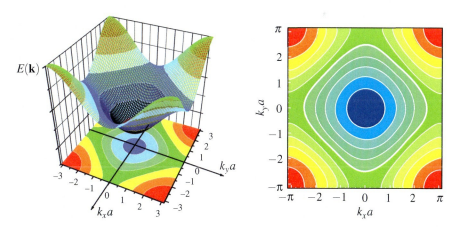

Fig. 3.7 Energy band for the square lattice in the tight-binding approximation. 3D representation of $E(\mathbf{k})$ (*left*) and constant energy curves obtained as projections on the ($k_x a$, $k_y a$) plane (*right*)

---

[3] See Problem 3: *Band Structure of* $YBa_2Cu_3O_7$.

b. Nearly Free Electrons in Two Dimensions

We now consider the 2D generalisation of the band structure calculation, using a periodic potential that is simplified to its first Fourier components in the x and y directions:

$$H = \frac{\mathbf{p}^2}{2m_0} + 2V_{2\pi/a}\left(\cos\frac{2\pi x}{a} + \cos\frac{2\pi y}{a}\right). \quad (3.36)$$

When the potential is zero, the surface $\varepsilon_0(k_x, k_y) = \hbar^2 k^2/2m_0$ is a simple paraboloid of revolution, and the constant energy curves are therefore circles. If the potential is weak enough to use a perturbation method, the eigenvalues $E(k_x, k_y)$ only differ from $\varepsilon_0(k_x, k_y)$ for values of $\mathbf{k}$ close to a Bragg plane. We thus see that the circles with constant energy E_i are unchanged unless they intersect a Bragg plane, i.e., unless they hit the first Brillouin zone. However, when the free electron circle of energy E_i cuts a Bragg plane for $\mathbf{k} = \mathbf{k}_i$, a forbidden energy zone (band gap) appears around E_i in the direction $\mathbf{k}_i$. This gives rise to a distortion of the constant energy curve, which deviates from the circle as shown in Fig. 3.8, with ever greater deviation as the matrix element of the potential increases. These perturbations will occur for all intersections with the Bragg planes, as happened in the 1D case.

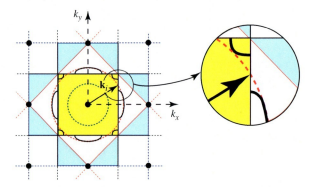

Fig. 3.8 Constant energy curves for a square lattice when the electrons are assumed to be nearly free. Close to $\mathbf{k}_i$, the free electron circle is distorted

> This suggests identifying the orthogonal bisecting planes of all the vectors of the reciprocal lattice taken from the origin and generalising the idea of the Brillouin zone. While the first Brillouin zone corresponds to the region of $\mathbf{k}$ space such that, moving out from $\mathbf{k} = 0$, one crosses no orthogonal bisecting plane of reciprocal lattice vectors, the second, third, and higher order Brillouin zones correspond to the region of reciprocal space whose points are obtained by crossing only one, two, or more orthogonal bisecting planes of reciprocal lattice vectors, respectively. These different Brillouin zones are all rather special primitive cells of the reciprocal lattice. Any of them can be used to represent the band structure. This can be illustrated in the 1D case, where the first Brillouin zone is $(-\pi/a, \pi/a)$, the second comprises two segments, viz., $(\pi/a, 2\pi/a)$ and its mirror image $(-2\pi/a, -\pi/a)$, and so on. We may represent the folded

3.2 Band Structure of Solids

band structure in the first Brillouin zone, but we may also represent the first energy band in the first Brillouin zone, the second in the second Brillouin zone, and so on. In such a representation, which corresponds to the bold lines in Fig. 3.5a, it is easier to recognise the free electron parabola. This unfolded representation in several Brillouin zones is thus more useful for representing a band structure obtained using the nearly-free electron approximation. The geometric construction of the Brillouin zones for the square plane lattice is shown in Fig. 3.9.

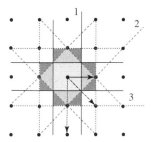

Fig. 3.9 First (*center square*), second (*light grey*), and third (*dark grey*) Brillouin zones for the square plane lattice. The Bragg lines corresponding to the first, second, and third neighbours are indicated by 1, 2, and 3

c. Forbidden Energy Band?

If $E(k)$ is represented in a particular direction $O\mathbf{k}$ of the (k_x, k_y) plane, we obtain something like the result shown in Fig. 3.6 for the chain, i.e., a band gap appears at the point belonging to the first Brillouin zone. However, the energy position of this band gap is not the same in all directions of the (k_x, k_y) plane. This can be understood from Fig. 3.10, where we have represented the results of a numerical calculation carried out for different values of the perturbing potential $V_{2\pi/a}$. Note that, as in the 1D case, for a given vector $\mathbf{k}_y$, a band gap appears at $k_x = \pi/a$. However, if we look at the band structure as a whole, we see that, for a weak potential V_1 (see Fig. 3.10a), the maximum of the lower band is higher than the minimum of the upper band. In this case there are therefore some states at all energies, and hence no band gap, something that could not happen in the 1D case. A true band gap can nevertheless be obtained by considering stronger potentials, as can be seen in Fig. 3.10b.

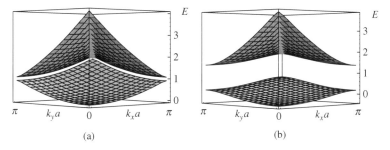

Fig. 3.10 The folded band structure $E_\mathbf{k}$ of the 2D model (3.36), for weak (**a**) and strong (**b**) periodic potentials. The energy E is given in units of $\varepsilon_0 = (\hbar^2/2m_0)(\pi/a)^2$

The absence of a band gap for weak potentials can also be illustrated in the representation of Fig. 3.8, which shows the different Brillouin zones. When the circle of constant energy E_i cuts the first Brillouin zone, some of its circular arcs will be in the second Brillouin zone. This means that there are states of energy higher than E_i in the first Brillouin zone and some states of energy lower than E_i in the second Brillouin zone. It thus shows that the two energy bands represented in the first and second Brillouin zone actually overlap. There is therefore no band gap between these two bands. However, note that, if the perturbing potential is strong, the removal of degeneracy at the edge of the Brillouin zone may be sufficient for the constant energy curves to remain entirely within the first Brillouin zone, as happens in the LCAO approximation (see Fig. 3.7). In this case, a band gap will appear.

Note that this result, obtained here by comparing the energies of the first two bands, can be repeated for the other bands corresponding to higher energy levels. The result depends in each case on the amplitude of the corresponding Fourier component of the potential. The main conclusion here is that, *in two or three dimensions, only a strong enough periodic potential will be able to generate a band gap*, in contrast to the 1D case, where an arbitrarily weak potential will be able to do that. This observation will be important when we come to discuss the metal or insulating nature of certain 2D or 3D compounds.[4]

3.2.3 Three-Dimensional Band Structures

There is no great difficulty in going from two to three dimensions, except that a description of the $E(\mathbf{k})$ surfaces would now require four dimensions. However, the constant energy surfaces can easily be represented in a 3D space.

a. Constant Energy Surfaces

It is straightforward to generalise the tight-binding approximation to the case of a cubic monatomic solid. By analogy with the 2D square lattice, each atom has six nearest neighbours located at $\mathbf{d} = \pm a\mathbf{x}, \pm a\mathbf{y}, \pm a\mathbf{z}$, and the energy band is given by

$$\boxed{E_{\mathbf{k}} = E_v - t_0 - 2t_1(\cos k_x a + \cos k_y a + \cos k_z a)} \,. \tag{3.37}$$

---

[4] When there is no band gap, the effects of the band structure can nonetheless be observed through the optical response (see, e.g., Problem 2: *Reflectance of Aluminium* and Problem 5: *Optical Response of Monovalent Metals*). The change in the total energy of the electronic states pertaining to a band can in some cases influence the atomic structure of a metal (see, e.g., Problem 4: *Electronic Energy and Stability of Alloys* and Problem 6: *One-Dimensional TTF-TCNQ Compounds*).

3.2 Band Structure of Solids

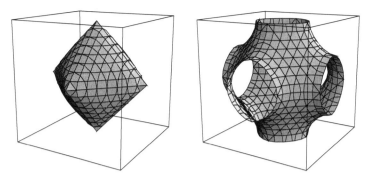

Fig. 3.11 Constant energy surfaces for the simple cubic lattice, in the first Brillouin zone: $E = E_v - t_0 - 2t_1$ (*left*) and $E = E_v - t_0$ (*right*)

Figure 3.11 shows several constant energy surfaces. We see that, by increasing the energy, we move from a simple sphere to more complex shapes.

As only few chemical elements form a simple cubic lattice, we consider instead the body- or face-centered cubic structures. In the body-centered cubic lattice, each atom has eight nearest neighbours at a distance $a\sqrt{3}/2$ and the displacement vectors to these eight neighbours are given by

$$\mathbf{d} = (\pm\mathbf{x}\pm\mathbf{y}\pm\mathbf{z})\frac{a}{2}. \tag{3.38}$$

In the face-centered cubic structure, the distance between nearest neighbours is $a\sqrt{2}/2$, and each atom of the conventional unit cell has twelve nearest neighbours, given by the displacement vectors

$$\mathbf{d} = (\pm\mathbf{x}\pm\mathbf{y})\frac{a}{2}, (\pm\mathbf{x}\pm\mathbf{z})\frac{a}{2}, (\pm\mathbf{y}\pm\mathbf{z})\frac{a}{2}. \tag{3.39}$$

The band structures are now determined by a similar argument to the one used in the last section. The only difference concerns the sum over the vectors **d** in (3.33), which is now specified by the different possibilities given by (3.38) or (3.39). The coefficients of the wave functions are still given by (3.30), and we then obtain

$$\boxed{E_{\mathbf{k}} = E_v - t_0 - 8t_1 \cos\frac{k_x a}{2} \cos\frac{k_y a}{2} \cos\frac{k_z a}{2}}, \tag{3.40}$$

for the body-centered cubic structure, and

$$\boxed{E_{\mathbf{k}} = E_v - t_0 - 4t_1 \left(\cos\frac{k_x a}{2}\cos\frac{k_y a}{2} + \cos\frac{k_x a}{2}\cos\frac{k_z a}{2} + \cos\frac{k_y a}{2}\cos\frac{k_z a}{2}\right)}, \tag{3.41}$$

for the face-centered cubic structure.

In most of the cases studied here (square lattice, simple cubic lattice, body-centered cubic lattice), the total band width, that is, the difference between the maximum and minimum of $E_\mathbf{k}$, is equal to $2zt_1$, where z is the number of nearest neighbours of an atom. There follows an empirical rule which is generally satisfied, with a few exceptions: *the more neighbours there are, the broader the band will be.* Any remarks on the nearly-free electron approximation go over from two to three dimensions.

b. Representing the 3D Band Structure

For a 3D solid, the relations $E(\mathbf{k})$ cannot be fully represented in a 3D space. However, an incomplete representation which allows a full appreciation of the properties of the band structure can be obtained by graphing $E_n(k)$ for values of k along particular axes of the reciprocal lattice. The preferred axes for such a depiction are the axes of symmetry of the first Brillouin zone. Consider for example the case of germanium (Ge) with structure (Ar) $3d^{10}4s^24p^2$, which crystallises into the diamond cubic structure with lattice constant $a = 5.66$ Å, shown in Fig. 2.7b. The Bravais lattice has the face-centered cubic structure. The first Brillouin zone of its

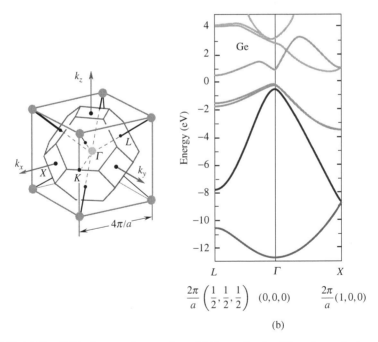

Fig. 3.12 (a) First Brillouin zone of germanium. (b) Energy bands of the valence electrons of Ge represented in the first Brillouin zone for the two directions ΓL and ΓX of (a)

3.2 Band Structure of Solids

body-centered cubic reciprocal lattice is illustrated in Fig. 3.12a. The curves $E(k)$ in the directions of the axes ΓL and ΓX are shown in Fig. 3.12b for the energy bands corresponding to the electrons in the $4s^2 4p^6$ orbitals.

c. Density of States

The density of states serves the same purpose in the 2D and 3D cases as it did for the linear chain. The aim is to be able to transform any sum of an arbitrary function g over values of **k** into an integral over the energies:

$$F = \sum_{\mathbf{k}} g(E_{\mathbf{k}}) = \frac{N_n}{\Omega_{BZ1}} \int_{BZ1} d^d k \, g(E_{\mathbf{k}}) = N_n \int dE \, g(E) D(E) \,. \quad (3.42)$$

Here the vectors **k** range over the first Brillouin zone, of volume Ω_{BZ1}. The replacement of the sum by an integral in the second equality is valid in the thermodynamic limit $N_n \to \infty$ for which the allowed values of **k** become continuous, and the third equality defines the *density of states per atom $D(E)$*. Finally, the volume element in **k** space is the one corresponding to the appropriate dimension $d = 2$ or $d = 3$.

The density of states $D(E)$ can in principle be calculated as in Chap. 1. However, in most cases, and in particular for the lattices we have been considering, an analytic evaluation is not possible. Figure 3.13 shows the density of states for the square and simple cubic lattices. Note the step-shaped singularity at the edge of the band for the square structure, and a square root singularity for the cubic lattice. These singularities are of the same type as those obtained for a free electron gas close to $\mathbf{k} = 0$. The singularity in the middle of the band, in the 2D case, is a logarithmic divergence which corresponds to the saddle points of the constant energy surfaces in Fig. 3.7. In the 3D case, $D(E)$ has gradient discontinuities. These singularities in the band structure are called Van Hove singularities.

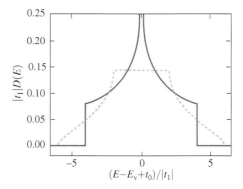

Fig. 3.13 Density of states per atom for the square lattice (*continuous curve*) and the simple cubic lattice (*dashed curve*) in the LCAO approximation

3.3 Metals, Insulators, and Semiconductors

In the last section, we discussed the eigenstates of an electron subject to a periodic potential in a crystal. In real crystals, there is a huge number of electrons, of the order of the Avogadro number, and this must clearly be taken into account for any question involving the electronic properties of the solid. With the approximation made here, which considers the electrons as independent, we know all the electron states, and the methods of statistical physics for fermions outlined in Sect. 3.3.1 can be used to determine the thermodynamic properties of the crystal. Note that, when the periodic potential is zero, we recover the situation for a free electron gas, as treated in statistical physics, presented here as a model for the metallic state. Applying Fermi–Dirac statistics to the Bloch states of the crystal, we will be able to identify what fundamentally distinguishes insulating solids from metals and show that, to a first approximation, the properties of metals generally arise from the properties of free electrons (Sect. 3.3.2). Finally, in Sect. 3.3.3, we will be in a position to discuss insulating solids with a small band gap, namely, semiconductors.

3.3.1 Statistical Physics of Fermions

The thermodynamic properties at temperature T of a system of independent fermions with quantised energy levels $E_n(\mathbf{k})$ can be determined from the grand canonical partition function

$$Z_G(T,\mu) = \prod_{n,\mathbf{k}} \left\{ 1 + e^{-\beta\left[E_n(\mathbf{k}) - \mu\right]} \right\}^2 , \qquad (3.43)$$

where we use the standard notation $\beta = 1/k_B T$, and μ is the chemical potential which determines the average number of electrons. Note that we have taken into account the spin degeneracy of the states $E_n(\mathbf{k})$ by including the square of each factor in the product. We now consider the grand canonical potential

$$A(T,\mu) = -k_B T \ln Z_G = -2k_B T \sum_{n,\mathbf{k}} \ln\left\{1 + e^{-\beta\left[E_n(\mathbf{k}) - \mu\right]}\right\} . \qquad (3.44)$$

The various thermodynamic quantities are obtained by taking derivatives of A. For example, the average number of electrons is

$$N(T,\mu) = -\frac{\partial A(T,\mu)}{\partial \mu} . \qquad (3.45)$$

At thermodynamic equilibrium, the average occupation number of an energy level E is given by the Fermi function (see Fig. 3.14a)

$$f(E) \equiv \frac{1}{1 + e^{\beta(E-\mu)}} . \qquad (3.46)$$

3.3 Metals, Insulators, and Semiconductors

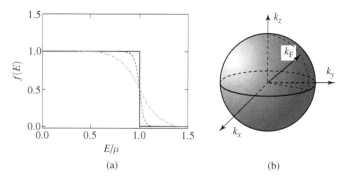

Fig. 3.14 (a) Fermi occupation function in the zero temperature limit (*continuous curve*), for $\beta\mu = 40$ (*dashed curve*), and for $\beta\mu = 10$ (*dot-dashed curve*). (b) Fermi sphere bounding the set of occupied states in wave vector space for a free electron gas

Note that, at zero temperature, this function is equal to 1 for $E < \mu$ and zero for $E > \mu$. This means that, at $T = 0$, all states are occupied up to the value of the chemical potential μ, which is called the *Fermi energy* E_F. The thermodynamic equilibrium state is simply the ground state in which all electron states of lower energy are occupied, taking into account the Pauli exclusion principle.

This has a simple illustration in the case of free electrons. Their eigenfunctions are plane waves (3.9), characterised by a wave vector $\mathbf{k}$, momentum $\mathbf{p} = \hbar\mathbf{k}$, and purely kinetic energy $\varepsilon(\mathbf{k}) = \hbar^2 \mathbf{k}^2 / 2m_0$. Note that, in the space of wave vectors $\mathbf{k}$, the surface of constant energy E_F which bounds the set of all occupied states is in this case a sphere of radius $k_F = (2m_0 E_F/)^{1/2}/\hbar$. This so-called *Fermi sphere* is shown in Fig. 3.14b.

Considering a box of volume $\Omega = L^3$ with the usual periodic boundary conditions, the quantisation condition leads to one state per unit of volume $(2\pi)^3/\Omega$ in wave vector space. If N_e is the total number of electrons in volume Ω, we obtain $k_F = (3\pi^2 N_e/\Omega)^{1/3}$, viz.,

$$E_F = \lim_{T \to 0} \mu = \frac{\hbar^2}{2m_0}\left(\frac{3\pi^2 N_e}{\Omega}\right)^{2/3}. \tag{3.47}$$

Note that the quantities E_F and k_F are completely determined by the density of electrons per unit volume.

Before considering the band structure of a real solid, we observe that, since the expression for $A(T, \mu)$ contains a sum over the energies of the eigenstates, it can be replaced by an integral in a representation involving $D_\Omega(E)$, the total electron density of states of the band structure per unit volume and per spin direction:

$$A(T, \mu) = -2\Omega k_B T \int_0^\infty \ln\left[1 + e^{-\beta(E-\mu)}\right] D_\Omega(E) dE. \tag{3.48}$$

In particular, note that the density of states for free electrons has a simple expression if we observe that the states with energies between E and $E + dE$ have wave vectors in the volume $4\pi k^2 dk$, where $dE = \hbar^2 k dk/m_0$. The density of states for free electrons in three dimensions thus has the form

$$D_3(E) = \frac{1}{4\pi^2}\left(\frac{2m_0}{\hbar^2}\right)^{3/2} E^{1/2}. \tag{3.49}$$

The analogous argument for free electrons in one or two dimensions leads to a density of states with the simple expression

$$D_d(E) = C_d \left(\frac{2m_0}{\hbar^2}\right)^{d/2} E^{(d-2)/2}, \tag{3.50}$$

where C_d is equal to $1/2\pi, 1/4\pi$ and $1/4\pi^2$, respectively, for $d = 1, 2$, and 3.

3.3.2 Real Solids: Conductor or Insulator?

We now consider a real solid with a periodic potential, described by band theory. Fermi statistics applies for the electron states, and in the ground state, at zero temperature, the electrons occupy levels up to some maximum. To determine this maximum, consider for simplicity the case of a crystal with monatomic primitive cell. In a solid with N_n primitive cells, one band contains N_n values of **k**, and each of these states can hold two electrons, owing to the two possible spin orientations. As a consequence, *one band can contain up to $2N_n$ electrons*.

Consider a monovalent element such as sodium (Na) whose band structure was described in Chap. 1. In this case, bands due to deep electronic levels (see Fig. 1.9b) are all occupied, whereas the energy band resulting from the valence level is only occupied by N_n electrons, and so is only half filled. For example, if we consider a band obtained using the LCAO approximation for a linear chain, a representation of the ground state which minimises the total energy is shown in Fig. 3.15. The states are filled up to the Fermi energy E_F which corresponds to the middle of the LCAO band.

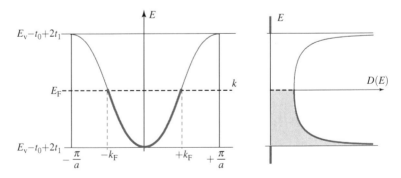

Fig. 3.15 Occupation of quantum states for a monovalent solid in its ground state, in a 1D representation. Occupied states are represented by *bold lines*. Here the Fermi energy is $E_F = E_v - t_0$

3.3 Metals, Insulators, and Semiconductors

It can be seen from Fig. 3.15 that, for electrons close to E_F, unoccupied states are available at arbitrarily low energy differences, as for the free electron gas. These electrons can thus absorb electromagnetic energy at arbitrarily low frequencies. This is indeed what would be expected in a *metallic conductor*.

Now consider a divalent chemical element. We then have $2N_n$ electrons to distribute over the same number of quantum states, and the band is therefore completely filled. Hence in this case the only accessible electron states lie in a higher band whose states will still be vacant. The electrons cannot be excited with a low frequency electromagnetic wave, exactly as would be expected in the case of an *insulator*. We thus arrive at the following two propositions:

Partly filled band $\Longleftrightarrow$ metallic conductor.

All bands completely filled (or empty) $\Longleftrightarrow$ insulator.

> **Question 3.2.** Consider the band structure of Ge described in Sect. 3.2.3b and illustrated in Fig. 3.12. Check that the ground state of Ge at zero temperature does indeed correspond to total occupation of a certain number of bands, and conclude that Ge is an insulator.

This analysis suggests that monovalent elements should all be metals, and this is indeed the case, while all divalent elements should be insulators, which is clearly false. In fact, all the alkaline earths, such as Be, Mg, and Ca, are metals. The flaw in the argument comes from the assumption that the bands arising from the different atomic levels have distinct energies. But we saw above that this will not always be the case in two or three dimensions, since the high energy bands can overlap when the atomic potential is weak enough.

We now examine the general situation with the help of Fig. 3.16. Imagine first a crystal with very large lattice constant, so that the hopping integrals are negligible. The electron energy spectrum of the solid will then simply be the atomic spectrum, as shown on the left of the figure. If the lattice constant is reduced, hopping integrals

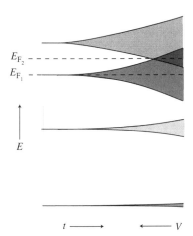

Fig. 3.16 Expected change in the electron levels in a solid when the hopping integrals t are gradually increased, i.e., when the atoms are moved closer together. This evolution can be obtained in the opposite direction if the strength of the atomic potential is gradually increased. The Fermi energies E_{F_1} and E_{F_2} correspond to monovalent and divalent solids, respectively

(abbreviated to t in the figure) will become significant, and each atomic level will broaden into a band. For a monovalent solid, the last occupied band will be partially occupied, with Fermi energy E_{F_1} in the middle of the band, whereas for a divalent solid, the last occupied band will be fully occupied. In this case, the Fermi energy E_{F_2} lies in a region where there is no quantum state, i.e., in a *band gap*.[5]

We thus recover the distinction between a metal and an insulator discussed earlier. However, another situation may arise if the lattice constant is reduced still further, thereby increasing t even more. In this case, the bands begin to overlap, as we have seen in Fig. 3.10a, and the corresponding band gaps will disappear, as shown on the right in Fig. 3.16. It is then energetically favourable to remove electrons from states close to the maximum of the lower band and place them close to the minimum of the higher band. Instead of a full band and an empty band, which would give an insulator, we find two partially filled bands, i.e., a metal. It is this situation with overlapping bands that arises in the alkaline earth elements, which are all metals.

Conversely, we may imagine gradually increasing the periodic potential V from an initially low value. As we saw in Sect. 3.2, a weak potential does not create a band gap in two or three dimensions, so we find ourselves in the situation corresponding to the right-hand side of Fig. 3.16. The weak potential situation typically prevails in dense polyvalent metals like Pb and In. When V increases, band gaps will eventually appear, as in the middle of the figure. This is what happens in diamond, silicon, germanium, and many semiconducting compounds, such as GaAs, GaP, InP, and so on. It should be noted that an increase in the periodic potential corresponds qualitatively to a decrease in the hopping integrals, and vice versa.

We may now formulate several conclusions regarding conduction properties. To begin with, *odd valence elements should always be metals, whereas even valence elements can be either metals or insulators*, depending on whether their bands overlap or not. Looking at the periodic table, we find that this conclusion is indeed correct in the vast majority of cases, generally with a metallic behaviour. The few exceptions (insulators with odd valence) concern substances like nitrogen or chlorine in the solid state at low temperatures. These exceptions are easy enough to explain: the basis of the crystalline solid is not in this case the isolated atom, but rather a molecule, viz., N_2 or Cl_2. The basis of the crystal thus carries an even number of electrons, and is therefore an insulator.

This conclusion can be extended without difficulty to non-elementary crystals, i.e., ones with a more complicated basis than a single atom. In the band theory framework, *if the basis contains an odd number of electrons, the result should be a metallic state, while if it contains an even number, the result may be either a metal or an insulator*.

---

[5] In an insulator, the Fermi energy is not defined by band occupation at $T = 0$. It must then be defined, as for semiconductors, from the chemical potential $E_F = \lim_{T \to 0} \mu$. We shall see below that E_F lies within the band gap.

3.3.3 Introduction to the Semiconductor Case

Note that according to the above definition a solid is only strictly an insulator at absolute zero temperature. Indeed, if all the bands are either filled or empty at zero temperature, a band gap will separate the last occupied state from the first empty state at zero temperature. However, for nonzero T, the Fermi factor implies that the last states $E \lesssim E_v$ of the lower band, usually called the *valence band*, are now only partially occupied, whereas the first states of the upper band $E \gtrsim E_c$, known as the *conduction band*, are partially occupied. The bands will thus be partially filled, and the material may be slightly conducting. It is when such conditions are fulfilled that we may define a *semiconductor* to be a material that can conduct significantly at room temperature.

> **Question 3.3.** In the case of Ge, whose band structure is depicted in Fig. 3.12, indicate the points in reciprocal space corresponding to E_v and E_c and the energy of the band gap.

For a pure (or intrinsic) semiconductor at $T \neq 0$, the number n_e of electrons occupying states in the conduction band must correspond exactly to the number n_h of missing electrons (or holes) in the valence band, in order to ensure electrical neutrality. This condition determines the position of the chemical potential for all T. Given the densities of states $D_c(E - E_c)$ and $D_v(E_v - E)$ of these two bands, we have

$$n_e = 2\int_{E_c}^{\infty} f(E)D_c(E - E_c)dE, \qquad n_h = 2\int_{-\infty}^{E_v}\left[1 - f(E)\right]D_v(E_v - E)dE, \quad (3.51)$$

where the factor of 2 accounts for the spin. It is easy to check that, provided that k_BT is much less than the width $E_g = E_c - E_v$ of the band gap, electrical neutrality requires the Fermi population factor to be very small for $E > E_c$ and very close to unity for $E < E_v$, whence the following approximations:

$$n_e = 2\int_{E_c}^{\infty} \exp\left[-\beta(E - \mu)\right]D_c(E - E_c)dE = N_c(T)\exp[\beta(\mu - E_c)],$$

$$n_h = 2\int_{-\infty}^{E_v} \exp\left[-\beta(\mu - E)\right]D_v(E_v - E)dE = N_v(T)\exp[\beta(E_v - \mu)], \quad (3.52)$$

where we have set

$$N_c(T) = \int_0^{\infty} 2D_c(E)\exp(-\beta E)dE, \qquad N_v(T) = \int_{-\infty}^{0} 2D_v(E)\exp(\beta E)dE. \quad (3.53)$$

Eliminating μ between relations (3.52), we thus obtain the general result

$$n_e n_h = N_c(T)N_v(T)\exp(-\beta E_g) = n_i^2. \quad (3.54)$$

In an intrinsic semiconductor, n_i is the electron density in the conduction band, equal to the missing density in the valence band. Note that, in this case, the chemical potential is then obtained by equating the right-hand sides of the two relations in (3.52), which yields

$$\mu = \frac{1}{2}(E_c + E_v) + \frac{1}{2}k_B T \ln\left[N_v(T)/N_c(T)\right]. \quad (3.55)$$

When the densities of states of the conduction and valence bands are equal, called the symmetric case, the chemical potential and hence also the Fermi energy lie exactly in the middle of the band gap (see Fig. 3.17). In this case, for bands in three dimensions with free electron density of states given by (3.49), we obtain

$$n_e = N_c(T)\exp(-\beta E_g/2), \quad (3.56)$$

where

$$N_c(T) = \frac{1}{2\pi^2}\left(\frac{2m_0 k_B T}{\hbar^2}\right)^{3/2} \int_0^\infty \exp(-u) u^{1/2} du. \quad (3.57)$$

The number of electrons transferred from the valence band to the conduction band varies exponentially with temperature, and is therefore very low as long as $k_B T \ll E_g$, as might have been expected. For example, when $E_g = 1$ eV at room temperature, i.e., $k_B T = 25$ meV, $n_e \approx 10^{-12}$ per primitive unit cell.

We thus discover that the ideal situation described above is difficult to obtain in practice, because it can be totally changed in real materials by the presence of sometimes uncontrollable impurities in comparable concentrations. We shall see in Chap. 4 how this situation has in fact been exploited by purifying certain semiconductors and then introducing selected impurities at controlled levels.

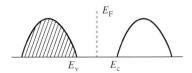

Fig. 3.17 Densities of states of the valence and conduction bands in a pure semiconductor. According to (3.55), E_F lies close to the middle of the band gap

3.3.4 The Special Case of Graphene

Graphene is an allotrope of carbon obtained from graphite by detaching a single crystal plane, in which the carbon atoms are positioned at the vertices of a honeycomb structure. As shown in Question 2.2, its Bravais lattice is hexagonal and the primitive cell contains two atoms in positions corresponding to two sublattices A and B. The latter only differ through the orientation of their three nearest neighbours, which belong to the complementary sublattice (see Fig. 3.18).

3.3 Metals, Insulators, and Semiconductors

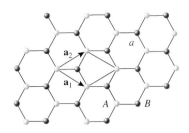

Fig. 3.18 Atomic structure, Bravais lattice ($\mathbf{a}_1, \mathbf{a}_2$), and primitive cell of the graphene sheet, showing the two sublattices A and B of carbon atoms

The structure is stabilised by the fact that the $2s^2 2p^2$ orbitals of a given carbon atom form three hybridised sp^2 covalent bonds oriented at an angle of 120° with the neighbouring carbon atoms. The corresponding electron energy levels constitute the saturated inner levels of the electronic structure of graphene. This is reminiscent of what happens in the benzene molecule, but replacing the hydrogen atoms by carbon atoms, whence the whole structure can be extended indefinitely.

As in benzene, the last electron of the carbon atoms is positioned in the $2p_z$ orbital, perpendicular to the plane of the graphene sheet. It is the delocalisation of this orbital that leads, by analogy with the π orbital of benzene, to the outer electron bands in the electronic structure of graphene. They can thus be described using the LCAO tight-binding approximation by considering one electron per carbon atom in the $2p_z$ orbital. The only complication with regard to the electronic structures so far investigated is that the primitive cell contains two atoms in the sublattices A and B, which leads to two energy bands.

Setting $|\varphi_j^A\rangle = |\varphi(\mathbf{r}-\mathbf{R}_j)\rangle$ and $|\varphi_j^B\rangle = |\varphi(\mathbf{r}-\mathbf{R}_j-\mathbf{d})\rangle$ for the eigenstates of energy E_0 associated with the 'atomic' Hamiltonians of the two atoms in the primitive cell, we seek the eigenstates of the total Hamiltonian in the form

$$|\psi_{\mathbf{k}}\rangle = \frac{1}{\sqrt{N}} \sum_j \left(\lambda_A |\varphi_j^A\rangle + \lambda_B |\varphi_j^B\rangle \right) e^{i\mathbf{k}\cdot\mathbf{R}_j} . \quad (3.58)$$

Assuming that the orbitals $|\varphi_j^A\rangle$ and $|\varphi_j^B\rangle$ are normalised and orthogonal, and considering only hopping integrals between nearest neighbours (j,A) and (j',B),

$$\langle \varphi_j^A | H | \varphi_{j'}^B \rangle = -t , \quad (3.59)$$

we determine the energies of the eigenstates $|\psi_{\mathbf{k}}\rangle$ of the Hamiltonian:

$$\boxed{E_{\mathbf{k}}^{\pm} = E_0 \pm t \left\{ 3 + 2 \left[\cos \mathbf{k}\cdot\mathbf{a}_1 + \cos \mathbf{k}\cdot\mathbf{a}_2 + \cos \mathbf{k}\cdot(\mathbf{a}_1-\mathbf{a}_2) \right] \right\}^{1/2} .} \quad (3.60)$$

Question 3.4. Projecting the Schrödinger equation with periodic potential onto the two atomic states $\langle \varphi_l^A |$ and $\langle \varphi_l^B |$, show that we obtain two coupled equations, the first of which is

$$\lambda_A E_0 - \lambda_B t \left(1 + e^{-i\mathbf{k}\cdot\mathbf{a}_1} + e^{-i\mathbf{k}\cdot\mathbf{a}_2} \right) = \lambda_A E . \quad (3.61)$$

Determine the second equation, which leads to Eq. (3.60).

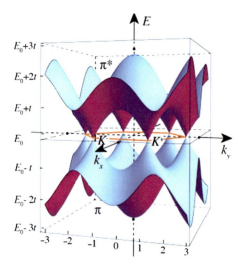

Fig. 3.19 Dispersion relation $E(\mathbf{k})$ of the π and π^* electronic bands of graphene, constructed from the p_z electronic orbitals of carbon. Note the six points at the vertices of the hexagonal first Brillouin zone where the valence and conduction bands touch one another. Near these points the dispersion relations have a conical shape for energies such that $|E - E_0| < t$. Image courtesy of Goerbig, M.O.: Laboratoire de Physique des Solides, Orsay, France

The two energy bands of (3.60) are shown in Fig. 3.19 for vectors in the hexagonal reciprocal lattice with lattice constant $4\pi/3a$, whose first Brillouin zone is the hexagon in the plane $E = E_0$. Note that these two binding π and antibinding π^* bands do not overlap, but touch one another solely at six points of energy $E = E_0$. These points of contact are obtained for vectors $\mathbf{k}$ such that the term in curly brackets in (3.60) vanishes. It can be checked that they correspond to the six vertices of the Brillouin zone. Since the primitive cell contains two electrons, the lower band is fully occupied at zero temperature and constitutes the valence band, while the upper band is completely unoccupied. The Fermi level is thus $E_F = E_0$, and *graphene can be considered to be a semiconductor with strictly zero band gap.*

Graphite, consisting of a stack of graphene sheets, has a similar electronic structure, with touching bands. However, in this case, the hopping integrals between graphene planes lead to overlaps between 3D orbitals, and graphite is a conductor at $T = 0$. It is thus a semi-metal. Note that graphene is also close to a semi-metal, since many electrons are thermally excited as soon as $T \neq 0$. As we shall see in Chap. 4, the rather special electronic structure of graphene leads to extremely unusual electronic transport properties.

3.3.5 Thermodynamic Properties of Metals

The thermodynamic properties of solids are determined by the quantum states populated at nonzero temperature. They thus provide valuable information about the excited states. Some such states correspond to collective vibrational modes of the atoms, known as phonons. They are present in all solids, whatever their electronic

3.3 Metals, Insulators, and Semiconductors

band structure. On the other hand, the excited states of electronic origin are radically different in metals and insulators, and this leads to a significant difference in the thermodynamic properties. One of the quantities that reflects these properties is the specific heat at constant volume $C_v = dU/dT$, which is the amount of energy dU required to cause a temperature rise of dT in the solid. It is thus determined by the change in internal energy associated with thermal population of excited states.

For example, for a semiconductor, the electron energy at a temperature T is increased above the ground state energy by the energy of the states excited from the valence band into the conduction band, viz., $U(T) - U(0) \approx n_e E_g$. In the case of a symmetric semiconductor, as considered in the last section,

$$C_v \approx E_g \frac{dn_e}{dT} = \frac{N_c E_g^2}{2 k_B T^2} \exp\left(-\frac{E_g}{2 k_B T}\right). \tag{3.62}$$

Since very few electrons are excited, due to the exponential factor, this contribution to the specific heat is negligible compared with the phonon contribution[6] which goes as T^3. Consequently, in semiconductors and insulators, the specific heat tells us little about electronic properties. We shall see in Sect. 3.4 that optical methods are better suited to measuring the width of the band gap. In contrast, in metals, many energy states close to E_F can be thermally excited, whence electron levels make a large contribution to the specific heat.

a. Specific Heat of Metals

The Fermi–Dirac distribution implies that the electron states over a width $\approx k_B T$ at the Fermi level are excited by an energy $\approx k_B T$. For a density of states $D_\Omega(E)$, the increase in the internal energy is therefore

$$\Delta U \approx k_B T \left[k_B T D_\Omega(E_F) \right], \tag{3.63}$$

which corresponds to a term linear in the temperature, i.e., $C_v \approx 2 k_B^2 T D_\Omega(E_F)$. The exact calculation is straightforward using the grand canonical potential. To first order in temperature, it leads to an electronic contribution to the specific heat that we shall denote by C_e, given by

$$\boxed{C_e(T) = \frac{\partial U}{\partial T} = \frac{2\pi^2 D_\Omega(E_F) k_B^2}{3} T}. \tag{3.64}$$

This term must be corrected by higher order terms expressed as powers of $k_B T/E_F$. This yields the Sommerfeld expansion, in which the first correction to the electronic density of states goes as $(k_B T/E_F)^2$ [3]. It is generally negligible for metals with no particular singularities in the density of states at the Fermi level. A metal is thus characterised by an *electronic contribution to the specific heat* that is linear in T at low temperatures, viz., $C_e = \gamma T$, where, according to (3.64), the coefficient γ can be used to determine the density of states at the Fermi level.

---

[6] See Problem 9: *Phonons in Solids*.

b. Pauli Susceptibility of Metals

Another quantity that clearly distinguishes between metals and insulators is the magnetisation induced by a magnetic flux density $\mathbf{B} = \mu_0 \mathbf{H}$ applied in some direction z. This removes the spin degeneracy of the Bloch states. The Zeeman term

$$H_{\text{Zeeman}} = -\boldsymbol{\mu}\mathbf{B} = 2\mu_B B \hat{s}_z, \qquad (3.65)$$

shifts the energy of the spin-up state $s_z = 1/2 = \uparrow$ relative to the energy of the spin-down state $s_z = -1/2 = \downarrow$. Consequently, the populations of the two spin states are no longer equal, and at equilibrium, the solid can acquire a magnetisation.

Note that at $T = 0$, for a fully occupied band, the energy balance is zero and the resulting magnetisation vanishes. In the band theory framework, the electron spins do not induce magnetic behaviour: *an insulator is essentially non-magnetic*. However, in a metal, some bands are not fully occupied. For these bands, the energies of all the spin-up states are increased in the presence of the field, whereas those of the spin-down states are lowered by the presence of the field. It follows that, at the Fermi level, the excess of spin down states constitutes a set spanning an energy width $2\mu_B B$, and hence $2\mu_B B D_\Omega(E_F)$ in number. The drop in energy induced by applying the magnetic field is therefore

$$\Delta U = -(\mu_B B)^2 D_\Omega(E_F).$$

The corresponding magnetisation is $M = -\partial \Delta U / \partial B = 2\mu_B^2 B D_\Omega(E_F)$, and the magnetic susceptibility is $\chi_P = \partial M / \partial B$, given by

$$\boxed{\chi_P = 2\mu_B^2 D_\Omega(E_F)}. \qquad (3.66)$$

In the corresponding Sommerfeld expansion, the first correction as a function of temperature is also of order $(k_B T / E_F)^2$. It is generally negligible, because $E_F > 1$ eV for most metals. We may conclude that band theory predicts a paramagnetic behaviour for the susceptibility of a metal which is essentially independent of T. This is commonly referred to as *Pauli paramagnetism*. Note that a measurement of the Pauli susceptibility can be used to estimate the density of states at the Fermi level.

c. Experimental Corroboration

The above discussion leads to several simple conclusions about the thermodynamic behaviour of solids in the framework of a band theory assuming independent electrons. There are only two possible situations:

- In an *insulator*, E_F is located in a band gap which separates totally occupied valence bands from the totally empty conduction bands. The density of states is zero at the Fermi level, and the electron contribution to the specific heat is negligible compared with the phonon contribution at any temperature. In addition, *insulators exhibit no spin magnetism*.

3.3 Metals, Insulators, and Semiconductors

- In a *metal*, one or more energy bands cross the Fermi level, thereby defining a Fermi surface comprising one or more sheets and representing all the last energy states occupied in **k** space. These electron states are easily excited thermally and contribute a term linear in T to the specific heat of the metal. Moreover, they are responsible for a *Pauli spin paramagnetism* that is independent of temperature.

The specific heat and the susceptibility are determined, for $T \ll T_F$, by the density of states at the Fermi level, using the simple relations (3.64) and (3.66). Note that a free electron model leads to similar relations for these two quantities C_e and χ_P, if we simply replace $D_\Omega(E_F)$ in these relations by $D_3(E_F)$, as calculated in (3.49) for free electrons.

For the specific heat, the T-linear electron contribution only dominates at low values of T, so measurements must be made at temperatures below a few kelvin for the phonon contribution to be negligible. The Pauli susceptibility, on the other hand, can be measured over a broad temperature range.

To make quantitative predictions of the thermodynamic properties of a solid, we must compute the density of states appearing in (3.64) and (3.66) to a certain level of accuracy, which generally proves to be a rather difficult task. However, one can use an experimental measurement of one or other of these properties in order to determine $D_\Omega(E_F)$. Measuring both provides a way of checking that band theory itself is applicable. Indeed, these two relations predict a ratio

$$\boxed{\lim_{T \to 0} \frac{T \chi_P(T)}{C_e(T)} = \frac{3 \mu_B^2}{\pi^2 k_B^2}}, \tag{3.67}$$

which depends only on universal constants and not on any specific property of the solid under investigation. Any deviation from this constant necessarily indicates that one or other of the approximations made in Chap. 1 is no longer valid.

Let us consider the monovalent alkali metals. The electron density in the highest occupied energy band is clearly one per atom. The data in Table 2.1 can be used to calculate the density per unit volume. Using (3.47) and (3.49), we deduce the density of states at the Fermi level $D_3(E_F)$, to be used when there is no crystal potential, i.e., for the case of a free electron gas. Furthermore, measurements of C_e and χ_P can be used to find two other values for $D_\Omega(E_F)$, which we shall denote by D_C and D_χ, respectively. Table 3.1 gives the values of the reduced quantities $d_C = D_C/D_3$ and $d_\chi = D_\chi/D_3$, together with the *Wilson ratio* R_W, which quantifies deviation from the prediction in (3.67):

$$\lim_{T \to 0} \frac{T \chi_P(T)}{C_e(T)}\bigg|_{\exp} = \frac{3 \mu_B^2}{\pi^2 k_B^2} R_W \iff R_W = d_\chi/d_C. \tag{3.68}$$

We find that the deviations of the density of states from the predictions of the free electron gas model are significant in most cases, although not excessive (there is no number of order 10 in the table). In every case, the Wilson ratio is close to unity. In a metal perfectly described by band theory, this ratio should be exactly unity.

Table 3.1 Reduced densities of states d_C and d_χ, together with the Wilson ratio R_W for the alkali metals

Metal	d_C	d_χ	R_W
Li	2.3	2.5	1.1
Na	1.3	1.7	1.3
K	1.2	1.5	1.3
Rb	1.3	1.6	1.2
Cs	1.5	1.7	1.1

Deviations are mainly due to interactions between electrons, which are neglected in this model. Note, however, that for more complicated metals, in particular those in the transition series (partially occupied 3d or 4d orbitals), the Wilson ratios can be much greater than unity, revealing the greater importance of interactions between electrons in these cases.

The classification into paramagnetic metal or non-magnetic insulator often proves accurate. However, there are many very interesting exceptions from the point of view of this book, in particular, ferromagnetic substances which may be either insulators or conductors, but which exhibit a spontaneous magnetic moment that is not predicted by band theory, or superconductors which are perfect conductors but which are not paramagnetic and whose specific heat is not linear in temperature. To understand these phenomena, we must abandon one or other of the approximations discussed in Chap. 1.

3.4 Experimental Determination of Band Structures

As we have seen above, thermodynamic methods and magnetic measurements can be used to estimate the density of states at the Fermi level in metals. But beyond this rather limited information, what experimental methods can be used to obtain more details regarding the density of states as a function of energy or the band structure of a solid? Here we shall discuss some experimental techniques for determining the band gaps in an insulator (optical absorption), the energy dependence of the density of states at the surface of conducting materials (tunneling effect and microscopy), and the dispersion curves $E(\mathbf{k})$ (angle-resolved photoemission spectroscopy).

3.4.1 Optical Absorption in Insulators

If we are only concerned with the value of the band gap in an insulator, spectroscopic methods are relatively simple and effective. Indeed, atomic energy levels are usually determined by absorption or emission spectroscopy. Similarly, if we shine a light beam at frequency ν on an insulating solid, this beam will pass right through the solid without absorption if the photon energy $h\nu$ is lower than the band gap

3.4 Experimental Determination of Band Structures

energy E_g. Otherwise there will be absorption of photons from the beam. The ratio of the outgoing to incident light intensity depends on the thickness x of the sample according to

$$I(x)/I_0 = e^{-\alpha x},$$

whence the optical absorption coefficient α may be determined. This absorption becomes significant when $h\nu = E_g$. The optical absorption threshold thus gives the width of the band gap in various semiconductors (see Fig. 3.20). The change in $\alpha(\nu)$ for different values of ν also gives information about the band structure. Indeed, the existence of many interband transitions at certain energies, notably close to the saddle points of $E(\mathbf{k})$, leads to singularities in the absorption curve. This is illustrated in Fig. 3.20.

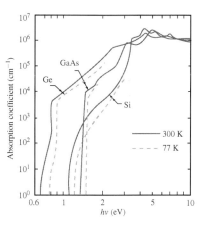

Fig. 3.20 Optical absorption coefficient α for various well known semiconductors, measured at 300 and 77 K as a function of the energy of the incident photons

3.4.2 Tunneling Effect and Scanning Tunneling Microscopy

Another very effective method for probing the density of states of a conducting material exploits the electronic tunneling effect. The basic idea is to determine the electron current transmitted through a thin insulating barrier between two conducting materials. This technique was initially used to study the electronic structure of superconducting materials (see Chap. 7).

a. Tunneling Effect

It is well known that incident electrons can tunnel through a potential barrier when the width d of the barrier is not too great. The transmission coefficient $\mathscr{T}$ of the barrier is

$$\mathscr{T} \sim \exp\left[-\frac{2d}{\hbar}\sqrt{2m_0(U_0 - E)}\right], \tag{3.69}$$

where U_0 is the height of the potential barrier and E is the energy of the incident electron [2]. The tunneling current is only significant if the width of the barrier is less than a few tens of Å.

If the barrier separates two normal metals, the chemical potentials are equal on either side at thermodynamic equilibrium, and there is no flow of electrons. However, if a bias V is applied across the metals, the potential energy $-eV$ lowers the energy levels on the $+$ side relative to those on the $-$ side, and there will be electron transport (see Fig. 3.21). We see that the electrons in metal 1 with energies between E_F and $E_F - eV$ can tunnel into states of the same energy which have become accessible in metal 2. If $D_1(E)$ and $D_2(E)$ are the densities of states of metals 1 and 2, and if we take into account the occupation probabilities at nonzero temperature, the tunneling current density from 2 to 1 is given by

$$j_{21} = A\mathcal{T} \int_{-\infty}^{+\infty} D_1(E)f(E)D_2(E+eV)\bigl[1-f(E+eV)\bigr]dE, \quad (3.70)$$

where A is a constant and $f(E)$ is the Fermi–Dirac function. At nonzero temperature, there is also a smaller current in the opposite direction (see Fig. 3.21) given by

$$j_{12} = A\mathcal{T} \int_{-\infty}^{+\infty} D_1(E)\bigl[1-f(E)\bigr]D_2(E+eV)f(E+eV)dE. \quad (3.71)$$

The total current density is therefore

$$j = A\mathcal{T} \int_{-\infty}^{+\infty} D_1(E)D_2(E+eV)\bigl[f(E)-f(E+eV)\bigr]dE. \quad (3.72)$$

We thus see that applying a bias simply lowers the energy levels in metal 2, whence a group of electrons in metal 1 corresponding to occupied states with energies lying

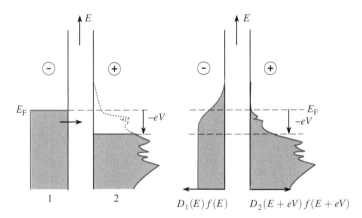

Fig. 3.21 *Left*: Energy levels at zero temperature in normal metals 1 and 2 with a positive bias V applied between 2 and 1. Electrons can only get from 1 to 2 by tunnelling. *Right*: Occupied states on either side of the barrier at nonzero temperature

3.4 Experimental Determination of Band Structures

in the range E_F to $E_F - eV$ are able to tunnel through to metal 2. If the latter is the material whose density of states we aim to determine, metal 1 is chosen to be a metal with a density of states that hardly varies with E at the Fermi level, so that $D_1(E)$ can be replaced by $D_1(E_F)$ to a good approximation. However, if $D_2(E)$ varies rapidly with E, the dependence of j on V will be nonlinear. Replacing $D_1(E)$ by $D_1(E_F)$ and setting $E' = E + eV$ in (3.72), it is straightforward to check that

$$\frac{dj}{dV} = A\mathcal{T}D_1(E_F) \int_{-\infty}^{+\infty} D_2(E') \frac{\partial f(E' - eV)}{\partial V} dE'. \tag{3.73}$$

At zero temperature, the derivative of the Fermi function is a δ function, and then

$$\frac{dj}{dV} = A\mathcal{T}eD_1(E_F)D_2(E_F + eV). \tag{3.74}$$

By measuring the dependence of dj/dV on V, we may thus deduce the density of states $D_2(E_F + eV)$.

b. Application to Scanning Tunneling Microscopy and Spectroscopy

A modern setup uses a very finely tapered metal tip for metal 1, with nanoscale dimensions. This is displaced in a controlled way just above the surface under investigation. The lateral displacement of the sample is achieved by means of a piezoelectric stage, whose x, y, z position is controlled by three applied voltages (see Fig. 3.22). The vertical position determines the tunnel current between the tip and sample for a given bias V between them. Using a feedback loop to fix the value of the tunnel current, one can then measure the change in z as the tip moves across the surface, thereby constructing a topographic image of the surface. This method has been used, for example, to produce images of alkane molecules deposited on graphite (see Fig. 2.1).

By measuring the tunnel current as a function of the bias at a given point on the surface with fixed height z, we obtain the density of states at this point as a function

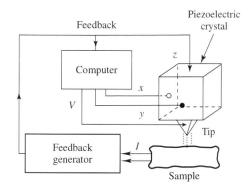

Fig. 3.22 Basic setup of a scanning tunneling microscope. The diagram only shows how the displacement of the piezoelectric crystal is controlled. The tunnel current between tip and sample can be used to control the z position. For fixed z, the dependence of the tunnel current on V can be used to find the local density of states

88   3 Electronic Structure of Solids: Metals and Insulators

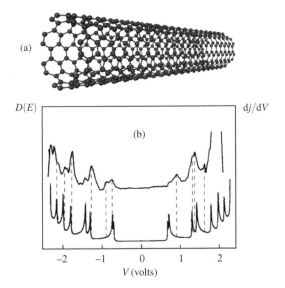

Fig. 3.23 (**a**) Carbon nanotube of diameter of about 10 Å. Image courtesy of Y. Ihara (Hokkaido University, Japan) (**b**) *Upper*: Density of states deduced from the tunnel characteristic dj/dV observed with a scanning tunneling microscope. *Lower*: LCAO calculation of $D(E)$ for a model nanotube. From Kim, P., et al.: Phys. Rev. Lett. **82**, 1225 (1999) ©American Physical Society. http://link.aps.org/doi/10.1103/physRevLett.82.1225

of the energy. An example is shown in Fig. 3.23, which displays the electronic density of states at a point of a carbon nanotube. The filamentary structure of this system induces Van Hove singularities in the density of states, similar to those calculated in Chap. 1. The experimental observations are compared with a calculation of the electronic structure using the LCAO approximation.

3.4.3 Angle-Resolved Photoemission Spectroscopy

In a photoemission experiment, one exploits the photoelectric effect. A photon with enough energy $h\nu$ removes an electron from an occupied state below the Fermi level, i.e., a conduction electron for a metal or a valence electron for an insulator or a semiconductor (see Fig. 3.24). This electron leaves the solid. Its initial state is a Bloch state of the crystal, while its final state is that of a free electron. By conservation of energy and momentum in the interaction process, the dispersion relation $\omega(\mathbf{k})$ of the Bloch states can be reconstructed.

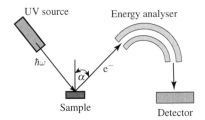

Fig. 3.24 Setup for a photoemission experiment (ARPES). UV photons excite electrons, which are then emitted with energy E in a direction α. By measuring E and α for the emitted electrons, the curves $E(\mathbf{k})$ can be deduced

3.4 Experimental Determination of Band Structures

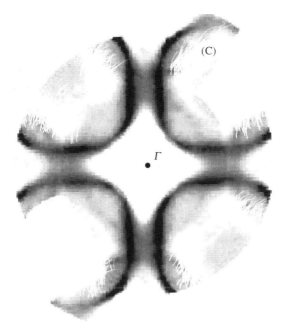

Fig. 3.25 Experimental determination by ARPES of the Fermi surface (C) of a CuO_2 plane, represented in several primitive cells of the reciprocal lattice

This technique, known as angle-resolved photoemission spectroscopy (ARPES), is widely used for 2D systems such as cuprate superconductors, whose structure was presented in Fig. 2.9b. In these compounds, the electronic structure of a CuO_2 plane at the surface of a single crystal can be investigated. The Fermi surface can thus be determined experimentally in several primitive cells of the reciprocal lattice, as shown for $Bi_2Sr_2CaCu_2O_8$ in the figure on p. 52. Part of the negative of this image is shown magnified in Fig. 3.25, in which the point Γ is taken as the origin of the reciprocal lattice.

Question 3.5. 1. Specify the Bravais lattice of the CuO_2 planes, together with the primitive cell and its reciprocal lattice.
2. From the shape of the Fermi 'curve', is it possible to say whether the electronic structure of the CuO_2 plane can be described by the nearly-free electron approximation or rather by a tight-binding approximation? Explain why.
3. Determine the number of charges per CuO_2 plane on the basis of the experimental Fermi surface.
4. Is this the expected number, if we assume that the atoms are in ionic form: Bi^{3+}, Sr^{2+}, Ca^{2+}, Cu^{2+}, and O^{2-}? If not, can you explain why there should be a difference?

Question 3.6. The second illustration in the preface shows a similar ARPES determination of the Fermi surface of another 2D system, viz., Sr_2RuO_4, in which the RuO_2 planes have the same structure as the CuO_2 planes in the cuprates. What can be concluded about the band structure of this compound?

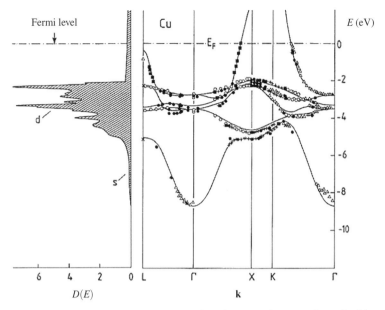

Fig. 3.26 Band structure of valence and conduction electrons of copper, determined by photoemission. The results of numerical simulations are shown by *continuous curves*. The corresponding density of states is shown *on the left*. From [6, p. 145] Springer © 1995. Experimental results from Courths, R., Hüfner, S.: Phys. Rep. **112**, 55 (1984). Band calculation from Eckardt, H., Fritsche, L., Noffke, J.: J. Phys. F **14**, 97 (1984)

The band structure of a 3D system can also be very accurately determined using ARPES. Figure 3.26 shows the experimental results for copper. The bands resulting from the 3d and 4s electrons of copper are shown for various orientations of the vector **k** in the reciprocal lattice. Since copper is fcc, Γ, X, K, L are the points of the first Brillouin zone given in Fig. 3.3b. The experimental results (points) agree very well with a numerical calculation of the band structure, which shows the quality of the methods currently available for calculating band structures. As we shall see later (see Chap. 8, and in particular Sect. 8.6), in magnetic systems for which local electron correlations become relevant, the approximations of Chap. 1 will no longer be applicable, and more elaborate methods are currently being developed.

3.5 Summary

Bloch's theorem in three dimensions gives the general form of the wave function of an electron subject to a periodic potential (Bloch function):

$$\psi_{n,\mathbf{k}}(\mathbf{r}) = e^{i\mathbf{k}\cdot\mathbf{r}} u_{n,\mathbf{k}}(\mathbf{r}), \quad \text{where} \quad u_{n,\mathbf{k}}(\mathbf{r}+\mathbf{R}_l) = u_{n,\mathbf{k}}(\mathbf{r}).$$

3.5 Summary

Each eigenstate is characterised by a vector $\mathbf{k}$ specified in the reciprocal lattice, and it is periodic under any translation along a vector $\mathbf{K}$ of the reciprocal lattice. This redundancy leads us to define the eigenstates in a primitive cell of the reciprocal lattice. An important consequence of this theorem is that it is the vector $\mathbf{k}$, called the crystal momentum or quasi-momentum, which plays the role of a conserved quantity in collisions with other particles.

It is interesting to consider the quantum states of an electron in a solid from the opposite standpoint to the one adopted in the tight-binding approximation, i.e., considering a free electron gas perturbed by a weak periodic potential. This approach applies for instance to the valence electrons which are the least tightly bound. The electron states most perturbed by the periodic potential are those whose crystal momentum $\mathbf{k}$ lies close to a Bragg plane:

$$2\mathbf{k} \cdot \mathbf{K} = \mathbf{K}^2 .$$

In this case, the degeneracy of the electronic levels is removed by the perturbing potential. The difference between the levels is $2V_{\mathbf{K}}$, twice the Fourier component of the perturbing potential at the wave vector $\mathbf{K}$.

This analysis brings out the role of the Bragg planes in the band structure of the solid, and leads us to specify the electronic states in the Brillouin zones, primitive cells of the reciprocal lattice bounded by the Bragg planes. In particular, the first Brillouin zone corresponds to those points of the reciprocal space closer to $\mathbf{k} = 0$ than to any other point of the reciprocal lattice.

It is then observed that, in one dimension, an arbitrarily weak potential introduces a band gap into the spectrum. On the other hand, in two or three dimensions, the periodic potential must have a certain minimal amplitude in order to generate a band gap. If the strength of the potential is increased, we recover a very similar band structure to the one obtained in the tight-binding approximation.

The full quantum state of a solid is obtained by filling the one-electron states with all the electrons of the given solid, while observing the rules of quantum statistical mechanics for fermions. We thus obtain a rather simple result: either the last occupied band (the one of highest energy) is only partially occupied, in which case we are dealing with a metal, or the last band is fully occupied, in which case the material is an insulator at $T = 0$.

In insulators, the last occupied levels of the valence band are separated from the first empty levels of the conduction band by a region of forbidden energy called the band gap. When this band gap is narrow enough ($\lesssim 1$ eV), a significant number of electronic states in the conduction band are populated by thermal excitation, leading to a weakly conducting state at room temperature. The material is then called a semiconductor.

Graphene is a 2D layer of carbon in which the outer p_z electrons form two electronic bands which touch one another at two points of the Brillouin zone. At $T = 0$, one is filled and the other empty, which corresponds to a very special electronic state of a semiconductor in which the band gap is strictly zero (or of a metal in which the Fermi surface reduces to just two points).

In the band theory framework, insulators are non-magnetic, with negligible electronic contribution C_e to the specific heat, whereas in metals, the electrons contribute a significant term linear in T which dominates at low temperatures. In addition, metals exhibit temperature-independent Pauli paramagnetism. These properties of the metal are determined by the highest energy occupied electronic states, whose energy E_F defines the Fermi energy. This means that the Fermi surface, which represents these states in reciprocal space, assumes a particularly important role, as does the density $D_\Omega(E_F)$ of these states. In particular,

$$C_e(T) = \frac{2\pi^2 D_\Omega(E_F)k_B^2}{3} T, \qquad \chi_P = 2\mu_B^2 D_\Omega(E_F),$$

and band theory predicts a universal constant for the ratio between these two quantities, which is indeed confirmed experimentally in many simple metals.

Several experimental methods can be used to study the electronic density of states or the band structure of solids. Optical absorption is a good way of determining band gaps, particularly for semiconductors. The electron tunneling effect between a nanoscale metal tip and a conducting surface gives the energy dependence of the electronic density of states at the location of the tip over the metal surface. These near-field microscopy methods can be used to study the topography of a surface at subnanometric scales, and hence to image atomic or molecular arrangements. One of the most powerful methods for studying the band structure of a solid is angle-resolved photoemission spectroscopy, which can directly determine the dispersion curves $E(\mathbf{k})$, and in particular the Fermi surface in the case of a metal.

3.6 Answers to Questions

Question 3.1

Let $\mathbf{a}_1$ and $\mathbf{a}_2$ be the vectors specifying the real lattice, and let us define their angle as $(\mathbf{a}_1, \mathbf{a}_2) = \alpha$. The 2D reciprocal lattice is given by

$$\mathbf{a}_1^* = \frac{2\pi \, \mathbf{a}_2 \wedge \mathbf{k}}{S}, \qquad \mathbf{a}_2^* = \frac{2\pi \, \mathbf{a}_1 \wedge \mathbf{k}}{S},$$

where $S = |\mathbf{a}_1 \wedge \mathbf{a}_2| = a_1 a_2 \sin \alpha$ is the area of the primitive unit cell, and $\mathbf{k}$ the normal to the $(\mathbf{a}_1, \mathbf{a}_2)$ plane (see Fig. 3.27). We thus have $(\mathbf{a}_2^*, \mathbf{a}_1^*) = (\mathbf{a}_1, \mathbf{a}_2) = \alpha$, and the reciprocal lattice has the same symmetry as the real lattice.

The rectangular lattice $(a\mathbf{x}, b\mathbf{y})$ has a rectangular reciprocal lattice with

$$\mathbf{a}_1^* = \frac{2\pi}{a}\mathbf{x}, \qquad \mathbf{a}_2^* = \frac{2\pi}{b}\mathbf{y}.$$

The reciprocal lattice of the hexagonal lattice is hexagonal with lattice constant $4\pi/a\sqrt{3}$, while the reciprocal lattice of the centered rectangular lattice is centered rectangular with lattice constant $2\pi/a\sin\alpha$. The first Brillouin zones are the Wigner–Seitz cells determined in Question 2.1.

Fig. 3.27 Reciprocal lattice vectors for a planar lattice

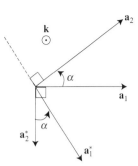

Question 3.2

The fcc primitive cell is shown in Fig. 3.28a. The basis of the diamond lattice contains two atoms of Ge at $(0,0,0)$ and $a(1/4,1/4,1/4)$, shown as white spheres in Fig. 3.28a. The reciprocal lattice is body-centered cubic with side $4\pi/a$. The first Brillouin zone is shown in Fig. 3.28b. There are two atoms per unit cell which donate each their four $(4s^2\ 4p^2)$ electrons to the bands issued from the atomic $(4s^2\ 4p^6)$ levels. These eight electrons therefore completely fill the first 4 bands (see Fig. 3.29). The others are empty at $T = 0$ and Ge is thus an insulator.

Question 3.3

The band gap separates the highest energy occupied levels at $k = 0$ and the lowest energy empty level at the point L with coordinates $(2\pi/a)(1/2, 1/2, 1/2)$. The energy difference between these levels is $E_g = 0.7$ eV. The energy width of this band gap

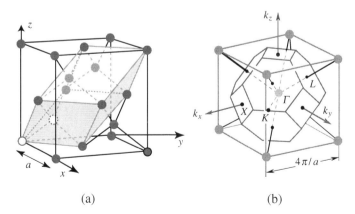

Fig. 3.28 Diamond structure. (**a**) Real space fcc unit cell with its two-atom basis. (**b**) Bcc reciprocal lattice and Brillouin zone

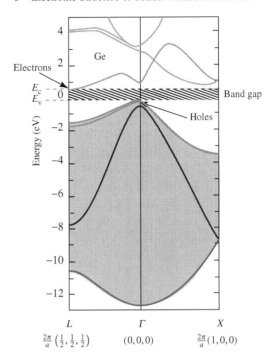

Fig. 3.29 Band structure of Ge. Filled valence bands are *shaded* and the band gap is *hatched*. The **k** space locations of the hole and electron states that can be populated by thermal excitation or doping are indicated

is therefore very small. At room temperature, the electronic states close to E_c with $\mathbf{k}_e \simeq (2\pi/a)(1/2,1/2,1/2)$ will be partly occupied and the states of energy E_v close to $\mathbf{k}_h \simeq 0$ will be partly unoccupied.

Question 3.4

Write

$$\left\langle \varphi_l^A \middle| H \sum_j \left(\lambda_A \middle| \varphi_j^A \right\rangle + \lambda_B \middle| \varphi_j^B \right\rangle \right) e^{i\mathbf{k}\cdot\mathbf{R}_j} = E \left\langle \varphi_l^A \middle| \sum_j \left(\lambda_A \middle| \varphi_j^A \right\rangle + \lambda_B \middle| \varphi_j^B \right\rangle \right) e^{i\mathbf{k}\cdot\mathbf{R}_j} .$$

Given the orthonormality of the functions $|\varphi_j^A\rangle$ and $|\varphi_j^B\rangle$, the right-hand term simplifies immediately to

$$\lambda_A \sum_j \left\langle \varphi_l^A \middle| H \middle| \varphi_j^A \right\rangle e^{i\mathbf{k}\cdot\mathbf{R}_j} + \lambda_B \sum_j \left\langle \varphi_l^A \middle| H \middle| \varphi_j^B \right\rangle e^{i\mathbf{k}\cdot\mathbf{R}_j} = \lambda_A E e^{i\mathbf{k}\cdot\mathbf{R}_l} .$$

The first sum reduces to the term $j = l$, while the second includes only hopping integrals from the site (l, A) to its three nearest neighbour B atoms, i.e., (l, B) and the two B atoms translated by $-\mathbf{a}_1$ and $-\mathbf{a}_2$. Cancelling the common factor $e^{i\mathbf{k}\cdot\mathbf{R}_l}$ then yields

3.6 Answers to Questions

$$\lambda_A E_0 - \lambda_B t\left(1 + e^{-i\mathbf{k}\cdot\mathbf{a}_1} + e^{-i\mathbf{k}\cdot\mathbf{a}_2}\right) = \lambda_A E.$$

In the same way, projecting onto $\langle\varphi_l^B|$, we obtain

$$\lambda_B E_0 - \lambda_A t\left(1 + e^{i\mathbf{k}\cdot\mathbf{a}_1} + e^{i\mathbf{k}\cdot\mathbf{a}_2}\right) = \lambda_B E.$$

These two equations have solution

$$(E_0 - E)^2 = t^2\left(1 + e^{-i\mathbf{k}\cdot\mathbf{a}_1} + e^{-i\mathbf{k}\cdot\mathbf{a}_2}\right)\left(1 + e^{i\mathbf{k}\cdot\mathbf{a}_1} + e^{i\mathbf{k}\cdot\mathbf{a}_2}\right),$$

which leads to the dispersion relations of the energy bands (3.60):

$$E_{\mathbf{k}}^{\pm} = E_0 \pm t\left\{3 + 2\left[\cos\mathbf{k}\cdot\mathbf{a}_1 + \cos\mathbf{k}\cdot\mathbf{a}_2 + \cos\mathbf{k}\cdot(\mathbf{a}_1 - \mathbf{a}_2)\right]\right\}^{1/2}.$$

Question 3.5

1. The 2D Bravais lattice is square with side a, and the primitive cell contains a Cu and 2O. The reciprocal lattice is plane square with lattice constant $2\pi/a$. The first Brillouin zone is square, centered on the origin and with vertices $(\pm\pi/a, \pm\pi/a)$. The axes are easily identified using the symmetry of the experimental Fermi surface (see Fig. 3.30).

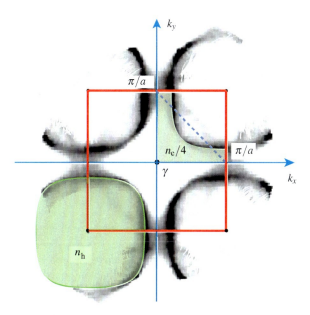

Fig. 3.30 Fermi 'curve' for the cuprate lattice. The square first Brillouin zone is displayed (see text). *Shaded areas* allow determination of the carrier content

2. The Fermi 'curve' (C) has large discontinuities at its intersections with the first Brillouin zone. The situation cannot therefore correspond here to one in which the electrons are nearly free. It is more like one with tight-binding, because the ionic potential gives large contributions at the edge of the zone. However, the simple LCAO model with hopping integrals between nearest neighbour Cu cannot explain this Fermi surface (see those plotted in Fig. 3.7). We must also introduce hopping integrals between next nearest neighbours.
3. The surface bounded in the plane by the Fermi curve (C) contains the occupied electronic states **k**. As the first Brillouin zone can accommodate 2 electrons per primitive cell, including spin, the number of electrons per primitive cell in the Brillouin zone is

$$n_e = 2 \times \frac{\text{Area within (C)}}{\text{Area of the Brillouin zone}}.$$

This ratio can be obtained from Fig. 3.30 by comparing the green area upper right with the area of one quarter of the Brillouin zone. By comparing the areas on either side of the diagonal of the square, it is clear that $n_e/4 < 2(1/8)$, that is $n_e < 1$. Greater accuracy is obtained by determining the area of the Fermi surface of holes, which is the green region bottom left. We find $n_h = 1.15 \pm 0.05 = 2 - n_e$. There are therefore 0.85 electrons per primitive cell in the CuO_2 plane in the partially occupied band. It is in fact this reduction of n_e below unity that leads to a metallic state. Indeed, for $n_e = 1$, the CuO_2 plane is an antiferromagnetic Mott insulator owing to the strong electron correlations at the Cu site.[7]
4. This means that there are fewer electrons per primitive cell in the CuO_2 plane than would be expected from the chemical formula $Bi_2Sr_2CaCu_2O_8$. Electrical neutrality thus implies an excess of negative charges outside the CuO_2 plane. This comes from the fact that, in the materials studied, the stoichiometry is not perfect. For example, there are interstitial oxygen atoms with concentration $x = (1 - n_e)/2$, and this modifies the doping in the CuO_2 planes. The chemical formula should be $Bi_2Sr_2CaCu_2O_{8+x}$. The possibility of changing the oxygen content x provides a controlled way of modifying the number of carriers in the CuO_2 planes.

Question 3.6

The figure on the cover of the book shows at the top the intensity of emitted photoelectrons as a function of their wave vectors k_x for different values of k_y. Peaks in the emitted intensity correspond to the last occupied levels and provide a direct way of visualising the Fermi surface of this compound in the first Brillouin zone via a projection onto the plane (k_x, k_y). This first Brillouin zone is a square in this plane, as for the cuprates. The Fermi surface is much more complicated than that of the cuprate in Fig. 3.25. Note that it is made up of three curves which can be generated

---

[7] This situation is described for La_2CuO_4 in Problem 17: *Electronic properties of La_2CuO_4*.

3.6 Answers to Questions

from any one of them by the four-fold symmetry of the square. This indicates that the band structure of this compound contains three bands which intersect the Fermi level and are therefore partly occupied.

For completeness, note that the crystal structure of Sr_2RuO_4 is obtained from a perovskite structure (see Chap. 2) and it is the same as that of La_2CuO_4.[8] If we consider the charges of the Sr^{2+} and O^{2-} ions, it is easy to deduce that the Ru are in a $4d^4$ configuration. However, in cuprates such as La_2CuO_4, where La is ionised to La^{3+}, the Cu are in a $3d^9$ configuration. In the ruthenate, it is not therefore surprising that several bands arising from the various d orbitals with similar energies should overlap near the Fermi level, in contrast to what happens with the cuprates.

---

[8] This structure is displayed in Problem 17: *Electronic Properties of La_2CuO_4*.

Chapter 4
Electron Transport in Solids

Contents

4.1	Drude Model for Transport in an Electron Gas: Relaxation Time and Collisions	101
	4.1.1 Electrical Conductivity	102
	4.1.2 Thermal Conductivity and the Wiedemann–Franz Law	105
4.2	Electron Transport in a Fermion Gas	107
	4.2.1 Electrical Conductivity	108
	4.2.2 Thermal Conductivity and the Wiedemann–Franz Law	110
4.3	Electrons in a Lattice: Dynamics of Bloch Electrons	111
	4.3.1 Group Velocity	111
	4.3.2 Acceleration in Reciprocal Space and Real Space	113
	4.3.3 Electronic Conductivity in a Crystal	115
4.4	Origin of Collisions	116
	4.4.1 Experimental Observation	116
	4.4.2 Scattering by Lattice Vibrations	118
	4.4.3 Collisions with Impurities and Defects	119
4.5	Electrons, Holes, and Dopants in Semiconductors	121
4.6	Electrons, Holes, and Transport in Graphene	126
4.7	Summary	132
4.8	Answers to Questions	133

All but one of the everyday objects illustrated here exploit physical processes with a common feature. As you read this chapter, you should be able to spot the odd one out, and explain why it differs, indicating for each object the key physical process underlying its use (answer in Sect. 4.8)

Our forebears were interested in metals for their malleability (at the forge) and good thermal conductivity (for making cooking vessels), and they have been used for centuries as a raw material for forging tools and weapons of various kinds. But their most characteristic feature, namely their good electrical conductivity (about 20 orders of magnitude greater than insulators), has only been recognised for a few centuries, since Coulomb and Ampère discovered the existence of electrical charge. In the preceding chapters, our discussion of allowed or forbidden energy bands for the electron levels in crystalline solids explains the origin of this qualitative difference between metals and insulators. However, we have not yet established a physical understanding for the limitation of conductivity in metals as described by Ohm's law. Indeed, we have seen that an electron in a metal is in a quantum state characterised by a Bloch wave function, which is a wave function extending throughout the space occupied by the metal. Since this state is stationary, one would expect an electron to be able to move freely throughout the volume of the metal without dissipation of energy, just as any electron in an atom remains around the nucleus in a stationary state, without loss of energy. *The idea of electrical resistivity is therefore quite foreign to the quantum description we have made of the electrons in a perfect metallic crystal.*

Note also that the descriptions used so far in the quantum mechanics or statistical physics textbooks concern stationary states or thermodynamic equilibrium, not directly relevant to charge transport in a material. The phenomenon of electrical conductivity is a problem of non-equilibrium thermodynamics. However, the departure from equilibrium here is not dramatic, provided that we consider only weak electric fields and currents. This phenomenon is thus described as a quasi-equilibrium transport regime.

In Sect. 4.1, we begin by describing the approach used by Drude in the 1900s to explain conductivity by treating the electrons as classical particles. We thus introduce the concepts of relaxation time and collisions. These concepts are developed in the more realistic context of a free fermion gas (see Sect. 4.2), then in the presence of a crystal potential (see Sect. 4.3). Section 4.4 discusses experimental observations that help us to determine the origin of collisions in real metals. Then in Sect. 4.5, we introduce the notion of *hole* which plays an important role in the case of semiconductors. Finally, the peculiar case of *graphene*, which has recently triggered intense research activity, is considered in Sect. 4.6.

4.1 Drude Model for Transport in an Electron Gas: Relaxation Time and Collisions

At thermodynamic equilibrium, there is no transfer of matter, heat, or charge within a continuous medium. However, in everyday life, such transfers are commonplace. While the underlying microscopic processes are generally very difficult to describe, there is nevertheless a simple situation where they are controlled by a slight departure from thermodynamic equilibrium. Indeed, if a macroscopic gradient is imposed on one of the intensive parameters characterising thermodynamic equilibrium, this will lead to a flow of the extensive quantity conjugate to it. For example, a temperature gradient ∇T will cause a heat flow (and hence a flow of energy $\mathbf{j}_{en}$), an electrical potential gradient (an electric field $\mathscr{E}$) will cause a flow of charge (and hence a current $\mathbf{j}$), and a chemical potential gradient $\nabla \mu$ will cause a flow of matter. In the limit of a very weak gradient, the resulting flow will be proportional to the gradient producing it. In the first two examples above, this corresponds to Fourier's law ($\mathbf{j}_{en} = -\kappa \nabla T$) and Ohm's law ($\mathbf{j} = \sigma \mathscr{E}$), which define the thermal conductivity κ and the electrical conductivity σ. In the rest of this chapter, we shall examine the microscopic origins of these transport processes in metals.

4.1.1 Electrical Conductivity

Following Drude's example at the beginning of the 20th century, we do not have to take into account quantum phenomena to obtain a qualitative understanding of the physical origins of electrical resistivity. Indeed, Drude treated the electron as a classical particle. When there is no electric field, the electrons have velocities $\mathbf{v}_i$, with randomly distributed directions in space, like the molecules in a gas, so their average velocity is zero, i.e., $\langle \mathbf{v}_i \rangle = 0$. Under the action of an electric field $\mathscr{E}$ applied at time $t = 0$, we know that an electron of velocity $\mathbf{v}_i$, mass m_0, and charge $-e$ will be accelerated, with

$$m_0 \frac{d\mathbf{v}_i}{dt} = -e\mathscr{E}. \tag{4.1}$$

All the electrons in the gas will therefore acquire a velocity component $\mathbf{v}_d$ called the *drift velocity*, corresponding to a uniformly accelerated motion in the direction opposite to the applied field:

$$\mathbf{v}_i = \mathbf{v}_i(0) + \mathbf{v}_d = \mathbf{v}_i(0) - \frac{et}{m_0}\mathscr{E}. \tag{4.2}$$

This would imply that $\mathbf{v}_d$ tend to infinity over a long period of time. However, we know by Ohm's law that a *static field produces a constant current*. The current density $\mathbf{j}$ is the electron flow per unit time and per unit area. It is given by

$$\mathbf{j} = -e \sum_i \mathbf{v}_i = -ne \langle \mathbf{v}_i \rangle . \tag{4.3}$$

Here the sum over i concerns all electrons crossing the unit area per unit time, n is the number of electrons per unit volume, and $\langle \mathbf{v}_i \rangle$ denotes their average velocity. The only way to obtain a stationary solution with constant flow rate is to introduce a frictional effect for the electron motion, whose role is to limit their drift velocity. To obtain a stationary solution with constant average velocity $\langle \mathbf{v}_i \rangle$, the friction term must be proportional to $\langle \mathbf{v}_i \rangle$, i.e., $-(m_0/\tau_e)\langle \mathbf{v}_i \rangle$, where τ_e has the physical dimensions of time:

4.1 Drude Model for Transport in an Electron Gas: Relaxation Time and Collisions

$$m_0 \frac{d\langle \mathbf{v}_i \rangle}{dt} = -e\mathcal{E} - m_0 \frac{\langle \mathbf{v}_i \rangle}{\tau_e}. \tag{4.4}$$

This leads to the stationary solution $\langle \mathbf{v}_i \rangle = -(e\tau_e/m_0)\mathcal{E} = -\mu_e\mathcal{E}$, where μ_e is the *electron mobility*, and finally,

$$\mathbf{j} = \frac{ne^2 \tau_e}{m_0} \mathcal{E}. \tag{4.5}$$

We thus recover *Ohm's law*:

$$\boxed{\mathbf{j} = \sigma \mathcal{E}}. \tag{4.6}$$

The electrical conductivity and resistivity are given by

$$\boxed{\sigma = \frac{ne^2 \tau_e}{m_0} = ne\mu_e = \frac{1}{\rho}}. \tag{4.7}$$

Note that the time τ_e introduced along with the friction hypothesis has a simple interpretation. Suppose that the electric field is cut off at time $t = 0$. The dynamical equation for the average velocity shows that this quantity decreases exponentially to 0, with time constant τ_e. This is thus the lapse of time required for the velocity distribution to return to equilibrium, and it is called the *relaxation time*. It therefore quantifies the rate at which the electron gas can reach thermodynamic equilibrium from a non-equilibrium state.

The relaxation time clearly originates from microscopic causes, which Drude attributed to the fact that the electrons, like the molecules of a gas, undergo *collisions*. The latter change the orientations of the electron velocities, but also their magnitudes. Indeed, collisions that only change the direction of the electron velocity without changing its kinetic energy should not contribute to dissipation. Drude thus assumed that the probability $p(\mathbf{v})$ that an electron have velocity $\mathbf{v}$ after a collision would be independent of its velocity before the collision, and hence that these collisions would destroy all memory of the increased drift velocity of the electrons. All the electrons will therefore undergo collisions that completely change their velocities. This is illustrated in Fig. 4.1, which shows the contribution of one electron to the drift velocity as time goes by.

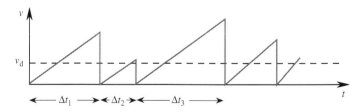

Fig. 4.1 Time dependence of the component of the velocity of a free electron induced by the presence of an electric field, according to the Drude model. The induced component is projected in the field direction (the acquired velocity is antiparallel to the field). The drift velocity of the electron is $\mathbf{v}_d$

Let dt/τ be the probability that an electron will undergo a collision during the time interval dt, and $\langle v(t) \rangle$ the average value of the velocity at time t. Then at time $t + dt$, the electrons fall into two groups:

1. The fraction dt/τ of the electrons which underwent a collision between t and $t + dt$. These make no contribution (to order dt) to the average velocity at time $t + dt$.
2. The fraction $(1 - dt/\tau)$ of the electrons which did not undergo a collision between t and $t + dt$. The velocities of these electrons increase to $\mathbf{v}_i(t + dt) = \mathbf{v}_i(t) - e\mathscr{E}dt/m_0$. Their contribution to the average velocity is therefore

$$\langle \mathbf{v}_i(t + dt) \rangle = \left(1 - \frac{dt}{\tau}\right)\left[\langle \mathbf{v}_i(t) \rangle - \frac{e\mathscr{E}dt}{m_0}\right], \tag{4.8}$$

or to order dt,

$$\langle \mathbf{v}_i(t + dt) \rangle - \langle \mathbf{v}_i(t) \rangle = -\frac{e\mathscr{E}dt}{m_0} - \frac{dt}{\tau}\langle \mathbf{v}_i(t) \rangle, \tag{4.9}$$

which leads back to (4.4).

We may thus identify τ with τ_e. In the box below it is shown that, in this model, τ_e is also the average time elapsed between two collisions of a given electron.

Let us find the value of τ_e for metallic copper, for which $\sigma = 6.54 \times 10^7 \; \Omega^{-1} \mathrm{m}^{-1}$ (or $\rho = 1.55 \; \mu\Omega\,\mathrm{cm}$). Given that copper crystallises into a face-centered cubic lattice with conventional unit cell of side $a = 3.61$ Å, and that the conduction band contains one electron per copper atom, it follows that $n = 4a^{-3} = 8.5 \times 10^{28} \; \mathrm{m}^{-3}$. Equation (4.7) yields $\tau_e = 3 \times 10^{-14}$ s, which indicates that an electron undergoes an enormous number of collisions per second.

Collision Probabilities

Let $p(t)$ be the probability that an electron undergoes no collision between times $t = 0$ and t. The probability of continuing until $t + dt$ without collision, viz., $p(t + dt)$, is the probability that the electron has not had a collision until t, and in addition that it has not had a collision between t and $t + dt$, whence

$$p(t + dt) = p(t)(1 - dt/\tau_e). \tag{4.10}$$

It follows that

$$dp = -p\,dt/\tau_e, \tag{4.11}$$

$$p(t) = \exp(-t/\tau_e). \tag{4.12}$$

The probability $P(t)dt$ that an electron undergoes no collision between 0 and t but undergoes a collision between t and $t + dt$ is then

$$P(t)dt = \exp(-t/\tau_e)dt/\tau_e. \tag{4.13}$$

4.1 Drude Model for Transport in an Electron Gas: Relaxation Time and Collisions

The probability density $P(t)$, described by an exponential distribution, can be used to calculate the average time between two collisions:

$$\langle t \rangle = \int_0^\infty t P(t) \mathrm{d}t = \tau_e . \tag{4.14}$$

Question 4.1. The Drude model can be extended to describe the response of a metal to an alternating electric field $\mathscr{E}(t) = \mathscr{E}(\omega) \operatorname{Re}\{e^{i\omega t}\}$. The associated current density can be used to define a complex conductivity

$$\sigma(\omega) = \sigma_1(\omega) - i\sigma_2(\omega) .$$

Making Drude's hypothesis that there is a relaxation time τ_e, determine $\sigma(\omega)$. Express $\sigma_1(\omega)$ as a function of $\sigma(0)$, and represent it graphically as a function of ω.

4.1.2 Thermal Conductivity and the Wiedemann–Franz Law

While an electric field $\mathscr{E}$, i.e., a potential gradient, produces a charge flow and hence an electron current $\mathbf{j}$, a temperature gradient ∇T induces an energy flow which can be characterised by an energy current density

$$\mathbf{j}_{\mathrm{en}} = -\kappa \nabla T . \tag{4.15}$$

The constant of proportionality κ is called the *thermal conductivity*.

a. Kinetic Model

We consider the energy transfer associated with the electronic system.[1] We assume that the same collisions underlie energy transfers in the electronic system and limit the drift velocity of the electrons when an external electric field is applied. If an electron undergoes a collision at a point, it either acquires or gives up an amount of energy that is determined by the temperature at the given point. It is this temperature that defines the average velocity acquired by the electrons that undergo a collision at this point. Take the temperature gradient in the metal to be along the x axis. As above, we assume that the electrons thermalise at each collision. Let $n(\mathbf{v})$ be the number of electrons per unit volume with velocity $\mathbf{v}$. Considering an area $\mathrm{d}S$ of the metal perpendicular to the x axis, the electrons with velocity $\mathbf{v}$ crossing this area between times t and $t + \mathrm{d}t$ lie in a volume $v_x \mathrm{d}t \mathrm{d}S$, so that their number is $n(\mathbf{v}) v_x \mathrm{d}t \mathrm{d}S$

---

[1] In solids, the vibrational modes of the atoms (*phonons*) also contribute to thermal conductivity. However, at low temperatures, very few phonons are excited and, in metals, the thermal conductivity is dominated by electrons.

Fig. 4.2 Volume containing all electrons of velocity **v** crossing the surface element dS during time dt. If dt = τ_e, then because of the temperature gradient, these electrons undergo a change in energy as given by (4.16)

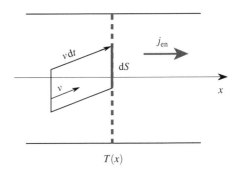

(see Fig. 4.2). If we take the time dt to be the time τ_e between two collisions, each of these electrons of velocity **v** undergoes an energy change of

$$\Delta u = v_x \tau_e \frac{du}{dx} = v_x \tau_e \frac{du}{dT}\frac{dT}{dx} = v_x \tau_e c \frac{dT}{dx}, \tag{4.16}$$

where $u(x)$ is the electron energy at x and c is the specific heat per electron.

The energy current along the x axis, which is the rate of flow of energy per unit area of the metal and per unit time τ_e, is thus obtained by integrating over **v**, which leads to

$$\mathbf{j}_{en} = -\int d^3\mathbf{v} n(\mathbf{v}) v_x v_x \tau_e c \frac{dT}{dx}\mathbf{x} = -\frac{1}{3}\left[\int d^3\mathbf{v} n(\mathbf{v}) v^2\right]\tau_e c \frac{dT}{dx}\mathbf{x},$$

or

$$\mathbf{j}_{en} = -\frac{1}{3} n \bar{v}^2 \tau_e c \frac{dT}{dx}\mathbf{x}, \tag{4.17}$$

where $\bar{v} = \langle v^2 \rangle^{1/2}$ is the root mean squared value of the velocity. This can be used to specify the average distance travelled between two collisions, known as the *mean free path* of the electron, viz., $\ell_e = \bar{v}\tau_e$. We thus arrive at an expression for the thermal conductivity:

$$\boxed{\kappa = \frac{1}{3} C_e \bar{v} \ell_e}, \tag{4.18}$$

where $C_e = nc$ is the electron specific heat per unit volume.

b. Wiedemann–Franz Law and Maxwell–Boltzmann Statistics

With his classical approach, Drude used Maxwell–Boltzmann statistics to interpret the empirical law due to Wiedemann and Franz, according to which κ/σ is proportional to T and independent of the metal (see the results in Table 4.1).

The energy equipartition theorem can be used to estimate the mean squared velocity at temperature T from $\langle v^2 \rangle = 3k_B T/m_0$, where k_B is Boltzmann's constant.

4.2 Electron Transport in a Fermion Gas

Table 4.1 Electrical and thermal conductivities at 0°C, and Lorenz constant $\mathscr{L}$ for various common metals. From [8]

Metal	σ [$10^7\,\Omega^{-1}\mathrm{m}^{-1}$]	κ [10^2 watt m^{-1}K^{-1}]	$\mathscr{L}$ [10^{-8} watt ΩK^{-2}]
Na	2.34	1.35	2.10
Cu	6.45	3.85	2.18
Ag	6.6	4.18	2.31
Be	3.6	2.3	2.34
Mg	2.54	1.5	2.16
Al	4.0	2.38	2.18
Pb	0.52	0.35	2.46
Bi	0.093	0.085	3.30
Pt	1.02	0.69	2.47

In addition, for a classical perfect gas of pointlike particles, the heat capacity per particle is $c = (3/2)k_B$, and $C = (3/2)nk_B$. As a consequence, κ is given by

$$\kappa = \frac{3}{2}nk_B\frac{k_B T}{m_0}\tau_e = \frac{3}{2}\frac{n}{m_0}\tau_e k_B^2 T . \tag{4.19}$$

If we assume that the same particles, subject to the same collisions, determine σ and κ, (4.7) and (4.19) can be used to eliminate n and τ_e. This leads to a simple relation between κ and σ:

$$\frac{\kappa}{\sigma} = \frac{3}{2}\left(\frac{k_B}{e}\right)^2 T = \mathscr{L}T . \tag{4.20}$$

This linear dependence of the ratio κ/σ on T is independent of the metal. The constant of proportionality is called the Lorenz number:

$$\mathscr{L} = (3/2)(k_B/e)^2 = 1.12 \times 10^{-8}\ \mathrm{W}\Omega/\mathrm{K}^2 . \tag{4.21}$$

This law due to Wiedemann and Franz is confirmed experimentally, with a larger value of the constant $\mathscr{L}$ than the theoretical value but of the same order of magnitude (see Table 4.1). This result was long regarded as a proof that Drude's hypothesis as formulated in the context of classical mechanics was actually correct.

4.2 Electron Transport in a Fermion Gas

At the time, Drude could legitimately consider the electron gas as a gas of classical particles, but with the advent of quantum mechanics, this approach clearly become obsolete insofar as the average velocity of the electrons cannot be given by the kinetic theory of gases. We know that the Pauli exclusion principle and Fermi–Dirac statistics lead to an electron kinetic energy which is determined by the Fermi energy rather than by $k_B T$. Moreover, we have seen that the classical approach suggests that all the electrons contribute to the electrical conductivity, whereas we know that only those electrons close to the Fermi level are likely to take part in energy exchanges since they are close in energy to unoccupied states.

4.2.1 Electrical Conductivity

We treat the free electron gas as an ensemble of fermions whose eigenstates are plane waves. All the occupied states in wave vector space, which correspond to n electrons per unit volume, lie within the Fermi sphere of radius k_F such that

$$\frac{4\pi}{3} k_F^3 \frac{2}{(2\pi)^3} = n, \tag{4.22}$$

which leads to the expression (3.47) for the Fermi energy:

$$E_F = (\hbar^2/2m_0)(3\pi^2 n)^{2/3}.$$

Plane waves are electron states with infinite spatial extent and precisely defined momentum. To describe electron transport, we must adopt a spatially confined representation of the electron, which can be done by considering the plane wave packet with wave vectors around some **k** value. Such a wave packet is spatially localised, at the expense of some uncertainty in the value of the wave vector. It has a group velocity $\mathbf{v}_g = \langle \mathbf{p} \rangle /m_0$ such that

$$\boxed{\mathbf{v}_g = \frac{1}{\hbar} \nabla_\mathbf{k} E(\mathbf{k}) = \frac{\hbar \mathbf{k}}{m_0}}. \tag{4.23}$$

When there is no applied force, for each occupied state with wave vector **k** there is an occupied state with wave vector −**k**, and the average velocity of the electron gas is strictly equal to zero, even if the electrons are moving at high speed.

In a quantum picture, the dynamics of an electron subjected to an applied force that varies slowly in time and space compared with the spatial characteristics of the wave packet can be described by a semi-classical equation which specifies the time evolution of the wave packet [2]:

$$\boxed{m_0 \frac{d\mathbf{v}_g}{dt} = \frac{\hbar d\mathbf{k}}{dt} = \mathbf{F}}. \tag{4.24}$$

For an applied electric field, (4.24) becomes

$$\frac{d \langle \mathbf{p} \rangle}{dt} = \frac{\hbar d\mathbf{k}}{dt} = -e\mathscr{E}. \tag{4.25}$$

4.2 Electron Transport in a Fermion Gas

The effect of the electric field is thus to increase the momentum of the electron linearly with time, as happened within the classical description. This therefore shifts the points representing the quantum states in **k** space in the opposite direction to the electric field vector. And this shift corresponds to the appearance of a flow of electrons, i.e., a current, in real space. Indeed, the states **k** on the shifted Fermi sphere are no longer paired as described above. Note that, as for a classical gas, the indefinitely accelerated electron motion once again contradicts Ohm's law. Here, too, we must introduce collision processes if we are to reach a stable situation in the presence of an electric field, where the Fermi sphere is shifted by

$$\boxed{\hbar \delta \mathbf{k} = -e\mathscr{E}\tau_e}. \tag{4.26}$$

Note that the number of occupied states within the Fermi sphere corresponds exactly to the n electrons per unit volume of the metal. With the displacement of the Fermi surface shown in Fig. 4.3, each is displaced by the above amount $\delta \mathbf{k}$, which corresponds to a drift velocity $\langle \mathbf{v} \rangle = \delta \mathbf{v}_g = -e\mathscr{E}\tau_e/m_0$, and hence a total current given by

$$\mathbf{j} = -ne\langle \mathbf{v}_i \rangle = \frac{ne^2 \tau_e}{m_0} \mathscr{E} = \sigma \mathscr{E}. \tag{4.27}$$

We thus recover a transport equation that is practically the same as the one derived by Drude. However, the quantity τ_e in this expression has a completely different meaning. Indeed, for a classical gas, the electrons can all undergo collisions whatever their kinetic energy, and τ_e is a collision time which affects all the electrons equally. Here, even though the effect of the field can be represented by a uniform shift of all the states, the equilibrium can only be maintained by collisions which

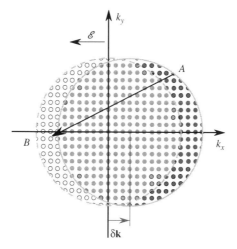

Fig. 4.3 Fermi surface at equilibrium and in the presence of an applied electric field. In the steady-state regime, the sphere is shifted by $\hbar \delta \mathbf{k} = -e\mathscr{E}\tau_e$ from the equilibrium position. This stationary state is obtained by collision processes which cause the electrons corresponding to the states A (*right-hand* Fermi surface) to come back to B (*left-hand* Fermi surface)

redistribute the electrons between these states and prevent the distribution from continuing its progression towards the right of Fig. 4.3. According to the Pauli exclusion principle, the collisions can only transfer electrons from occupied states to unoccupied states. If k_F is the radius of the Fermi surface when there is no field, these collisions can therefore only bring states with $|\mathbf{k}| > k_F$ on the right back to states with $|\mathbf{k}| < k_F$ on the left of the figure. Only collision processes involving electrons with energy close to E_F can therefore contribute to establishing the stationary state and the corresponding redistribution of the velocities. *Pauli's principle thus restricts the kind of collision that can occur*, and the collision time τ_e only concerns electrons at the Fermi level, with speed $v_F = \hbar k_F/m_0$. Likewise, the mean free path concerns only these electrons, which can undergo collisions. The mean free path is given by

$$\boxed{\ell_e = v_F \tau_e(E_F)} . \tag{4.28}$$

Consider again the example of copper, where $\tau_e = 3 \times 10^{-14}$ s. The drift velocity is $\mathbf{v}_e = -(e\tau_e/m_0)\mathscr{E} \simeq 5 \times 10^{-3}\mathscr{E}$. Even for an electric field of 10^4 V/m, this implies that $v_e \sim 50$ m/s. It can be compared with the Fermi velocity $v_F = (3\pi^2 n)^{1/3}\hbar/m_0$, according to (4.22), or $v_F \simeq 1.5 \times 10^6$ m/s, with the value of n determined in Sect. 1.1.1. The drift velocity is thus very small compared with the Fermi velocity. This indicates that the displacement of the Fermi surface is at most $10^{-4}k_F$. The representation in Fig. 4.3 is therefore not at all to scale. Finally, we may estimate the mean free path $\ell_e \simeq 5 \times 10^{-8}$ m $\simeq 500$ Å, according to (4.28). In good metals, at room temperature, an electron only moves a few hundred times the interatomic separation before suffering a collision.

4.2.2 Thermal Conductivity and the Wiedemann–Franz Law

The kinetic arguments in Sect. 4.1.2a for expressing κ as a function of microscopic quantities are still valid, provided we use the parameters associated with the Fermi energy and the value of C_e obtained for a free electron gas (see Chap. 3):

$$C_e = \frac{\pi^2}{2} n k_B \frac{k_B T}{E_F} . \tag{4.29}$$

Using (4.18), this implies that

$$\kappa = \frac{1}{3} C_e v_F^2 \tau_e = \frac{\pi^2}{3} n \frac{k_B^2 T}{m_0} \tau_e . \tag{4.30}$$

In the context of Fermi–Dirac statistics, the ratio κ/σ is thus equal to

$$\boxed{\frac{\kappa}{\sigma} = \frac{\pi^2}{3}\left(\frac{k_B}{e}\right)^2 T = \mathscr{L} T} , \tag{4.31}$$

where the Lorenz number is exactly twice that obtained using classical statistics. This theoretical value $\mathscr{L} = 2.45 \times 10^{-8}$ wattΩ/K^2 is much closer to the experimental values of $\mathscr{L}$ given in Table 4.1. It is remarkable that the classical and quantum calculations give such close results. It is an amazing coincidence if we recall that the expressions for κ and σ are very different. Indeed, C_v is proportional to $k_B T$ for a fermion gas, whereas it is independent of T for a classical gas. Conversely, v^2 is equal to v_F^2 and is independent of temperature for fermions, whereas $\langle v \rangle^2$ is proportional to T for a classical gas. These two terms compensate one another quite accidentally, up to a factor of 2 in the expression for the Lorenz number!

4.3 Electrons in a Lattice: Dynamics of Bloch Electrons

Is this description of electron transport still valid in a crystal, in the presence of a periodic crystal potential? As in the case of free electrons, an electron in a crystal will have to be described by a wave packet, but this time by a packet of Bloch waves centered on a wave vector **k**, involving states $\mathbf{k} + \Delta \mathbf{k}$ belonging to the same electron energy band with index n. The wave packet thereby constructed will have spatial extent Δr related to its extent Δk in reciprocal space by $\Delta r \sim 1/\Delta k$.

4.3.1 Group Velocity

The wave vector **k** which specifies the Bloch state of energy $E_n(\mathbf{k})$ is not simply related to the momentum operator $\mathbf{p} = -i\hbar \nabla$, in contrast to the case of plane waves. Indeed, for the Bloch function

$$\psi_{n\mathbf{k}} = u_{n\mathbf{k}}(\mathbf{r}) e^{i\mathbf{k} \cdot \mathbf{r}}, \tag{4.32}$$

the momentum is given by

$$\langle \mathbf{p} \rangle_{n\mathbf{k}} = \langle \psi_{n\mathbf{k}} | \mathbf{p} | \psi_{n\mathbf{k}} \rangle = m_0 \langle \mathbf{v} \rangle_{n\mathbf{k}} \tag{4.33}$$

$$= \hbar \mathbf{k} - i\hbar \int u_{n\mathbf{k}}(\mathbf{r}) \nabla u_{n\mathbf{k}}(\mathbf{r}) \mathrm{d}^3 \mathbf{r} \neq \hbar \mathbf{k}. \tag{4.34}$$

However, the crystal momentum $\hbar \mathbf{k}$ plays a very special role, as we shall see. A simple calculation carried out in the box in next page relates the average velocity $\langle \mathbf{v} \rangle_{n\mathbf{k}}$ of the electron to the **k** dependence of the dispersion relations $E_n(\mathbf{k})$. In fact, we find that

$$\boxed{\langle \mathbf{v} \rangle_{n\mathbf{k}} = \frac{1}{\hbar} \nabla_{\mathbf{k}} E_n(\mathbf{k})}, \tag{4.35}$$

which naturally also applies to free electrons [see (4.23)]. In the 1D case, this expression reduces to $\langle \mathbf{v} \rangle = \hbar^{-1} \mathrm{d} E_n(k)/\mathrm{d} k$, and the velocity of an electron in a state

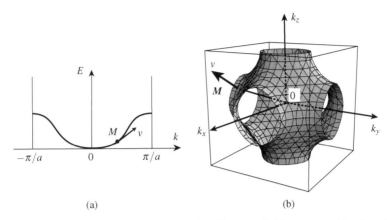

Fig. 4.4 Velocity of an electron in a Bloch state **k**, with energy $E(\mathbf{k})$, represented by the point M. (**a**) For the 1D band calculated using the LCAO approximation. (**b**) For a cubic crystal. A constant energy surface is shown

(n, k) is nothing other than the derivative of the dispersion curve with respect to k. This is illustrated in Fig. 4.4, for the simple band calculated using the LCAO approximation. The velocity is zero at the bottom of the band and at the point $k = \pi/a$ of the first Brillouin zone.

In two or three dimensions, the geometric interpretation of (4.35) is a little more complicated. It shows that the velocity $\langle \mathbf{v} \rangle_{n\mathbf{k}}$ lies along the *normal at the point* $\mathbf{OM} = \mathbf{k}$ *to the constant energy curve or surface constructed from* $E_n(\mathbf{k})$ (see Fig. 4.4b).

Note that, in a Bloch state which is a stationary eigenstate of the Hamiltonian, *the electronic velocity is well defined and time independent*, agreeing with the idea that the electron undergoes no collision on the crystal potential, the latter being entirely accounted for by the Hamiltonian. The periodic crystal potential does not scatter the electrons, but simply determines their velocity as given by (4.35). This quantum description invalidates Drude's hypotheses, according to which the collision time τ_e is determined by collisions of the electrons on the atomic nuclei.

Relation Between $\langle \mathbf{v} \rangle_{n\mathbf{k}}$ and $E_n(\mathbf{k})$

If $E_n(\mathbf{k})$ is the energy eigenvalue associated with the Bloch state (4.32), we may write

$$\left[\frac{p^2}{2m} + V(\mathbf{r}) \right] u_{n\mathbf{k}} e^{i\mathbf{k}\cdot\mathbf{r}} = E_n(\mathbf{k}) u_{n\mathbf{k}} e^{i\mathbf{k}\cdot\mathbf{r}}, \quad (4.36)$$

which reduces to

$$\left[\frac{\hbar^2}{2m}(\mathbf{k} - i\nabla)^2 + V(\mathbf{r}) \right] u_{n\mathbf{k}} = E_n(\mathbf{k}) u_{n\mathbf{k}}. \quad (4.37)$$

This corresponds to an effective Hamiltonian given by

$$H_\mathbf{k} = \frac{\hbar^2}{2m}(\mathbf{k} - i\nabla)^2 + V(\mathbf{r}), \quad (4.38)$$

4.3 Electrons in a Lattice: Dynamics of Bloch Electrons

which acts on the space of $u_{n\mathbf{k}}(\mathbf{r})$ functions and has eigenvalue $E_n(\mathbf{k})$. The effective Hamiltonian corresponding to $\mathbf{k}+\mathbf{q}$ can be written

$$H_{\mathbf{k}+\mathbf{q}} = \frac{\hbar^2}{2m_0}(\mathbf{k}+\mathbf{q}-i\nabla)^2 + V(\mathbf{r}) \tag{4.39}$$

$$= H_{\mathbf{k}} + \frac{\hbar^2}{m_0}(\mathbf{k}-i\nabla)\mathbf{q} + \frac{\hbar^2}{2m_0}q^2. \tag{4.40}$$

For small $\mathbf{q}$, a perturbation treatment to first order in $\mathbf{q}$ determines the energy of the corresponding eigenstate to be

$$E_n(\mathbf{k}+\mathbf{q}) = E_n(\mathbf{k}) + \left\langle u_{n\mathbf{k}} \left| \frac{\hbar^2}{m_0}(\mathbf{k}-i\nabla)\cdot\mathbf{q} + \frac{\hbar^2}{2m_0}q^2 \right| u_{n\mathbf{k}} \right\rangle + \cdots \tag{4.41}$$

$$= E_n(\mathbf{k}) + \frac{\hbar}{m_0}\langle u_{n\mathbf{k}}e^{i\mathbf{k}\cdot\mathbf{r}}|\mathbf{p}|u_{n\mathbf{k}}e^{i\mathbf{k}\cdot\mathbf{r}}\rangle \mathbf{q} + O(q^2) + \cdots \tag{4.42}$$

$$= E_n(\mathbf{k}) + \hbar\langle \mathbf{v}\rangle_{n\mathbf{k}}\cdot\mathbf{q} + \cdots. \tag{4.43}$$

Identifying with the Taylor expansion

$$E_n(\mathbf{k}+\mathbf{q}) = E_n(\mathbf{k}) + \mathbf{q}\cdot\nabla_{\mathbf{k}}E_n(\mathbf{k}) + \cdots, \tag{4.44}$$

we do then recover (4.35).

4.3.2 Acceleration in Reciprocal Space and Real Space

Under the effect of an applied force $\mathbf{F}$, the electron energy is modified. Assume that $\mathbf{F}$ varies slowly in space on the scale of the unit cell, and slowly in time on the scale of the frequencies associated with transitions between allowed energy bands. The work dW done on an electron of velocity $\mathbf{v}$ and charge $-e$ over a time interval dt modifies its energy $E_n(\mathbf{k})$ by changing the value of $\mathbf{k}$ and hence the crystal momentum, whence

$$dW = \mathbf{F}\cdot\mathbf{v}dt = dE_n(\mathbf{k}) \tag{4.45}$$

$$= \mathbf{F}\cdot\frac{1}{\hbar}\nabla_{\mathbf{k}}E_n(\mathbf{k})dt = \nabla_{\mathbf{k}}E_n(\mathbf{k})\cdot d\mathbf{k}, \tag{4.46}$$

or

$$\boxed{\hbar\frac{d\mathbf{k}}{dt} = \mathbf{F}}. \tag{4.47}$$

This is the acceleration theorem in reciprocal space. *In response to an external force that varies slowly in space and time, it is the derivative of the crystal momentum (rather than the derivative of the momentum) that equals the applied force.* In the presence of a magnetic field $\mathbf{B}$, it can be shown that (4.47) can be rewritten as

$$\frac{\hbar d\mathbf{k}}{dt} = -e(\mathcal{E}+\mathbf{v}\wedge\mathbf{B}) = \mathbf{F}. \tag{4.48}$$

Differentiating **v**, given by (4.35), with respect to time and using (4.47), we obtain

$$\frac{d\mathbf{v}}{dt} = \nabla_\mathbf{k}\mathbf{v} \cdot \frac{d\mathbf{k}}{dt} = \frac{1}{\hbar^2}\nabla_\mathbf{k}\big[\nabla_\mathbf{k} E_n(\mathbf{k})\big] \cdot \mathbf{F}, \qquad (4.49)$$

or

$$\boxed{\frac{dv_\alpha}{dt} = \sum_\beta \frac{1}{(m_\text{e})_{\alpha\beta}} F_\beta}, \qquad (4.50)$$

with

$$\boxed{\frac{1}{(m_\text{e})_{\alpha\beta}} = \frac{1}{\hbar^2}\frac{\partial^2 E_n(\mathbf{k})}{\partial k_\alpha\,\partial k_\beta}}. \qquad (4.51)$$

Equations (4.50) and (4.51) constitute the acceleration theorem in real space. Being consequences of (4.47), they are subject to the same restrictions regarding the slow spatiotemporal variation of **F**. Note that the tensorial relation between the acceleration and the force lead to the definition of a tensor $1/(m_\text{e})_{\alpha\beta}$ whose components have the dimensions of a reciprocal mass. This defines the *effective mass tensor* $(m_\text{e})_{\alpha\beta}$ at a point **k** of a given band n. By definition, the effective mass tensor is symmetric.

The idea of effective mass is of particular interest near an extremum $\mathbf{k}_0$ of an energy band. In this case, close to $\mathbf{k}_0$, if we set $\mathbf{k} = \mathbf{k}_0 + \Delta\mathbf{k}$, the energy is given to lowest order by

$$E_n(\mathbf{k}) - E_n(\mathbf{k}_0) \simeq \sum_{\alpha\beta}\frac{\hbar^2}{(m_\text{e})_{\alpha\beta}}\Delta k_\alpha\,\Delta k_\beta. \qquad (4.52)$$

In the particular case of a cubic crystal, at the center of the zone, if the energy is not degenerate, the constant energy surfaces are spheres and the effective mass is thus isotropic, given by m_e. The acceleration is then

$$m_\text{e}\frac{d\mathbf{v}}{dt} = \mathbf{F}, \qquad (4.53)$$

and the derivative $m_0 d\mathbf{v}/dt$ of the ordinary momentum is not equal to the external force. In this same special case, the crystal momentum and velocity are related by (4.35):

$$\mathbf{v} = \frac{\hbar\mathbf{k}}{m_\text{e}}. \qquad (4.54)$$

In general, the effective mass is positive near a band minimum and negative near a maximum. According to (4.50), a *negative effective mass* expresses the fact that the velocity acquired under the effect of F_β is in the opposite direction to the velocity that would be acquired by an electron subjected to F_β in vacuum. This somewhat paradoxical behaviour should come as no surprise, however. In the crystal, the electron feels both the applied electric potential and the crystal potential $V(\mathbf{r})$. This shows that, in a crystal, *the response of an electron is determined to a large extent by the potential associated with the ions of the crystal*. Even when the effective mass has the same sign as the mass m_0 of the free electron, the value of m_e/m_0 may differ significantly from unity. While in metals $m_e/m_0 \sim 1$, the same is not always true in semiconductors. For example, the effective mass at the bottom of the conduction band is $+0.067 m_0$ for GaAs and $+0.014 m_0$ for InSb.

These effective masses can be determined by studying the response of the electrons to an alternating electric field in the presence of a static magnetic field. The eigenmodes of the electron states give rise to so-called cyclotron resonances.[2]

4.3.3 Electronic Conductivity in a Crystal

The results obtained for the Bloch states can help us to determine the conductivity of a band. When there is no electric field, the total current density for a band is by definition

$$\mathbf{j} = -e \sum_{\mathbf{k}\text{ occupied}} v(\mathbf{k}) = -\frac{e}{\hbar} \sum_{\mathbf{k}\text{ occupied}} \nabla_\mathbf{k} E_n(\mathbf{k}) . \tag{4.55}$$

Since the functions $E_n(\mathbf{k})$ are even in $\mathbf{k}$, the total current vanishes. This is true for the case of a fully occupied band, for example. Let us now apply an electric field in the latter case. By the reciprocal space acceleration theorem, over a time interval dt, each wave vector $\mathbf{k}$ will change by $\Delta \mathbf{k} = -e\mathscr{E}\Delta t/\hbar$, and all wave vectors will thus be shifted by the same amount.

Given the definition of the first Brillouin zone and the periodicity in $\mathbf{k}$ space, the states leaving this zone are equivalent up to a vector $\mathbf{K}$ of the reciprocal lattice to those that become empty, and the band remains fully occupied (see Fig. 4.5). Overall

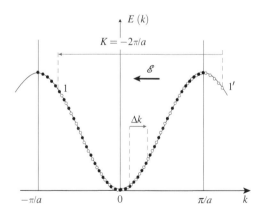

Fig. 4.5 Effect of an electric field $\mathscr{E}$ on the electron states of a filled band. The states in the absence of $\mathscr{E}$ (*empty dots*) are shifted by $\Delta \mathbf{k} = -e\mathscr{E}\Delta t/\hbar$ (*full dots*). The states shifted out of the first Brillouin zone are equivalent to those that are depopulated (1' is equivalent to 1). The band therefore remains entirely full and the total current is identically zero. The magnitude of Δk and the quantisation of the levels with respect to k have been exaggerated in the figure for better visibility

---

[2] See Problem 8: *Cyclotron Resonance*.

the electronic states of the band are conserved and **j** remains identically zero:

$$\mathbf{j} = -\frac{e}{\hbar} \sum_{\mathbf{k} \text{ full band}} \nabla_{\mathbf{k}} E_n(\mathbf{k}) \equiv 0, \quad (4.56)$$

even when an external electric field is applied. We have thus shown what was anticipated in Chap. 3, namely that *a fully occupied band does not contribute to the current*. Metals thus correspond to the case in which bands are only partially filled.

The conductivity of metals can then be understood in the same framework as free electrons. The Fermi surface, which is no longer spherical in an arbitrary crystal, is shifted when an electric field is applied, and the magnitude of the displacement is determined by the number of collisions involving electrons at the Fermi level.

4.4 Origin of Collisions

So far we have seen how to understand why the conductivity of metals should be finite. However, we still have no microscopic description of what causes collisions, even though we have some intuitive notions. In particular, we have seen that a crystal potential, resulting from the sum of the Coulomb potentials of the ions in the crystal lattice will not contribute to any collision process if it is strictly periodic. On the other hand, defects such as impurities will be able to scatter electrons and can therefore be a source of collisions. As always, it is experiment that leads the way to understanding and classifying the various collision processes. A key point here is the temperature dependence of the resistivity of metals. We shall describe briefly how the resistivity is measured and discuss the main results.

4.4.1 Experimental Observation

It is in principle an experimentally simple matter to measure the resistivity of a metal. The idea is just to pass a constant current through the metal and measure the voltage across the sample. However, when the conductivity is very high, some precautions are necessary, otherwise the measurement wires and the resistance of their contacts with the sample can distort the result. The method generally used is shown in Fig. 4.6. The idea is to decouple the wires supplying the current from those measuring the potential drop.

To investigate what causes conductivity, we may begin by considering a monovalent metal such as silver, for which the Fermi surface is almost spherical and the conduction band is similar to a band of free electrons. The results obtained on various silver samples with different levels of purity (see Fig. 4.7) show that the resistivity is affected by the presence of impurities. They also show a marked temperature dependence, with an almost linear variation at high temperatures and strong

4.4 Origin of Collisions

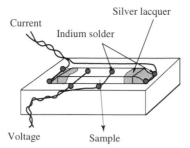

Fig. 4.6 Measuring the resistivity of a metal. Four contacts are made on the sample, which is placed on an insulating substrate. The two outer contacts are connected to the current supply. The voltage measured across the two others is used to determine the resistivity ρ. Slightly resistive contacts will not significantly affect the measurement

Fig. 4.7 Temperature dependence of the resistivity of silver samples with varying degrees of purity. Adapted from result published in [5]

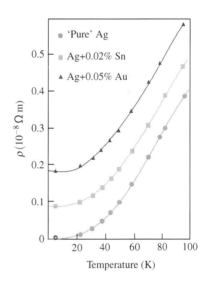

curvature at lower temperatures. These observations suggest that the resistivity takes the form

$$\rho(T) = \rho_0 + \rho_1(T) . \tag{4.57}$$

The first term is largely determined by the purity of the metal, while the second is practically independent of the sample. Compare these results with the expected expression for the resistivity of a metal:

$$\rho = \frac{1}{\sigma} = \frac{m_e}{ne^2} \frac{1}{\tau_e} , \tag{4.58}$$

where we have introduced the effective mass m_e to account for the band structure. The latter is in principle almost independent of temperature. The only modifications arise from thermal expansion of the metal, which does not exceed about 1%. Moreover, the number of valence electrons introduced by impurities with concentrations lower than 1% will barely alter the number of conduction electrons. It follows that the only quantity that varies with the purity and the temperature T is the collision rate $1/\tau_e$. The experimental results shown in Fig. 4.7 imply that this has approximately the form

$$\boxed{\frac{1}{\tau_e} = \frac{1}{\tau_1} + \frac{1}{\tau_2(T)}}.\qquad(4.59)$$

It thus looks as though two totally different collision processes are operating in parallel. One of these, described by τ_1, is independent of temperature, but depends on the purity of the material, while the other, described by τ_2, is exactly the opposite. Experiment thus leads to a simple general law: if different independent microscopic processes underlie the collisions of electrons at the Fermi level, the probabilities of the collisions are additive. The additivity of the collision rates contributing to the resistivity is called *Matthiessen's law*. Below, we consider one by one the various collision processes that can contribute to the resistivity.

4.4.2 Scattering by Lattice Vibrations

The temperature dependent term is insensitive to the presence of impurities and thus seems to be a characteristic of the metal without defects. It dominates the resistivity at high temperatures when the metal is pure. It thus seems that the underlying process is more effective at higher temperatures. We have seen that the Bloch states which take into account the periodic potential of the crystal cannot be scattered by the ions. However, any departure from periodicity, due to thermal agitation of the ions from their equilibrium position, for example, is likely to cause electron scattering. As the amplitude of these oscillations increases with temperature, such a collision term would indeed be expected to increase at high T, as shown experimentally. For this reason, the temperature dependent term is attributed to electron scattering by *atomic vibration modes*.

These lattice vibrations are collective harmonic displacements of the ions from their equilibrium positions. They propagate as waves and are called *phonons*. They are characterised by a wave vector $\mathbf{q}$, and by their energy $\hbar\omega$. The function $\omega(\mathbf{q})$ is the *phonon dispersion relation*.[3] An electron can thus be scattered from an initial state $\mathbf{k}$ to a final state $\mathbf{k}'$, while the energy difference between the states is balanced by absorption or emission of a phonon. This process can therefore modify the energy state of the electron system and lead to a state of thermodynamic equilibrium. The associated dissipation corresponds to the *Joule effect*. The slowing down of the electrons proceeds through excitation of vibrational modes which cause the metal to heat up. In principle, the energy associated with these vibrations increases as $k_B T$, leading to a term linear in T for $1/\tau_e$ and for the resistivity at high temperatures.

At low temperatures, only the low energy oscillatory modes of the atomic lattice are excited. They correspond to low frequency, hence long wavelength modes,

---

[3] See Problem 9: *Phonons in Solids*.

4.4 Origin of Collisions

which are much less efficient. Given the vibrational eigenmodes of the atoms in the crystal and their interaction with the electrons, we can calculate the resistivity associated with the electron–phonon collision process. Phonon modes are characterised by their maximal energy $k_B \theta_D$, and this defines a characteristic temperature θ_D called the *Debye temperature*. The resistivity due to the electron–phonon interactions in different metals has a temperature dependence which is approximately a characteristic function of T/θ_D. This has been modelled by Grüneisen (see Fig. 4.8).

Note that, since this temperature dependence varies little with the purity of the material, measurements of the resistivity of a metal wire are often used as a secondary thermometer, after calibrating by one or two fixed temperature points. In fact most precision electronic thermometers use platinum, a noble metal that barely oxidises and is easily purified.

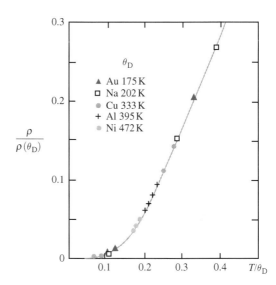

Fig. 4.8 Resistivities of different metals as a function of T/θ_D, where θ_D is the Debye temperature characteristic of the phonons in the given metal. Adapted from [7, p. 221], 4th edn., Wiley © 1996

4.4.3 Collisions with Impurities and Defects

Electrons will obviously be scattered by crystal defects such as impurities. The results in Fig. 4.9 do indeed show that the resistivity increases as a function of the concentration of impurities, and that it depends on the kind of impurity. An electron passing close to an impurity will be deflected by the interaction with the Coulomb potential of the impurity. In contrast to scattering by phonons, the electron energy is conserved along its trajectory. Once it has moved a long distance away, it will recover the same kinetic energy. Such a collision is said to be elastic, and only alters the direction of the electron velocity, not its magnitude. The trajectory and the

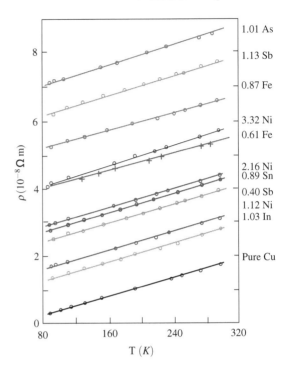

Fig. 4.9 Resistivity of copper and various alloys of the form Cu-M, where M is a chemical element with concentration given in atomic percentage *on the right* of the figure. Note that the residual resistivity ρ_0 associated with scattering by impurities increases with their concentration and depends on the chemical nature of the element inserted into the alloy. Adapted from Linde, J.: Ann. Phys. (Leipzig) **5**, 15 (1932)

probability of scattering are independent of temperature, which explains why the resistivity associated with this mechanism is independent of T. Note that impurities are not the only defects occurring in a crystal that can play the role of scattering center. The absence of an ion at a given crystal site, known as a vacancy, or an ion at some interstitial position in the crystal lattice both constitute structural defects that are extremely effective electron scatterers.

At very low temperatures, the resistivity of a metal is thus determined to a large extent by its purity. However, even for the purest metals that can be obtained, the resistivity is at best only 10^6 times lower than the resistivity at room temperature. The electron mean free path can then reach the millimeter range for copper, or a few centimeters for aluminium.

Question 4.2. 1. In Fig. 4.9, how does the contribution to ρ of the impurity M vary for different values of the concentration c?
2. Make a quantitative comparison of the effects of In, Sn, Sb, and As impurities on the resistivity.
3. Atomic copper has the configuration $3d^{10}4s$. In copper metal, only the $4s$ electron contributes to the conduction band, which has a quasi-spherical Fermi surface. In, Sn, Sb, and As give 3, 4, 5, and 5 electrons to the conduction band, respectively. The ionic cores thus have valences 3^+, 4^+, 5^+, and 5^+, respectively. Express the result of part 2 as a function of the valence of the impurity. Can you explain this experimental result qualitatively? What kind of calculation would be needed to give a quantitative explanation of these results?

4.5 Electrons, Holes, and Dopants in Semiconductors

Consider a situation in which a band is almost completely occupied, except for a few states at the top of the band. This may be the case, for example, in the valence band of a semiconductor, out of which an electron has been excited, or in the conduction band of a metal, when the Brillouin zone contains 'pockets' near the Fermi level. Under such conditions, in order to investigate electron transport problems, we shall see in Sect. 4.5a that, rather than considering all the electrons in the band, it is simpler to restrict attention to the unoccupied states of the band, called *holes*, which behave like positive charges. We shall then see in Sect. 4.5b that, in semiconductors, the electron or hole nature and the concentration of the charge carriers can be chosen by doping the semiconductor with the appropriate impurity.

a. Holes

When an electric field is applied to the electrons in an electronic band that is almost completely occupied, all the Bloch wave vectors representing the occupied states are displaced in the opposite direction to the applied field. The same is therefore true of the unoccupied states (see Fig. 4.10). The behaviour of the unoccupied states can be used to describe the global response of the band. Consider for instance a valence band containing just one empty state with quasi-momentum $\mathbf{k} = \mathbf{k}_e$. It is tempting to define a quasi-particle [*valence band filled except for a single empty state*], which we shall call a *hole*, whose behaviour will describe the behaviour of the band. The total current associated with the electron states in this band will define the current $\mathbf{j}_h$ associated with the hole:

$$\mathbf{j}_h = -e \sum_{\mathbf{k}\text{ occupied}} \mathbf{v}(\mathbf{k}) = -e \sum_{\text{filled band}} \mathbf{v}(\mathbf{k}) + e\mathbf{v}(\mathbf{k}_e) = e\mathbf{v}(\mathbf{k}_e), \quad (4.60)$$

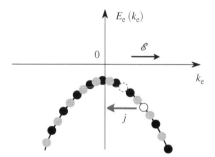

Fig. 4.10 Hole in a filled electron band. The effect of an electric field $\mathscr{E}$ is to shift all the electron states in the opposite direction (*dark* states shift into *light* states, while the *empty* state moves to the *empty dotted* state). The current **j** associated with the band is negative according to (4.61), because $v(k_e) = \hbar^{-1} dE_e/dk_e < 0$

with zero total current for the filled band (see Sect. 4.3.3). We thus have

$$\mathbf{j}_h = e\mathbf{v}(\mathbf{k}_e) = \frac{e}{\hbar}\nabla_{\mathbf{k}_e} E_e(\mathbf{k}_e), \qquad (4.61)$$

where $E_e(\mathbf{k}_e)$ is the dispersion relation of the valence electrons at the point $\mathbf{k}_e$.

Likewise, the total *momentum* of the filled band is zero because the electrons can be associated in pairs with opposite momenta. The total momentum of the band in the absence of the electron, that is, *the wave vector of the hole*, is therefore

$$\mathbf{k}_h = -\mathbf{k}_e. \qquad (4.62)$$

Finally, consider the total energy of the band. It is given by $E_T - E_e(\mathbf{k}_e)$, where E_T is the total energy of the filled band. By choosing E_T as the zero energy, we can define the energy of the hole as

$$E_h = -E_e(\mathbf{k}_e). \qquad (4.63)$$

This energy increases as the energy of the missing electron gets smaller. Note in particular that, under the action of an electric field, the change in energy of the band is in the opposite direction to that of the missing electron. The hole thus has an energy that varies in the opposite way to that of an electron under the effect of a field. A *positive electrical charge* must therefore be associated with the hole:

$$e_h = +e. \qquad (4.64)$$

With the above definitions, the current associated with the hole, as given by (4.61), is

$$\mathbf{j}_h = e\mathbf{v}(\mathbf{k}_e) = \frac{e}{\hbar}\nabla_{\mathbf{k}_e} E_e(\mathbf{k}_e) = \frac{e}{\hbar}\nabla_{\mathbf{k}_h} E_h(\mathbf{k}_h) = e\mathbf{v}_h(\mathbf{k}_h). \qquad (4.65)$$

The velocity of the hole is therefore simply the velocity of the missing electron at $\mathbf{k}_e$. Finally, since $\mathbf{k}_h = -\mathbf{k}_e$, the dynamical equation describing the time dependence of $\mathbf{k}_h$ and $\mathbf{v}_h$ in the presence of applied electric and magnetic fields is

$$\hbar\frac{d\mathbf{k}_h}{dt} = -\hbar\frac{d\mathbf{k}_e}{dt} = e(\mathcal{E} + \mathbf{v}_h \wedge \mathbf{B}). \qquad (4.66)$$

This is indeed the equation of motion of a positive charge of velocity $\mathbf{v}_h$ in a state with wave vector $\mathbf{k}_h$. The effective mass tensor for the hole can then be defined as

$$\frac{dv_{t\alpha}}{dt} = \sum_\beta \frac{1}{(m_h)_{\alpha\beta}} F_\beta, \quad \text{where} \quad \frac{1}{(m_h)_{\alpha\beta}} = \frac{1}{\hbar^2}\frac{\partial^2 E_h}{\partial k_{t\alpha} \partial k_{t\beta}}. \qquad (4.67)$$

It has the opposite sign to the effective mass tensor of the missing electron

$$(m_h)_{\alpha\beta} = -(m_e)_{\alpha\beta}. \qquad (4.68)$$

Equations (4.62) to (4.68) thus provide the rules for moving from the electron representation to the hole representation (see Fig. 4.11).

4.5 Electrons, Holes, and Dopants in Semiconductors

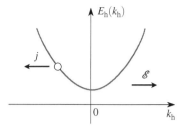

Fig. 4.11 Energy, wave vector, and velocity associated with the band in Fig. 4.10 as represented using the hole language. We do indeed have $j < 0$, because according to (4.65), $j_h = ev_h(k_h) = (e/\hbar)dE_h/dk_h < 0$

The notion of hole is particularly important when treating semiconductors with a fully occupied valence band at $T = 0$. Electron states in the conduction band and holes in the valence band are populated when $T \neq 0$.

Question 4.3. In the case of Ge, which has the band structure shown in Fig. 3.12, what are the vectors **k** and energies corresponding to excited electrons and holes at nonzero temperature. Estimate their effective masses using Fig. 3.12.

A light wave of higher energy than the width of the band gap can also be used to excite an electron of the valence band by an optical transition from a state $\mathbf{k}_e$ to a state of the same wave vector in the conduction band. An electron–hole pair can thus be created even at low temperature.

Question 4.4. An electron–hole pair is created in the conduction and valence bands of a semiconductor. What will be the total current associated with these carriers in the presence of an applied electric field?

b. Doped Semiconductors

As we saw in Chap. 3, in a pure or intrinsic semiconductor, the concentration of electrons and holes excited at room temperature is rather low. However, this situation can be altered by introducing impurities into the semiconductor in a controlled way. For example, consider the case of silicon, a tetravalent element, into which pentavalent phosphorus atoms are introduced. This amounts to breaking the translation symmetry of the crystal, by placing a P at an Si site. At this particular site, the Coulomb potential will thus be higher, because there is an extra charge in the atomic nucleus. For the first four valence electrons of the P, the situation is similar to that of silicon. They are involved in sp^3 bonds with the Si and fill the electron states corresponding to those at the top of the valence band, thereby taking the place of the missing silicon electrons.

However, the last electron donated by the P should be transferred to a state in the conduction band of silicon, in an energy level E_c. But it is subjected to the Coulomb potential due to the excess charge $+e$ of the phosphorus nucleus, which will generate bound electron levels, as in a hydrogen atom.

The main difference from the well known case of the hydrogen atom is that this electron is not moving around in the vacuum, but rather in solid silicon. Its eigenstates must be constructed from unoccupied electron states of the conduction band of silicon, which are Bloch waves with a rather low effective mass m_e ($\simeq 0.3 m_0$ for silicon). In addition, this long range Coulomb potential produces an electric field which acts on all the nuclei and electrons in the silicon. Hence, in this material, which is basically an insulator, an applied electric field produces an electrostatic dipole polarisation of the electronic orbitals around the Si nuclei. This polarisation will in turn create a depolarising field, i.e., a field opposing the applied electric field. This is the effect that leads to the dielectric constant ε of the insulator. It follows that the Coulomb electric field produced by the extra charge of the P is reduced at large distances by this factor ε. The net result of all this is that we have an excess electron with effective mass m_e, subjected to a similar potential to that of the hydrogen atom but reduced by the factor ε:

$$V_c(|\mathbf{r}-\mathbf{r}_i|) = -\frac{e^2}{4\pi\varepsilon\varepsilon_0 |\mathbf{r}-\mathbf{r}_i|} \,. \tag{4.69}$$

The ground state of this hydrogen-like atom is thus a bound state whose binding energy measured from the bottom of the silicon conduction band is obtained by replacing e^2 by e^2/ε and m_0 by m_e, whence

$$E_P = -\frac{E_1}{\varepsilon^2} \frac{m_e}{m_0} \,, \tag{4.70}$$

where $E_1 = 13.6$ eV is the binding energy of the hydrogen atom. Likewise, the associated wave function is a $1s$ orbital with Bohr radius equal to

$$a_P = a_1 \varepsilon m_0 / m_e \,. \tag{4.71}$$

With $\varepsilon \simeq 10$ and $m_e \simeq 0.3 m_0$ for silicon, we obtain $E_P \simeq -40$ meV and $a_P \simeq 16$ Å. The fact that a_P is greater than the lattice constant of the silicon lattice justifies with hindsight the approximation wherein silicon is treated as an effective medium for this extra electron.

Furthermore, at zero temperature, the binding energy E_P is high enough for the extra phosphorus electron to occupy this electron state of energy

$$E_d = E_c + E_P \,, \tag{4.72}$$

for which the wave function is a highly dilated Bohr orbit. On the other hand, the binding energy is low enough to ensure that $k_B T \simeq |E_P|$ at room temperature. The electron is thermally excited into the states of the conduction band and thus contributes to electron transport.

4.5 Electrons, Holes, and Dopants in Semiconductors

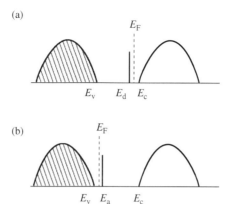

Fig. 4.12 Band structures of (**a**) *n*-doped and (**b**) *p*-doped semiconductors. At zero temperature, the Fermi level lies between the donor level E_d and the conduction band in the *n*-doped case, and between the valence band and the acceptor level E_a in the *p*-doped case

All this can be formalised by extending the statistical physics arguments of Chap. 3 to the band structure shown schematically in Fig. 4.12a. Here, apart from the electron levels of the silicon valence and conduction bands, we have introduced the level E_d of the extra phosphorus electron, which we call the *donor level* $E_d = E_c + E_P$. This level is not spin degenerate, because it is forbidden to put two electrons on the corresponding Bohr orbit, the Coulomb interaction energy between the two electrons on this orbit being greater than E_P.[4]

At $T = 0$, all the electron states of the valence band and also the level E_d of the phosphorus atoms are occupied, and the chemical potential $\mu = E_F$ lies between E_d and E_c. When $T \neq 0$, the electrical neutrality condition becomes

$$n_e = n_h + n_{d0}, \qquad (4.73)$$

where n_{d0} is the concentration of donors that have lost their electron (and are hence ionised to P^+). The latter is given as a function of the concentration n_d of donor phosphorus atoms by

$$n_{d0} = n_d \big[1 - f(E)\big].$$

The relation (4.73) can be used to determine the exact position of μ at an arbitrary temperature.

For temperatures such that $k_B T > |E_P| = E_c - E_d$, it can be checked that μ goes below E_d and that this ends up exciting practically all the electrons of the donor level into the conduction band, with

$$n_h \ll n_e \approx n_d.$$

---

[4] The consequences of this Coulomb interaction are examined in Problem 7: *Insulator–Metal Transition*.

Substituting P atoms for Si atoms therefore creates a semiconducting compound with majority electron doping, which is said to be *n*-doped. Note that the electron and hole densities are always related by (3.54).

In the same way, some Si atoms could be replaced by atoms of a trivalent element like B, which is an electron acceptor, introducing an acceptor level E_a above the top of the valence band. One can thus obtain a *p*-doped compound, i.e., doped with holes, with the electronic structure illustrated in Fig. 4.12b.

Note that the presence of positive charges can be detected directly by electron transport experiments with a magnetic field $B = \mu_0 H_a$ perpendicular to the current. In the plane perpendicular to the field, the Laplace force deflects the carrier trajectories in the same direction whether they be positively or negatively charged. It follows that charges accumulate in the direction perpendicular to the current, and in the steady-state regime, this produces an electric field transverse to the direction of the current. The associated potential difference has the sign of the majority carriers in the material. This effect, called the *Hall effect*, can be used to determine the density and sign of the carriers in a semiconductor.

The industrial applications of semiconductors stem from this ability to control the density and *n* or *p* type of the carriers by doping. This in turn led to the development of microelectronics, where diodes and transistors were constructed by making junctions between *n*- and *p*-doped semiconductors. Another important step was to make transistors by applying an electric field to a semiconducting layer in order to control the level of doping, as we shall see for the specific example of graphene.

4.6 Electrons, Holes, and Transport in Graphene

As we saw in Chap. 3, graphene has the band structure of a zero gap semiconductor. We shall see here how the particular shape of the dispersion curves $E(\mathbf{k})$ shown in Fig. 3.19 thus leads to rather novel properties.

a. Consequences of the Graphene Band Structure

To begin with, consider the dispersion relations (3.60), taking E_0 as the zero energy. The experimental value $t \simeq 3$ eV is high, and we may restrict to eigenstates of energy $|E| \ll t$ if we are interested in low energy properties at experimentally accessible temperatures. Indeed, for neutral graphene, the conduction electrons and holes appear for energies close to $E_F = 0$. The corresponding momenta lie near the six vertices of the Brillouin zone, to be reduced to two points K^+ and K^- with wave vectors $\mathbf{K}^\varepsilon$ with $\varepsilon = \pm$, called Dirac points (see box below). The four others are obtained from K^+ and K^- by translations along a vector of the reciprocal lattice $(\mathbf{a}_1^*, \mathbf{a}_2^*)$, as shown in Fig. 4.13a. Setting

$$\mathbf{k} = \mathbf{K}^\varepsilon + \mathbf{q},$$

4.6 Electrons, Holes, and Transport in Graphene

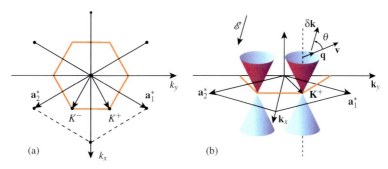

Fig. 4.13 (a) Hexagonal reciprocal lattice of graphene and its first Brillouin zone. The primitive cell ($\mathbf{a}_1^*, \mathbf{a}_2^*$) is shown, as are the two Dirac points K^+ and K^- of this cell. (b) The conical dispersion relation $E(\mathbf{k})$ near the Dirac points is shown inside the same primitive cell. This is a zoom on Fig. 3.19 in the vicinity of the Dirac points. The real space velocity $\mathbf{v}$ of an electron with wave vector $\mathbf{K}^+ + \mathbf{q}$ is indicated

we may consider an expansion of (3.60) to first order in $\mathbf{q}$:

$$E_{\mathbf{k}} = \pm \frac{3}{2} at |\mathbf{q}| \:. \tag{4.74}$$

Figure 4.13b shows these dispersion relations, which are conical close to the Dirac points. The velocity of an electron in a state $E_{\mathbf{k}}$ is given by

$$\langle \mathbf{v_k} \rangle = \frac{1}{\hbar} \nabla_{\mathbf{k}} E_{\mathbf{k}} = \pm \frac{3at}{2\hbar} \mathbf{n} = \pm v_F \mathbf{n} \:, \tag{4.75}$$

where $\mathbf{n}$ is the unit vector normal to the constant energy circle at $\mathbf{K}^\varepsilon + \mathbf{q}$, which thus lies along $\mathbf{q}$ (see Fig. 4.13b). The electron velocity is aligned with its quasi-momentum $\mathbf{q}$ and has magnitude $v_F = 3at/2\hbar$, which is independent of the electron state. Note that the zero curvature of the constant energy surface implies that *the effective mass vanishes*. Moreover, the dispersion relation (4.74) can be rewritten in the form

$$E_{\mathbf{K}^\varepsilon + \mathbf{q}} = \pm \hbar v_F |\mathbf{q}| \:. \tag{4.76}$$

This dispersion relation provides a simple way of calculating the density of energy states $D_S(E)$ per unit area and spin direction, when $|E| \ll t$. The number $D_S(E)dE$ of states with energies lying between E and $E + dE$ is the number of electron states contained in the area $2\pi q dq$ of reciprocal space, viz.,

$$D_S(E)dE = 2\pi q dq \times (2\pi)^{-2} \times 2 \:,$$

taking into account the existence of the two branches at points K^+ and K^-. We obtain

$$D_S(E) = \frac{|E|}{\pi \hbar^2 v_F^2} \:. \tag{4.77}$$

This density of states, which vanishes at $E = 0$, increases linearly with energy, a distinctly unusual behaviour compared with standard metals or semiconductors. A detailed calculation using the general form of the dispersion curves (3.60) gives Van Hove singularities for $D_S(E)$ when $E = \pm t$, energies at which the dispersion curves have saddle points, and a constant limiting value for the extrema of the energy $E = \pm 3t$, where a parabolic dispersion relation and finite effective mass are obtained (see Fig. 4.14).

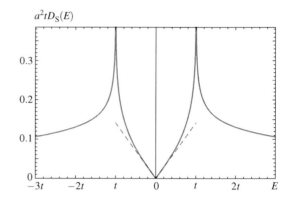

Fig. 4.14 Density of states per unit area for graphene, obtained numerically from the dispersion curves given in (3.60). For $|E| < t$, we recover the linear relation (4.77) close to the Dirac points. Figure courtesy of Montambaux, G.: Laboratoire de Physique des Solides, Orsay, France

Dirac Points and Relativistic Physics

The similarity between (4.76) and the dispersion relation for photons, together with the vanishing effective mass, has inspired an analogy between the equations of motion of the electrons in graphene around the points K^+ and K^- and those of massless quantum particles. The speed v_F given in (4.75), which is independent of the electron energy, plays the role of the speed of light. For this reason, the points K^+ and K^- are commonly referred to as Dirac points today. For the same reason, the surfaces representing the energy dispersion relation are called Dirac cones. The states in the conduction band then represent particles, while those of the valence band are identified with antiparticles.

Note that the degeneracy between the two conical valleys introduces a complication which we have hidden in (4.76), where the sign $\pm$ for the energy is obviously not related to the one used to characterise the Dirac points K^ε.

This analogy with 2D relativistic quantum physics can be taken a long way. As in this last case, where the particle spin is represented by 2×2 Pauli matrices, in the case of graphene, these matrices take into account the sublattices A and B. In this representation, the Hamiltonian yielding the conical dispersion relations can then be written

$$\hat{H} = \hbar v_F |\mathbf{q}| \hat{\xi} \ .$$

The operator $\hat{\xi}$, with eigenvalues ± 1, describes the projection of the sublattice 'spin' in the direction of propagation of the electron in the graphene. It is analogous to helicity in particle physics. Helicity is the projection of the real spin of the particle onto its direction of propagation. This analogy leads us to consider massless charged particles, something unknown in particle physics, since the only known massless particles, photons or neutrinos, are electrically neutral.

b. Electron Transport

At $T = 0$, there are in principle no carriers, and one therefore expects zero conductivity for graphene. However, for $T \neq 0$, electrons with surface density n_e are excited in the conduction band, with a resulting hole surface density n_h in the valence band, and graphene becomes a conductor. Given the electron–hole symmetry of the band structure, charge conservation implies that $n_e = n_h$ and that the chemical potential remains fixed at $\mu = E_F = 0$.

> **Question 4.5.** Show that the number of carriers thermally excited varies as T^2 and that the same is true of the electronic specific heat.

Insofar as the dispersion relation is not quadratic, and since the effective mass is zero, the Drude relation is not applicable, in contrast to the case of standard semiconductors. We must therefore consider a more general expression for the electrical conductivity, taking into account the total current of the occupied electron states, in the form

$$\mathbf{j} = -2\sum_{\mathbf{q}} e\mathbf{v}(\mathbf{q}) f(E(\mathbf{q})), \tag{4.78}$$

where $f(E)$ is the Fermi function and the factor of 2 is due to spin degeneracy. This current is clearly zero when there is no electric field. If we assume, as discussed in this chapter, that the electrons are subjected to viscous friction, in the presence of a field $\mathscr{E}$, the stationary solution implies a shift $\delta \mathbf{k}$ of the electronic states given by

$$\hbar \delta \mathbf{k} = -\tau_e e \mathscr{E} = \hbar \delta \mathbf{q}, \tag{4.79}$$

where τ_e is the collision time, assumed to be the same for both electrons and holes, since the problem is symmetric. Taking into account the displacement $\delta \mathbf{q}$ of the Fermi surface in (4.78), we obtain

$$\mathbf{j} = -2e \sum_{\mathbf{q}} \mathbf{v}(\mathbf{q} + \delta \mathbf{q}) f[E(\mathbf{q})] = -2e \sum_{\mathbf{q}} \mathbf{v}(\mathbf{q}) f[E(\mathbf{q} - \delta \mathbf{q})]$$

$$= 2e \sum_{\mathbf{q}} \mathbf{v}(\mathbf{q}) \frac{df[E(\mathbf{q})]}{dE} (\nabla_{\mathbf{q}} E) \cdot \delta \mathbf{q} = 2e\hbar \delta \mathbf{q} \sum_{\mathbf{q}} v_x^2(\mathbf{q}) \frac{df[E(\mathbf{q})]}{dE}.$$

Equation (4.79) was used in the last equality, taking $\mathbf{x}$ to be a unit vector in the direction $\mathbf{j} \parallel \mathscr{E} \parallel \delta \mathbf{q}$. For temperatures such that $k_B T \ll t$, we can restrict the dispersion relations to their conical part given by (4.76). Setting $\theta = (\mathbf{q}, \delta \mathbf{q})$ as shown in Fig. 4.13b, this leads to

$$\mathbf{j} = -2e^2 \tau_e v_F^2 \mathscr{E} \sum_{\mathbf{q}} \cos^2 \theta \frac{df[E(\mathbf{q})]}{dE}. \tag{4.80}$$

The sum can be replaced by an integral over energy, after integrating over the angular orientation of **q** at given energy E, viz., $\int_{-\pi}^{\pi} \cos^2\theta (d\theta/2\pi) = 1/2$. We thus obtain

$$\mathbf{j} = e^2 \tau_e v_F^2 \mathscr{E} \int_{-\infty}^{\infty} -\frac{df}{dE} D_S(E) dE . \tag{4.81}$$

The 2D conductivity can then be written in the form

$$\sigma = \frac{2}{\pi \hbar^2} e^2 \tau_e \int_0^\infty -\frac{df}{dE} E dE = \frac{2}{\pi \hbar^2} e^2 \tau_e k_B T \int_0^\infty -\frac{df}{dx} x dx , \tag{4.82}$$

where $x = E/k_B T$. The integral has the form

$$\int_0^\infty \frac{x dx}{\cosh^2 x} = \ln 2 ,$$

so

$$\boxed{\sigma = 2\frac{\ln 2}{\pi} \frac{e^2}{\hbar^2} \tau_e k_B T} . \tag{4.83}$$

We thus expect zero conductivity at $T = 0$, and *growing linearly with temperature*. Note that, in two dimensions, the resistivity $\rho = 1/\sigma$ corresponds to a current density **j** per unit width of material (not per unit area), so ρ has the dimensions of a resistance (and the symbol $\Omega/\square$ is sometimes used to express the units of ρ).

c. Doping by Field Effect

It is certainly an interesting fact that graphene is metallic at room temperature because carriers are thermally excited. However, for applications in electronics, it is essential to be able to construct active components producing some specific response to an applied voltage. One of the most widely used approaches with semiconductors is to make a metal oxide semiconductor capacitive structure, called a MOSFET (MOS field effect transistor), to control the doping of the semiconductor through an electric field obtained by applying a voltage to the metal gate.

This effect can be illustrated by the equivalent structure used for graphene, shown schematically in Fig. 4.15a. In this case, the gate is a doped Si substrate, oxidised at the surface to produce a film of SiO_2 with a thickness d of a few hundred nanometers. The graphene layer constitutes the second plate of the capacitor. When a voltage V_G is applied across the gate and the graphene, the charges appearing on the capacitor plates dope the graphene. If ε is the dielectric constant of the oxide layer, the capacitance $C = \varepsilon\varepsilon_0/d$ per unit area of the junction induces a charge

$$\boxed{n = \frac{\varepsilon\varepsilon_0}{ed} V_G} . \tag{4.84}$$

4.6 Electrons, Holes, and Transport in Graphene

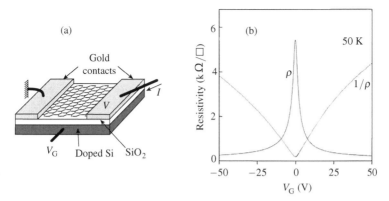

Fig. 4.15 (a) Semiconductor–oxide–graphene structure used to dope the graphene sheet by varying the voltage V_G applied to the semiconducting gate. (b) Dependence of the resistance and conductivity of graphene on V_G. Figure courtesy of A.K. Geim, from results published by Morozov, S.V., et al., Phys. Rev. Lett. **100**, 016602 (2008)

The surface density n of the carriers induced in graphene by an electric field thus varies linearly with the gate voltage V_G. We therefore observe, e.g., in Fig. 4.15b, that the resistivity of graphene falls off rapidly as a function of the gate voltage.

Question 4.6. 1. Which experimental observations in Fig. 4.15b can be explained by the approach discussed in Sects. 4.6b and c? Which cannot be explained? Compare the density of carriers induced experimentally by the field effect with the one obtained in Question 4.5 at thermodynamic equilibrium at 50 K ($\varepsilon = 4$, $d = 300$ nm).
2. What becomes of the chemical potential μ of the carriers in this experiment?

Note that the observations made with the atomic layer of graphene are similar to those carried out on silicon MOSFET transistors, which are made by depositing a metal film on a much thicker semiconducting layer whose carrier density is controlled by the voltage across the metal gate. In the case of graphene, the change in the resistance induced by a voltage $V_G = 10$ V is of the order of a factor of 10, which is a significant transistor effect. This type of structure has been used over the past few years to demonstrate many novel physical properties in graphene, suggesting applications for graphene or graphene multilayers in nanotechnology in the near future.

Question 4.7. What experiment could determine the sign of the injected carriers? How could one check experimentally whether the graphene energy spectrum does indeed correspond to that of the Dirac cones?

4.7 Summary

Electrical conductivity is a non-equilibrium electron transport phenomenon. An electric potential gradient induces charge transfer which, in a gas of non-interacting classical or quantum particles, should lead to an accelerated, hence non-linear motion of the electrons as a function of the applied electric field $\mathscr{E}$. Ohm's law, which says that $\mathbf{j} = \sigma_e \mathscr{E}$, can thus only be explained if there is a frictional force due to the collisions suffered by the electrons, whereby the electron velocities can reach a steady state equilibrium distribution. The relaxation time τ_e of the velocity distribution is the average time between collisions for a given electron in the classical Drude model. We thus define the electron conductivity σ_e and the mobility μ_e by

$$\sigma = \frac{ne^2 \tau_e}{m_0} = ne\mu_e .$$

In the more realistic case of a free fermion gas, the effect of an electric field is to translate the points representing the electron states in the space of quasi-momenta $\mathbf{k}$ by an amount $\hbar \delta \mathbf{k} = -e\mathscr{E}\tau_e$, where τ_e is now the collision time of the electrons whose representative states lie on the Fermi surface, i.e., have an energy close to E_F. For these electrons, the electron mean free path is thus $\ell_e = v_F \tau_e$.

In a metal, the electrical and thermal conductivities of electronic origin, σ and κ, respectively, are generally limited by the same collision processes, and this leads to the Wiedemann–Franz law:

$$\frac{\kappa}{\sigma} = \frac{\pi^2}{3}\left(\frac{k_B}{e}\right)^2 T = \mathscr{L}T .$$

The group velocity of an electron wave packet built up from Bloch states of quasi-momentum $\mathbf{k}$ is given directly by the dispersion curves $E_n(\mathbf{k})$ of the associated electronic band. In three dimensions, it is normal to the corresponding constant energy surface in $\mathbf{k}$ space, and is given by

$$\langle \mathbf{v} \rangle_{n\mathbf{k}} = \frac{1}{\hbar} \nabla_\mathbf{k} E_n(\mathbf{k}) .$$

The effect of a force on a Bloch state of quasi-momentum $\mathbf{k}$ is to shift the point representing the state by an amount $\hbar d\mathbf{k}/dt = \mathbf{F}$.

In real space, the electron acceleration is related to the force by

$$\frac{dv_\alpha}{dt} = \sum_\beta \frac{1}{(m_e)_{\alpha\beta}} F_\beta ,$$

where $(m_e)_{\alpha\beta}$ is the effective mass tensor, related to the curvature tensor of the constant energy surfaces in reciprocal space.

Experimental observations show that various collision processes contribute to the resistivity: collisions with impurities for which τ_e does not depend on temperature, and interactions with vibrational modes of the crystal lattice, for which τ_e falls off at high temperatures.

In semiconductors, two types of carrier contribute to the resistivity, namely, the electron states at the bottom of the conduction band and the unoccupied states at the top of the valence band, which are called holes. The latter can be treated as positively charged carriers, with the velocity of the missing electron, but with energy, quasi-momentum, and effective mass opposite to those of the missing electron.

The density of the carriers, electrons or holes, can be controlled in a semiconductor by doping it with heterovalent substitution impurities. For example, phosphorus is an electron donor in silicon, whereas boron is an acceptor. The excess electron of the donor is trapped at low temperatures by the Coulomb potential of the donor, on a hydrogen-like atomic orbital constructed from Bloch states of the conduction band. The effective Coulomb potential is reduced by the dielectric constant ε of the material. As a consequence, this electron state has a lower binding energy than the states of the conduction band, being reduced to

$$E_P = -\frac{E_1}{\varepsilon^2} \frac{m_e}{m_0},$$

while its dilated Bohr orbit has radius

$$a_P = a_1 \varepsilon m_0 / m_e.$$

The Fermi level lies between the donor level and the conduction band, but at high temperatures, this donor state is thermally excited toward the electronic states of the conduction band. In this way, n-doped semiconductors can be obtained, in which the majority carriers are electrons. In a similar manner, an acceptor produces a bound hole state at low temperatures, with slightly higher energy than the top of the valence band. This hole is excited into the valence band at high temperature, thereby producing a p-doped semiconductor.

Junctions between n- and p-doped semiconductors can be exploited to make diodes and transistors, which have spurred the massive expansion of microelectronics.

The novel band structure of graphene, with its two Dirac points and an electronic density of states that varies linearly with energy, leads to original electron transport properties for this 2D system. In particular, a significant change in the carrier density can be induced by applying an electric field perpendicular to the graphene sheet. This structure is analogous to MOSFET structures obtained with standard semiconductors, and is expected to give rise to specific applications.

4.8 Answers to Questions

Odd Man Out on Page 100

Most of the everyday objects illustrated in the figure make use of *electrical or thermal conductivity of electrons in a metal*:

- The *thermometer* uses both effects: thermal conductivity of the metal tip which thermalises at the temperature of the tested substance, and electron transport in a metal or semiconductor, exploiting the temperature dependence of a resistance, for example, to indicate the temperature. Electrical conductivity is clearly as crucial in the underlying measurement and display circuitry.
- The *iron* exploits both effects, one to heat a resistance by electrical conductivity and the Joule effect, the other to heat the iron shoe which is itself metallic.
- The *lamp and fuses* use electrical conductivity and the Joule effect to heat the filament or melt the metal in the fuse to cut the current.
- The *credit card* stores data in magnetic form. This data is read by an electronic chip which uses electronic circuits and hence exploits electron transfer for read and write operations. Recorded data is only useful if it can be read.
- The *kettle* heats a liquid by thermal conductivity of electronic origin in the metal element at the bottom of the container.
- The *outdoor heater* uses the infrared radiation produced by a chemical reaction, namely oxidation of butane or propane, to heat objects receiving this radiation. The metal surface of the reflector reflects infrared radiation back down. However, there is no electron transport here. Moreover, this reflection serves only to avoid energy loss in the upward direction. The heater works, but in a less energy efficient way, without the reflector.

We may conclude that the *outdoor heater is the odd man out* in this set of everyday objects.

Question 4.1

With the Drude hypothesis, the equation of motion of an electron is

$$m\frac{d\mathbf{v}_e}{dt} = e\mathscr{E} - \frac{m\mathbf{v}_e}{\tau_e},$$

where $\mathbf{v}_e$ is the drift velocity and τ_e the collision time. With $\mathscr{E} = \mathscr{E}(\omega)e^{i\omega t}$ and $\mathbf{v}_e = \mathbf{v}_e(\omega)e^{i\omega t}$, we easily obtain

$$i\omega m \mathbf{v}_e(\omega) = e\mathscr{E}(\omega) - \frac{m\mathbf{v}_e(\omega)}{\tau_e},$$

and hence,

$$\mathbf{j}(\omega) = ne\mathbf{v}_e(\omega) = \frac{ne^2 \mathscr{E}(\omega)}{m}\frac{1}{i\omega + 1/\tau_e} = \sigma(\omega)\mathscr{E}(\omega),$$

where

$$\sigma(\omega) = \frac{\sigma(0)}{1+i\omega\tau_e}, \qquad \sigma(0) = \frac{ne^2\tau_e}{m},$$

so that

$$\sigma_1(\omega) = \frac{\sigma(0)}{1+\omega^2\tau_e^2}, \qquad \sigma_2(\omega) = \frac{\sigma(0)\omega\tau}{1+\omega^2\tau_e^2}.$$

The function $\sigma_1(\omega)$ is shown graphically in Fig. 4.16.

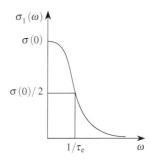

Fig. 4.16 Graph of the function $\sigma_1(\omega)$

Question 4.2

1. The contribution of impurities to the resistivity depends linearly on their concentration c. In the case of Ni or Fe, this can be checked by plotting the difference $\Delta\rho$ between the resistivities of the alloy and the pure metal as a function of c (see Fig. 4.17).
2. It is easy to determine $\Delta\rho$ for In, Sn, Sb, and As impurities, and normalise it by the concentration c, to give $\rho_i = \Delta\rho/c$.
3. ρ_i is plotted as a function of Z in Fig. 4.18a. We observe that the resistivity due to impurities grows rapidly with Z. This happens because the relaxation rate τ_e^{-1} is determined by the probability of Bloch electronic states being scattered by the potential associated with the impurity. As the Cu ions are in a 1^+ state, the extra Coulomb potential introduced by an impurity is determined by the charge difference between the ion associated with the impurity and Cu^+, that is, $\Delta Z = Z - 1$. The probability of an electron being scattered by the extra Coulomb

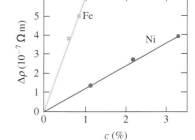

Fig. 4.17 Difference $\Delta\rho$ between the resistivities of an alloy and the pure metal as a function of the concentration c

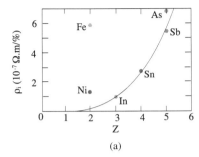
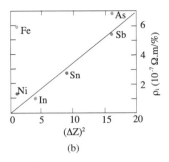

Fig. 4.18 (a) ρ_i as a function of Z. (b) ρ_i as a function of $(\Delta Z)^2$

potential $\Delta Z e^2/4\pi\varepsilon_0 r$ introduces the square of a matrix element of this potential. The scattering probability must therefore go as $(\Delta Z)^2$. It can be checked that the experimental results do indeed obey such a law (see Fig. 4.18b). A quantitative calculation of the resistivity due to impurities could be done using the Born approximation, as for collision cross-sections between two particles [21, Chap. 6]. Note that Fe and Ni impurities behave anomalously. These are magnetic impurities, for which this approach is not appropriate.

Question 4.3

The effective masses are obtained from the *curvatures of the bands*. For free electrons, an energy $E = 1$ eV $= 1.6 \times 10^{-19}$ J would correspond to a wave vector $k = \hbar^{-1}\sqrt{2m_0 E} = 5.4 \times 10^9$ m$^{-1} = 0.49(2\pi/a)$. It can be checked from Fig. 3.29 that the curvature of the conduction band in the (111) direction is close to what is found for free electrons. There is a corresponding effective mass $m_e \sim m_0$. For the valence bands, the last band of holes (doubly degenerate) has higher curvature: on the right, 1 eV corresponds to $0.25(2\pi/a)$, so that $m_h \sim 0.25 m_0$. The second band of holes, with a width of about 8 eV, has a much greater curvature, and thus corresponds to 'lighter' holes, with $m_h \sim 0.08 m_0$.

Question 4.4

The optical excitation of an electron by a direct transition from a state $\mathbf{k}_e$ in the valence band (VB) to a state with the same wave vector in the conduction band (CB) leaves a hole of wave vector $\mathbf{k}_h = -\mathbf{k}_e$ in the valence band. If we apply an electric field $\mathscr{E}$ to this system, we show that the electric currents due to electron and hole drift are additive, for a standard band diagram like the one in Fig. 4.19 where the effective masses m_e and m_h of the electrons and holes are positive and isotropic.

According to the real space acceleration theorem, the motion of an electron in the presence of a field $\mathscr{E}$ is given by

4.8 Answers to Questions

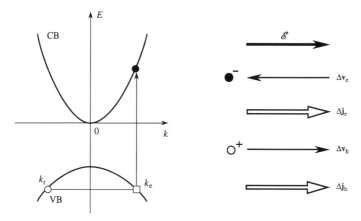

Fig. 4.19 Excitation of an electron–hole pair by direct optical transition. CB conduction band. VB valence band (*left*). Electron and hole drift currents $\Delta \mathbf{j}_e$ and $\Delta \mathbf{j}_h$ in the presence of an applied electric field $\mathscr{E}$ (*right*)

$$m_e \frac{d\mathbf{v}_e}{dt} = -e\mathscr{E},$$

and the velocity and current acquired over a time lapse Δt are

$$\Delta \mathbf{v}_e = -\frac{e\Delta t}{m_e}\mathscr{E}, \qquad \Delta \mathbf{j}_e = -e\Delta \mathbf{v}_e = +\frac{e^2 \Delta t}{m_e}\mathscr{E}.$$

For the hole,

$$m_h \frac{d\mathbf{v}_h}{dt} = e\mathscr{E}, \qquad \Delta \mathbf{v}_h = \frac{e\Delta t}{m_h}\mathscr{E}, \qquad \Delta \mathbf{j}_h = \frac{e^2 \Delta t}{m_h}\mathscr{E}.$$

The electric currents due to electron and hole drift thus lie in the same direction as $\mathscr{E}$, and are therefore additive. This amounts to saying that the conduction band and the valence band each contribute to the total current, one by an amount $\Delta \mathbf{j}_h$, and one by an amount $\Delta \mathbf{j}_e$. Figure 4.19 shows this result schematically.

This result is quite general, i.e., the total current is the sum of the contributions from the different bands, which are described in 'electron language' or 'hole language', depending on what is most convenient.

Question 4.5

We have

$$n_e = n_h = \int_0^{3t} f(E)D_S(E)dE \simeq 2(\pi \hbar^2 v_F^2)^{-1} \int_0^\infty Ef(E)dE,$$

since, for $k_B T \ll t$, we can use (4.76) for the density of states, and sum to infinity:

$$n_e = n_h = \frac{2}{\pi}\left(\frac{2k_B T}{3at}\right)^2 \int_0^\infty f(x)x\,dx \,.$$

The number of carriers does indeed go as T^2, and we may estimate its order of magnitude per unit area, given that $\int_0^\infty f(x)x\,dx = R(2) = \pi^2/12$:

$$n_e = n_h = \frac{2\pi}{27}\left(\frac{k_B T}{at}\right)^2 . \tag{4.85}$$

For graphene, $a = 1.42$ Å and $t = 3$ eV, which yields

$$n_e \approx 8 \times 10^{10} \text{ cm}^{-2} \text{ at } T = 300 \text{ K} \,. \tag{4.86}$$

The specific heat can be evaluated approximately by noting that the energy of this set of electrons which scales with T^2 is increased by $k_B T$, and the internal energy increases then as T^3, and the specific heat as T^2. More precisely,

$$U = 2\int_0^\infty Ef(E)D_S(E)dE = 2(\pi\hbar^2 v_F^2)^{-1}\int_0^\infty E^2 f(E)dE \,,$$

so

$$U = 2(k_B T)^3(\pi\hbar^2 v_F^2)^{-1}\int_0^\infty f(x)x^2\,dx \,,$$

and with $I(x) = \int_0^\infty f(x)x^2\,dx = 4R(3)/3 \approx 1.8$, we obtain

$$C = \frac{dU}{dT} = 10.8k_B(k_B T)^2(\pi\hbar^2 v_F^2)^{-1} . \tag{4.87}$$

Note that these quadratic dependences expected for the number of carriers and the specific heat bear no relation to what happens in standard metals or semiconductors. These are direct consequences of the singular electronic structure of graphene. As we shall see later, they have never been observed so far.

Question 4.6

1. Experiment shows that the resistivity decreases when either a negative or a positive voltage is applied, i.e., when either holes or electrons are created. This is to be expected if the band structure is symmetric. Note that the field effect is very efficient experimentally, because the conductivity σ increases very quickly and almost linearly as a function of V_G.
   Using $\varepsilon = 4$ and $d = 300$ nm, we obtain the density of carriers created:

$$n = 7 \times 10^{10}|V_G| \text{ (cm}^{-2}/\text{V}) \,. \tag{4.88}$$

4.8 Answers to Questions

For comparison, according to (4.85), the expected content of thermally excited carriers at 50 K is $n = 2.2 \times 10^9$ (cm^{-2}), so an applied voltage of $|V_G| \simeq 32$ meV would suffice to create an equivalent number of carriers. Such a gate voltage should double the conductivity, but this is only observed in Fig. 4.15b for a voltage two orders of magnitude greater. It thus looks as though (4.85) does not apply, since for $V_G = 0$, it corresponds to far fewer thermally excited carriers than are actually observed experimentally.

This certainly means that there are carriers due to charged defects causing extrinsic chemical doping. But this kind of doping, if uniform and hence of definite sign, would break the electron–hole symmetry. For example, electron doping should be reduced to zero by applying a suitable negative voltage. One would thus expect to see the minimum of $\sigma(V_G)$ shift toward the right, quite the opposite of what is observed! The fact that the conductivity curve does not break the electron–hole symmetry can only be explained by an inhomogeneous situation on the graphene layer, which must comprise some positively charged and some negatively charged regions, with average close to zero. The spatial extension of these regions induced by the distribution of defects at the interface with the SiO$_2$ layer prevents the recombination of electrons and holes.

Figure 4.20 shows experimental curves for $\sigma = 1/\rho$ at various temperatures, similar to the one in Fig. 4.15b obtained at 50 K. It shows that these conductivities, and in particular the minimum of the conductivity at $V_G = 0$, vary little with T. Moreover, the carriers associated with defects and those injected by field effect seem to behave independently. Indeed, we may suppose that

$$\rho = \rho_1(V_G) + \rho_0, \quad (4.89)$$

and thereby deduce an experimental value of $\sigma_1(V_G) = 1/\rho_1(V_G)$. The result is shown in Fig. 4.20. It is found that $\sigma_1(V_G)$ is proportional to V_G and almost independent of T. This is not obvious a priori, and is hence a highly relevant experimental observation. We shall discuss the implications below.

2. The dependence of n on V_G corresponds to injection of electrons ($V_G > 0$) in the conduction band or holes ($V_G < 0$) in the valence band of graphene. It thus corresponds to a shift in the chemical potential for different values of V_G.
   At $T = 0$, if the dispersion equation (4.76) applies, and if we set $E_F = \hbar v_F q_F$, the number of electrons with energy $E < E_F$ per unit area is $2 \times 2 \times \pi q_F^2/(2\pi)^2 = n$. The relation between E_F and the doping is therefore

$$E_F = \hbar v_F q_F = \hbar v_F (\pi n)^{1/2}. \quad (4.90)$$

This dependence of E_F on V_G, and the corresponding occupation of the Dirac cones on the doping, are both illustrated in Fig. 4.20.

Note that we have taken into account the effect of the carrier number on the conductivity and the Fermi level, but we have not yet considered the relevance of the collision time τ_e, which is the key factor limiting conductivity. Of course it comes into the calculation of Sect. 4.6c, which leads to the expression (4.84) for

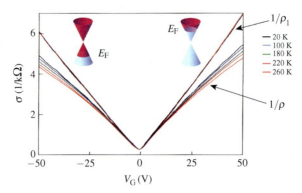

Fig. 4.20 Experimental curves for $\sigma = 1/\rho$ vs. gate voltage at different temperatures (see text). Figure courtesy of A.K. Geim, from results published by Morozov, S.V., et al., Phys. Rev. Lett. **100**, 016602 (2008)

σ when there is no field-effect doping. Note, however, that when there is doping, (4.82) remains perfectly valid, provided that we take the Fermi function with the chemical potential $\mu = E_F$ given by the above result. In this case, the calculation proceeds easily for $k_B T \ll E_F$, because in this case, $-df/dE$ can be replaced by a Dirac delta $\delta(E - E_F)$ and we have

$$\sigma = e^2 \tau_e v_F^2 D_S(E_F) = \frac{1}{\pi} \frac{e^2}{\hbar^2} \tau_e |E_F| \,. \tag{4.91}$$

The fact that the experimental conductivity is almost independent of T implies that τ_e must also be temperature independent and hence determined rather by defects than by phonons. In a conventional metal, the Drude relation for σ would naturally lead to a conductivity proportional to n, and hence proportional to V_G. But according to the expression obtained above, the conical dispersion relation of graphene leads rather to $\sigma \propto \tau_e E_F \propto \tau_e n^{1/2} \propto \tau_e V_G^{1/2}$. One would then expect $\sigma \propto V_G^{1/2}$, whereas experiment shows that $\sigma \propto n \propto V_G$! This can only be explained if τ_e goes as $\sqrt{n} \propto \sqrt{V_G}$, and this is counter-intuitive, since it implies that the more we dope, the less the defects will scatter electrons! (Are the defects screened by the carrier density?)

These experimental results are still under intensive investigation to determine which defects cause collisions. Could it be charged defects on the interface layer of SiO_2 which arise with doping? And we still need to understand what increases the conductivity at $V_G = 0$.

Even if defects limit conductivity, the latter remains high. We can define an electron mobility μ_e by writing σ in the form $\sigma = n e \mu_e$, and compare with μ_e for doped semiconductors. The experimental value obtained for graphene is $\mu_e \simeq 200{,}000$ cm^2/V s, which is huge at room temperature when compared with the cleanest semiconductors we can produce.

Question 4.7

As for semiconductors, the carrier sign can be ascertained by measuring the sign of the Hall effect, i.e., the transverse voltage induced in the presence of a current in graphene when a magnetic field is applied perpendicularly to the graphene plane. The experimental results in Fig. 4.21a show that the Hall voltage changes sign for $V_G = 0$. The reciprocal of the Hall voltage, called the Hall constant $(\rho_{xy})^{-1} = nq/B$, can be used to determine nq, where q is the carrier charge, which is indeed found experimentally to be proportional to V_G, except for $V_G \sim 0$, as we saw above.

Cyclotron frequency measurements can be used to determine the form of the dispersion relations. Recall that cyclotron modes are oscillatory eigenmodes of electrons in an applied magnetic field **B**, determined from the area of the electron orbits. In semiconductors, the cyclotron frequency $\omega_c = eB/m_c$ is used to obtain a cyclotron mass m_c related to the effective mass of the carriers.[5] It is given quite generally by

$$m_c = \frac{\hbar^2}{2\pi} \frac{dS}{dE}\bigg|_{E=E_F}, \tag{4.92}$$

where S is the area of the electron orbit on the surface of energy E in reciprocal space. In the case of the Dirac cones,

$$S(E) = \pi [q(E)]^2 = \pi \left(\frac{E}{\hbar v_F}\right)^2, \tag{4.93}$$

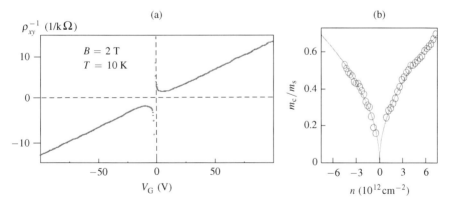

Fig. 4.21 (a) Hall effect conductance versus gate voltage showing the linear increase in carrier content of either sign except near the neutral point. (b) Cyclotron resonance mass versus carrier content for graphene. Figures courtesy of A.K. Geim, from results published by Novoselov, K.S., et al., Nature **438**, 197 (2005)

---

[5] See Problem 8: *Cyclotron Resonance*.

and then

$$m_c = \frac{E_F}{(v_F)^2} = \frac{\hbar(\pi n)^{1/2}}{v_F}. \tag{4.94}$$

The experimental results in Fig. 4.21b confirm the $n^{1/2}$ dependence, i.e., going as $V_G^{1/2}$. The Dirac cone dispersion relations are thus perfectly corroborated. Note that m_c is not at all the effective mass, which is zero for graphene.

The point about these experiments is that they help us to distinguish the effects of collision time and carrier number on transport phenomena. With cyclotron resonance, the magnetic field used must nevertheless be strong enough to ensure the condition $\omega_c \tau_e \gg 1$. This generally requires low temperature experiments to increase τ_e. For graphene, τ_e is independent of temperature, as noted from Fig. 4.20, and the mobility is high enough to allow cyclotron frequency measurements at $T = 300$ K, like the ones shown in Fig. 4.21b.

In order to determine the dispersion relations $E(\mathbf{k})$ directly, another method described in Chap. 3 is angle-resolved photoemission spectroscopy (ARPES). In 2D systems, $E(\mathbf{k})$ can be determined directly for occupied states by analysing the energies of electrons emitted in one direction by incident photons. For graphene, as shown in Fig. 4.22, the linear form of $E(\mathbf{k})$ is observed directly at the Dirac point for the π electron band. This experiment requires the surface to be homogeneous over the extent of the incident photon beam, viz., 40 μm, and could not be carried out for graphene deposited on SiO_2 owing to the disorder induced by the substrate. As mentioned in the solution to Question 4.6, this disorder-induced doping varies over an area comparable with the beam area. The results of Fig. 4.22 were obtained on a graphene multilayer deposited by epitaxy on the face of an SiC single crystal made up entirely of carbon atoms. The graphene layers formed here do not stack up

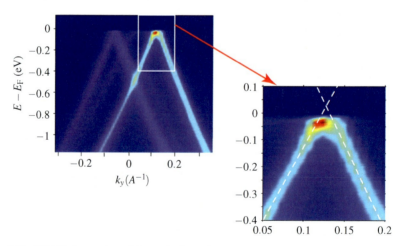

Fig. 4.22 ARPES observation of the conic dispersion relation of graphene (see text). Image courtesy of C. Berger, from results published by Sprinkle, M., et al., Phys. Rev. Lett. **103**, 226803 (2009)

4.8 Answers to Questions

as in graphite and are only weakly coupled. The experiment shows that they behave like isolated graphene sheets. It can be seen from Fig. 4.22 that the dispersion is indeed linear, and on the zoom, that the point of maximal intensity corresponding to the Fermi level lies less than 20 mV from the point $E = 0$ of the dispersion curve. According to (4.88), this shows that the doping induced by the SiC substrate is less than 10^{10} holes/cm^2 over the beam area. Also on Fig. 4.22, note the existence of a second, less intense Dirac cone. It corresponds to a graphene sublayer rotated with respect to the surface layer.

Chapter 5
Introduction to Superconductivity

Contents

5.1	Conditions for Superconductivity	147
	5.1.1 Persistent Currents	147
	5.1.2 Critical Field and Critical Current	150
5.2	Difference Between a Perfect Conductor and a Superconductor: The Meissner Effect	152
	5.2.1 Effect of a Magnetic Field on a Perfect Conductor	152
	5.2.2 Meissner Effect: Exclusion of the Magnetic Field	154
	5.2.3 London Equations and Penetration Depth	155
5.3	A Macroscopic Quantum Effect	158
	5.3.1 Macroscopic Wave Function and Current	158
	5.3.2 Quantisation of Magnetic Flux	159
	5.3.3 Measuring the Flux Quantum	160
	5.3.4 Josephson Effect	162
	5.3.5 SQUID	165
5.4	Summary	167
5.5	Answers to Questions	168

The levitation effect raises many questions. Peter on the right is practically minded and imagines how this phenomenon might be put to use, while Claire and David, who have read up to Chap. 7, try to visualise the strange mechanism that governs the behaviour of the electrons in the superconducting compound

5.1 Conditions for Superconductivity

We have seen how the conductivity of a metal is limited by collisions, which determine the electron mean free path. After a collision, an electron completely loses all memory of its quantum state as specified by its quasi-momentum **k**. It is thus impossible to follow a Bloch state over any distance much greater than the mean free path. To understand the microscopic origin of these collisions, one had to measure the conductivity of very pure metals at low temperatures. This was made possible by the work of the physicist Kammerlingh Onnes, who specialised in the liquefaction of gases and opened the way to the use of cryogenic fluids. In 1911, he succeeded in liquefying ^{4}He, at a temperature of 4.2 K. He then suggested using the low temperatures created in this way to study the low temperature conductivity of pure metals. Quite unexpectedly, he discovered that the conductivity of mercury increased by several orders of magnitude at temperatures below 4.18 K. This discovery of superconductivity remained a mystery for more than 40 years. With hindsight, it is clear that the prerequisites of quantum mechanics had not yet been established. Of course, this did not prevent the physicists of the first half of the twentieth century from gradually getting a hold on the fundamental manifestations of this phenomenon *through ideas based entirely on experimental observations*.

As mentioned in the Preface, we shall present superconductivity here in a way that closely follows historical progress, to show how scientific research actually proceeds in practice. In Sect. 5.1, we indicate how existence of a superconducting state can only be established by demonstrating the occurrence of *persistent currents*. In Sect. 5.2, we show that this phenomenon cannot simply be attributed to the conductivity of a perfect metal. We can then discuss an effect discovered by Meissner in 1933, namely *the total expulsion of magnetic flux from a superconducting material*. We go on to describe the phenomenological explanation of this novel magnetic manifestation of superconductivity, due to the London brothers in 1935. In Sect. 5.3, we suggest how the basic ideas of quantum mechanics allow one to establish the quantisation of the magnetic flux, on the basis of these two phenomena. Experimental observations then justify the idea that *electrons are paired in the superconducting state*.

5.1 Conditions for Superconductivity

5.1.1 Persistent Currents

It was by trying to understand what limits the conductivity of a pure metal at low temperatures that Onnes discovered that this conductivity becomes almost infinite at low T in some metals. In his first published experiments, Onnes observed that, below a temperature of 4.18 K, a mercury wire has a resistance that plummets from 0.1 Ω to well below 10^{-4} Ω. This sudden change in the resistance at a critical temperature T_c cannot of course be taken to show that the resistance has completely vanished, but it does raise the question as to how the acccuracy of electrical resistance

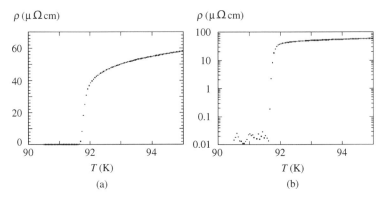

Fig. 5.1 Temperature dependence of the resistivity of a sample of the cuprate $YBa_2Cu_3O_7$. (**a**) The resistivity is observed to drop abruptly at 91.7 K in an experiment similar to the one shown in Fig. 4.6. (**b**) The measured voltage is plotted on a logarithmic scale. A residual voltage of experimental origin limits the accuracy of the measurement. Figures courtesy of Rullier-Albenque, F.: SPEC, CEA. Saclay, France

measurements might be improved. In the typical setup of Fig. 4.6, the results are clearly limited by the accuracy of the voltmeter (see Fig. 5.1). For a given sample and voltmeter, the most naïve idea would be to increase the current. It is then observed that the resistance of a normal metal reappears once the current becomes too large. It would thus seem that there is a *critical current density* $\mathbf{j}_c$ above which the superconducting state cannot exist.

How then can we determine whether the superconductor has nonzero resistance? To this end, it seems more interesting to ask whether we can demonstrate the existence of dissipation in a superconductor in which there is a current less than $\mathbf{j}_c$. Indeed, we may consider a closed loop of conducting material of inductance L and set up a current in it. If the loop has resistance R, the current I satisfies

$$L\frac{dI}{dt} + RI = 0,$$

which implies that

$$I = I(0)\exp\left(-\frac{Rt}{L}\right), \tag{5.1}$$

in the case where a current $I(0)$ has been established in this loop at time $t = 0$. Any resistance, no matter how small, will thus determine the time constant for the current to fall off. But how could such an experiment be realised in practice? How can a current be made to circulate in a closed loop, and how can we measure the way it varies in time?

There is no question of using a current generator, since such a device introduces normal metal conductors and hence dissipation, and it would have to be disconnected in order to close the loop! Fortunately, classical electromagnetism tells us that a current can be created by induction by applying a magnetic field. We may thus apply a magnetic field $\mathbf{H}_a = \mathbf{B}_a/\mu_0$ in a direction perpendicular to the plane of the loop, for a very short time, by means of some external source. Lenz's law tells us that the current induced in the conductor will oppose the time variation of the magnetic flux Φ through the loop due to $\mathbf{B}_a(t)$. We may thus write

$$-\frac{d\Phi}{dt} = -S\frac{dB_a}{dt} = RI + L\frac{dI}{dt},$$

where S is the area spanned by the loop. If the variation of B_a is produced in a very short time compared with the time constant L/R of the loop, R can be neglected during the time over which the current is being set up, which implies that

$$\frac{d(B_a S + LI)}{dt} = 0.$$

We thus find that, if R is small, the total magnetic flux, which is the sum of the flux due to the applied magnetic induction B_a and LI due to the induced current, will be constant inside the loop. For a change ΔB_a in the magnetic induction, the induced current is

$$I = -\frac{S\Delta B_a}{L}.$$

It only remains to measure the time variation of I and we may determine R using (5.1). If R is really zero, the current will persist indefinitely. We thus see that, if the superconductor is indeed a perfect conductor, it will be able to accommodate a *persistent current*.

In practice, there is little hope of measuring the time dependence of I. In fact, it is preferable to observe B. To increase sensitivity, it is easy to understand that the most effective procedure will be to apply an external field B_a when the conductor is resistive, i.e., at a temperature $T > T_c$. In this case, the resistance in the normal state is big enough for the current to drop to zero rather quickly. The loop is then cooled below T_c in the presence of the applied field, which in principle should change nothing.[1] The applied field B_a is subsequently reduced to zero over a very short time

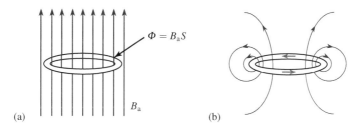

Fig. 5.2 (a) Cooled metal loop in the normal state in the presence of an applied magnetic field. (b) The field is switched off when the metal has been cooled to the perfectly conducting state. According to Lenz's law, the current induced in the loop is such as to oppose the change in the magnetic flux through the loop. The current persists indefinitely in zero field if the conductor is perfect

---

[1] We shall see in Sect. 5.2 that changes in B nevertheless occur within the superconducting loop.

lapse, thereby inducing a current $I = SB_a/L$. The lines of the magnetic flux density are illustrated in Fig. 5.2 before cooling and after switching off the magnetic field.

If the initial time $t = 0$ is taken to be the moment when B_a has dropped to zero, the magnetic flux induced by the current I is equal to SB_a, and decreases exponentially in time with time constant L/R. We thus measure the time dependence of B at the center of the loop in order to determine this time constant and hence R. A very accurate experiment was carried out in 1961 with a superconducting loop, and the value of B was determined by measuring the Larmor resonance frequency of the nuclear spins of the protons in a sample placed at the center of the loop. This frequency can be measured extremely precisely, and no change was observed over two years. This led to an upper estimate for the resistivity of a superconducting wire of $10^{-26}\,\Omega\,\text{m}$, a value at least 18 orders of magnitude lower than that of copper at room temperature. *Superconductivity therefore corresponds to a new and profoundly original physical state.* For comparison, note that the resistivity of the best conducting materials in their normal state is barely 20 orders of magnitude lower than those of the most perfect known insulators.

5.1.2 Critical Field and Critical Current

We saw above that the fact that the resistance of a superconductor is zero led naturally to the idea of using magnetic fields to study the behaviour of superconductors, and this right from the moment that superconductivity was first discovered. The above experiment suggests applying the strongest possible fields H_a in order to increase the sensitivity of the measurements. But in fact it was observed instead that superconductivity disappears for strong fields: if a current $j \ll j_c$ is driven through a superconducting wire, it is observed that, beyond a certain value H_c of the applied field, a potential difference appears between the measurement contacts. This shows that *a magnetic field stronger than H_c destroys the superconducting state*.

The temperature dependence of H_c was thus investigated, thereby determining the domain of existence of the superconducting phase. The experimental results shown in Fig. 5.3 can be represented approximately by the empirical law

$$H_c\left(\frac{T}{T_c}\right) = H_c(0)\left[1 - \left(\frac{T}{T_c}\right)^2\right]. \tag{5.2}$$

This limitation by an applied field shows that *superconductivity is essentially a magnetic phenomenon*, other manifestations of which will be discussed below.

But what exactly is the relevance of this critical current for the existence of the superconducting state? In other words, does the disappearance of the superconducting state in the presence of a high enough current point toward some novel physical effect, or is it related to the field induced by this current? It is easy to see that, for a

5.1 Conditions for Superconductivity

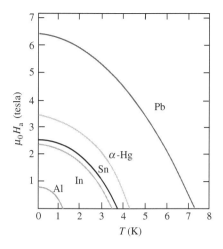

Fig. 5.3 Curves representing the field H_c destroying superconductivity as a function of the temperature T for different superconducting metals. The curves are approximately modelled by (5.2)

zero applied field, a current I in a superconducting wire of radius a (see Fig. 5.4a) will induce a tengential field $H = I/(2\pi a)$ at its surface. If $H > H_c$, the superconducting phase will be destroyed at the surface, and the superconducting region could only survive for a radius smaller than a. All the current would then have to flow within this superconducting core, and this would generate an even stronger field. This implies that the superconductivity must disappear suddenly and completely whenever $I = 2\pi a H_c$. This observation, first made by Silsbee, suggests that *the relevant physical quantities limiting the existence of the superconducting phase are the temperature T_c and the field $H_c(0)$*.

In the geometry of Fig. 5.4a, this defines the simple relation

$$I_c = 2\pi a(H_c^2 - H_a^2)^{1/2}$$

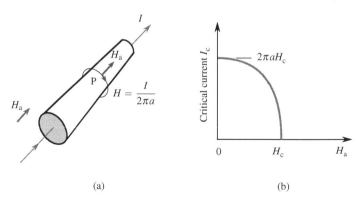

Fig. 5.4 (**a**) Setup for measuring the critical current in a cylindrical wire, as a function of an external field H_a applied along the axis of the wire. (**b**) Observed dependence of I_c on H_a. Adapted from [18, p. 84]

between the critical current and the critical field (see Fig. 5.4b). Paradoxically, this result, which can be checked experimentally, would appear to show that the critical current density j_c is inversely proportional to the radius rather than the cross-sectional area of the superconductor.[2]

Note that, for the materials discovered in the first half of the twentieth century, the critical temperature T_c never exceeded 20 K, while H_c was limited to 0.2 tesla. It was this second limit that blocked the way to any possible application of superconductivity for creating permanent magnetic fields, because the latter were then limited to values well below what could be achieved with soft iron magnets.

5.2 Difference Between a Perfect Conductor and a Superconductor: The Meissner Effect

5.2.1 Effect of a Magnetic Field on a Perfect Conductor

As superconductors are perfect conductors, it is important to understand the expected effect of a field on a perfect conductor. When there is no collision term, the acceleration of the electrons in the presence of an electric field $\mathscr{E}$ is

$$m_0 \frac{d\mathbf{v}}{dt} = -e\mathscr{E},$$

and if there are n electrons per unit volume, the current density is given by $\mathbf{j} = -ne\mathbf{v}$. This implies the first of the London equations:

$$\frac{d\mathbf{j}}{dt} = \frac{ne^2}{m_0}\mathscr{E}. \quad (5.3)$$

Maxwell's equations can be used to evaluate the behaviour of this perfect conductor in the presence of a magnetic induction $\mathbf{B}$:

$$\mathbf{curl}\,\mathscr{E} = -\frac{d\mathbf{B}}{dt}, \quad \mathbf{curl}\,\mathbf{B} = \mu_0 \mathbf{j} + \varepsilon_0 \mu_0 \frac{d\mathscr{E}}{dt}.$$

The first Maxwell equation can be combined with the first London equation (5.3) to give

$$\frac{m_0}{ne^2} \mathbf{curl}\left(\frac{d\mathbf{j}}{dt}\right) = -\frac{d\mathbf{B}}{dt}. \quad (5.4)$$

In the steady state $\mathscr{E} = 0$, otherwise $\mathbf{j}$ would grow indefinitely, according to (5.3). In addition, we consider only fields and currents that vary slowly in time, thus excluding for example the response to an electromagnetic wave. In this case, $\mathscr{E}$ and $d\mathbf{j}/dt$

---

[2] We shall see that in reality currents only flow at the surface of the wire, over some thickness λ (see Sect. 5.2). The critical current density is thus in this surface layer $j_c \approx H_c/\lambda$.

5.2 Difference Between a Perfect Conductor and a Superconductor

are both small, and since $d\mathscr{E}/dt$ is second order compared with $d\mathbf{j}/dt$, the displacement currents are negligible in the second Maxwell equation. The latter therefore simplifies to $\mathbf{curl\,B} = \mu_0 \mathbf{j}$. Substituting into (5.4), we then obtain

$$-\frac{d\mathbf{B}}{dt} = \frac{m_0}{ne^2\mu_0}\mathbf{curl\,curl}\left(\frac{d\mathbf{B}}{dt}\right),$$

and hence

$$\frac{d}{dt}\left(\mathbf{B} + \frac{m_0}{ne^2\mu_0}\mathbf{curl\,curl B}\right) = 0. \tag{5.5}$$

Since $\mathbf{curl\,curl B} = \mathbf{grad\,div B} - \Delta\mathbf{B}$ and $\mathbf{div B} = 0$, and setting

$$\boxed{\lambda_L^2 = \frac{m_0}{ne^2\mu_0}}, \tag{5.6}$$

where λ_L has the dimensions of length, it follows that

$$\frac{d(\mathbf{B} - \lambda_L^2\Delta\mathbf{B})}{dt} = 0. \tag{5.7}$$

For a conductor filling a half-space bounded by a plane surface perpendicular to the x axis and subjected to an applied magnetic induction $\mathbf{B_a}(t)$ parallel to its surface, the above equation leads to

$$\frac{dB(x)}{dt} = \frac{dB_a}{dt}\exp\left(-\frac{x}{\lambda_L}\right),$$

implying that, at a great distance from the surface as compared with λ_L, the magnetic induction $B(x)$ is time independent. In other words, *a change in B_a induces surface currents which oppose any change in the magnetic induction inside the perfect conductor* (see Fig. 5.5).

The value of $B(x)$ inside the metal is entirely determined by the field there when it was cooled below T_c, and it does not vary in time for a perfect conductor. This is illustrated in Fig. 5.6a. This result tells us that, under an applied field H_a, the magnetic induction within a perfect conductor will differ depending on whether

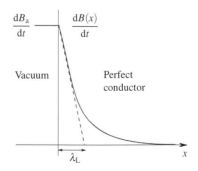

Fig. 5.5 Variation of $dB(x)/dt$ inside a perfect metal obtained when a variation dB_a/dt is imposed on the applied field. It goes to zero exponentially over a distance λ_L which corresponds to the thickness of the layer containing the surface currents

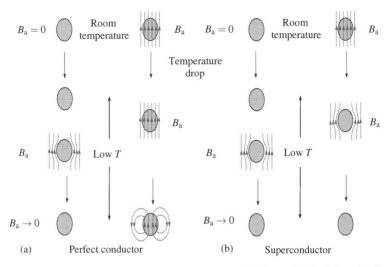

Fig. 5.6 (**a**) Magnetic field lines for a perfect conductor cooled in zero field, with application of a magnetic induction B_a for $T < T_c$, and reduction of B_a to zero (*left*). Application of B_a above T_c, followed by cooling in the presence of B_a and suppression of B_a (*right*). (**b**) The same sequences for a superconductor. In this case the field is excluded below T_c, whatever the magnetothermal history of the sample. The state obtained at low temperature is reversible. Adapted from [18, pp. 18–20]

the metal was cooled below T_c in the presence or the absence of this magnetic field. Note that λ_L is entirely determined by n, the number of conduction electrons per unit volume in the metal, which is of the same order of magnitude as for conventional metals.

For copper, $n = 8.5 \times 10^{28}$ electrons/m^3, and according to (5.6), this implies that $\lambda_L \simeq 200$ Å. The surface currents preventing any change in the magnetic induction in the metal only exist over a distance containing a few hundred atomic layers.

5.2.2 Meissner Effect: Exclusion of the Magnetic Field

It was not until 1933, some 22 years after the discovery of superconductivity, that Meissner and Ochsenfeld were able to demonstrate experimentally that a superconductor is not the same thing as a perfect conductor. Indeed they observed that, not only is the magnetic induction constant, but it is in fact zero inside a macroscopic superconductor, and this whatever magnetothermal path is taken (see Fig. 5.6), i.e.,

$$\boxed{B = 0}.$$

This observation, since called the Meissner effect, confirms that superconductivity is a novel magnetic effect of the electron gas, but it also shows that it is a fully *reversible* phenomenon. *The electron gas undergoes a transition at T_c to a new and more stable thermodynamic state.*

5.2 Difference Between a Perfect Conductor and a Superconductor

This observation leads to a series of considerations that we shall develop further in Chap. 6, and which will allow us to characterise the behaviour of the electronic system on the basis of experimental observations.

Note that the magnetisation in the bulk of a superconductor is given by

$$\mathbf{B} = \mu_0(\mathbf{H} + \mathbf{M}),$$

and then

$$\boxed{\mathbf{M} = -\mathbf{H}}. \tag{5.8}$$

The magnetic susceptibility $\chi = \tilde{\chi}/\mu_0$ of a superconductor is diamagnetic, with $\tilde{\chi} = -1$. A superconductor is thus a *perfect diamagnetic system*. This is illustrated in Fig. 5.7, where we have plotted the measured magnetisation of a superconductor for different values of the applied magnetic field. One of the easiest manifestations of this diamagnetism to demonstrate experimentally is the *levitation* of a magnet above a superconductor, as illustrated on p. 146. The exclusion of the magnetic field by the superconductor corresponds to a repulsive interaction between the magnet and the superconductor, leading to an equilibrium position for the magnet at a height h such that the force due to the magnetic interaction exactly balances the weight of the magnet.

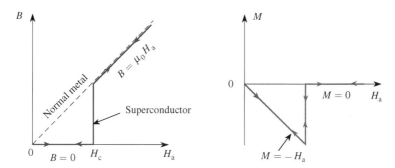

Fig. 5.7 Dependence of B and M in a superconductor on the field H_a

5.2.3 London Equations and Penetration Depth

Given the state of knowledge in the 1930s, it was just as difficult to understand the vanishing of the resistivity as the disappearance of the magnetic induction in the bulk of the superconductor. However, it was observed that the vanishing of $\mathbf{B}$ in the bulk of the material could be explained by setting the argument of the derivative equal to zero in Eq. (5.7). The phenomenological relation

$$\mathbf{B} - \lambda_L^2 \Delta \mathbf{B} = 0 \tag{5.9}$$

was thus used by the London brothers from 1935 to account for the Meissner effect. Note that (5.9) results from the need to account for observation, and cannot be justified on a priori grounds. However, it implies that it is **B** rather than d**B**/dt that goes to zero exponentially over the distance λ_L from the surface of the superconductor. The field is excluded only from the heart of the bulk superconductor. The existence of a penetration depth λ_L for the magnetic induction was demonstrated experimentally by measurements made on low dimensional samples, i.e., metal films and powders. This shows that persistent currents arise in a thickness λ_L at the surface and exactly compensate for the applied field in the bulk of the superconductor.

> **Question 5.1.** Consider a lead cylinder of inner radius $r_i = 10$ μm and outer radius $r_o = 20$ μm. In zero applied field, lead becomes superconducting at $T_c = 7.23$ K. Its critical field is $H_c = 0.08$ tesla. The density of superconducting electrons at zero temperature in lead is $n_s = 2 \times 10^{28}$/m^3. This cylinder is cooled in a magnetic induction field $B_a = \mu_0 H_a = 3.3$ mtesla applied parallel to its axis, starting when $T > T_c$ and continuing to a temperature $T \ll T_c$. The source of the magnetic field is maintained after cooling. What is the magnetic field between 0 and r_i? What is the magnetic field inside the cylinder between r_i and r_e? If there are currents in the cylinder, locate them and find their direction. How could the magnitude of these currents be estimated?

In fact the two London equations which best describe the experimental properties of superconductors in this phenomenological way are

$$\boxed{\lambda^2 \mu_0 \frac{d\mathbf{j}}{dt} = \mathscr{E}}, \tag{5.10}$$

and

$$\boxed{\lambda^2 \mu_0 \mathbf{curl\,j} = -\mathbf{B}}. \tag{5.11}$$

The first is just (5.3), and accounts for the absence of resistivity. The second London equation replaces (5.4) here, and implies (5.9), i.e., the Meissner effect, with the help of the second Maxwell equation.

These equations are barely more justifiable[3] than (5.9). They replace Ohm's law for the superconductor, and do not help to establish the value of λ as λ_L in (5.6). The penetration depth is thus defined experimentally. Measurements show that λ *increases with temperature*, and diverges at T_c, leading to a certain continuity with the normal metal state, in which the field penetration is perfect.

---

[3] It is worth noting how important empirical models can be for scientific progress. A full formulation is often only possible much later, when many conceptual obstacles have been removed. And even though today the BCS theory provides an explicit understanding of the observed relationships, the simplified representation due to the London brothers remains an extremely useful guide to understanding many experimental observations.

5.2 Difference Between a Perfect Conductor and a Superconductor

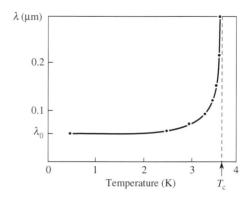

Fig. 5.8 Temperature dependence of the penetration depth measured in tin. Contradicting the simple London model of (5.6), it varies with the temperature and diverges for $T = T_c$, as shown by (5.12). From Schawlow, A.L., Devlin, G.E.: Phys. Rev. **113**, 120 (1959). With the kind permission of the American Physical Society. http://link.aps.org/doi/10.1103/PhysRev.113.120

The measured temperature dependence is often represented in an empirical form (see Fig. 5.8):

$$\lambda = \lambda_0 \left[1 - \left(\frac{T}{T_c} \right)^4 \right]^{-1/2}. \tag{5.12}$$

For most metals, the value of λ_0 measured at low temperatures is of the same order of magnitude as λ_L, but generally larger. To account for these observations, one commonly used approach is to use the *two-fluid model* for the electrons in a superconductor. The superconducting electrons, whose density n_s varies with the temperature, coexist with normal electrons of density $n - n_s$. Only the superconducting electrons contribute to the persistent currents, so by analogy with (5.6), we write

$$\lambda^2 = \frac{m_0}{n_s e^2 \mu_0}. \tag{5.13}$$

Comparing (5.12), (5.13), and (5.6), we thus obtain

$$\lambda_0^2 = \frac{m_0}{n_s(0) e^2 \mu_0}, \qquad n_s(T) = n_s(0) \left[1 - (T/T_c)^4 \right]. \tag{5.14}$$

In this model, the density of superconducting electrons thus decreases as the temperature increases, vanishing at $T = T_c$, when the normal state is reached.

Note that this two-fluid model can be used to predict the response of a superconductor to an electric field at frequency ω. The complex alternating conductivity $\sigma_s(\omega)$ of the superconducting electrons adds to $\sigma_n(\omega)$ for the normal electrons, determined in Question 4.1.

Question 5.2. 1. Determine the imaginary part of $\sigma_{2s}(\omega)$ using the London equations.
2. What response would you expect for $\sigma_{1s}(\omega)$ at zero frequency? [It can be found as the limit of $\sigma_1(\omega)$ for a metal when $\tau_e \to \infty$.]
3. At what frequency ω_c are the amplitudes of the two currents j_n and j_s equal? Assuming that $n_s(0)$ is the total electron number n per unit volume, determine ω_c at $T = 0.9 T_c$. (Take $\tau_e \sim 10^{-12}$ s.)
4. Can this approach be applied for $T \to 0$?

5.3 A Macroscopic Quantum Effect

How do we explain the lack of electrical resistivity? Clearly, defects are not going to disappear at T_c, and superconductivity is observed in particularly impure alloys, e.g., lead–indium alloys, where the residual resistivity is high in the normal state. The fact that currents persist over macroscopic lengths suggests that the electron flow is not perturbed over such lengths. There are almost as many obstacles as electrons, and yet the electrons go through undisturbed, as if these obstacles were just not there! This is as striking a paradox as the discovery that electrons orbit around atomic nuclei without radiating electromagnetic energy.

The solution is clearly of the same kind as the one proposed by Bohr to resolve the latter puzzle. *The superconducting state of the electron gas can only be a stationary quantum state of the electronic system.* The main difference with the Bohr atom is that this quantum state involves a *macroscopic* number of electrons, so if it is to be characterised by a wave function, this must necessarily involve a collective state of these electrons.

5.3.1 Macroscopic Wave Function and Current

The wave function of this macroscopic quantum state must be a multielectron wave function, and hence exceedingly complex. It is not useful a priori to try to describe the probabilities of finding the various electrons, but the macroscopic wave function can be reduced to one denoted by $\psi(\mathbf{r})$ that delivers *the spatial variation of the probability density of the superconducting carriers*,[4] viz., $|\psi(\mathbf{r})|^2$. By doing this, we preserve the key feature of a wave function, which is to include a *phase factor*, with

$$\psi(\mathbf{r}) = \phi(\mathbf{r})\exp\left[i\theta(\mathbf{r})\right]. \tag{5.15}$$

As a general rule, we shall be concerned with situations in which the superconducting carrier density $\phi(\mathbf{r})^2 = \phi^2$ is spatially uniform in the bulk of the superconductor. However, we shall see in Chap. 6 that we must expect a spatial dependence of this density in the vicinity of the interface between a superconductor and any other material.

In the presence of a magnetic field with vector potential $\mathbf{A}$, the momentum operator $\mathbf{p} = -i\hbar\nabla$ for a particle of mass m and charge q is [2, Chap. 15]

$$\mathbf{p} = m\mathbf{v} + q\mathbf{A}, \tag{5.16}$$

---

[4] Can the quantities $\phi(\mathbf{r})$ and $\theta(\mathbf{r})$ be defined at the same point of space? In the case of a one-particle wave function, the state and energy are defined, but the phase cannot be simultaneously measured. However, in a macroscopic system, the density and phase can be defined simultaneously, because the relevant uncertainty relation is $\Delta n_s \Delta\theta > 1$. Since n_s is macroscopic, the two quantities can be defined accurately enough, and can thus be treated semi-classically.

5.3 A Macroscopic Quantum Effect

where **v** is the velocity operator at point **r**. The electric current density at point **r** is the product of the charge and the probability current, viz.,

$$\mathbf{j}(\mathbf{r}) = q\mathrm{Re}(\psi^* \mathbf{v}\psi) = \frac{q}{m}\mathrm{Re}\left\{\psi^*[-i\hbar\nabla - q\mathbf{A}(\mathbf{r})]\psi\right\},$$

$$\mathbf{j} = -i\frac{\hbar q}{2m}(\psi^*\nabla\psi - \psi\nabla\psi^*) - \frac{q^2\mathbf{A}}{m}|\psi|^2.$$

With the expression for the wave function given above, this implies

$$\boxed{\mathbf{j} = \frac{q\phi^2}{m}(\hbar\nabla\theta - q\mathbf{A})}. \tag{5.17}$$

Note that, unlike θ and **A**, the current **j** is a measurable physical quantity. It is therefore independent of the choice of gauge for the vector potential, whence a change of gauge for **A** must lead to a change in $\theta(\mathbf{r})$. In particular, if $\mathbf{B} = 0$ in the bulk of a superconductor, the current **j** will be associated with a gradient in the phase of the wave function obtained in the gauge for which $\mathbf{A} = 0$.

5.3.2 Quantisation of Magnetic Flux

The simple fact that a superconductor constitutes a macroscopic quantum state has important implications, because the phase $\theta(\mathbf{r})$ of the wave function cannot vary in an arbitrary way in space. In particular, continuity of the wave function implies that the change in $\theta(\mathbf{r})$ along any closed path within a superconductor must be a multiple of 2π:

$$\boxed{\oint_{(C)} \nabla\theta(\mathbf{r})\cdot d\boldsymbol{\ell} = 2n\pi}. \tag{5.18}$$

Substituting the value of $\nabla\theta(\mathbf{r})$ given by (5.17) into (5.18), we obtain

$$\oint_{(C)} \left(\frac{m}{q\phi^2}\mathbf{j} + q\mathbf{A}\right)\cdot d\boldsymbol{\ell} = nh. \tag{5.19}$$

Since $\Phi = \oint \mathbf{A}\cdot d\boldsymbol{\ell}$ is the magnetic flux through the closed path (C), we may define a quantised quantity Φ^* with the dimensions of a flux, called the *fluxoid*:

$$\Phi^* = \Phi + \frac{m}{q^2\phi^2}\oint_{(C)} \mathbf{j}\cdot d\boldsymbol{\ell} = nh/q. \tag{5.20}$$

We thus consider a contour (C) inside a connected superconductor (see Fig. 5.9a). If we choose a contour that is everywhere distant from the surface of the superconductor, i.e., at a distance much greater than λ, then $\mathbf{B} = 0$ and $\mathbf{j} = 0$ imply that $n = 0$, and *the phase of the wave function is uniform*.

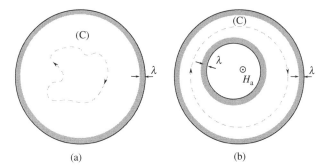

Fig. 5.9 (**a**) Contour (C) inside a connected superconductor. (**b**) Contour (C) around a hole in a non-simply-connected superconductor. The flux Φ of the magnetic induction in (C) reduces to the flux of the applied field $\mu_0 H_a$ through the hole, provided the latter has a diameter much greater than λ

For a non-simply-connected superconductor, consider a contour (C) around a hole (see Fig. 5.9b). If (C) is taken a long distance away from the edges of the hole compared with λ, then **j** is zero and, according to (5.20), we obtain

$$\Phi = n\Phi_0, \qquad \Phi_0 = h/q. \tag{5.21}$$

If we consider a ring superconductor with a hole of dimensions much greater than λ, the flux Φ is practically equal to the flux of the external field through the hole. *The magnetic flux inside the superconducting ring is quantised, and the flux quantum is* Φ_0. Note that this quantisation of the flux is only obtained because of the Meissner effect, and would not hold for a perfectly conducting metal or for a superconductor with thickness less than λ. In the latter case, there is still a quantised quantity for any integration contour, but it is the fluxoid Φ^*, which is not the magnetic flux and is more difficult to measure.[5]

5.3.3 Measuring the Flux Quantum

The quantisation of the flux can be demonstrated experimentally by measuring the flux in a superconducting tin cylinder as a function of the magnetic field in which the tin tube has been cooled. As can be seen from Fig. 5.10, the trapped flux varies in steps, and is a multiple of about 2×10^{-15} weber. This is two times smaller than the value $h/e = 4.14 \times 10^{-15}$ weber expected for an electron.

---

[5] In an almost perfect metal with very low resistivity, the mean free path can be long enough to ensure that the electron wave functions are coherent on a macroscopic scale. Effects associated with this quantum coherence can then be detected, appearing as periodic oscillations in certain macroscopic quantities, such as the magnetic susceptibility or the resistance of the ring. These observations bear no relation to a quantisation of the magnetic flux.

5.3 A Macroscopic Quantum Effect

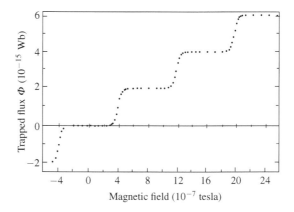

Fig. 5.10 Flux trapped in a superconducting tin cylinder, measured by determining the magnetic field on the cylinder axis. The value of the flux trapped in the cylinder is plotted as a function of the magnetic field in which the tin tube was cooled. From Goodman, W.L. et al.: Phys. Rev. B **4**, 1530 (1971). With the kind permission of the American Physical Society. http://link.aps.org/doi/10.1103/PhysRevB.4.1530

This means that the charge q of the superconducting carriers is not $-e$ but $-2e$. Hence, *the electrons are paired in the superconducting state*. The flux quantum is therefore

$$\Phi_0 = \frac{h}{2e} = 2.07 \times 10^{-15} \text{ weber} . \tag{5.22}$$

In (5.16), (5.17), (5.18), (5.19), and (5.20), we must therefore take

$$q = -2e . \tag{5.23}$$

It is natural to take the mass m of these electron pairs to be

$$m = 2m_0 . \tag{5.24}$$

Finally, comparing the curl of (5.17), viz.,

$$\mathbf{curl\, j} = -\frac{4e^2 \phi^2}{m} \mathbf{B} ,$$

with (5.11), viz.,

$$\mathbf{curl\, j} = -\frac{n_s e^2}{m_0} \mathbf{B} ,$$

it follows that the modulus ϕ^2 of the wave function must be equal to half the density of electron pairs

$$\phi^2 = n_s/2 . \tag{5.25}$$

So by measuring the flux quantum, it is established that, in a superconductor, the carriers are pairs of electrons. This is a common feature of all the electronic superconductivity phenomena so far observed. We shall examine the origin of these electron pairs in more detail in Chap. 7.

Question 5.3. Reconsider the lead superconducting ring of Question 5.1. Suppose it is cooled below T_c in an applied field $B_a = \mu_0 H_a = 3.3$ µtesla. Answer the same questions as in Question 5.1.

Let us note that with (5.13) for λ^2, and (5.23), (5.24), and (5.25), the relation (5.17) between the current and phase of the superconducting wave function can be rewritten as[6]

$$\boxed{\lambda^2 \mu_0 \mathbf{j} = -\frac{\Phi_0}{2\pi} \nabla\theta - \mathbf{A}}. \tag{5.26}$$

5.3.4 Josephson Effect

A further spectacular manifestation of the macroscopic quantum nature of the superconducting state is revealed by the characteristics of the tunneling effect between two superconductors. We have already seen in Chap. 3 how electrons can transfer by quantum tunneling between two metals separated by a thin vacuum layer. An analogous situation can be produced by introducing a thin insulating barrier between two superconductors, as shown in Fig. 5.11a. However, there is nevertheless an important difference, as we shall see, between a tunnel junction between two normal metals and a tunnel junction between two superconductors. As discussed in Chap. 3, a tunneling current will only be set up if there is a potential difference between two metals. However, in the case of two superconductors, the quantum coherence effects

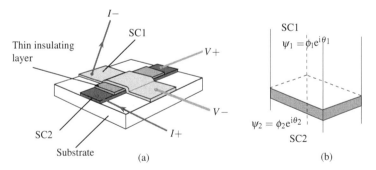

Fig. 5.11 (a) Junction between two superconducting metals (called a Josephson junction). The superconductor and insulator films can be deposited by sputtering and masking. In some cases, the surface of the superconducting metal SC2 can simply be oxidised to form the insulating barrier. (b) Schematic representation of the junction in (a). There is a step in the phase of the wave function between SC1 and SC2

---

[6] Note that, for a connected sample, Eq. (5.26) resumes into

$$\lambda^2 \mu_0 \mathbf{j} = -\mathbf{A}$$

and we directly obtain then an elegant representation of the two London equations, which are recovered by differentiating with respect to time or by taking the curl.

5.3 A Macroscopic Quantum Effect

are such that *a current of paired electrons can pass through an insulating barrier even if the potential difference is zero.*

Below we shall explain how the flow of superconducting pairs is related *to the phase of the macroscopic wave function of the superconducting electrons*. This is called the *Josephson effect*, named after the British physicist who predicted it in 1962 (Nobel prize in 1973). This will only occur in general for very thin insulating barriers, much thinner than those leading to a tunneling current between normal metals.

A schematic view of the junction is given in Fig. 5.11b, where we consider two superconductors (1) and (2), assumed identical to simplify. In this case, each superconductor is described by a macroscopic wave function

$$\psi_1 = \left(\frac{n_{s1}}{2}\right)^{1/2} e^{i\theta_1}, \qquad \psi_2 = \left(\frac{n_{s2}}{2}\right)^{1/2} e^{i\theta_2}, \qquad (5.27)$$

where n_{s1} and n_{s2} are the superelectron densities in the two superconductors.

If the insulating barrier is thin enough for the electron pairs to cross it by tunnelling, there will be a coupling between the two superconductors. This tunneling effect can be described by an effective coupling K between the two superconductors, whose value will depend on the physical characteristics of the barrier. In particular, it will decrease exponentially with the thickness. Suppose also that a constant potential difference V is applied across the two superconductors. The Schrödinger equations for the two wave functions ψ_1 and ψ_2 characterising the two superconductors can be written in the form

$$\left. \begin{array}{l} i\hbar \dfrac{\partial \psi_1}{\partial t} = E_0 \psi_1 + K \psi_2, \\[6pt] i\hbar \dfrac{\partial \psi_2}{\partial t} = (E_0 - 2eV) \psi_2 + K \psi_1, \end{array} \right\} \qquad (5.28)$$

where E_0 is the energy per electron pair of the ground state of each of the superconductors when there is no coupling and no electric potential V. The energy of superconductor (2) is $E_0 - 2eV$, since the effective charge of the superconducting carriers is $-2e$. For $K = 0$, we do indeed obtain the equations for the two isolated superconductors. The terms in $K\psi_2$ and $K\psi_1$ describe the coupling between the two superconductors due to the fact that the wave function associated with one superconductor does not suddenly drop to zero in the other.

Using (5.27), we have

$$\frac{\partial \psi_1}{\partial t} = \left(\frac{1}{2n_{s1}} \frac{dn_{s1}}{dt} + i \frac{d\theta_1}{dt}\right) \psi_1. \qquad (5.29)$$

Substituting (5.29) and the analogous relation for $\partial \psi_2/\partial t$ into (5.28), we obtain

$$\begin{array}{l} i\hbar \left(i\dfrac{d\theta_1}{dt} + \dfrac{1}{2n_{s1}} \dfrac{dn_{s1}}{dt} \right) - E_0 = K \left(\dfrac{n_{s2}}{n_{s1}}\right)^{1/2} e^{i(\theta_2 - \theta_1)}, \\[10pt] i\hbar \left(i\dfrac{d\theta_2}{dt} + \dfrac{1}{2n_{s2}} \dfrac{dn_{s2}}{dt} \right) - E_0 + 2eV = K \left(\dfrac{n_{s1}}{n_{s2}}\right)^{1/2} e^{i(\theta_1 - \theta_2)}. \end{array} \qquad (5.30)$$

The imaginary parts of these equations can be written

$$\frac{dn_{s1}}{dt} = \frac{2K}{\hbar}(n_{s1}n_{s2})^{1/2}\sin(\theta_2 - \theta_1) = -\frac{dn_{s2}}{dt}. \tag{5.31}$$

The current density

$$J = -e\left[\left(\frac{dn_{s2}}{dt}\right) - \left(\frac{dn_{s1}}{dt}\right)\right]$$

in the junction can thus be written

$$\boxed{J = J_c \sin(\theta_2 - \theta_1)}, \tag{5.32}$$

setting

$$J_c = \frac{4eK}{\hbar}(n_{s1}n_{s2})^{1/2}.$$

The real parts of the equations in (5.30) are

$$-\hbar\frac{d\theta_1}{dt} - E_0 = K\left(\frac{n_{s2}}{n_{s1}}\right)^{1/2}\cos(\theta_2 - \theta_1),$$
$$-\hbar\frac{d\theta_2}{dt} - E_0 + 2eV = K\left(\frac{n_{s1}}{n_{s2}}\right)^{1/2}\cos(\theta_2 - \theta_1). \tag{5.33}$$

If the two superconductors of the junction are identical $n_{s1} = n_{s2}$, the right-hand sides of (5.33) are equal, and equating the left-hand sides, we obtain

$$\boxed{\frac{d(\theta_2 - \theta_1)}{dt} = \frac{2eV}{\hbar}}. \tag{5.34}$$

The two equations (5.32) and (5.34) relating the current and the voltage across the junction are the two Josephson equations.

a. Current–Phase Relation

The first relation (5.32), known as the current–phase relation, tells us that the current crossing the insulating barrier is directly related to the phase difference $\theta_2 - \theta_1$ between the wave functions of the two superconductors. It shows that *a current can cross the barrier when there is no applied voltage.*

For $V = 0$, according to the second Josephson relation (5.34), the phase difference and hence the current J are time independent. The existence of a finite current through the junction can be understood by analogy with (5.17), which shows that a gradient $\nabla\theta$ of the phase of the superconducting wave function always implies a current of electron pairs. Here the phase does not vary continuously, but suddenly in going from superconductor (1) to superconductor (2). In a certain sense, (5.32) is like a generalisation of (5.17) (with zero magnetic field here).

5.3 A Macroscopic Quantum Effect

In practice, a current is injected through the junction by means of a current source. In this case, the phase difference $\theta_2 - \theta_1$ between the superconductors adjusts in such a way as to satisfy (5.32), which is only possible for an injected current less than the maximal value J_c, called the *critical current* of the junction. It is proportional to the coupling coefficient K and depends sensitively on the characteristics of the junction.

b. Voltage–Phase Relation

When V is nonzero, the second Josephson relation (5.34), called the voltage–phase relation, shows that if a constant, finite voltage V is applied to the junction, it will induce a linear time dependence of the phase difference, which leads, according to (5.32), to an alternating current. So a constant voltage will generate an alternating current with frequency ν_J proportional to V:

$$\nu_J = \frac{2e}{\hbar} V. \tag{5.35}$$

The full set of physical consequences that follow from the two relations (5.27) and (5.28) are commonly referred to as the *Josephson effect*.[7] This effect, which is a direct consequence of the existence of a macroscopic wave function, has important applications. It has provided a very precise way of measuring the ratio e/h. Conversely, using (5.35), a precision oscillator can be made by applying a well calibrated voltage, or a voltage can be determined with unequalled accuracy by measuring the frequency ν_J. Indeed, the Josephson junction is used today to specify the voltage standard.

5.3.5 SQUID

The quantisation of flux inside a superconducting loop, as discussed in Sect. 5.3.3, is a novel manifestation of superconductivity that can be put to use. Indeed, if a superconducting ring is subjected to a magnetic field B, a very accurate measurement of B can be made by determining the quantum number corresponding to the number of flux quanta inside the ring. Recall that, for a ring of diameter 10 μm, the flux Φ_0 is just the flux of the Earth's magnetic field, i.e., about 0.2 gauss or 2×10^{-5} tesla. However, it is difficult to change this number of quanta directly by changing the field applied to a simple superconducting loop, and nor is it easy to measure the number of flux quanta.

On the other hand, if two Josephson junctions are inserted into the superconducting ring, as shown in Fig. 5.12, the flux in the loop is no longer quantised. Indeed, due to the insulating barriers, no continuous contour can now be defined in the superconductor.

---

[7] Some features can be understood by attempting Problem 11: *Direct and Alternating Josephson Effects in Zero Magnetic Field*.

Fig. 5.12 Interferometer with two Josephson junctions, as used in the DC-SQUID, which runs on a direct current I

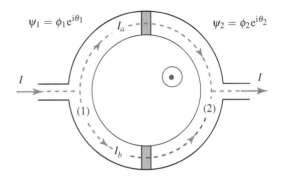

The Josephson junctions force a current of superconducting pairs to move around the loop in the absence of any applied voltage. By applying a magnetic field through this closed circuit, we may act on the phase difference $\theta_2 - \theta_1$ between the two superconducting circuits on either side of the junctions. In this way, we set up an interferometry experiment between the currents in the two branches of the loop, whose phase relation depends on the applied magnetic field B.

> **Question 5.4.** Assuming that the two Josephson junctions are the same, show that the maximum total current $I_{\max}$ that can go around the loop varies sinusoidally with the applied magnetic field, where
>
> $$I_{\max} = I_0 \cos\left(\pi \left|\frac{\Phi}{\Phi_0}\right|\right).$$
>
> This is what is observed experimentally, as shown in Fig. 5.13. Can you explain why the oscillations disappear beyond a certain value of the applied field?

This kind of interferometer can be used to measure the magnetic flux Φ through the loop with a much higher resolution than the flux quantum. As the latter is very small, experimental devices developed on the basis of Josephson junctions, called SQUIDs (superconducting quantum interference device) were used to make highly sensitive magnetometers in the 1980s. These were commercialised in the 1990s, resulting in the large scale dissemination of these magnetic measurement techniques. Several applications will be discussed in Chap. 10.

Fig. 5.13 Current through an interferometer containing two Josephson junctions as a function of the applied field. From Jaklevic, R.C., Lambe, J., Mercereau, J.E., Silver, A.H.: Phys. Rev. A **140**, 1628 (1965). With the kind permission of the American Physical Society (APS). http://link.asp.org/doi/10.1103/PhysRev.140.A1628

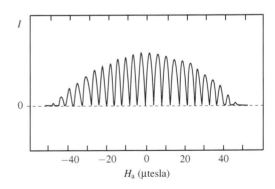

5.4 Summary

Superconductivity is a state of the electron gas of a metal with zero resistivity, manifesting itself through the presence of persistent currents. This state occurs below a critical temperature T_c, but only for an applied magnetic field smaller than a certain critical field H_c. The temperature dependence of the critical field is approximately

$$H_c(T/T_c) = H_c(0)\left[1 - \left(\frac{T}{T_c}\right)^2\right].$$

The fact that there is a critical field H_c also implies a bound on the current that can pass through the superconductor (the critical current).

A superconductor is not simply a perfect conductor. The difference is manifested via the Meissner effect, which shows that the magnetic induction is zero in the bulk of a superconducting material, which is therefore a perfect diamagnetic material. In a perfect metal, the magnetic induction is uniform, but it is determined by the magnetothermal history of the material. In a superconductor, on the other hand, the zero magnetic induction is reversible, and corresponds to a specific thermodynamic state of the electron gas.

The phenomenological London equations explain the Meissner effect, and show that the state of zero magnetic induction results, in the presence of an applied field $H_a < H_c$, from the occurrence of persistent currents at the surface of the superconductor. The latter circulate in a surface layer of thickness λ, called the penetration depth. This is given by

$$\lambda^2 = \frac{m_0}{n_s e^2 \mu_0}.$$

In this expression, the experimental quantity n_s is the number of superconducting electrons per unit volume. It vanishes at the superconductor–metal transition.

The superconducting state is a quantum state of the electron gas whose coherence is preserved at the macroscopic scale. The probability density of the superconducting carriers is given by a wave function

$$\psi(\mathbf{r}) = \phi(\mathbf{r})\exp i\theta(\mathbf{r}),$$

where the phase factor $\theta(\mathbf{r})$ plays an important role, because it determines the persistent currents through

$$\mathbf{j} = \frac{q\phi^2}{m}(\hbar\nabla\theta - q\mathbf{A}).$$

The macroscopic quantum nature of this system manifests itself through the quantisation of the magnetic flux through a macroscopic superconducting loop. The flux through a superconducting loop is quantised with

$$\Phi = n\Phi_0,$$

where the measured flux quantum is

$$\Phi_0 = h/2e .$$

This shows that the current carriers in a superconductor are pairs of electrons and allows one to rewrite the relation between the current and phase of the wave function as

$$\lambda^2 \mu_0 \mathbf{j} = -\frac{\Phi_0}{2\pi} \nabla\theta - \mathbf{A} .$$

Another manifestation of the macroscopic quantum nature of the wave function in the superconducting state is the Josephson effect, observed when we consider the tunneling current of electron pairs through a thin insulating layer between two superconductors. The Josephson equations

$$J = J_c \sin(\theta_2 - \theta_1) , \qquad \frac{d(\theta_2 - \theta_1)}{dt} = \frac{2eV}{\hbar} ,$$

relate this current and the voltage across the junction to the phase difference between the wave functions of the two superconductors on either side of the junction.

Josephson junctions are used to make SQUIDs, quantum interferometers that can measure the magnetic flux through a macroscopic loop to much higher accuracy than the flux quantum.

5.5 Answers to Questions

Question 5.1

The applied field is less than the critical field H_c, so the cylinder is superconducting. The magnetic flux through the tube is held at its initial value (see Fig. 5.14a). The value of n_s is used to determine the London penetration depth from

$$\lambda^2 = \frac{m_0}{n_s e^2 \mu_0} .$$

This leads to $\lambda = 376$ Å. Since the thickness of the tube is $r_e - r_i = 10$ μm $\gg \lambda$, the Meissner effect cancels the magnetic induction in the region $r_i < r < r_e$, apart from the regions of thickness λ next to the surface (see Fig. 5.14b).

Currents are localised in the ring-shaped regions of thickness λ at the surface of the tube. The current at the outer surface creates a field within the tube which opposes the external field. On the inner surface of the tube, the currents are opposite, and tend to maintain the flux inside the cylinder. To carry out a detailed calculation, we must solve $\mathbf{B} - \lambda^2 \Delta \mathbf{B} = 0$ in cylindrical coordinates and then determine the currents using $\mu_0 \mathbf{j} = \text{curl } \mathbf{B}$

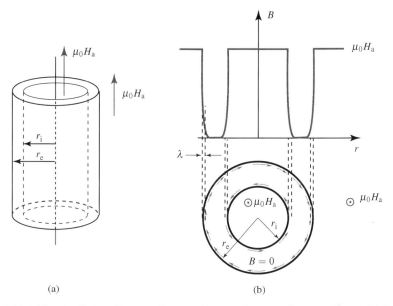

Fig. 5.14 (a) Lead cylinder of inner radius $r_i = 10$ μm and outer radius $r_o = 20$ μm. (b) Cross-section showing magnetic induction and currents when the ring is cooled in an applied field of 3.3 mtesla

Question 5.2

1. The London equation

$$\lambda^2 \mu_0 \frac{d\mathbf{j}}{dt} = \mathscr{E}$$

implies

$$\mathbf{j}(\omega) = \frac{\mathscr{E}(\omega)}{i\omega\lambda^2\mu_0},$$

and then

$$\sigma_s(\omega) = -\frac{i}{\omega\lambda^2\mu_0},$$

so

$$\sigma_{2s}(\omega) = \frac{1}{\lambda^2\omega\mu_0} = \frac{n_s(T)e^2}{m\omega}.$$

2. For a superconductor, $\sigma_{1s}(\omega)$ is infinite at zero frequency. For a perfect conductor, the conductivity $\sigma_1(\omega)$ can be obtained by letting τ_e tend to ∞. The function in Fig. 4.16 thus has a width that tends to zero. This implies that $\sigma_{1s}(\omega)$ is a delta function $\delta(\omega)$. Note that the area under $\sigma_1(\omega)$ is independent of τ_e:

$$\int_0^\infty \sigma_1(\omega)d\omega = \frac{ne^2}{m}\int_0^\infty \frac{\tau_e}{1+\omega^2\tau_e^2}d\omega = \frac{\pi}{2}\frac{ne^2}{m},$$

so that
$$\sigma_{1s}(\omega) = \frac{\pi}{2}\frac{n_s(T)e^2}{m}\delta(\omega).$$

Therefore,

$$\boxed{\sigma_s(\omega) = \frac{n_s(T)e^2}{m}\left[\frac{\pi}{2}\delta(\omega) - \frac{i}{\omega}\right], \quad \sigma_n(\omega) = \frac{n_e(T)e^2}{m}\frac{\tau_e}{1+\omega^2\tau_e^2}(1-i\omega\tau_e).}$$

3. We have
$$|j_n| = \frac{n_e e^2}{m}\frac{\tau_e}{\sqrt{1+\omega^2\tau_e^2}}|\mathscr{E}|,$$

and, for $\omega > 0$,
$$|j_s| = \frac{n_s e^2}{m\omega}|\mathscr{E}|.$$

Hence,
$$\left|\frac{j_n}{j_s}\right| = \frac{n_e}{n_s}\frac{\tau_e\omega}{\sqrt{1+\omega^2\tau_e^2}}.$$

Letting $X = n_s(T)/n_e(T)$, we will have $|j_n/j_s| = 1$ for $\omega_c^2\tau_e^2 = X^2/(1-X^2)$. From the T dependence of the penetration depth λ given in the two fluid model by (5.14), we obtain an estimate

$$X = n_s/n_e = 1 - (T/T_c)^4.$$

For a temperature $T = 0.9T_c$, this leads to $\omega_c\tau_e \sim 0.36$, which corresponds to a frequency $\nu_c \sim 0.6 \; 10^{11}$ Hz for $\tau_e = 10^{-12}$ s. One really needs to reach such high frequencies even close to T_c to feel the dissipation of the 'normal' electron current.

4. For $T \to 0$, there will be no dissipation whatever the frequency, according to this simple two-fluid model. This is hardly possible, because the electronic system has excited states. The two-fluid model does not take these excitations into account. The BCS calculation predicts that there will be a gap in the excitations and absorption will occur for $\hbar\omega > 2\Delta$.

Question 5.3

The flux $B_a = \mu_0 H_a = 3.3$ µtesla inside the cylinder is

$$\Phi = B_a(\pi r_i^2) = 1.03 \times 10^{-15} \text{ weber/m}^2.$$

This is close to half the flux quantum Φ_0. Consequently, since the flux inside the tube is quantised, we will only be able to trap a flux $\Phi = 0$ or $\Phi = \Phi_0$ in the tube.

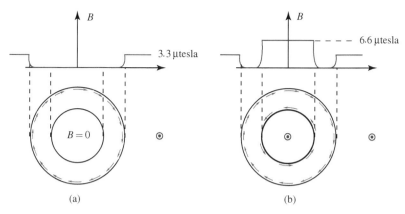

Fig. 5.15 Superconducting ring cooled with an applied axial field of 3.3 µtesla. (**a**) $B \equiv 0$ for $0 < r < r_e$. (**b**) $B = 6.6$ µtesla for $0 < r < r_i$

The field between 0 and r_i will thus be either 0 or 6.6 µtesla, with a probability close to 1/2 depending on the experiment (see Fig. 5.15).

Inside the cylinder walls, the field obviously remains zero, except in the surface regions of thickness λ, where it must match the value outside the superconducting walls. The current at the outer surface produces a field inside the tube which always opposes the external field. For the case shown in Fig. 5.15a, where $B \equiv 0$ for $0 < r < r_e$, there is no current on the inner face of the tube. For the case shown in Fig. 5.15b, where $B = 6.6$ µtesla for $0 < r < r_i$, the current flows in the same direction as in Fig. 5.14b.

Note that the current on the outer face of the tube is 1,000 times smaller than for the case illustrated in Fig. 5.14b. However, the current on the inner face of the tube for the case shown in Fig. 5.15b is 500 times smaller than for the case illustrated in Fig. 5.14b.

Question 5.4

If θ_1, θ_2, and θ'_1, θ'_2 are the phases of the wave functions of superconductors 1 and 2 on either side of the two junctions, as shown in Fig. 5.16, while I_0 is the critical current of the two junctions, assumed identical, we have

$$I = I_0 \sin(\theta_2 - \theta_1) + I_0 \sin(\theta'_2 - \theta'_1). \tag{5.36}$$

In the absence of an applied field, the phase of the superconducting wave functions is uniform on both superconducting sections (1) and (2), so that $\theta_1 = \theta'_1$ and $\theta_2 = \theta'_2$ obviously yield $I = 2I_0 \sin(\theta_2 - \theta_1)$.

For $H_a \neq 0$, the relation (5.26) between current and phase applies and can be integrated within the superconductor along the contour C shown in Fig. 5.16 to yield

Fig. 5.16 Interferometer with two Josephson junctions in a DC-SQUID, which operates with a direct current I. Here the sides of the upper junction are labelled 1 and 2, and those of the lower junction 1′ and 2′. The contour C does not include the junctions

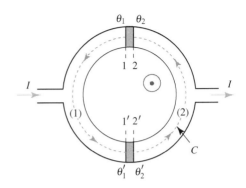

$$\mu_0 \lambda^2 \oint_C \mathbf{j} \cdot d\mathbf{l} + \oint_1^{1'} \mathbf{A} \cdot d\mathbf{l} + \oint_{2'}^{2} \mathbf{A} \cdot d\mathbf{l} = -\frac{\Phi_0}{2\pi} \oint_C \nabla\theta \cdot d\mathbf{l} \quad (5.37)$$

$$= \frac{\Phi_0}{2\pi}(\theta_1 - \theta_1' + \theta_2' - \theta_2) \quad (+n\Phi_0)$$

The first integral vanishes because $\mathbf{j}$ is zero in the bulk of the superconductor. The other two integrals on the left-hand side represent the flux Φ through the loop if we neglect the circulation of $\mathbf{A}$ within the insulating barriers. We thus have

$$\Phi = \frac{\Phi_0}{2\pi}(\theta_1 - \theta_1' + \theta_2' - \theta_2) \quad (+n\Phi_0), \quad (5.38)$$

or

$$\theta_2' - \theta_1' = \theta_2 - \theta_1 + 2\pi\frac{\Phi}{\Phi_0} \quad (+2\pi n),$$

and, setting $\delta = \theta_2 - \theta_1$, (5.36) gives

$$I = I_0 \sin\delta + I_0 \sin\left(\delta + 2\pi\frac{\Phi}{\Phi_0}\right),$$

$$I = 2I_0 \cos\pi\frac{\Phi}{\Phi_0} \sin\left(\delta + \pi\frac{\Phi}{\Phi_0}\right). \quad (5.39)$$

The total critical current $I_{\max}$ thus varies periodically with the flux in the loop:

$$\boxed{I_{\max} = 2I_0 \cos\left(\pi\frac{\Phi}{\Phi_0}\right).}$$

This current clearly results from interference between the currents through the two junctions, which is why this DC-SQUID is called a two-junction interferometer. Such a setup allows a precise determination of the number of flux quanta through the loop.

The experimental results shown in Fig. 5.13 do indeed reveal such a periodicity. Note, however, that the current in the interferometer is not strictly periodic. This

5.5 Answers to Questions

happens because we have considered ideal junctions here, with a spatially uniform phase difference on the junction. This would only be realistic for a junction with a very small cross-sectional area, or for a negligible flux of H_a through the junction. The geometry of the junctions is responsible for the decrease in I_{max} when Φ increases.

For a single junction of finite cross-section, the non-uniformity of the phase difference across the surface of the junction produces interference between the currents at the center and at the edges of the junction.[8] The critical current then depends sinusoidally on the flux in the junction.

Note: On the left-hand side of (5.37), we neglected the integrals over the part of the circuit through the two junctions. In fact the expression (5.26) used for the current versus the phase difference is not strictly accurate in the presence of a magnetic field, as it is not gauge invariant. Strictly speaking, we must replace $\theta_2 - \theta_1 = \delta$ in (5.36) by

$$\delta = \theta_2 - \theta_1 - \frac{2\pi}{\Phi_0} \oint_1^2 \mathbf{A} \cdot \mathbf{dl},$$

to obtain a gauge invariant expression for the current.[8]

It can be checked that, on the right-hand side of (5.37), the sum of the phases transforms to

$$\delta' - \frac{2\pi}{\Phi_0} \oint_{1'}^{2'} \mathbf{A} \cdot \mathbf{dl} - \delta - \frac{2\pi}{\Phi_0} \oint_2^1 \mathbf{A} \cdot \mathbf{dl} \quad (+2\pi n).$$

We then recover precisely those terms neglected in the integral over the interferometer loop, so that

$$\delta' = \delta + 2\pi \frac{\Phi}{\Phi_0},$$

which is a rigorous proof of the result in (5.38).

---

[8] See Problem 12: *Josephson Junction in a Magnetic Field*.

Chapter 6
Thermodynamics of Superconductors

Contents

6.1 Thermodynamics of Bulk Superconductors 177
    6.1.1 Free Energy of the Superconducting State 178
    6.1.2 Entropy and Specific Heat 179
6.2 Thin Films and Coherence Length 181
    6.2.1 Thermodynamics of Thin Films... 181
    6.2.2 Coherence Length ξ 184
    6.2.3 Mixed State: A Simple Model 185
6.3 Two Types of Superconductivity 188
    6.3.1 Type I Superconductors 189
    6.3.2 Simple London Model for a Vortex 189
    6.3.3 Type II Superconductivity 191
    6.3.4 Applications of Type II Superconductors 194
6.4 Summary 195
6.5 Answers to Questions 196

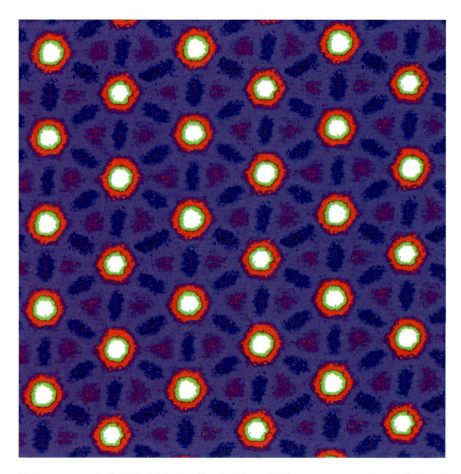

Abrikosov vortex lattice. The Meissner effect is only perfect in some superconductors. In the so-called type II superconductors, with a high applied field, the thermodynamic equilibrium state is one in which normal metallic regions appear within the superconducting material. Tubes of normal material are arranged in a hexagonal lattice, called an Abrikosov lattice, named after its discoverer (Nobel prize in 2003). The exceptional image above was obtained by scanning tunneling microscopy (STM). Image courtesy of Davis, J.C. and coworkers, Cornell University, Ithaca, NY. Scale 3,000 Å×3,000 Å

While the historical discovery of superconductivity sprang from the observation of a zero resistance state, we have seen that the most novel manifestation of this phenomenon is the Meissner effect. The existence of a diamagnetic transition is today the main experimental signature sought to establish the presence of superconductivity in new materials. In 1986, it was indeed the observation by Bednorz and Müller of the Meissner effect in $La_{2-x}Ba_xCuO_4$, after the original observation of a drop in the resistance at temperatures below 35 K, which was taken as incontrovertible proof of the superconductivity of this compound. This observation, rewarded by the Nobel prize in 1988, initiated the study of a whole range of similar copper oxides which make up the family of high-T_c cuprate superconductors.

The fact that a superconductor excludes a magnetic field has many physical consequences. This fact alone implies the existence of persistent currents, and leads to the understanding that superconductivity is essentially a magnetic phenomenon, at thermodynamic equilibrium, due to the condensation of the electron gas in a novel macroscopic quantum state at low temperatures. This state must therefore have a *lower free energy than the electron gas in a normal metal*.

In Sect. 6.1, we discuss the first thermodynamic consequences of this observation, and in particular *the existence of a critical field* for superconductivity. The experimental investigation of the free energy in the presence of an applied field will help us to identify certain physical quantities which characterise the superconducting state. In Sect. 6.2, we show how the thermodynamic properties of thin superconducting films lead to the notion of a coherence length for superconducting pairs and the so-called *mixed state*. A brief description of type I and type II superconductors and vortices is undertaken in Sect. 6.3.

6.1 Thermodynamics of Bulk Superconductors

In the presence of a magnetic field, the free energy of a material is lowered if it is paramagnetic and raised if it is diamagnetic: *the exclusion of a magnetic field by a superconductor corresponds to an increase in its free energy*. It follows immediately that the Meissner effect cannot hold out indefinitely if the field is increased because, for a given value of the field, the increase in free energy due to exclusion of the field will compensate for the drop in free energy associated with condensation of the electron gas in the superconducting state. The free energy of the normal state, which hardly depends on the magnetic field, thus becomes lower for a strong enough applied field. The very existence of the Meissner effect therefore implies that the electron gas must return to its normal state above some thermodynamically critical value H_c. We thus see that, by investigating the temperature dependence of H_c, we will obtain valuable insight into the condensation energy in the superconducting state.

6.1.1 Free Energy of the Superconducting State

In a magnetic system, the thermodynamic potential can be taken as the free enthalpy, or Gibbs free energy [1]:

$$G = U - TS + PV - \mu_0 \mathbf{H}_a \cdot \mathbf{M}, \quad (6.1)$$

where U is the internal energy. The intensive thermodynamic quantities T, $-P$, and $\mu_0 \mathbf{H}_a$ are the conjugate variables with respect to G of the extensive variables S, V, and $\mathbf{M}$, respectively. So for constant T and P, we have

$$dG = -\mu_0 \mathbf{M} \cdot d\mathbf{H}_a, \quad (6.2)$$

in such a way that the dependence of the free energy $G_s(T, H_a)$ of the superconductor on the applied field is given by the area under the magnetisation curve (see Fig. 6.1):

$$G_s(T, H_a) - G_s(T, 0) = \int_0^{H_a} -\mu_0 \mathbf{M} \cdot d\mathbf{H}_a = \int_0^{H_a} \mu_0 H_a dH_a, \quad (6.3)$$

$$G_s(T, H_a) - G_s(T, 0) = \frac{\mu_0}{2} H_a^2. \quad (6.4)$$

Likewise, the change in free energy related to the Pauli paramagnetism of the electrons in a metal ($M = \tilde{\chi}_P H_a$) is given by $G_n(T, H_a) - G_n(T, 0) = -\mu_0 \tilde{\chi}_P H_a^2/2$. When the applied field is equal to the critical thermodynamic field H_c, the normal and superconducting phases have the same free energy, viz.,

$$G_s(T, H_c) = G_s(T, 0) + \frac{\mu_0}{2} H_c^2 = G_n(T, 0) - \frac{\mu_0}{2} \tilde{\chi}_P H_c^2 = G_n(T, H_c).$$

The difference in free energy between the superconducting and normal states is therefore

$$G_s(T, 0) - G_n(T, 0) = -\frac{\mu_0}{2}(1 + \tilde{\chi}_P) H_c^2.$$

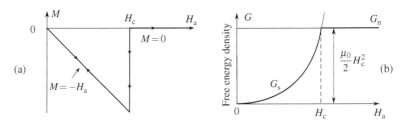

Fig. 6.1 (a) Magnetisation of a superconductor as a function of H_a, exhibiting the Meissner effect. (b) Free energy density of a superconductor and the normal metal as a function of H_a

6.1 Thermodynamics of Bulk Superconductors

Since $\tilde{\chi}_P \ll 1$ for the normal metal, the energy of condensation of the electron gas in the superconducting state is given to a good approximation by

$$G_s(T,0) - G_n(T,0) = -\frac{\mu_0}{2}H_c^2. \tag{6.5}$$

Hence if we neglect the change in the free energy of the normal state with field, we can express the free energy of the superconducting state in the form

$$\boxed{G_s(T,H_a) - G_n(T,0) = -\frac{\mu_0}{2}\left(H_c^2 - H_a^2\right).} \tag{6.6}$$

This quadratic dependence on the field is shown in Fig. 6.1.

6.1.2 Entropy and Specific Heat

Given the free energy and the temperature dependence of H_c, we may immediately deduce the following expression for the entropy:

$$S_s = -\frac{dG_s}{dT} = S_n + \mu_0 H_c \frac{dH_c}{dT}, \tag{6.7}$$

which implies that the entropy of the superconducting state is less than that of the normal state, because H_c decreases with T. We do therefore find that *the superconducting state is more ordered than the normal state*. This entropy difference suggests that there is a latent heat $L = T(S_n - S_s)$ at the normal/superconducting transition in the presence of a magnetic field. This is a first order thermodynamic transition. On the other hand, experiment shows that H_c goes to zero at T_c with a finite slope for $H_c(T)$ (see Fig. 5.3), and according to (6.7), this means that $S_s = S_n$ at T_c. There is no latent heat, and the transition is second order in zero field. However, there is a discontinuity in dS/dT and hence also in the specific heat $C_s = TdS_s/dT$:

$$\frac{C_s - C_n}{T} = \mu_0 \frac{d(H_c dH_c/dT)}{dT} = \mu_0 \left[\left(\frac{dH_c}{dT}\right)^2 + H_c \frac{d^2 H_c}{dT^2}\right]. \tag{6.8}$$

For $T = T_c$, the critical field H_c vanishes, and it follows that the discontinuity in C_n at T_c is given by

$$C_s(T_c) - C_n(T_c) = \mu_0 T_c \left(\frac{dH_c}{dT}\right)^2_{T=T_c}.$$

It is worth pausing for a moment to consider the experimental measurements of the specific heat, since they can be achieved to much higher accuracy than measurements of $H_c(T)$, and according to (6.8), provide a handle on the temperature derivative of this quantity. Suppose in particular that we accept the quadratic form of $H_c(T)$ given by (5.2), viz.,

$$H_c(T) = H_c(0)\left[1 - \left(\frac{T}{T_c}\right)^2\right]. \tag{6.9}$$

By (6.7), it is easy to show that

$$S_s - S_n = -2\mu_0 \left[\frac{H_c(0)}{T_c}\right]^2 \left[1 - \left(\frac{T}{T_c}\right)^2\right] T,$$

and hence,

$$\frac{C_s - C_n}{T} = -2\mu_0 \left[\frac{H_c(0)}{T_c}\right]^2 \left[1 - 3\left(\frac{T}{T_c}\right)^2\right].$$

Experiment shows that C_s/T vanishes when $T \to 0$, which implies a relation between the characteristic quantities of the normal and superconducting states[1]:

$$\frac{C_n}{T} = 2\mu_0 \left[\frac{H_c(0)}{T_c}\right]^2 = \gamma.$$

The quadratic form (6.9) for $H_c(T)$ can thus be used to deduce the following relation between the specific heats of superconductor and normal metal:

$$\frac{C_s}{T} = 3\gamma \left(\frac{T}{T_c}\right)^2. \tag{6.10}$$

The discontinuity in the specific heat at the transition would therefore be given by $(C_s - C_n)/T_c = 2\gamma$.

The expected changes in $S/\gamma T_c$ and $C_s/\gamma T_c$ are shown in Fig. 6.2. Experiment shows that these results are not quite right. For example, in Fig. 6.3, it is observed that the discontinuity in C/T at the transition is slightly less than 2γ. Moreover, the linear dependence on $(T/T_c)^2$ is not corroborated. This suggests that the form (6.9) for $H_c(T)$ is not such a good approximation. Paradoxically, the most significant physical difference regarding this discrepancy does not result from the observed difference at T_c, but rather from the low temperature behaviour, which is exponential. We shall understand the full importance of this experimental result in Chap. 7. Indeed, it led to the suggestion that there might be a band gap in the spectrum of energy states of a superconductor.

> **Question 6.1.** Explain the temperature dependence of the specific heat observed in the normal state.

---

[1] As discussed in Chap. 3, in a normal metal, the entropy S_n and specific heat C_n due to the electrons are given by $S_n = \gamma T = C_n$, where $\gamma \propto D_\Omega(E_F)$.

6.2 Thin Films and Coherence Length

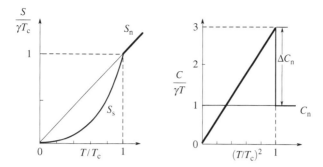

Fig. 6.2 Entropy and specific heat as a function T/T_c for the normal metal and superconducting state, obtained using (6.9) for $H_c(T)$

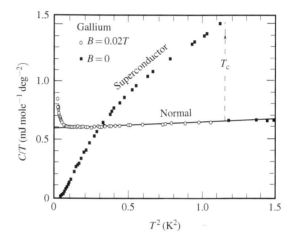

Fig. 6.3 Measurements of C/T as a function of T^2, carried out on metallic gallium in the superconducting state and in the normal state restored by applying a field $H_a > H_c$. The dependence on T^2 expected on the basis of the approximate quadratic form for $H_c(T)$ is not corroborated. From [7, p. 365], 8th edn., Wiley © 2005

6.2 Thin Films and Coherence Length

So far we have only discussed bulk superconductors, and we have focused on the characteristic behaviour in the bulk of such materials, without worrying too much about the surface region in which the persistent currents are to be found. But the experimental understanding of superconductivity suggests that we should look more closely at thin samples, in particular to investigate the penetration depth and its temperature dependence. To get some idea of the kind of calculations involved, it will be useful to consider thin films.

6.2.1 Thermodynamics of Thin Films

Consider for example a film of thickness $2a$ perpendicular to the x axis, subjected to an applied field H_a in the z direction. By symmetry, the field within the superconductor only has a component in the z direction, and the London equations imply $\lambda^2 d^2 B/dx^2 = B(x)$, according to (5.8).

The general solution of this differential equation is

$$B(x) = B_1 \exp\left(-\frac{x}{\lambda}\right) + B_2 \exp\left(\frac{x}{\lambda}\right).$$

With the boundary conditions $B = \mu_0 H_a$ at $+a$ and $-a$, we obtain

$$B(x) = \mu_0 H_a \frac{\cosh(x/\lambda)}{\cosh(a/\lambda)}, \qquad (6.11)$$

and if $a \gg \lambda$, this does indeed correspond to a profile in which the field decreases exponentially over a length λ on the two faces of the film and vanishes at the center (see Fig. 6.4).

Note that, in a film, the surface currents can be determined from $\mu_0 \mathbf{j} = \text{curl } \mathbf{B}$, whence they must be oriented in the y direction, with

$$\mu_0 j_y = -\frac{dB}{dx} = -\frac{\mu_0 H_a}{\lambda} \frac{\sinh(x/\lambda)}{\cosh(a/\lambda)}. \qquad (6.12)$$

This profile is shown in Fig. 6.5.

In order to measure the penetration depth λ, it is useful to reduce a/λ as far as possible. The important point here is that, if $a \approx \lambda$, the diamagnetism is no longer perfect, as it would have been in a bulk superconductor. Naturally, it would be nice to be able to measure the spatial variation of B across the film directly. But this is no easy matter, even with the techniques and instruments of modern physics.

As we shall see in Chap. 10, one quantity that can be simply measured is the magnetisation of the film, which is the spatial average of $(1/\mu_0)B(x) - H_a$, i.e.,

$$M = \int_{-a}^{a} \left[\frac{B(x)}{\mu_0} - H_a\right] \frac{dx}{2a},$$

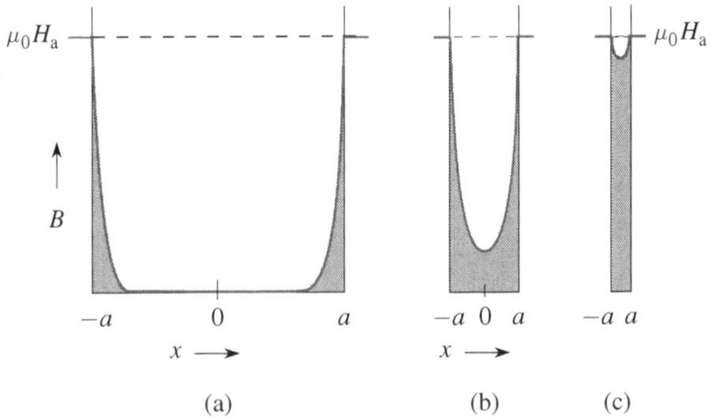

Fig. 6.4 Penetration of a field (**a**) in a thick layer $a > \lambda$, (**b**) in a thin film $a \sim \lambda$, (**c**) and in an ultrathin film $a < \lambda$

6.2 Thin Films and Coherence Length

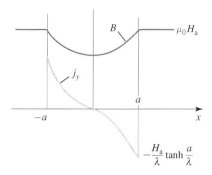

Fig. 6.5 Magnetic induction and surface current profiles in the y direction for a thin film

and using (6.11), it is easy to check that

$$M = H_a \left(\frac{\lambda}{a} \tanh \frac{a}{\lambda} - 1 \right). \tag{6.13}$$

If $a \gg \lambda$, we do indeed recover $M = -H_a$. By measuring M as a function of the film thickness a, we can thus determine λ. Knowing the material making up the film, the free energy of condensation $G_s(T,0)$ in the superconducting state is the same as the one measured for bulk samples of this metal, expressed by (6.5) as a function of the critical field H_c. On the other hand, if $a \sim \lambda$, the free energy $G_{sf}(T,H_a)$ of the film increases less than it would in a bulk sample in the presence of a field H_a, since the field is barely excluded now. As in (6.3), we clearly have

$$G_{sf}(T,H_a) - G_s(T,0) = \int_0^{H_a} -\mu_0 M \cdot dH_a = \frac{\mu_0 H_a^2}{2} \left(1 - \frac{\lambda}{a} \tanh \frac{a}{\lambda}\right).$$

Hence, substituting in the expression (6.5) for $G_s(T,0)$,

$$G_{sf}(T,H_a) - G_n(T,0) = \frac{\mu_0 H_a^2}{2} \left(1 - \frac{\lambda}{a} \tanh \frac{a}{\lambda}\right) - \frac{\mu_0 H_c^2}{2}. \tag{6.14}$$

The free energy of the superconducting film will only equal that of the normal state when the right-hand side of this equation is zero, i.e., when H'_c satisfies

$$H'^2_c \left(1 - \frac{\lambda}{a} \tanh \frac{a}{\lambda}\right) = H_c^2. \tag{6.15}$$

Since $H'_c > H_c$, this shows that, in a thin film, the superconducting state can be maintained for larger fields than H_c. In the limit $a \ll \lambda$, a truncated expansion of (6.15) leads to

$$H'_c = \frac{\lambda \sqrt{3}}{a} H_c.$$

The critical field H'_c of the film can be made significantly larger than H_c if $a \ll \lambda$, as has been confirmed by measurements on thin films of cadmium or tin. This result shows that the original limitations on the use of superconductivity due to the low values of H_c can be overcome by using films or wires of thickness less than λ. The thermodynamic properties of a thin superconducting tube are also altered in the presence of a quantised magnetic flux.[2]

6.2.2 Coherence Length ξ

We shall see that the effect described above can be used to raise a very basic question. We have seen that the free energy required to exclude the magnetic field can be decreased by reducing the dimensions of the superconductor. As a consequence, since the superconducting state is a thermodynamic equilibrium state, why should a sample not divide itself up spontaneously into normal and superconducting regions when a field is applied? *By accepting to lose condensation energy in the normal regions, this would allow the material to considerably reduce the increase in free energy due to exclusion of the field in the superconducting regions.* For example, for a superconductor in the form of a layer of thickness $2d \gg \lambda$, it is easy to imagine a structure built up from superconducting layers of thickness $2a \sim \lambda$ alternating with normal layers of thickness $2b$ such that, in an applied field, the free energy is less than that of the bulk sample (see Fig. 6.6). The point is just to make b extremely small in order to gain some free energy!

Experimentally, many situations have been observed where the Meissner effect is not perfect, and much effort has been expended in the search for explanations, although other possibilities related in particular to the inhomogeneity and shape of the samples were initially favoured (see Chap. 10). However, some bulk superconducting metals exhibit a perfect Meissner effect which disappears suddenly for $H = H_c$. In these cases, no lamellar structure was observed, suggesting that we may have made some mistake in the above analysis, and that some crucial physical element has escaped our notice. Since thin films of the same metals exhibit superconductivity above H_c, as mentioned above, *there must be an energy cost that opposes the formation of interfaces between superconductor and normal metal.*

In the above argument, we assumed that there could be a sharp interface between normal and superconducting states. The fact that we are considering a crystal with

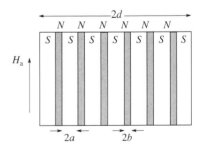

Fig. 6.6 Structure built up from alternating thin films of superconductor and normal metal, with thicknesses $2a$ and $2b$, respectively, which can in principle reduce the free energy in an applied magnetic field

---

[2] See Problem 10: *Thermodynamics of a Thin Superconducting Cylinder.*

6.2 Thin Films and Coherence Length

a lattice spacing of a few angstrom units will certainly impose a physical limit on the thickness of such an interface. But this would not really be a sufficient limitation because the penetration depth is macroscopic, and this would still leave room for macroscopic thicknesses b of normal regions. Indeed, we can today make artificial structures that retain the properties of a metal while comprising only two, or even just one atomic monolayer!

However, it seems reasonable to ask whether electrons can transfer easily within the same crystal from a collective quantum state to an independent electron state over a very short distance. In other words, *over what distance can the density of superconducting electrons go from n_s to zero?* This suggests reexamining the very hypotheses that lead to the London equation close to the surface of a superconductor. Indeed, it was assumed there that n_s would remain constant right up to the surface. But the notion of electron pairs suggests that this could not be possible. An electron close to a surface has a lower probability of finding a partner than one in the bulk of the superconductor.

The superconducting state, a coherent macroscopic quantum state, can only be set up over a spatial distance at least as long as the average distance between paired electrons. We thus allow n_s to vary from n_s to zero over a finite distance ξ, called the *coherence length* of the superconducting state (see Fig. 6.7). This means that the interface has a minimal thickness and an energy cost, which explains the existence of the Meissner effect.

Indeed, close to an interface, the free energy per unit area increases by approximately $\sim \xi\mu_0 H_c^2/2$ compared with the bulk superconductor, because the density of superconducting electrons is lower. However, it is reduced by the penetration of the magnetic field by an amount of the order of $\sim \lambda\mu_0 H_a^2/2$. While the first term dominates, there is no energy advantage in forming a lamellar state. However, if the field is increased, such a structure might appear before H_a reaches H_c. We shall now make this explicit by developing the above model.

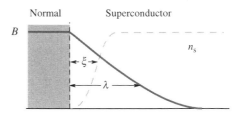

Fig. 6.7 Penetration depth λ and coherence length ξ at a metal–superconductor interface

6.2.3 Mixed State: A Simple Model

We can now investigate whether the structure specified above can have a lower free energy than the bulk superconducting phase. Insofar as the normal metal–superconductor interface requires a minimal thickness of the order of ξ, there is no longer any real problem in imagining that the thickness b of the normal layers might be infinitesimal (the existence of ξ is tantamount to imposing a minimal thickness of ξ for b). The structure thus looks like a series of thin films, with period $2a$. The free energy of such a film is then given by (6.14) increased by the loss of condensation energy in the interface of thickness ξ that is $(\xi/a)\mu_0 H_c^2/2$.

The reduction in free energy $G_{\text{ms}}(T, H_a)$ per unit volume of this mixed state compared with the normal state is therefore given by

$$G_{\text{ms}}(T,H_a) - G_n(T,0) = \frac{\mu_0 H_a^2}{2}\left(1 - \frac{\lambda}{a}\tanh\frac{a}{\lambda}\right) - \frac{\mu_0 H_c^2}{2}\left(1 - \frac{\xi}{a}\right), \quad (6.16)$$

and this should be compared with the value obtained for a perfect Meissner effect, as given in (6.6):

$$\Delta G_M = G_s(T,\mathbf{H}_a) - G_n(T,0) = -\frac{\mu_0}{2}(H_c^2 - H_a^2), \quad (6.17)$$

$$G_{\text{ms}}(T,H_a) - G_n(T,0) = -\frac{\mu_0 H_a^2}{2}\frac{\lambda}{a}\tanh\left(\frac{a}{\lambda}\right) + \frac{\mu_0 H_c^2}{2}\frac{\xi}{a} + \Delta G_M. \quad (6.18)$$

It follows that the lamellar structure is only thermodynamically favoured over a perfect Meissner effect if

$$-H_a^2\frac{\tanh\alpha}{\alpha} + H_c^2\frac{\xi}{\lambda\alpha} < 0,$$

where we have set $\alpha = a/\lambda$. This can also be written

$$\left(\frac{H_a}{H_c}\right)^2 \frac{\lambda}{\xi}\tanh\alpha > 1.$$

Since $\tanh\alpha < 1$, this will be impossible when $\lambda/\xi < 1$, for any $H_a \le H_c$. When $\lambda < \xi$, *the Meissner effect therefore persists until the thermodynamic critical field is reached.*

On the other hand, *a lamellar solution will be possible if $\lambda > \xi$*. The width of the layers giving the lowest free energy is obtained by minimising G_{ms} with respect to a, i.e., solving $dG_{\text{ms}}/da = 0$. Cancelling a factor of $\mu H_a^2\lambda/2$ in (6.18), we obtain

$$-\left(\frac{1}{a^2}\right)\left[\frac{\xi}{\lambda}\left(\frac{H_c}{H_a}\right)^2 - \tanh\left(\frac{a}{\lambda}\right)\right] - \left[a\lambda\cosh^2\left(\frac{a}{\lambda}\right)\right]^{-1} = 0,$$

which simplifies into

$$\tanh\alpha - \frac{\alpha}{\cosh^2\alpha} = \frac{\xi}{\lambda}\left(\frac{H_c}{H_a}\right)^2. \quad (6.19)$$

The function on the left-hand side is a monotonic increasing function of α which tends asymptotically to 1. There is therefore no solution unless the right-hand side is less than 1, i.e., for $H_a < H_c$, there will only be a solution if $\xi < \lambda$, agreeing with

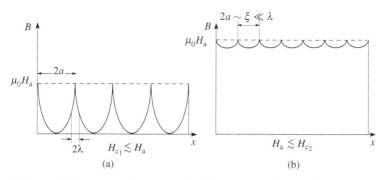

Fig. 6.8 Change in the magnetic induction in lamellar structures obtained for different values of the field H_a. The normal region is assumed to have infinitesimal thickness. (**a**) For $H_a \gtrsim H_{c_1}$, the distance between layers is much greater than λ. The change in the magnetic induction in each layer is given by (6.11) and shown in Fig. 6.4. (**b**) For $H_a \lesssim H_{c_2}$, the layer thickness tends to ξ. For the case depicted here, $\lambda \gg \xi$ and the field penetrates almost uniformly

what we have already seen above. Even in this case, a solution will only exist for fields above a critical field H_{c_1} defined by

$$H_a > H_c \left(\frac{\xi}{\lambda}\right)^{1/2} = H_{c_1}. \tag{6.20}$$

We may thus conclude that *there is a total Meissner effect for $0 < H < H_{c_1}$*, and that beyond H_{c_1}, *a lamellar structure can appear*. For $H_a = H_{c_1}$, α is infinite. On the other hand, as soon as $H_a > H_{c_1}$, α drops off and interfaces form with very long period to begin with. As H increases, the density of interfaces steadily increases (see Fig. 6.8).

This structure will clearly remain in much larger fields than H_c, until the free energy in (6.18) becomes equal to that of the normal state. This happens for a second critical value H_{c_2} of the field such that

$$G_{ms}(T, H_{c_2}) - G_n(T, 0) = \frac{\mu_0 H_{c_2}^2}{2}\left[1 - \frac{\lambda}{a}\tanh\left(\frac{a}{\lambda}\right)\right] - \frac{\mu_0 H_c^2}{2}\left(1 - \frac{\xi}{a}\right) = 0, \tag{6.21}$$

$$\left(\frac{H_{c_2}}{H_c}\right)^2 \left(1 - \frac{\tanh \alpha}{\alpha}\right) - \left(1 - \frac{\xi}{\lambda \alpha}\right) = 0. \tag{6.22}$$

The period of the layers is given at H_{c_2} by (6.19), viz.,

$$\tanh \alpha - \frac{\alpha}{\cosh^2 \alpha} = \frac{\xi}{\lambda}\left(\frac{H_c}{H_{c_2}}\right)^2. \tag{6.23}$$

Equations (6.22) and (6.23) thus imply the simple relation

$$\tanh \alpha = \frac{H_c}{H_{c_2}}.$$

In the limit $\lambda \gg \xi$, a truncated expansion of (6.23) then leads to

$$\frac{H_c}{H_{c_2}} = \frac{3\xi}{2\lambda},$$

which corresponds to $H_{c_2} \gg H_c$.

The main result here is that, when $\xi < \lambda$, *a state in which superconducting regions alternate with normal regions is energetically favoured over the perfect Meissner state* for a range of magnetic field strengths between H_{c_1} and H_{c_2}.

Note that, for $\lambda \gg \xi$, the period of the layers is of the same order of magnitude as ξ when H_a comes close to H_{c_2}. Under such conditions the magnetic field penetrates almost everywhere in the superconductor, while the change in B can become relatively small in the lamellar structure (see Fig. 6.8b).

The calculation exposed above brings out the main physical phenomena, but it artificially favours a lamellar structure for the field penetration. There is no justification for this choice, which is based on the form chosen for the sample, and takes into account its surface geometry. One would be unlikely to choose the same form for the interfaces in the case of a cylindrical sample! The thermodynamic argument does indeed suggest that the state obtained has a lower free energy than the one corresponding to the Meissner effect, but it in no way guarantees that this will be the state of minimal energy. If the sample is not assumed to have any particular geometry, there is only one privileged direction, that of the applied field, and we should expect rather to observe a penetration of the field in the form of 'field tubes'. This is what happens in reality, and we shall describe it in more detail in the next chapter.

6.3 Two Types of Superconductivity

The above analysis of experimental observations showed that we require another characteristic length, the coherence length ξ, in order to describe the behaviour of superconductors in an applied magnetic field. As indicated at the end of Sect. 6.2.2, the formation of an interface between the superconducting state and the normal state has an energy cost of the order of

$$\frac{\mu_0}{2}(\xi H_c^2 - \lambda H_a^2)$$

per unit interfacial area. It is this interface energy that determines the resulting state of thermodynamic equilibrium. Two types of behaviour are then possible depending on the value of the ratio λ/ξ, which is generally denoted by κ. For $\kappa = \lambda/\xi \lesssim 1$, a Meissner effect is observed up to H_c, and the normal state is restored above H_c. We then have a *type I superconductor*. On the other hand, for $\xi \lesssim \lambda$, the Meissner effect is only observed up to H_{c_1}. Above H_{c_1}, a mixed state is established, and this only disappears above a second critical field value H_{c_2}. We then speak of a *type II superconductor*.

6.3.1 Type I Superconductors

Let us note to begin with that, for a type I superconductor, the interface energy between a superconductor and a normal metal is always positive. This situation forbids the creation of interfaces within the superconductor and leads to a perfect Meissner effect. In this case, the density of superconducting pairs is low close to the surface of the superconductor, in a region of thickness equal to the penetration depth of the magnetic field (see Fig. 6.9).

This does not correspond to the London limit considered in Chap. 5, for which the density of superconducting carriers was assumed constant right up to the surface. *The phenomenological London model does not therefore apply under the conditions for which it was designed.* This explains why the value λ_0 of the penetration depth at zero temperature is not λ_L. The fact that the density n_s of superconducting carriers is reduced over a much greater distance than λ clearly suggests that $\lambda_0 > \lambda_L$. This was first understood by Pippard on the basis of ultrahigh frequency absorption experiments, which led him to introduce the notion of coherence length. As we shall see later, the two lengths λ and ξ depend on the properties of both the superconducting state and the normal state of the metal.

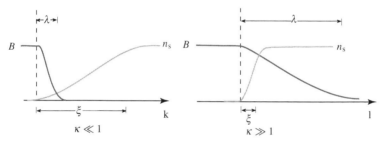

Fig. 6.9 Lengths λ and ξ at the interface between superconductor and normal metal, for type I (*left*) and type II (*right*) superconductors

6.3.2 Simple London Model for a Vortex

As in the case of the layers discussed above, a simple model for describing a vortex can be obtained by considering, in the infinite superconducting material, a cylindrical core of normal material of radius ξ along the z axis of the magnetic field. This amounts to assuming a discontinuous change in n_s. Applying the London equations in the superconducting material, we can thus determine the magnetic field distribution around the normal core. We have

$$\lambda^2 \Delta \mathbf{B}(\mathbf{r}) - \mathbf{B}(\mathbf{r}) = 0.$$

Taking into account the cylindrical symmetry of the problem, this can be written

$$\lambda^2 \frac{d^2 B_z(r)}{dr^2} - B_z(r) = 0. \tag{6.24}$$

As the field vanishes in the bulk of the superconductor, i.e., for $r \to \infty$, this equation has the solution

$$B_z(r) = B_0 K_0(r/\lambda), \qquad \text{for } r > \xi, \tag{6.25}$$

where $K_0(x)$ is the modified Bessel function of the second type, which is given approximately by

$$K_0(x) \approx \left(\frac{\pi}{2x}\right)^{1/2} \exp(-x), \qquad \text{for } x \to \infty,$$

and

$$K_0(x) \approx -\ln x, \qquad \text{for } x \ll 1.$$

Note that $\int_0^\infty K_0(x) x \, dx = 1$.

In the normal core of the vortex, the constant field is, by continuity,

$$B_z(r) = B_0 K_0(\xi/\lambda), \qquad \text{for } r < \xi. \tag{6.26}$$

This field is maintained by the persistent currents circulating tangentially around the core of the vortex which, from $\mu_0 \mathbf{j} = \text{curl } \mathbf{B}$, can be written

$$j_\perp(r) = -\frac{B_0}{\lambda} K_0'(r/\lambda). \tag{6.27}$$

The spatial dependencies of B_z and $j_\perp$, which both decrease exponentially, are shown in Fig. 6.10.

Question 6.2. Show that the value of B_0 can be determined generally by taking into account flux quantisation inside a contour containing the core of the vortex. Show that, in the case of an extreme type II superconductor for which $\lambda \gg \xi$, if the flux through this contour is a flux quantum Φ_0, the constant B_0 is independent of ξ and given by

$$B_0 = \frac{\Phi_0}{2\pi\lambda^2}. \tag{6.28}$$

For $\xi \ll \lambda$, the core of the vortex reduces to a singularity, because $\xi \to 0$. A simple way of describing the properties of the vortex mathematically is then to insert a $\delta(\mathbf{r})$ source term for the field into the second London equation (5.11) which becomes

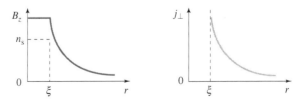

Fig. 6.10 Spatial dependencies of the superconducting carrier density, the magnetic field, and the current circulating around the core of the vortex in the London model

6.3 Two Types of Superconductivity

$$\lambda^2 \mu_0 \, \text{curl} \, \mathbf{j} + \mathbf{B} = \Phi_0 \delta(\mathbf{r}) \mathbf{z} \, . \tag{6.29}$$

It can be checked that this equation does indeed yield the solutions obtained in (6.25) and (6.27) for the fields and currents, and also the flux quantisation condition, but with an unphysical divergence of the field and current for $r < \xi$. To account for the physical reality, we simply equate the field value in this case with that for $r = \xi$, i.e.,

$$B_z = \frac{\Phi_0}{2\pi \lambda^2} K_0(\xi/\lambda) \, , \qquad \text{for } r < \xi \, .$$

This simplified model of the vortex can be used to determine the energy of the vortex per unit length, interactions between vortices, critical fields, and so on.[3]

6.3.3 Type II Superconductivity

As early as 1950, physicists working with Landau in Russia were developing similar thermodynamic arguments to the ones we presented in Sect. 6.2, but more realistic. In particular, in the expression for the free energy, Ginzburg and Landau allowed for a spatial variation of the superconducting pair density, which defines the coherence length. Their analysis does not lead to simple analytic solutions. This is why we have chosen here to develop more approximate models which nevertheless include the main physical effects. They show in a similar way to what was said above that superconductivity is type I when

$$\kappa = \frac{\lambda}{\xi} < \frac{1}{\sqrt{2}} \, ,$$

and that, when this does not hold, the energy required to create a vortex becomes negative[4] above a lower critical field

$$\boxed{H_{c_1}(T) = \frac{\Phi_0}{4\pi \mu_0 \lambda^2} \ln\left(\frac{\lambda}{\xi}\right) = \frac{\xi}{\lambda\sqrt{2}} \ln\left(\frac{\lambda}{\xi}\right) H_c(T)} \, . \tag{6.30}$$

---

[3] This can be seen in Problem 13: *Magnetisation of a Type II Superconductor*.
[4] This almost linear relation between H_{c_1} and ξ/λ may seem somewhat surprising, since the lamellar model led to $H_{c_1} = H_c(\xi/\lambda)^{1/2}$. The difference comes about because, for the geometry of a vortex, the energy of formation of the normal region per unit length is proportional to

$$\frac{\mu_0}{2} \left(\xi^2 H_c^2 - \lambda^2 H_a^2 \right) \, ,$$

since the cross-sectional area of the normal core of the vortex goes as ξ^2 and the field penetrates a cylinder of cross-sectional area proportional to λ^2.

For $H < H_{c_1}$, the Meissner effect occurs, but since $\lambda > \xi$, this situation is close to that prevailing for the London model. Indeed, when $\kappa \gg 1$, i.e., $\lambda \gg \xi$, the density of superconducting carriers barely changes over the penetration depth, and hence $\lambda_0 \simeq \lambda_L$.

As long as $H \lesssim H_{c_1}$, the field penetrates via widely spaced vortices, as shown in Fig. 6.11a. These vortices look very similar to those described in Sect. 6.3.2, but in this more realistic case, the spatial dependencies of $n_s(r)$, $B_z(r)$, and $j_\perp(r)$ do not exhibit discontinuities, as shown in Fig. 6.11b. Since the spacing between vortices is much greater than λ, the current vanishes between them. The flux corresponding to one vortex is quantised, and exactly equal to the flux quantum Φ_0.

Abrikosov succeeded in minimising the free energy in the presence of a field, and thereby demonstrated that *the lattice formed by these vortices is hexagonal*, as observed experimentally in the image on p. 176.

As H_a increases, the size of the primitive cell of this lattice decreases, and when the magnetic field reaches the upper critical field value

$$\boxed{H_{c_2}(T) = \frac{\Phi_0}{2\pi \mu_0 \xi^2} = \frac{\lambda\sqrt{2}}{\xi} H_c(T)}, \qquad (6.31)$$

the field penetrates uniformly and the metal becomes normal again.

Question 6.3. In principle, the ratio λ/ξ can be increased indefinitely. Show using a thermodynamic argument that the value of H_{c_2} is nevertheless bounded.

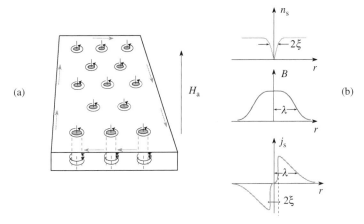

Fig. 6.11 Penetration of the field in a type II superconductor. (**a**) Penetration by tubes of field, arranged in a hexagonal lattice. The flux is Φ_0 per primitive cell. (**b**) Vortex structure close to H_{c_1}. Radial dependencies of the superconducting electronic density n_s, the magnetic induction B, and the supercurrents **j**, indicating the lengths λ and ξ

6.3 Two Types of Superconductivity

When $H_a \lesssim H_{c_2}$, as we have already seen for the lamellar model, the vortex spacing becomes lower than λ, and the vortices can no longer be considered independent, in contrast to those for fields close to H_{c_1}, as described in Fig. 6.11b. Close to H_{c_2}, the situation is similar to that described in Fig. 6.8b for the lamellar model. A map of the magnetic field still reveals a hexagonal lattice, but with lattice constant ever closer to ξ. *The flux per primitive cell of this lattice remains at the same value*, i.e., Φ_0, but there is no more than a spatial undulation of the magnetic field, whose amplitude is steadily attenuated until it disappears altogether for $H_a = H_{c_2}$. This disappearance occurs more gradually than in the lamellar model. Indeed, the result obtained in Sect. 6.2.3 gives a discontinuity in B at H_{c_2}, and this no longer exists in Abrikosov's exact solution, nor in the experimental observations (see Chap. 10).

Question 6.4. Can you determine the magnetic field that was used experimentally to obtain the image of the vortex lattice in the figure on p. 176?

The phase diagram (H, T) of a type II superconductor is shown in Fig. 6.12. Ginzburg and Abrikosov received the Nobel prize in 2003 for this work.

A non-uniform current distribution is also produced in Josephson junctions in the presence of a magnetic field.[5]

Question 6.5. Is there a simple way to find out whether a material is type I or type II? Would it be possible via a levitation experiment, for example, if you were given a superconducting pellet with area S and three small magnets with much smaller dimensions than S and with magnetisations $\mu_0 M_0$ equal to 0.01, 0.03, and 0.1 tesla, together with a cryogenic fluid that could be used to cool the pellet below T_c. How would you find out whether the material was type I or type II?

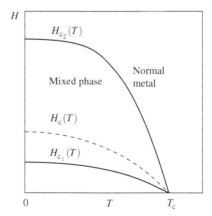

Fig. 6.12 Phase diagram for a type II superconductor

---

[5] See Problem 12: *Josephson Junction in a Magnetic Field*.

6.3.4 Applications of Type II Superconductors

The existence of type II superconductors provides a way of using superconductivity to produce high magnetic fields. One obvious reason is that one can significantly increase the magnetic field range over which the superconducting state persists. A current can thus be circulated without losses in a superconducting coil to create uniform fields H_a close to H_{c_2}. In the geometry of a solenoid comprising many layers of wire, one inner turn of the coil is subjected to a total field H_a which is well above the field created by the current passing through the turn itself, so the limiting factor will indeed be H_{c_2}.

But the use of type II superconductors provides another advantage. We have seen that, in a type I superconductor, whatever the geometry of the material, the current will only circulate in a thickness λ, with a critical current density determined by H_c/λ. In an extreme type II superconductor, close to H_{c_2}, the vortices have a spacing of the order of ξ, and although the microscopic currents are modulated over this distance, the transport current in the superconductor circulates throughout the bulk of the material. So for a given critical current density, the current through a wire of diameter d is larger than for a type I superconductor by a factor of the order of d/λ. As $\lambda \approx 500$ Å, the current gain is very high, viz., $\approx 2,000$ for a wire measuring 1 mm across.

Note on the other hand that these favourable conditions nevertheless lead to a certain level of technical complexity due to the effect of the electric current and magnetic field on the vortex lattice. In particular, the inner wires of the coil are subjected to a field H_a that is perpendicular to the current circulating in the wire. The associated Lorentz force tends to displace the vortices perpendicularly to these two directions. As the vortices contain a core of normal material, this displacement leads to dissipation, and the material is no longer strictly speaking superconducting! In practice, these vortex displacements only occur easily if the material is exceedingly clean, this allowing rapid attainment of thermodynamic equilibrium. On the other hand, when there are defects that can trap the vortices and prevent them from being displaced by the Lorentz force, a superconducting state without dissipation can be maintained. Considering all the various requirements, it turns out that the technical optimisation of the wires for superconducting coils is much more complex than one might have imagined.

The transition from idea to actual application inspired a technical ingenuity that would have been difficult to anticipate. The photograph in Fig. 6.13 is an electron microscope image of a wire made from an NbTi alloy of the kind used to make superconducting coils that can produce high persistent fields of the order of 8 tesla. As can be seen, it comprises several thousand NbTi filaments a few microns in diameter inserted within a copper matrix. The arrangement of the NbTi filaments is similar to that used in a thick rope. They are distributed in thin strings that are twisted in the direction perpendicular to the plane of the image. These strings are themselves twisted together to form the rope structure that constitutes the wire.

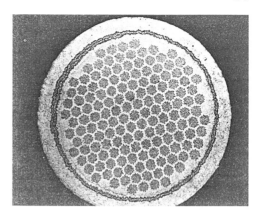

Fig. 6.13 Cross-section of a wire comprising 6,000 superconducting NbTi filaments in a copper matrix. The NbTi multifilament structure was designed to optimise the thermodynamic stability of the fields and currents in extreme type II conditions when the applied field is close to H_{c_2}. With the kind permission of D.R.

6.4 Summary

The superconducting state has a lower free energy than the normal state. The exclusion of the magnetic field from a superconductor leads to an increase in the free energy. The Meissner effect thus implies the existence of a thermodynamical critical field H_c for which these two effects balance out. Knowing only the experimental temperature dependence of H_c, the following thermodynamic quantities can be determined: the free energy G_s, entropy S_s, and specific heat C_V, which characterise the superconducting phase. In particular,

$$G_s(T, H_a) - G_n(T, 0) = -\frac{\mu_0}{2}\left(H_c^2 - H_a^2\right).$$

If we consider a thin superconducting film of thickness smaller than the penetration depth λ, the Meissner effect is not perfect and the increase in free energy in the presence of a field H_a is less than the value given above for a bulk superconductor. Superconductivity therefore persists for fields above H_c.

This observation implies that, in order to minimise its free energy in the presence of a magnetic field, a superconductor should divide up into normal and superconducting regions. The fact that the Meissner effect exists shows that this subdivision has a nonzero energy cost, and that the normal metal–superconductor interface cannot be very abrupt. We are compelled to introduce a new characteristic length ξ called the coherence length, which characterises the average distance between two paired electrons in the superconducting state.

Calculating the thermodynamic free energy (illustrated for the lamellar model), it can be shown that there are two types of superconductor:

1. Type I superconductors, for which $\lambda < \xi/\sqrt{2}$ and there is a perfect Meissner effect up to H_c.
2. Type II superconductors, for which $\lambda > \xi/\sqrt{2}$ and there is a Meissner effect up to a lower critical field

$$H_{c_1}(T) = \frac{\Phi_0}{4\pi\mu_0\lambda^2}\ln\left(\frac{\lambda}{\xi}\right) = \frac{\xi}{\lambda\sqrt{2}}\ln\left(\frac{\lambda}{\xi}\right)H_c(T).$$

Beyond H_{c_1} the magnetic field penetrates in the form of flux tubes called vortices. In this mixed phase, the vortices are arranged into a hexagonal lattice. The density of the lattice increases with increasing field, until the superconductivity is destroyed and the field penetrates uniformly for an upper critical field value

$$H_{c_2}(T) = \frac{\Phi_0}{2\pi \mu_0 \xi^2} = \frac{\lambda\sqrt{2}}{\xi} H_c(T).$$

The existence of this mixed phase provides a way of making superconducting wires that can be used to produce intense magnetic fields.

6.5 Answers to Questions

Question 6.1

With a larger magnetic field value than $H_c(0)$, the normal state is restored. We observe that, for $T > 0.3$ K, the specific heat has the form

$$C_n/T = 0.6 + 0.057 T^2.$$

The first term gives the coefficient γ of the electronic specific heat, while the second is a contribution to the measured specific heat going as T^3. It is produced by the vibrational degrees of freedom of the atoms in the metal, i.e., phonons.[6]

We also observe at low T an increase in the specific heat that does not exist for zero field in the superconducting state. There are therefore excited states which appear when a field is applied. These are thus of magnetic origin and not associated with the metallic state. They would appear to be due to paramagnetic impurities exhibiting a Curie susceptibility. It is easy to check that the internal energy of such an ensemble of paramagnetic entities contributes in the following way to the specific heat [1]:

$$C = \frac{dU}{dT} = Nk_B(\mu_B B/k_B T)^2 \left[\cosh(\mu_B B/k_B T)\right]^{-2}.$$

Since $\mu_B B \lesssim k_B T$ under the experimental conditions, the corresponding contribution of such terms to C_n increases as T^{-2} at low T.

Question 6.2

The quantisation condition for the flux through a contour (C) around the vortex is

$$\Phi_0 = \oint_{(C)} \lambda^2 \mu_0 \mathbf{j} \cdot d\mathbf{l} + \iint_S \mathbf{B} \cdot \mathbf{n} dS.$$

---

[6] See Problem 9: *Phonons in Solids*.

6.5 Answers to Questions

If we consider a circular contour of radius $R \to \infty$, then $|j| \to \exp(-R)$, and the first integral vanishes for large R. Only the second contributes to the flux, which can be determined by taking an integration surface extending over the whole plane. Then, using (6.24) and (6.25), we can write

$$\Phi_0 = \int_0^\xi B_0 K_0(\xi/\lambda) 2\pi r \, dr + \int_\xi^\infty B_0 K_0(r/\lambda) 2\pi r \, dr \, .$$

This can be used to determine B_0 by straightforward integration:

$$\Phi_0/B_0 = 2\pi \lambda^2 \left[K_0(\xi/\lambda) \frac{\xi^2}{2\lambda^2} + \int_{\xi/\lambda}^\infty K_0(x) x \, dx \right] \, .$$

For an extreme type II superconductor, $\xi/\lambda \ll 1$, the first term corresponds to the flux in the core of the vortex and tends to 0, while the integral is equal to 1. We thus recover the proposed expression

$$\boxed{B_0 = \frac{\Phi_0}{2\pi \lambda^2}} \, .$$

Question 6.3

In a field H_a, the drop in free energy of the normal state associated with the Pauli susceptibility of the metal is

$$G_n(T, H_a) - G_n(T, 0) = -\frac{1}{2} \mu_0 \chi_P H_a^2 \, .$$

We neglected this term at the beginning of Chap. 6 when discussing the Meissner effect. However, for high fields close to H_{c_2}, this term can no longer be ignored. When

$$-\frac{1}{2} \mu_0 \chi_P H_a^2 < G_s(0,0) - G_n(0,0) = -\frac{1}{2} \mu_0 H_c^2 \, ,$$

the normal state will necessarily be reestablished. This therefore leads to an upper bound for any critical field, and in particular for H_{c_2}, called the *Pauli limit*:

$$\boxed{H_{c_2}^m < H_c / \sqrt{\chi_P}} \, .$$

Question 6.4

Flux quantisation implies that, in a primitive cell of the vortex lattice of area S,

$$\mu_0 H_a S = \Phi_0 \, .$$

If a is the measured dimension of the hexagonal cell, we have $S = a^2\sqrt{3}/2$, and hence

$$\mu_0 H_a = \frac{2\Phi_0}{a^2\sqrt{3}}.$$

From the figure, we obtain $a = 550$ Å, so

$$\boxed{\mu_0 H_a = 1.06 \text{ tesla}}.$$

Question 6.5

If the material is type I, the Meissner effect is total, provided that $H < H_c$. The properties of the magnet–pellet system are the same whatever its thermomagnetic history. *The levitation will thus be the same whether the magnet is approached by the cold pellet or the magnet is placed on the pellet and subsequently cooled.* The first case tests screening, the second the Meissner effect.

However, if one of the magnets has a magnetisation $M_0 > H_c$, this reversibility may not be observed. It might levitate when brought near to the pellet and remain at a distance from the pellet such that the field on the pellet is less than H_c. On the other hand, if the magnet is placed on the hot pellet and then cooled, part of the pellet may remain in its non-superconducting state, and the force due to the residual Meissner effect may not be sufficient to oppose gravity. In this case, *the pellet would not levitate*.

If the material is type II, there is in principle no difference with a type I material as long as $M_0 < H_{c_1}$, but a significant difference with a type I material is to be expected if $M_0 > H_{c_1}$. When the magnet is placed on the pellet and the system subsequently cooled, the field will penetrate in the form of vortices when T goes below T_c. In this case, the diamagnetism is significantly reduced and the levitation will be less high than when the pellet has been cooled in zero field. For this reason, it is not always so easy to distinguish a type I material with a magnet such that $M_0 > H_c$ and a type II material with $M_0 > H_{c_1}$.

For a type II material and $M_0 < H_{c_1}$, the identical behaviour to a type I material and $M_0 < H_c$ only becomes effective *at thermodynamic equilibrium*. Indeed, for a type II material, H_{c_1} varies with T and vanishes at T_c. When the magnet is placed on the pellet and the latter is then cooled down to $T_0 \ll T_c$, the field continues to penetrate for $T \lesssim T_c$, and must be excluded in the form of vortices when T goes down to T_0. If the superconductor is very pure, this exclusion occurs very quickly, and at T_0 the vortices are effectively excluded if $M_0 < H_{c_1}$. There is reversibility and the same situation occurs with a type I superconductor. On the other hand, *if the material is not pure*, the exclusion does not occur, because the vortices are trapped by defects (*vortex pinning*) and the levitation is reduced, even if $M_0 < H_{c_1}(T_0)$.

The situation is thus generally irreversible for a type II superconductor, whatever the value of M_0, while it is reversible for a type I material when $M_0 < H_c$.

6.5 Answers to Questions

If the vortex pinning is effective, the magnetisation of the type II superconductor may even be positive and lead to an attraction of the magnet by the superconducting material (see Fig. 10.7, curve E). In this case one cannot strictly speak of levitation of the magnet over the superconductor, but rather of attraction and trapping of the magnet at a well defined distance. This is important for applications in which one wants to levitate a superconducting material over magnets, while ensuring an attraction and stability of the object above the magnetic material. This is indeed the case for a levitated train, where an on-board superconducting coil levitates over magnetic rails.

Chapter 7
Microscopic Origins of Superconductivity

Contents

7.1	Conventional Metal Superconductors	203
	7.1.1 Superconducting Metals and Alloys	203
	7.1.2 Attractive Interaction Due to Phonons	205
	7.1.3 The Band Gap in the Superconducting State	208
7.2	Cooper Instability	210
	7.2.1 Electrons in the Conduction Band of a Semiconductor	212
	7.2.2 Extra Electrons in a Metal	213
7.3	BCS Theory: Experimental Evidence	214
	7.3.1 Ground State and Band Gap	215
	7.3.2 Excited States in the BCS Theory	216
	7.3.3 Electron Tunneling Determination of the Gap	217
	7.3.4 Critical Temperature and BCS Theory	220
	7.3.5 Coherence Length	221
7.4	High-T_c Superconductors	223
	7.4.1 Cuprates	223
	7.4.2 Other Families of Superconductors	225
7.5	Summary	226
7.6	Answers to Questions	227

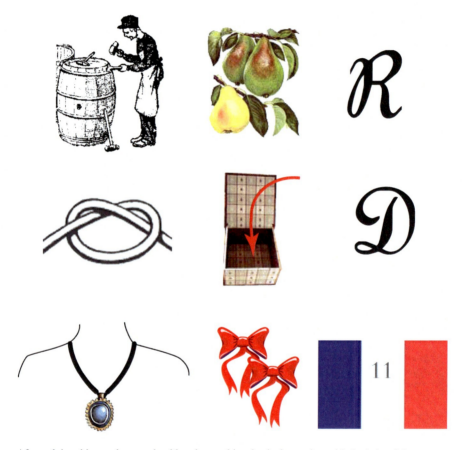

After solving this puzzle, you should understand its physical meaning with the help of the present chapter. You should also be able to work out what Claire and David are thinking about in the picture on p. 146

By measuring the flux quantum, it was established that the carriers are electron pairs in a superconductor. This is a key feature of all electron superconductivity phenomena so far observed. Note, however, that the discussion of the model based on Cooper pairs in Chap. 5 seriously distorts the actual chronology of these discoveries. London did indeed realise as early as 1950 that superconductivity would have to be treated as a macroscopic quantum effect, and envisaged flux quantisation, but he assumed that the superconducting carriers had charge $-e$. The point is that the experiment testing the flux quantisation hypothesis was not carried out until 1961, by which time the existence of Cooper pairs was well accepted.

In Sect. 7.1, we describe how the microscopic origin of superconductivity was established for metals and conventional alloys. To begin with, we consider the properties of simple metal superconductors. The *experimental* observations we describe led step by step to the essential elements of the puzzle that were then assembled in 1957 by Bardeen, Cooper, and Schrieffer (BCS) into a theoretical edifice that could explain the origins of superconductivity in conventional metallic systems.

Section 7.2 examines the conditions required for an attractive interaction between two electrons to produce a bound state in the conduction band of an insulator, and in a free electron gas. We will then be able to describe the calculation with which Cooper demonstrated in 1956 the instability of the electron gas in the presence of an attractive interaction between electrons, no matter how weak it may be.

Section 7.3 presents without derivation the main consequences of the BCS theory, which have led to a fairly complete explanation of the experimental properties of superconductivity in metals and alloys.

The originality of the new superconductors with high critical temperature, discovered in 1986 by Bednorz and Müller, is outlined in Sect. 7.4. We shall then be able to understand the intense activity in this area, not only on the fundamental level but also from the point of view of applications.

7.1 Conventional Metal Superconductors

We begin in Sect. 7.1.1 with a presentation of the known metal superconductors. The thermodynamic considerations presented in Chap. 6 led us to seek a ground state for the electron system that had lower energy than a free electron gas. We shall see in Sect. 7.1.2 how the involvement of the atomic nuclei in the superconducting state was established experimentally and led naturally to the idea of an attractive interaction between electrons, associated with the Coulomb interactions with the atomic nuclei. Experiment also revealed other novel properties of metal superconductors, and in particular the existence of a band gap in the excitations of the electronic system (Sect. 7.1.3).

7.1.1 Superconducting Metals and Alloys

One of the first questions raised right after the discovery of superconductivity by Kammerlingh Onnes was the issue of whether this phenomenon was particular to certain specific systems, or whether it had a more general relevance. The conductivity of metals was thus systematically studied at low temperatures. New cryogenic techniques had to be developed, in particular using ^{3}He to achieve temperatures below 1 K, before it could be demonstrated that many metals were in fact superconducting for a wide range of different values of T_c. Examining the periodic table shown in Fig. 7.1, we may make the following observations:

1. There are some rather surprising exceptions to superconductivity, including good metals like the noble metals Cu, Ag, and Au, but also the alkaline earth metals Li, Na, and K.
2. Metals with low conductivity at room temperature, such as the $4d$ or $5d$ transition metals, are among the best superconductors.
3. Semiconductors and some semi-metals in groups III to VI become conducting and superconducting under pressure.
4. The magnetic $3d$ and lanthanide metals are not superconducting.

Among the elementary metals, the one with the highest critical temperature is niobium (Nb) with $T_c = 9.2$ K. On the other hand, some have very low critical temperatures, e.g., tungsten (W), with $T_c \simeq 0.01$ K, and their superconductivity was only discovered when they could be produced in a pure enough form. But does this suggest that superconductivity is limited only to pure metals? Far from it! Specific alloys and compounds even exhibit higher critical temperatures than pure metals (see Table 7.1). Note, however, that alloys are generally type II superconductors, whereas most pure metals are type I.

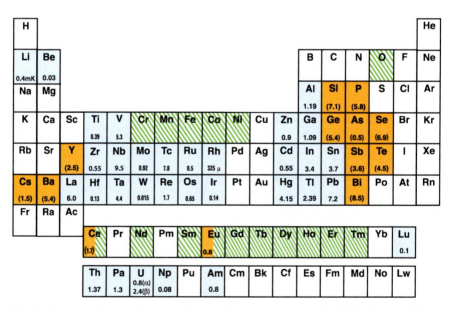

Fig. 7.1 Periodic table showing those elements that are superconducting (*full colours*) and those that are magnetic (*green hatching*). *Grey*: Elements that become superconducting at their critical temperature under standard pressure. *Orange*: Elements that become superconducting under pressure, with observed value of T_c in brackets

Table 7.1 Superconductivity of some metal compounds and alloys

Compound	T_c (K)	Compound	T_c (K)
Nb_3Sn	18.05	V_3Ga	16.50
Nb_3Ge	23.0	V_3Si	17.10
Nb_3Al	17.5	UCo	1.70
Nb_3Au	11.5	Ti_2Co	3.44
NbN	16.0	La_3In	10.40
MoN	12.0	InSb	1.90

Without even mentioning more recent discoveries of new compounds exhibiting superconductivity, made since the 1980s, it is quite clear that this phenomenon is far from being the exception. It is therefore essential to understand not only its microscopic origin, but also in the opposite case the fundamental explanation as to why some metals are not superconducting.

7.1.2 Attractive Interaction Due to Phonons

We have seen that T_c characterises the condensation energy of the electron system in the superconducting state, i.e., the drop in energy that occurs when the electron gas becomes superconducting. Some effort has been made to determine whether the superconducting transition is associated with a change in crystal structure or in the lattice parameter. However, careful crystallographic studies using X rays or neutrons have never revealed any evidence of such a structural change, and this suggests that *these processes are essentially electronic*. In a system of independent electrons like those considered in Chap. 1, no such effect could occur. We must therefore introduce a hitherto neglected negative energy contribution into the Hamiltonian of such a system. Experimentally, the obvious thing to do in order to find the origin of this contribution is to look for microscopic factors affecting the critical temperature. But to reach a clear conclusion, a given factor has to be modified experimentally while keeping all other factors the same, in order to see whether any change in T_c occurs as a consequence. For example, we may ask whether T_c varies with the density of carriers in the normal metal. This can be altered by applying a pressure. It is easy to see that the pressure will also change the hopping integrals between the orbitals of nearest neighbouring atoms, and hence also the band structure (see, for example, what happens for the semiconductors mentioned above, which become metallic). It is thus difficult to change one physical parameter without changing others too.

a. The Isotope Effect

It turns out that the mass of the ions can be changed quite easily without changing their charge, and therefore without changing the Coulomb potentials affecting the electrons. This is done by exploiting the fact that some elements of the periodic table exist in the form of several different stable isotopes. An example is mercury (Hg), which has five stable isotopes, with masses between 198 and 204, and also tin (Sn). One byproduct of the nuclear power industry has been the possibility of producing stable isotopes artificially, and improving techniques for isotope separation. Since the 1950s, it has thus been possible to prepare isotopically enriched samples of Hg and Sn, which have been used to show that the critical temperature of superconductivity decreases as the atomic mass M of the ions is increased. Accurate measurements of the critical temperature for different values of M for a given metal show that the dependence is (see Fig. 7.2)

$$\boxed{T_c = kM^{-\alpha}, \qquad \alpha \simeq 0.5}. \tag{7.1}$$

This isotope effect constitutes an irrefutable demonstration that *the ions are involved in the stabilisation processes of the superconducting state.*

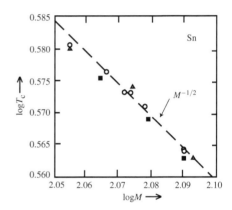

Fig. 7.2 Dependence of the critical temperature of tin on the atomic mass (log–log graph). *Dashed line:* $1/\sqrt{M}$ dependence. Different symbols correspond to measurements made by different groups. From [6, p. 245] Springer © 1995

b. Electron–Phonon Interaction

With the simplifying hypotheses of Chaps. 1 and 3, we assumed that the ions held fixed positions in the crystal. However, we have already seen that their oscillations about their equilibrium positions are responsible for a contribution to the electrical and thermal conductivities (see Chap. 4). The motions of the ions can be decomposed into eigenmodes called phonons,[1] whose energy dispersion relation defines the Debye frequency ω_D, or equivalently, the Debye temperature $\theta_D = \hbar\omega_D/k_B$. It is the interaction between electrons and phonons that leads to a temperature dependent contribution to the electron relaxation time. The ions thus play an important role in the physical properties of a metal. But can they help to create a different ground state energy for the electron system?

---

[1] See Problem 9: *Phonons in Solids.*

7.1 Conventional Metal Superconductors

Note that the existence of an attractive Coulomb interaction between the electrons and ions is such that, when an electron propagates, it creates a distortion of the crystal lattice. It attracts towards it those positive ions that sit close to its trajectory. The distortion produced in the environment of the trajectory will persist for a certain time because ionic displacements occur at low frequencies (see Fig. 7.3). Another electron will thus gain free energy if it stays in the region distorted by the first electron rather than in a non-distorted region. This corresponds to an attractive interaction between the two electrons.

A simple picture that may help to understand this fact considers two heavy beads lying on a horizontal elastic membrane. One bead distorts the membrane. The total gain in potential energy and distortion energy of the membrane is greater for two nearby beads than for two widely separated beads. The result is an effective attractive interaction between the beads, which thus prefer to be stuck one next to the other.

The case under consideration here is dynamic and hence much more complicated than the static model of the beads. Indeed, an electron moves with speed close to the Fermi speed. When it passes close by the ions, the latter begin to move, but since the collective modes of the ionic motions have a low Debye frequency ω_D, the amplitude of their displacement will only reach its maximum after a time of the order of $2\pi/\omega_D$. We can thus understand that the maximum energy gain for another electron will only be obtained roughly speaking when the distance between the two electrons is of the order of $2\pi v_F/\omega_D$, and this specifies a characteristic distance between the electrons in a pair. This in turn can be used as a physical basis for estimating the coherence length ξ, at least to within an order of magnitude. This idea was put forward by Fröhlich as early as 1950, and was supported by the discovery of the isotope effect. Indeed, the gain in energy due to the distortion induced by the electron is determined by the energy $\sim \hbar \omega_D$ of the atomic vibrational modes. Since these harmonic modes are determined by the repulsive interaction forces K between the ions, their frequencies, and hence also ω_D, are proportional to $(K/M)^{1/2}$, which provides an explanation for the isotope effect.

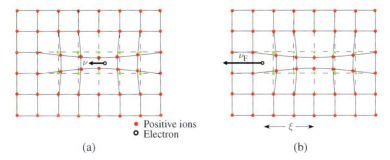

Fig. 7.3 Ion displacements induced by an electron in a solid. (**a**) A slowly moving electron produces a distortion which follows its trajectory. (**b**) If the electron moves quickly, the distortion lags behind. This is the situation when electrons move at the Fermi velocity in a metal. A second electron can follow the distortion caused by the first

The discovery of the isotope effect thus suggests that *there is some kind of attractive effective interaction between electrons induced by the vibrational modes of the ions*, and that this interaction plays a key role in the mechanisms of superconductivity. We shall see in Sect. 7.2 how an attractive interaction between electrons can induce a bound state between electrons.

7.1.3 The Band Gap in the Superconducting State

The thermodynamic measurements discussed in Chap. 6 brought out the difference in energy between the superconducting ground state and the normal metal state. However, it is just as important to understand the excited states of the superconductor. We can find out about the spectrum of excited states by investigating the thermodynamic properties at low temperatures. More direct information can be obtained by studying the absorption of photons. These experiments show that there is a band gap between the ground state and excited states of the electron system.

a. Specific Heat at Low Temperature

We return here to the specific heat measurements already mentioned in Chap. 6. They showed that the temperature dependence of the critical field was not perfectly quadratic. Recall that CdT is by definition the energy dU that must be supplied to increase the temperature of the electron gas by dT. At low temperatures, the specific heat can thus be used to test the energy distribution of the excited states of the system. For a normal metal, all the states close to the Fermi level are accessible and a great deal of energy must be supplied to significantly change the occupations of the levels. In the superconducting state, the specific heat is found experimentally to be much lower (see Fig. 6.3), indicating that few excited states are actually accessible. With the quadratic form of $H_c(T)$ given in (6.9), we expect C_s/T to have a T^2 dependence, but a much slower variation is in fact found experimentally.

Plotting the experimental results of Fig. 6.3 as a function of $1/T$, we find that (see Fig. 7.4)

$$\boxed{C_s = a\exp(-\Delta/k_B T)}. \tag{7.2}$$

This looks like what would be expected for the electron contribution to the specific heat of an insulator. As discussed in Chap. 3, the population factor of the levels at the bottom of the conduction band is proportional to $\exp\left[-(E_c - E_F)/k_B T\right]$, which would lead to a specific heat going as $\exp(-E_g/2k_B T)$, where E_g is the width of the band gap [see (3.62)].

These specific heat measurements, carried out on superconductors in 1954, thus suggest that there is *a band gap of width Δ in the electronic level structure of the superconducting state* (see Fig. 7.4). The energy Δ is found experimentally to lie close to $1.5k_B T_c$ for various metals (V, Sn, Al, and others).

7.1 Conventional Metal Superconductors

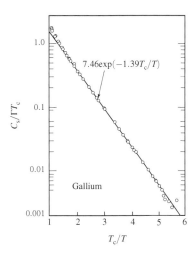

Fig. 7.4 Measurements of the specific heat C_s of metallic gallium. The experimental values of C_s (experimental points shown in Fig. 6.3, obtained for the lowest values of T) are plotted on a logarithmic scale as a function of $1/T$, and are found to vary as $\exp(-\Delta/k_B T)$. This shows that there is a band gap in the elementary excitations of the superconducting state. From [7, p. 365], 8th edn., Wiley ©2005

b. Microwave Absorption

This similarity with insulators naturally encouraged the search for spectroscopic methods that could determine the band gap. We know that a semiconductor is transparent to photons with lower energy than the band gap energy E_g, and absorbs them by creating electron–hole pairs if $h\nu$ is greater than E_g. We saw in Chap. 3 that the optical absorption or reflectivity can be used to determine the band gap of a semiconductor. Similar experiments were thus carried out on metal superconductors in 1955. From the values of Δ suggested by the measurements of C, the band gap is expected in the microwave frequency range (10 K $\sim$ 1 meV $\sim$ 10 cm^{-1}).

The experiments shown in Fig. 7.5 demonstrate the existence of a band gap. The microwave absorption is zero for low frequencies and increases suddenly above a certain frequency that can be used to determine the width of the band gap. The

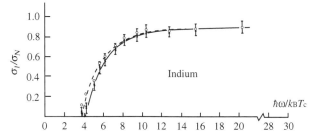

Fig. 7.5 Frequency dependence of microwave absorption in superconducting indium for $T \ll T_c$. The absorption normalised by the value in the normal state is plotted as a function of the frequency normalised at the critical temperature. An absorption threshold is observed for $\hbar\omega \sim 3.8 k_B T_c$. From Ginsberg, D.M., Tinkham, M.: Phys. Rev. **118**, 990 (1960) Standard American Physical Society Copyright. http://link.aps.org/doi/10.1103/PhysRev.118.990

energy of the microwave absorption gap is found to be twice the value Δ obtained via specific heat measurements. This reflects the fact that a photon can only be absorbed in the superconducting state if the photon energy is high enough to break a superconducting electron pair, and hence excite two electrons by making them 'normal'. The measured gap therefore corresponds to twice the energy Δ per electron needed to populate a thermally excited state. We also understand that this pair binding energy 2Δ is related to the condensation energy in the superconducting state, and hence to $k_B T_c$.

7.2 Cooper Instability

The electron–phonon interaction in a solid can lead to an indirect attractive interaction between two electrons. This may therefore induce a state in which the electrons are bound together in pairs. In quantum mechanics, we know that a potential well does not always yield a bound state, i.e., a particle is not necessarily confined spatially by a potential well. This leads us to ask under what conditions an attractive interaction between two electrons in a solid will in fact yield a bound state. Let us examine the cases where these two electrons are in the conduction band of a semiconductor or at the Fermi level of a metal. We consider the simple situation where, in the absence of any interaction between the electrons, the eigenstates are plane waves.

The Hamiltonian for two interacting electrons is thus

$$H = \frac{1}{2m_0}(\mathbf{p}_1^2 + \mathbf{p}_2^2) + V(\mathbf{r}_1 - \mathbf{r}_2), \tag{7.3}$$

where $\mathbf{r}_{1,2}$ are the positions of the two electrons, $\mathbf{p}_{1,2}$ are their momenta, and m_0 is their mass. The interaction potential $V(\mathbf{r})$ tends to zero at infinity.

By the usual transformation where $\mathbf{R}$ is the position of the center of mass,

$$\mathbf{R} = \frac{1}{2}(\mathbf{r}_1 + \mathbf{r}_2), \qquad \mathbf{r} = \mathbf{r}_1 - \mathbf{r}_2, \tag{7.4}$$

the total wave function factorises to

$$\psi(\mathbf{R},\mathbf{r}) = e^{i\mathbf{K}\cdot\mathbf{R}}\phi(\mathbf{r}), \tag{7.5}$$

where the relative wave function $\phi(\mathbf{r})$ satisfies

$$\left[-\frac{\hbar^2}{m_0}\Delta_\mathbf{r} + V(\mathbf{r})\right]\phi(\mathbf{r}) = E\phi(\mathbf{r}), \tag{7.6}$$

with $\Delta_\mathbf{r}$ the Laplacian with respect to $\mathbf{r}$.

7.2 Cooper Instability

The wave function $\phi(\mathbf{r})$ can be expanded relative to the basis of plane waves

$$\langle \mathbf{r} | \mathbf{k} \rangle = (\Omega)^{-1/2} e^{i\mathbf{k}\cdot\mathbf{r}} ,$$

with energy eigenvalues $\hbar^2 \mathbf{k}^2 / m_0 = 2\varepsilon_{\mathbf{k}}$. Here Ω is the volume over which periodic boundary conditions are imposed. The wave function thus becomes

$$|\phi\rangle = \sum_{\mathbf{k}'} \phi_{\mathbf{k}'} |\mathbf{k}'\rangle . \tag{7.7}$$

We now expand $V(\mathbf{r})$ in a Fourier series over the same states

$$V(\mathbf{r}) = \frac{1}{\Omega} \sum_{\mathbf{k}'} V_{\mathbf{k}'} e^{i\mathbf{k}'\cdot\mathbf{r}} . \tag{7.8}$$

Substituting (7.7) and (7.8) into the expression in (7.6), we obtain

$$\sum_{\mathbf{k}'} (2\varepsilon_{\mathbf{k}'} - E) \phi_{\mathbf{k}'} |\mathbf{k}'\rangle + \frac{1}{\Omega} \sum_{\mathbf{k}'} \sum_{\mathbf{k}''} V_{\mathbf{k}''} e^{i\mathbf{k}''\cdot\mathbf{r}} \phi_{\mathbf{k}'} |\mathbf{k}'\rangle = 0 .$$

If we now project this equation onto an eigenstate of the basis with wave vector $\mathbf{k}$, by multiplying on the left by the bra $\langle \mathbf{k} |$, only those terms such that $\mathbf{k}' + \mathbf{k}'' = \mathbf{k}$ will survive in the double sum. This will therefore lead to

$$(2\varepsilon_{\mathbf{k}} - E) \phi_{\mathbf{k}} + \frac{1}{\Omega} \sum_{\mathbf{k}'} V_{\mathbf{k}-\mathbf{k}'} \phi_{\mathbf{k}'} = 0 . \tag{7.9}$$

The usual Schrödinger equation (7.6) is thus replaced by the set of integral equations (7.9) written for all possible $\mathbf{k}$. Such equations are often more difficult to handle than differential equations, and the formulation (7.9) of the problem is rarely used in general. However, in this case, it will prove to be very useful.

We are concerned here with the case where the interaction between electrons is short range. When the interaction is a delta function $\delta(\mathbf{r})$ with zero range, $V_{\mathbf{k}}$ is constant in momentum space, with

$$V_{\mathbf{k}} = -V_0 , \tag{7.10}$$

where V_0 is positive. In this particular case, (7.9) can be written

$$(2\varepsilon_{\mathbf{k}} - E) \phi_{\mathbf{k}} - \frac{V_0}{\Omega} A = 0 , \tag{7.11}$$

where $A = \sum_{\mathbf{k}'} \phi_{\mathbf{k}'}$ can be eliminated by summing $\phi_{\mathbf{k}}$ over all the $\mathbf{k}$, whereupon

$$\boxed{\frac{V_0}{\Omega} \sum_{\mathbf{k}} \frac{1}{2\varepsilon_{\mathbf{k}} - E} = 1} . \tag{7.12}$$

This equation can be used to determine whether the energy E of the electron pair can be lowered in the presence of the interaction V_0, in which case the ground state is a bound electron pair.

7.2.1 Electrons in the Conduction Band of a Semiconductor

Suppose to begin with that two extra electrons are injected into a semiconductor at $T = 0$. In these cases, the available states $|\mathbf{k}\rangle$ are those in the conduction band, and their energy levels are given by $\varepsilon_\mathbf{k} = \hbar^2 \mathbf{k}^2 / 2m_0$ as measured from the bottom of the conduction band. Replacing the sum over $\mathbf{k}$ in (7.12) by an integral over the kinetic energy $\varepsilon_\mathbf{k}$ of the two particles, we obtain

$$V_0 \int_0^\infty \frac{D_d(\varepsilon)}{2\varepsilon - E} d\varepsilon = 1, \tag{7.13}$$

where the density of states is given by $D_d(\varepsilon) = C'_d \varepsilon^{d/2-1}$ for electrons in dimension d ($d = 1, 2, 3$), and $C'_d = C_d (2m_0/h^2)^{d/2}$ was determined in (3.50). The existence of a bound electron pair corresponds to a solution of (7.13) with $E < 0$.

Note that the integral in (7.13) diverges at the upper bound, except for $d = 1$. This reflects the fact that the form (7.10) for the interaction corresponds to an attraction with zero range. To remove this anomaly, we could consider a potential with nonzero range. In the present case, where the attraction between the electrons is transmitted via phonons, the interaction is only significant if the energy difference between the two electrons is lower than the typical energy $\hbar\omega_D$ of a phonon. We saw a qualitative illustration of this in Sect. 7.1.2. This suggests limiting the interaction to states with lower energy than $\hbar\omega_D$, and hence replacing the upper bound at ∞ in the integral of (7.13) by $\hbar\omega_D$. We thereby obtain

$$V_0 C'_d \int_0^{\hbar\omega_D} d\varepsilon \frac{\varepsilon^{d/2-1}}{2\varepsilon - E} = 1. \tag{7.14}$$

A considerable effort is required to treat the phonon dynamics in a more rigorous way, while the main results can in fact already be obtained from (7.14).

The condition for the existence of a bound electron pair[2] corresponds to a solution of (7.14) with $E < 0$. Note that the left-hand side of (7.14) is a function of E which tends to zero as $-1/E$ when $E \to -\infty$. In dimension $d = 3$, it remains finite when $|E| \to 0$ and takes the value $V_0 C'_d \sqrt{\hbar\omega_D}$. Consequently, a solution $E < 0$ and hence a bound state *only exist if the attractive potential is stronger than the critical value* $V_{0c} = 1/(C'_d \sqrt{\hbar\omega_D})$. In one or two dimensions, however, the integral in

---

[2] Electrons can be injected into the conduction band of a semiconductor by illuminating it with a laser at an energy above the band gap. The above calculation shows that bound pairs, called bipolarons, can only exist under certain specific conditions.

7.2.2 Extra Electrons in a Metal

Now consider the case of a metal, described as a free electron gas. All states with $|\mathbf{k}| < k_F$ are therefore occupied. We may ask once again whether *two extra interacting electrons* added to the electron gas will form a bound pair. Compared with the previous situation, the Pauli exclusion principle stipulates that the eigenstates of the electron pair can only be constructed from unoccupied plane wave states above the Fermi level, and the $\phi_{\mathbf{k}}$ are only nonzero if $|\mathbf{k}| > k_F$. By restricting the sum over $\mathbf{k}$ to $|\mathbf{k}| > k_F$ in (7.9), the condition (7.12) for the existence of a bound state changes to

$$\frac{V_0}{\Omega} \sum_{|k|>k_F} \frac{1}{2\varepsilon_{\mathbf{k}} - E} = 1. \tag{7.15}$$

When there is no interaction between the extra electrons, they will have an energy $2E_F$. The existence of a bound electron pair thus requires a solution of (7.15) with $E < 2E_F$ (rather than with $E < 0$ as in the previous case). Going to an integral over the energies, (7.15) is equivalent to

$$V_0 \int_{E_F}^{E_F+\hbar\omega_D} d\varepsilon \frac{D(\varepsilon)}{2\varepsilon - E} \approx V_0 D(E_F) \int_{E_F}^{E_F+\hbar\omega_D} d\varepsilon \frac{1}{2\varepsilon - E} = 1. \tag{7.16}$$

In the second expression, we have neglected the energy dependence of the density of states over the given energy range, because the typical values of $\hbar\omega_D$ are such that $\hbar\omega_D \ll E_F$.

Equation (7.16) is easy to understand. Note the resemblance between this equation and the case $d = 2$ of (7.14). We thus expect there to be a bound state whatever the value of V_0. Indeed, calculating the integral in (7.16) and setting $E = 2E_F - \delta$, we find a solution with *binding energy* δ given for small V_0, i.e., $V_0 D(E_F) \ll 1$, by

$$\delta = 2\hbar\omega_D \exp\left[-\frac{2}{V_0 D(E_F)}\right]. \tag{7.17}$$

This is the *Cooper effect*: whatever the value of the attractive potential, in the presence of a gas of independent electrons, two electrons will always form a bound pair, which fully justifies the name *Cooper pair*.

From (7.11), it is easy to see that $\phi_{\mathbf{k}} = \phi_{-\mathbf{k}}$, so we also have $\phi(\mathbf{r}) = \phi(-\mathbf{r})$, or $\phi(\mathbf{r}_2 - \mathbf{r}_1) = \phi(\mathbf{r}_1 - \mathbf{r}_2)$. The spatial part of the wave function is even under exchange of the two electrons. The Pauli principle thus requires the spin part of the total wave function to be odd, which means that the bound state is a *spin singlet*. We conclude that two extra electrons subjected to an attractive interaction in a free electron gas will form a singlet bound state with energy below $2E_F$.

Fig. 7.6 Spatial representation of a Cooper pair. The distortions shown by *orange* and *green shading* are produced by the electrons at r_2 and r_1, respectively

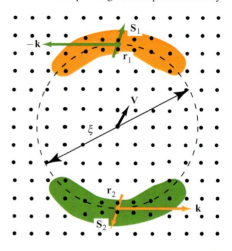

Since $\phi_{\mathbf{k}} = (AV_0/\Omega)[2(\varepsilon_{\mathbf{k}} - E_F) + \delta]^{-1}$ decreases rapidly for $\varepsilon_{\mathbf{k}} - E > \delta/2$, the pair formed in this way is built up from plane wave states $\mathbf{k}$ close to E_F.

The spatial representation of the pair depicted in Fig. 7.6 shows that each electron in the pair moves in association with the delayed distortion of the lattice produced by the motion of the other electron.[3] We shall see below that this wave function can be used to find the spatial extent ξ of the pair.

Note that the binding energy δ of the pair given by (7.17) depends exponentially on the reciprocal of the coupling V_0, and cannot by expanded as a power series in V_0. This shows that there is no way of generating this paired electron state in which the attractive interaction V_0 between the electrons can be treated as a simple perturbation of the non-interacting electron system.

7.3 BCS Theory: Experimental Evidence

The calculation carried out by Cooper in 1956 refers to an unrealistic situation in which the electron gas is assumed to be free from interactions, in which only the two 'added' electrons actually interact. In a real metal, a macroscopic number of electrons with energies close to the Fermi level actually interact together, and the above argument will apply to many electron pairs. We may thus anticipate a global instability of the set of electrons on the Fermi surface, leading to the formation of an infinite number of Cooper pairs. Naively, one might think of an ensemble of pairs like the one in Fig. 7.6, which could be viewed as independent bosons. This would correspond to a picture in which the gas of independent electrons is like a set of

---

[3] The representation in Fig. 7.6, which provides a good picture of what an isolated Cooper pair might look like, corresponds to a pair of electrons with orbital angular momentum, and thus breaks rotation symmetry. When there is no interaction that can remove the orbital degeneracy, the ground state of the pair is strictly speaking a superposition of such states with opposite rotations.

7.3 BCS Theory: Experimental Evidence

individualistic dancers on the dance floor who, at temperature T_c, gain internal energy by pairing off into couples dancing a tango. Note that this state would correspond to independent bosons, of the kind arising in Bose condensation discussed in the box below. However, the superconducting state is in fact very different, because the density of superconducting pairs is so high that the average distance between pairs is less than the coherence length ξ.

This is what Bardeen, Cooper, and Schrieffer realised in 1957, when they constructed a ground state comprising an infinite number of pairs, such that the pairs actually lose their individual identities. At any particular moment, an electron is paired with another electron, but the pairs change round all the time, in such a way that the macroscopic ensemble is coherent in the quantum sense. An analogy along the above lines would be a square dance, in which each male dancer must have a female partner at any moment of time, but change partner in a correlated way as time goes by. In this way, the dancers constitute a coherent ensemble on the dance floor, while the long range coherence of the electron pairs is induced by the interaction between the electrons. (In the square dance, coherence results from the rules for the dance learnt beforehand by the participants!)

It is beyond the scope of the present book to give a full demonstration of the BCS theory, which was rewarded by the Nobel prize in 1972. However, Cooper's calculation provides the basic theoretical arguments required to understand the BCS results and the way they account for experimental observations. One success of the BCS theory was *the prediction of a band gap*, together with its value.

7.3.1 Ground State and Band Gap

The BCS theory can be used to show that the ground state of the electron system is a quantum state involving a macroscopic number of electron pairs. It goes well beyond Cooper's calculation, because it can determine the energies of the excited states, and it has the advantage of showing that that there is a band gap, since the first thermally excited state of the macroscopic electron system lies at an energy 2Δ above the ground state. An energy of 2Δ must be supplied to excite an electron pair, i.e., to break up a pair and transform the electrons into 'non-superconducting' carriers (see Fig. 7.7).

In the weak coupling limit $V_0 D(E_F) \ll 1$, Δ is given by

$$\Delta = 2\hbar\omega_D \exp\left[-\frac{1}{V_0 D(E_F)}\right]. \tag{7.18}$$

As we saw in the Cooper calculation, the ground state is constructed from electron states with energies close to the Fermi energy E_F (in an energy band $\hbar\omega_D$). The energy of condensation in the superconducting state, which is related to the thermodynamic critical field by $E_c = -\mu_0 H_c^2/2$, is determined in the BCS theory to be

Fig. 7.7 Ground state and excited states of a superconductor

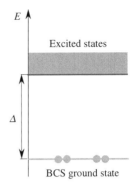

$$E_c = -\frac{1}{2}D(E_F)\Delta^2 = -2(\hbar\omega_D)^2 D(E_F)\exp\left[-\frac{2}{V_0 D(E_F)}\right]. \quad (7.19)$$

Note that E_c and Δ have a similar form to the binding energy δ calculated by Cooper for an interacting electron pair in a gas of independent electrons [see (7.17)]. In both cases, we obtain an exponential function of the reciprocal of the coupling. The absence of resistivity and the existence of the Meissner effect can be understood in this quantum ground state.

7.3.2 Excited States in the BCS Theory

At zero temperature, the BCS theory can determine the spectrum of excited states of the system. For all practical purposes, these can be treated as independent 'normal' electron states. The density of excited states is given by

$$D_s(E) = \begin{cases} D(E_F)\dfrac{E - E_F}{\left[(E - E_F)^2 - \Delta^2\right]^{1/2}}, & \text{for } E > E_F + \Delta, \\ 0, & \text{for } E_F < E < E_F + \Delta. \end{cases} \quad (7.20)$$

This function is shown in Fig. 7.8.

Note that, while the excited states can be considered as independent electron states, the same cannot be said for the ground state. The latter is a collective state of the electron system in which the electrons are strongly correlated, since it is made up of electron pairs. In the schematic representation of Fig. 7.7, electrons can only be thermally excited out of the ground state in pairs, so the gap is indeed equal to 2Δ. Regarding the excited states, we may consider that the gap appears by shifting the set of accessible states between E_F and $E_F + \Delta$ in the normal metal beyond $E_F + \Delta$.

The BCS calculation can be carried out *as a function of temperature*. It transpires that, not only is the population of excited states given by a Fermi factor, but the gap closes as the temperature increases, and vanishes at T_c, guaranteeing continuity with

7.3 BCS Theory: Experimental Evidence

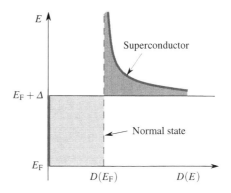

Fig. 7.8 Density of states of excited levels. The vertical scale is much expanded, so the density of states $D(E)$ in the normal state appears constant $\simeq D(E_F)$. The states that can be reached in the normal metal (*light shading*) are shifted beyond the gap (*dark shading*) in the superconductor

the normal state. The critical temperature is found to be directly related to the value of Δ at zero temperature by

$$\boxed{2\Delta(0) = 3.52 k_B T_c}. \tag{7.21}$$

To begin with we discuss experiments to check the prediction of the energy spectrum of the superconductor, before considering the validity of the theoretical determination of T_c on the basis of (7.21).

> **Question 7.1.** Can BCS theory explain the experimental values of the critical field in Fig. 5.3? What can be deduced about the coefficient γ of the electronic specific heat for these metals? What about the Pauli limit for H_{c_2}?

7.3.3 Electron Tunneling Determination of the Gap

We have seen that, in order to observe the gap and the excited states in the superconducting state, we could use measurements of specific heat or reflectivity in the microwave range. The former only inform us about an integral of the density of thermally occupied states, while the latter are difficult to set up experimentally, because measurements must be made as a function of the microwave frequency. To do a detailed study of the energy spectrum of the excited states, tunneling effect measurements prove to be extremely effective.

We saw in Chap. 5 that *tunneling of superconducting electron pairs* through a thin insulating barrier gives rise to the Josephson effect, exploited to make SQUIDs. Here we consider *the tunneling of electrons between a normal metal and a superconductor*. This kind of junction can be made by depositing thin films as shown in Fig. 5.11. In this case, one of the layers is a normal metal, while the other becomes superconducting for $T < T_c$.

As we saw in Chap. 3 regarding scanning tunneling microscopy, the current–voltage characteristic of a tunnel junction can inform about the density of states of the two metals as a function of energy over an energy range eV determined by the tunnel voltage V. In general, metals 1 and 2 are such that $D_1(E)$ and $D_2(E)$ vary little with the energy in their normal state, and according to (3.74),

$$j_{NN} = A \mathcal{T} D_1(E_F) D_2(E_F) eV, \qquad (7.22)$$

which is an Ohmic characteristic, independent of temperature.

Consider what happens if metal 2 becomes superconducting for $T \ll T_c$. Since the excited levels are separated by Δ from the ground state, no electron can be transferred to the superconductor as long as $eV < \Delta$, because there is no accessible state. The tunneling current is therefore zero and rises suddenly when $eV > \Delta$, because the number of accessible states also goes up suddenly (see Fig. 7.9). The current–voltage characteristic $j(V)$ can thus be used to determine Δ for $T \ll T_c$. It only became possible to carry out such experiments in 1960, when the techniques for depositing thin films were finally developed.

The tunneling conductance dj/dV accurately gives the function $D_{2s}(E_F + eV)$ rounded off by the temperature dependence of the Fermi factor [see (3.73)]. This was achieved with great accuracy by Giaver (Nobel prize with B. Josephson in 1973), who showed in 1962 that the tunneling characteristics agree extremely well with those deduced from the BCS density of states (7.20) (see Fig. 7.10). The temperature dependence of Δ as measured by tunneling also agrees extremely well with the predictions of the BCS theory (see Fig. 7.11).

Question 7.2. The Abrikosov vortex lattice imaged on p. 176 was observed by scanning tunneling microscopy at the surface of a single crystal of the compound NbSe$_2$, for which $T_c = 7.2$ K. The colour code represents the value of the tunneling conductance at each point of the crystal surface for $V = 1$ mV. What characteristic length is represented by the diameter of the spots in the image? What observations of this image might seem unexpected? The detailed modifications of the tunneling effect in the mixed state[4] will help to understand some of these observations.

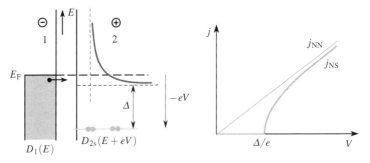

Fig. 7.9 Tunneling between a normal metal (1) and a superconductor (2) with a positive applied voltage V. An electron can be injected by tunneling into an excited state of the superconductor if $eV > \Delta$. The current–voltage characteristic $j(V)$ is used to determine Δ

---

[4] Examined in Problem 13: *Magnetisation of a Type II Superconductor*.

7.3 BCS Theory: Experimental Evidence

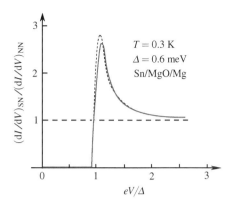

Fig. 7.10 Tunneling conductance as a function of V for an Mg/MgO/Sn junction at $T = 0.3$ K. The experimental results (*continuous curve*) agree extremely well with the theoretical prediction from the BCS density of states (*dashed curve*). From Giaver, I., Hart, H.R., Megerle, K.: Phys. Rev. **126**, 941 (1962). With the kind agreement of the American Physical Society (APS). http://link.aps.org/doi/10.1103/PhysRev.126.941

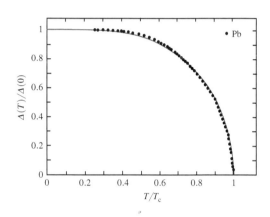

Fig. 7.11 Temperature dependence of the gap energy for lead, measured by tunneling. These results agree well with the prediction of the BCS theory (*continuous curve*). Adapted from Gasparovic, R.E., Taylor, B.N., Eck, R.E.: Solid State Com. **4**, 59 (1966) Elsevier copyright

The BCS theory shows that, when $V < 0$, the behaviour is totally symmetrical. As soon as $e|V| > \Delta$, an electron can be transferred to the normal metal, leaving an excitation equivalent to a hole in the superconductor, i.e., an electron which no longer has a partner. This is identical to the excitation corresponding to a solitary injected electron, so we obtain an analogous representation to the one obtained for a

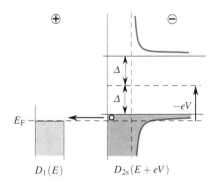

Fig. 7.12 Electron tunneling. Description of the excited states of a superconductor using a semiconductor model (7.23). An electron is transferred to the normal metal when $e|V| > \Delta$, releasing an excited state in the superconductor that resembles a hole

semiconductor (see Fig. 7.12), for which $D_s(E)$ is symmetrical with respect to the Fermi level:

$$D_s(E) = \begin{cases} D(E_F) \dfrac{|E - E_F|}{\left[(E - E_F)^2 - \Delta^2\right]^{1/2}}, & \text{for } |E - E_F| > \Delta, \\ = 0, & \text{for } |E - E_F| < \Delta. \end{cases} \quad (7.23)$$

In this representation, the superconducting ground state is not described. Only the excited states are considered.

> The superconducting state cannot be described as a metallic state of independent electrons in which a gap appears at T_c and leads to a new one-electron state described by the semiconductor-type band structure of Fig. 7.12. The point is that the energy involved in the appearance of such a band gap would correspond to a transfer below $E_F - \Delta$ of the normal electron states which would lie between $E_F - \Delta$ and E_F. It can be shown that the drop in energy associated with such a transformation is given by
>
> $$E_c^* = \int_{-\infty}^{E_F} E\bigl[D(E) - D(E_F)\bigr] dE = -\frac{C}{2} D(E_F) \Delta^2,$$
>
> where $C > 1$. The drop in energy in the superconducting state given by (7.19) is smaller as it involves the interaction energy between electrons in the BCS ground state, which is not taken into account in this independent electron band gap.
>
> A metal–insulator transition that can be described by the appearance of such a band gap occurs in some metals when a lattice distortion leads to a doubling of the primitive cell. This kind of transition, called a Peierls transition for a metal in one dimension, can only occur if the gain in electron energy is greater than the cost in elastic energy needed to cause the distortion.[5]

7.3.4 Critical Temperature and BCS Theory

As we have just seen, the validity of the BCS solution is well attested by experimental determination of the gap and its temperature dependence. But can the theory explain the measured values of T_c? Note that (7.21), which tells us that $\Delta = 1.76 k_B T_c$, is also perfectly satisfied for many metals. Moreover, the fact that T_c is proportional to $\hbar \omega_D$ obviously leads back to the isotope effect, insofar as the Debye frequency goes as $M^{-1/2}$. As we said earlier, it is this observation that suggests that the electron–phonon interaction underpins the attractive interaction in conventional metal superconductors. Combining (7.18) and (7.21), the BCS results thus imply

$$\boxed{k_B T_c = 1.14 \hbar \omega_D \exp\left[-\frac{1}{V_0 D(E_F)}\right]}, \quad (7.24)$$

in terms of the parameters ω_D, V_0, and $D(E_F)$ which characterise the normal metal. The last two parameters appear inside the exponential in the expression for T_c, which explains the differences *in order of magnitude of T_c between different metals*.

---

[5] Such a case is described in Problem 6: *One-Dimensional TTF-TCNQ Compounds*.

However, any error in these quantities, which must be determined by integrating over the Fermi surface, will lead to a large error in the estimate for T_c. In fact there is not a single metal for which it has been possible to determine T_c directly by calculation before obtaining its value experimentally. It is generally easier to justify the measured values a posteriori!

In order not to seem too negative about this point, we should nevertheless note that the various factors appearing in the expression for Δ and hence also in the expression for T_c can provide a qualitative understanding of the criteria for the existence of superconductivity, and also the differences in the values of T_c across the periodic table.

To begin with, the coupling between electrons is only attractive if the Coulomb repulsion between them is weak compared with the electron–phonon coupling. The latter will be stronger the more the electrons feel the crystal potential due to the ions. Now we know that the potential of the crystal lattice can only significantly modify the electronic band structure as compared with a free electron model in those metals for which the Fermi surface intersects one or more Brillouin zones. In this case, a modification of the energies of the electron states is induced close to vectors **k** corresponding to these intersections (see Chap. 3).

Such an effect will not generally occur for *monovalent metals*, for which the Fermi surface is included within the first Brillouin zone and is generally close to spherical. In such metals, the electron–phonon interaction is weak and so the term associated with phonons in the resistivity is rather small. The fact that superconductivity is not detected in such systems therefore suggests that *the electron–phonon coupling is not sufficient* to dominate the Coulomb repulsion.

On the other hand, T_c is higher for *polyvalent metals*, for which the Fermi surface *intersects many Brillouin zones*, and for which the electron–phonon coupling is therefore strong, which corresponds to *high resistivity* in the metallic state. This therefore perfectly explains the fact that good superconductors are often rather poor conductors.

Finally, the maximal values of T_c are obtained for the *transition metals* in the $4d$ and $5d$ series, for which the two terms V_0 and $D(E_F)$ are large, which is indeed expected on the basis of (7.24). Indeed, in these systems, the overlaps between the d orbitals are relatively small, and this leads to *narrow bands*, and hence *high electronic densities of states*.

To end this section, note that the fact that the magnetic metals are not superconducting suggests some kind of incompatibility between magnetism and superconductivity.

7.3.5 Coherence Length

We have seen that the wave function of the Cooper pair is built up from plane wave states taken over an energy range δ above E_F. The corresponding plane wave packet has wave vectors with magnitudes between k_F and $k_F + \Delta k$, where

$$\Delta(\hbar^2 k^2/2m_0) = \hbar^2 k_F \Delta k/m_0 \simeq \delta \ .$$

Corresponding to this extension $\Delta k = \delta/\hbar v_F$ in momentum space is an extension in position space given by $\Delta r = 1/\Delta k = \hbar v_F/\delta$. This spatial extent of the Cooper pair therefore provides an estimate of the coherence length ξ. The more exact BCS calculation leads to a similar form for ξ, in which δ is replaced by the superconducting gap Δ, that is

$$\boxed{\xi = \frac{\hbar v_F}{\pi \Delta}} \ . \tag{7.25}$$

This is valid for a pure metal system. Since Δ decreases with temperature, it follows that ξ increases and diverges at the critical temperature.

Dirty Superconductors

In reality, the two lengths λ and ξ depend on the properties of the normal state of the metal. In particular, if the metal is not pure, i.e., if the mean free path is short in the normal state, the coherence length gets shorter, while the penetration depth increases. These changes in λ and ξ take place without affecting the microscopic properties of the superconducting state, i.e., the condensation energy and gap. Introducing impurities in controlled amounts into a metal thus leads to an increase in the ratio λ/ξ, and it is then possible to move gradually from type I to type II superconductivity. We shall see an experimental example in Chap. 10. Note also that the temperature dependences of λ and ξ are similar and that the ratio $\kappa = \lambda/\xi$ does not therefore depend on temperature.

Superconductivity and Bose Condensation

In the known superconductors, the fact that the electron pairs are in *spin singlet states* is sometimes said to mean that the pairs behave as bosons. In simple metals, superconductivity is then likened to Bose condensation, a phenomenon predicted at low temperatures for systems of non-interacting bosons (see any textbook on statistical physics). Superfluidity, that is, the absence of viscosity at low temperatures, which is observed in the boson ^{4}He, is also similar to Bose condensation. However, strictly speaking, one cannot really speak of Bose condensation in either case, because the bosons under consideration here interact rather strongly together. This is already the case with ^{4}He, owing to the significant Van der Waals forces between the atoms in the liquid phase. As far as superconductivity in metals is concerned, there is not really any transition between a state in which the bosons are non-interacting and a condensed boson state. The electron pairs identified with bosons in this case are just not present in the metal at temperatures above T_c, and the pairing only occurs at the superconducting transition. In addition, in the BCS ground state, the electron pairs are intimately 'mixed', the average distance between paired electrons being greater than the interatomic distance. The comparison with Bose condensation is not therefore justified. Bose condensation in dilute gases, where the atoms are only weakly interacting, has been observed (for the first time in 1995, with rubidium gas). These gases are confined by laser beams, which permit to considerably reduce the kinetic energy of the atoms and thereby cool them to very low temperatures (a few tens of nK).

7.4 High-T_c Superconductors

By the 1970s, BCS theory seemed to provide a satisfactory explanation of superconductivity, while theoretical estimates suggested that the maximal value of T_c would be intrinsically limited to 25 K. The only effort in those days was therefore to try to reach this upper bound in technically usable compounds, so that liquid hydrogen could be used as cryogenic fluid, with a liquefaction temperature of $T = 20$ K at standard pressure.

The discovery by Bednorz and Müller in 1986 of copper oxide-based materials, the *cuprates*, with critical temperatures $T_c \approx 40$ K, was a genuine shock to the scientific community, and they were awarded the Nobel prize in 1988, an exceptionally short time after their discovery. Since then, the quest for new superconductors has snowballed and further compounds of the same family have been discovered, followed by other families with values of T_c exceeding all expectations. In many cases, the applicability of the BCS theory was put in question. Here we shall review some aspects of these discoveries.

7.4.1 Cuprates

In the year following the discovery of the compound $La_{2-x}Ba_xCuO_4$ by Bendorz and Müller, many other superconducting compounds of the same family were found, the current record being 135 K at standard pressure and $\approx$150 K at high pressure. Since T_c has gone above the liquefaction temperature of nitrogen at standard pressure (77 K), the *prospects for applications* have increased significantly. But the *fundamental questions* raised by this discovery are themselves exceedingly important.

Indeed, these new high-T_c superconductors are lamellar compounds comprising *copper oxide sheets*, separated by planes of transition ions or other oxides. We see immediately that we are dealing with quite different materials from the conventional elemental metals discussed so far. These are rather ionic crystals, like many other transition metal oxides, which are generally insulators. The very fact that some cuprates are metallic is not obvious!

One novel feature of these compounds is of chemical origins, because some of these planes play the role of charge reservoirs. The oxygen content can often be modified by simple thermal treatment. For example, the archetypal cuprate superconductor $YBa_2Cu_3O_7$, with $T_c = 92$ K, comprises two types of copper oxide planes (see Fig. 7.13). The CuO_2 planes at the center of the unit cell, separated by a layer of Y, are chemically highly stable. In contrast, the oxygen content of the base plane of the cell, made up of linear CuO chains, can be reduced by thermal treatment in vacuum.

These CuO chains can be totally deoxidised to obtain a composition $YBa_2Cu_3O_6$. This compound is insulating, its valence band being associated with the bands formed from the $Cu(3d)$–$O(2p)$ orbitals of the CuO_2 planes. To increase the oxygen

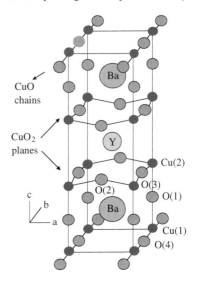

Fig. 7.13 Crystal structure of $YBa_2Cu_3O_7$ ($T_c = 92$ K). The conducting planes are the central CuO_2 planes, which correspond to the Cu(2) sites. The base planes containing CuO chains play the role of charge reservoirs. By partially suppressing the O(4) oxygens of the CuO chains, the number of carriers (holes) is reduced in the CuO_2 planes. The compound $YBa_2Cu_3O_6$ is an antiferromagnetic insulator

content and thereby return to the compound $YBa_2Cu_3O_7$, the idea is to introduce electronegative O^{2-} ions into the base plane. These extra oxygen ions capture electrons from the CuO_2 planes and thereby produce holes in the Cu(3d)–O(2p) valence band. This energy level system is valid for practically all the cuprate superconductors. The *hole density n_h can thus be continuously varied* in the CuO_2 planes of these materials by changing their oxygen content, and it is observed that T_c is maximum for $n_h \sim 0.15$ holes per Cu in the CuO_2 planes.

The density of free carriers which determines the metallic behaviour of these superconducting compounds is therefore *much lower* than in metal alloys. Their Fermi energies, and hence also the value of v_F, are both low. Consequently, the fact that T_c and hence Δ are high implies immediately, according to (7.25), that the coherence length ξ of the material is very low. *The superconductivity of the cuprates is thus always type II.* Note that the conductivity and superconductivity are both *highly anisotropic*. Taken together, these properties lead to novel behaviour of the vortex system, which has a different phase diagram to conventional type II superconductors (see Chap. 10). The low value of ξ (~ 20 Å) also leads to signficant practical difficulties when making junctions between superconductors, which must be structurally perfect on the scale of ξ.

Are These Superconductors Exotic?

Since the Debye temperature is of the same order of magnitude as one finds in metal alloys, and since the density of states is lower, it seems hard to imagine on the basis of (7.24) that the electron–phonon coupling can be the fundamental cause for such a high value of T_c. Various approximations of the BCS calculation need to be reexamined, including the weak coupling approximation $V_0 D(E_F) \ll 1$. Moreover, in a 2D metal, $D(E)$ exhibits Van Hove singularities, so that it cannot be treated as constant over the energy range $\hbar \omega_D$, and the band structure must be fully taken into consideration.

Furthermore, the CuO_2 planes have the property of being magnetic when not doped, the Cu ions ($3d^9$) carrying spin 1/2. This is true for $YBa_2Cu_3O_6$. In this case, as we shall see in the following chapters, *it seems difficult to neglect the interactions between electrons at the copper sites*, which are responsible for the magnetic behaviour. Many experiments have shown that the cuprates have strange magnetic properties in their metallic state. Various possibilities have been invoked to explain their high-T_c superconductivity. In some models, there is more talk of Bose condensation, because some authors consider that the pairs may actually exist above T_c.

One key experimental observation was the demonstration that, in the cuprates, the superconductor gap is not uniform in momentum space. It vanishes in the so-called nodal directions ($\pm \pi/a, \pm \pi/a$) of the wave vector in the Brillouin zone, and is maximum for the antinodal directions ($0, \pm \pi/a$). The gap thus has a similar symmetry to the d atomic wave functions, and changes sign under a rotation through 90°. This special d-type symmetry of the superconductor gap is today the strongest evidence for the exotic origin of the superconductivity of the cuprates. It leads to the novel properties of the Josephson junctions and also, in specific configurations of the junctions, to the existence of a vortex with a flux $\Phi_0/2$, as shown in the experiment illustrated in the figure on p. 296.

For many physicists, this observation provides strong evidence that electron pairing has magnetic origins. Despite intense effort since the cuprates were discovered, which has generated considerable progress on both the experimental and theoretical fronts, there is still no general agreement within the scientific community. This highlights the difficulty of the problem, which remains one of the main challenges of solid state physics.

7.4.2 Other Families of Superconductors

Following the discovery of the cuprates, new families of compounds with values of T_c above 25 K were identified, including alkaline earth compounds with fullerene C_{60}, discussed in Chap. 2 (see Fig. 2.10), or the compound MgB_2.[6] In the latter two cases, electron–phonon coupling and the BCS theory seem applicable, but with several special features to explain the high values of the critical temperature T_c.

On the other hand, since 1980, many other families of superconducting compounds with low T_c have been investigated. Some 1D or 2D organic salts, called Bechgaard salts,[7] and compounds of metals in the lanthanide series, called heavy fermion compounds, seem also to be exotic superconductors, like the cuprates. More recently, lamellar compounds of the transition elements, such as the ruthenate Sr_2RuO_4, with the electronic structure illustrated on the cover page of this book, or the cobaltates Na_xCoO_2, have been attracting much interest. Regarding the latter, the cobalt atoms in the CoO_2 planes are arranged in a hexagonal structure, and doping obtained by electron transfer from intercalated Na atoms. In 2003, it was found that these metallic compounds become superconducting, with $T_c = 4.5$ K, only when the CoO_2 planes are pushed apart by insertion of H_2O between the Na and CoO_2 planes.

---

[6] See Problem 16: *Magnesium Diboride A New Superconductor?*

[7] Those have similar molecular structure to that displayed in Problem 6: *One-Dimensional TTF-TCNQ Compounds.*

Finally, in 2008, new families of lamellar compounds comprising square planes of Fe, e.g., the Fe pnictides, where the FeAs$_2$ films are separated by planes which may provide electron or hole charge transfer, turn out to be superconducting with a maximal T_c of 55 K! The physical chemistry of 2D transition metal compounds will surely continue to deliver quite unexpected superconductivity properties.

Magnetism plays a key role in the electronic structure of these compounds, as for the cuprates, as we shall see in Chap. 8 after introducing some elementary notions of magnetism. In most of these compounds, a magnetic origin is often suspected for the coupling between electrons that leads to pairing.

7.5 Summary

Many metallic elements are superconducting, with values of T_c ranging over several orders of magnitude. The best conductors, in particular the noble metals, are not. The discovery of the isotope effect, i.e., the dependence of T_c on the atomic mass, shows that the ions are involved in the processes stabilising the superconducting state. The Coulomb interaction between electrons and ions, i.e., with their vibrational degrees of freedom known as phonons, leads to an indirect attractive interaction between electrons.

Specific heat and microwave absorption experiments reveal the existence of a band gap above the ground state of the superconductor. The width of the band gap corresponds to the energy 2Δ that must be provided to break up an electron pair and make them 'normal'.

Under what conditions can an attraction between two electrons lead to a bound state? In the case of a 3D insulator, two electrons in the conduction band will only form a bound state if the attractive potential exceeds a certain minimum value. On the other hand, in a metal, two electrons in attractive interaction at the Fermi level lead to a bound state whatever the strength V_0 of the interaction. This result due to Cooper suggests an instability of the electron ensemble near the Fermi level at low temperatures. The BCS theory shows that this new ground state of the electron gas comprises spin singlet superconducting pairs. This ground state is separated from the excited states by an energy

$$\Delta = 2\hbar\omega_D \exp\left[-\frac{1}{V_0 D(E_F)}\right] = 1.76 k_B T_c \,.$$

The excited states can be observed by tunnel effect in N-I-S junctions, where electrons or holes are injected into the superconductor from the normal metal. The particles injected in this way propagate in states separated by more than Δ from the Fermi level. The density of states of these excitations can be described by a semiconductor model with

$$D_s(E) = \begin{cases} D(E_F)\dfrac{|E-E_F|}{\left[(E-E_F)^2 - \Delta^2\right]^{1/2}}, & \text{for } |E-E_F| > \Delta \,, \\ 0, & \text{for } |E-E_F| < \Delta \,. \end{cases}$$

The BCS theory can explain all the properties of the conventional metal superconductors described so far, and in particular the broad range of values of T_c and the fact that the good metals are not superconducting. But having understood the superconductivity of metals and simple alloys, new superconducting compounds were discovered which apparently cannot be described as a form of superconductivity in which phonons are responsible for electron pairing.

Among these, many compounds have a lamellar atomic or molecular structure and electronic properties with distinct 2D features. Lamellar copper oxides, called *cuprates*, are without doubt the materials that have done most to shake the scientific community, since superconductivity occurs with the highest values of T_c yet obtained (as high as 150 K). Beyond the applications that become feasible with such high critical temperatures and the high associated critical fields, there is considerable debate over the origins of this kind of superconductivity.

Many new superconducting compounds have been discovered since the 1980s, and in many cases the origin of the superconductivity remains poorly understood. It will be a great scientific challenge to push our understanding beyond the BCS theory. Industrial applications of this physical phenomenon, something nobody doubts in the long term, will require a very tight control on the processes for synthesising these materials on a large scale.

7.6 Answers to Questions

Puzzle on Page 202

The pictures should lead you to the following sentence:

Cooper pairs are not independent bosons,

Cooper - pears - R - knot - in - D - pendent - bows - onze (French).

In the superconducting state, the electron–phonon interaction is responsible for the pairing of the electrons. In the macroscopic quantum state of the BCS theory, many pairs condense out at the same time, and the characteristic size of a pair, as determined by the coherence length ξ, is much greater than the average distance between electrons. In this sense, the pairs of dimension ξ are totally entangled with one another. While an isolated Cooper pair might possibly be considered as a boson, the Cooper pairs in the BCS state are in a highly correlated state, and therefore *cannot be considered as non-interacting bosons*. Note that the pairing of the Cooper pairs only occurs below T_c, and that *there are no states resembling bosons above the critical temperature T_c*.

This situation should be contrasted with Bose–Einstein condensation, where non-interacting bosons exist at all temperatures. In their ground state, these independent bosons are in the same quantum state. The occupation number of this ground state becomes an extensive quantity below the Bose condensation temperature T_B. The bosons are thus independent particles in the quantum sense at all temperatures, but move in a quantum mechanically coherent way for $T < T_B$.

Note: In the figure on the first page of Chap. 5, Claire and David are looking for explanations of superconductivity. David is trying to understand how the elastic deformation of the lattice can induce an attractive interaction between electrons, responsible for the formation of Cooper pairs. Claire is more intrigued by the collective phenomenon leading to quantum coherence over macroscopic distances. The electrons seem to her to execute a coordinated danse, rather like square dancers.

Question 7.1

The condensation energy of the superconducting state at zero temperature is given by the BCS theory as

$$G_s(0,0) - G_n(0,0) = -\frac{1}{2}D(E_F)\Delta^2 = -\frac{1}{2}\mu_0 H_c^2.$$

This suggests therefore that the temperature dependence of H_c may be related to the temperature dependence $\Delta(T)$ of the gap, these two quantities vanishing at $T = T_c$. At $T = 0$, this leads to

$$\frac{\mu_0 H_c(0)}{\Delta} = \left[\mu_0 D(E_F)\right]^{1/2}.$$

Now according to the BCS theory,

$$\Delta = 1.76 k_B T_c.$$

Using the relation between the electronic density of states at the Fermi level and the electronic specific heat, viz.,

$$\gamma = \frac{2\pi^2}{3} k_B D(E_F),$$

we obtain

$$\boxed{\left[\frac{\mu_0 H_c(0)}{T_c}\right]^2 = 0.47 \mu_0 \gamma}.$$

The experimental curves are approximately homothetic, with an almost constant ratio $H_c(0)/T_c$, which corresponds to an electronic specific heat and density of states that barely depend on the metal. This is indeed the case for the metals Al, In, Sn, Hg, and Pb considered in Fig. 5.3. In fact, the above relation applies perfectly with the experimental values of γ. The 3d and 4d transition metals have a higher density of states, leading to a higher ratio between the field $H_c(0)$ and the critical temperature.

7.6 Answers to Questions

The Pauli limit for H_{c_2} determined in Question 6.3, viz.,

$$H_{c_2}^m < H_c/\sqrt{\chi_P},$$

can be found by taking the value of $H_c(0)$ given by the BCS theory:

$$\mu_0 H_c(0) = \Delta \left[\mu_0 D(E_F)\right]^{1/2},$$

and

$$\chi_P = 2\mu_0 \mu_B^2 D(E_F).$$

We immediately obtain

$$\mu_0 \mu_B H_{c_2}^m = \Delta/\sqrt{2},$$

which yields, using $\Delta = 1.76 k_B T_c$ given by BCS,

$$\boxed{\mu_0 H_{c_2}^m = 1.84 T_c \quad (\text{tesla/K})}.$$

For example, for V_3Ga, which is a superconducting compound[8] at the relatively high $T_c = 14$ K, this would give the Pauli limit $\mu_0 H_{c_2}^m = 25.7$ tesla.

It is clear from Fig. 7.14 that the experimental curve for $H_{c_2}(T)$ in V_3Ga is not parabolic. For the quadratic shape, the tangent at $T = T_c$ cuts the vertical axis at $2H_c(0)$. The tangent to the experimental curve would give $\mu_0 H_{c_2}^*(0) = 28.5$ tesla, which is higher than the experimental value $\mu_0 H_{c_2}^{\exp}(0) \sim 21$ tesla. The parabola

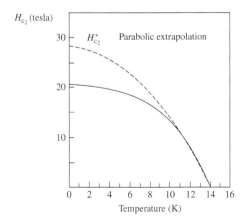

Fig. 7.14 Experimental curve for $H_{c_2}(T)$ in V_3Ga, showing that it saturates at low temperature with respect to a parabolic extrapolation

---

[8] See Problem 14: *Electronic Structure and Superconductivity of $V_3 Si$*.

fitted to the experimental results for $T \sim T_c$ is shown in Fig. 7.14. This saturation of the experimental curve $H_{c_2}(T)$ is undoubtedly related to the existence of the Pauli limit.

Question 7.2

The image on the first page of Chap. 6 was obtained at the surface of NbSe$_2$ by scanning tunneling microscopy for $V = 1$ meV. The gap predicted by the BCS theory is

$$\Delta = 1.76 k_B T_c \approx 1.05 \text{ meV}, \quad \text{for } T_c = 7.2 \text{ K}.$$

We thus see that the tunnel conductance must be close to zero in the superconducting state since the characteristic is taken for V slightly less than the value of the gap. This therefore corresponds to the blue continuum background. But in the core region of the vortices, the tunnel conductance increases and tends to the value for the normal metal. The white spots, of diameter about 7 mm, are thus close in size to 2ξ, and we obtain

$$\xi = \frac{7}{2} \times \frac{3{,}000}{130} \text{ Å} = 80 \text{ Å}.$$

Chapter 8
Magnetism of Insulators

Contents

8.1   Magnetic Behaviour of Solids 234
8.2   Magnetism of Atoms 236
     8.2.1   Hydrogen Atom 236
     8.2.2   Multielectron Atoms with Filled Shells 237
     8.2.3   An Atom with a Partially Filled Shell: Carbon 237
     8.2.4   Atoms with Partially Filled Shells: Hund Rules 240
8.3   Paramagnetism of an Ensemble of Isolated Ions 243
8.4   Ordered Magnetic States 244
     8.4.1   Interatomic Exchange Interaction 245
     8.4.2   Heisenberg Model 247
     8.4.3   Ferromagnetism: Molecular Field Approximation 249
8.5   Antiferromagnetism and Ferrimagnetism 252
8.6   From Insulator Magnetism to Metallic Magnetism 255
     8.6.1   Mott–Hubbard Insulator 256
     8.6.2   Mott Transition and Doped Mott–Hubbard Insulators 258
     8.6.3   Magnetism and Superconductivity 259
     8.6.4   Metallicity and Magnetism in a Band Approach 261
8.7   Summary 265
8.8   Answers to Questions 266

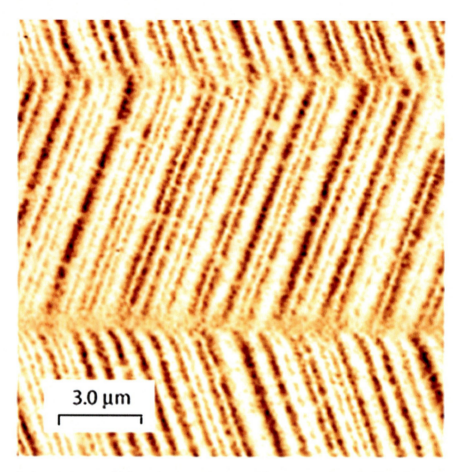

An important means of information storage is to record data on magnetic media. The information recorded on a digital audio tape (DAT) can be visualised by magnetic force microscopy (MFM, see Chap. 10). On this tape, the magnetisation of the grains in a magnetic film deposited on a polymer substrate has been imposed by a read head that moves much more quickly than the winding rate of the tape, and which can write the successive tracks on the tape. Here a 1 or a 0 corresponding to an upward magnetisation (*black*) or a downward magnetisation (*white*) is imposed on a region of area 0.25×10 μm^2. Data is written onto hard disks in a similar way, with a data storage density that is currently 400 times higher (1 bit on 25×250 nm^2). Image courtesy of Miltat, J., Thiaville, A.: Laboratoire de Physique des Solides. Orsay, France

8 Magnetism of Insulators

Magnetic fields are understood by classical electromagnetism as resulting from macroscopic currents circulating in conductors. However, this description does not help us to understand the existence of magnetic fields spontaneously induced by materials like iron or cobalt when there are no macroscopic currents imposed by external sources. Only quantum mechanics can explain electronic states corresponding to permanent currents at the microscopic scale. We have just seen an example in the case of superconductors. Likewise, the electron orbital angular momentum and spin are the purely quantum entities which underlie the magnetism of materials.[1]

In this chapter we begin by explaining the quantum origins of spontaneous magnetic moments in materials, and in particular ferromagnetism, using a model which applies to insulating materials for which the electron magnetic moments interact via an exchange interaction between near neighbours. We shall see in Chap. 9 that the presence of magnetic domains in the materials around us involves a very low energy compared with the exchange interaction responsible for magnetism. These low energy phenomena nevertheless govern most magnetic material behaviours which are of great importance for their technical uses. In Chap. 10, we shall discuss experimental methods used for measurements and imaging of magnetic materials. Chapter 11 deals with the response of magnetic materials to alternating fields, on the basis of which we can discuss magnetic resonance methods. Finally, in Chap. 12, we consider the thermodynamic properties of magnetic materials.

Note that the methods used in Chaps. 1 and 3 to describe electronic bands in solids, and in particular the approximation which consists in only considering the *averaged* interaction between electrons, led to the consideration of only two types of solid: metals exhibiting Pauli magnetic susceptibility, and non-magnetic insulators. *There is thus no room for spontaneous magnetism in such a description.* The magnetic properties of superconductors stem from the existence of an attractive interaction between electrons which is induced by the phonons of the atomic lattice, and this attractive interaction leads to a ground state in which the electrons are no longer independent, and where persistent currents can be established on a microscopic scale. To be able to describe the magnetic states of matter, we must therefore take into account the *interactions between electrons.*

After a brief review of the various kinds of magnetic behaviour observed in materials (Sect. 8.1), we shall illustrate how interactions between electrons (direct Coulomb interactions or indirect interactions induced by the Pauli principle) explain atomic magnetism, i.e., the existence of atoms with an electronic magnetic moment (Sect. 8.2). We then investigate how ions carrying independent electronic magnetic moments lead to Curie magnetism (Sect. 8.3). In Sect. 8.4, we shall see how the

---

[1] Note that atomic nuclei also carry spins, and have magnetic moments that give rise to Curie magnetism. However, the nuclear magnetic moments are very small compared with that of the electron and make an altogether negligible contribution to the magnetic properties of the solid, which are thus entirely electronic in origin. The nuclear magnetism can only be directly detected by a magnetometer in quite exceptional circumstances (materials with weak electronic magnetism at very low temperatures), but is easier to observe using resonance methods (see Chap. 11).

overlap between electron atomic orbitals induces a coupling between the electronic magnetic moments that is responsible for ordered magnetic states observed at sufficiently low temperatures. A minimal approach for dealing with the interactions between electrons will be outlined in Sect. 8.6 in order to understand how magnetic and metallic behaviour might coexist.

8.1 Magnetic Behaviour of Solids

The vast majority of natural compounds exhibit very weak magnetism. A highly sensitive magnetometer is needed to measure the magnetic susceptibility in this case. Measurements show that these compounds are generally *diamagnetic*. This arises because most materials are insulators made up of ions (or molecules) with *saturated electronic shells* whose orbitals barely overlap with those of nearest neighbours. In the models of electronic structure in Chaps. 1 and 3, the electronic states differ little from those of isolated ions (or molecules) and give rise to similar magnetic behaviour to that observed in the corresponding liquid or gaseous states. For ions (or molecules) with saturated electronic shells, the electrons occupy electronic levels whose spins and orbital angular momenta cancel. In this case, the only source of magnetism lies in modifications to the electronic orbitals when an external field is applied. The classical analogy between an orbital and an electronic current can be used to understand that the current induced by applying an external field will oppose this field, simply from Lenz's law, and this situation corresponds to a diamagnetic magnetisation. The diamagnetic susceptibility is small and *temperature independent (see Fig. 8.1), since it is a quantum mechanical property of the electronic ground state*. Indeed, the excited states are hardly populated thermally, because they are generally several eV away. This diamagnetism ($\tilde{\chi}_d \simeq -10^{-5}$) is negligible compared with values for superconductors ($\tilde{\chi} = -1$). Note that $\tilde{\chi}_d$ is directly related to the size of the electronic orbital in the direction perpendicular to the applied field, and will be particularly relevant for electrons involved in molecular states that

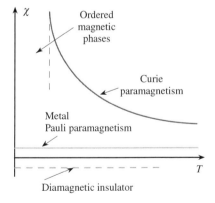

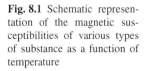

Fig. 8.1 Schematic representation of the magnetic susceptibilities of various types of substance as a function of temperature

8.1 Magnetic Behaviour of Solids

extend over several atoms. This happens, for example, in the benzene rings of aromatic molecules.

In metals, the inner electron shells also overlap very slightly and lead to a similar contribution. In addition to this diamagnetic susceptibility from the inner shells, there is another due to the electrons in the partially filled bands which exhibit *Pauli spin paramagnetism* [see (3.66)]:

$$\chi_P = 2D_\Omega(E_F)\mu_B^2, \quad (8.1)$$

which is also almost independent of the temperature insofar as the conduction band is broad and has no fine structure on the scale $k_B T \simeq 25$ meV. If the density of states $D_\Omega(E_F)$ is small, the total resulting susceptibility $\chi = \chi_d + \chi_P$ remains diamagnetic, as happens for example for copper. For metals in the transition series, e.g., the $4d$ series, the bands are narrow and the densities of states higher than for the noble metals. The resulting magnetic susceptibility is then paramagnetic.

The most significant magnetic effects arise in solids containing ions whose atomic shells are not saturated, and which carry a *magnetic moment*. In this case, strong paramagnetism is observed, increasing according to $\chi \simeq C/T$ as the temperature decreases. This is *Curie paramagnetism*, which arises from the removal of degeneracy of the electron spin states when a field is applied. This paramagnetism of independent magnetic moments corresponds to a reduction in the statistical disorder of the orientations of the moments in the presence of a field. It is also of purely atomic origin.

However, it is found experimentally that, in many cases, the observed paramagnetism deviates significantly from this law, especially at low temperatures. In particular, at low temperatures, *spontaneous magnetisation* is often observed, and the solid retains a nonzero magnetic moment even if the applied magnetic field is switched off:

$$\lim_{H_a \to 0} M(H_a) \equiv M_0 \neq 0. \quad (8.2)$$

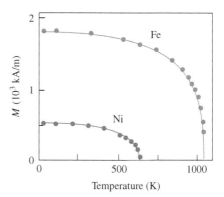

Fig. 8.2 Spontaneous moments of Ni and Fe as a function of temperature. These are metals with high Curie temperature T_C

The existence of a state with spontaneous magnetisation is due at least in part to an ordering of the magnetic moments. We have here a collective phenomenon of the system of electronic moments induced by interactions between the moments of neighbouring atoms. This *collective magnetism* only occurs in the solid phase. The best known case is no doubt *ferromagnetism*, which corresponds to a parallel alignment of all the spins, at least on average. The typical temperature dependence of the spontaneous magnetisation $M_0(T)$ is shown in Fig. 8.2 for nickel and iron. Spontaneous magnetisation appears at the *Curie temperature* T_C.

8.2 Magnetism of Atoms

Before considering the origins of the ordered magnetism which arises in certain solids, let us briefly review some features of the electronic structure of atoms.

8.2.1 Hydrogen Atom

Consider first the electronic energy levels in the simplest atom, namely hydrogen (see Fig. 8.3a) [2, Chap. 11]. Note that there is a high level of degeneracy in the levels. Each level with principal quantum number $n = 1, 2, 3, \ldots$, corresponds to a state of energy $E_n = -E_1/n^2$ and is degenerate with respect to different values of the orbital quantum number l (with $l = 0, 1, \ldots, n-1$). Each of these levels is itself $2l+1$ times degenerate. If we include the spin degeneracy, the total degeneracy of a given energy level is $2n^2$. Note that, since the hydrogen atom has only one electron,

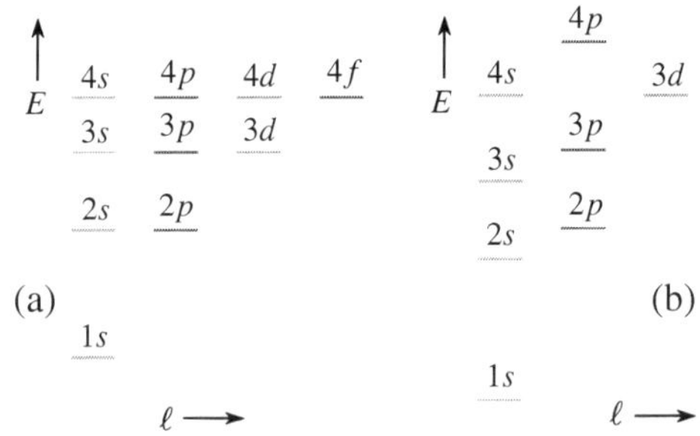

Fig. 8.3 (a) Energy level diagram for the hydrogen atom. The angular momentum quantum numbers are given using the traditional notation, where s corresponds to $l = 0$, p to $l = 1$, and so on. The energy is $E_n = -E_1/n^2$, where $E_1 = 13.6$ eV. The ground state is $1s$. (b) Energy level diagram for a multielectron atom. The energy of an electron state depends on the quantum numbers n and l

its ground state is clearly the 1s level, with angular momentum $l = 0$, while the spin is $S = 1/2$. In this case, the magnetic moment of the atom is

$$\mu = 2\mu_B S = \mu_B. \tag{8.3}$$

8.2.2 Multielectron Atoms with Filled Shells

For a multielectron atom with Z electrons, the Coulomb potential of the nucleus of charge Z is $-ZV_c(\mathbf{r})$. If we only take this potential into account, the corresponding electron levels are similar to those of the hydrogen atom with $E_{n,Z} = -Z^2 E_1/n^2$. However, the Coulomb interactions between electrons will modify these levels, and in particular, they will remove the degeneracy of levels with the same principal quantum number but different values of l. The energy of a level increases with its orbital angular momentum (see Fig. 8.3b). Only the orbital degeneracies (due to the spherical symmetry of the Coulomb potential) and spin degeneracies (due to the rotational symmetry in spin space) will remain. The degeneracy of a level in Fig. 8.3b is thus $2(2l+1)$.

This partial lifting of degeneracy, revealed directly through the structure of the periodic table, can be understood in a highly qualitative way by the following argument. Electrons in s levels have wave functions concentrated close to the nucleus and therefore feel the repulsive effect of other electrons only rather weakly. In contrast, an electron in a p level (and even more so in d or f) with the same principal quantum number is on average further from the nucleus and thus feels the repulsive effect of other electrons to a greater extent, which increases its energy compared with the s level. Note that, in this level diagram, the Z electrons occupy the lowest levels allowed by the Pauli exclusion principle in the ground state. If the number of electrons is such that all the low energy states can be filled, the atomic shells are saturated and the atom has no spin or orbital magnetic moment.

8.2.3 An Atom with a Partially Filled Shell: Carbon

To understand the origin of the magnetic moments of atoms with partially filled shells, consider the case of carbon, which has 6 electrons. Recall that, if $V_{at}(\mathbf{r})$ is the atomic potential including the Coulomb potential of the nucleus and the mean value of the repulsion between electrons, the energy levels of this multielectron atom are solutions of

$$H_0 = \frac{p^2}{2m_0} + V_{at}(\mathbf{r}).$$

The eigenstates correspond to levels 1s, 2s, 2p, and so on. In the ground state of the atom, two electrons occupy the 1s level and two the 2s level, while the last

Fig. 8.4 p_x, p_y, p_z orbitals

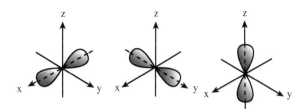

two are in the $2p$ state. The pairs in the $1s$ and $2s$ levels must be in configurations with total spin zero, owing to the Pauli principle. It remains to determine the exact configuration of the two p electrons which occupy the six-fold degenerate energy level E_{2p}. A basis for these eigenstates is formed by the three radial wave functions p_x, p_y, and p_z oriented along the x, y, and z axes, shown in Fig. 8.4. These states have the same energy, as they can be obtained from one another by rotations in space.

To construct the two-electron eigenstates, a first possibility is to put the two electrons in the same orbital, so the spatial part of the wave function would be $|p_x p_x\rangle = |p_x\rangle_1 |p_x\rangle_2$, where α and β in the notation $|\alpha\beta\rangle$ indicate the states of the first and second electrons, respectively. According to the Pauli exclusion principle, the total wave function of these two electrons must be antisymmetric under exchange of the two particles ($\mathbf{r}_1 \leftrightarrow \mathbf{r}_2$). As this spatial wave function is symmetric, the spin wave function must be antisymmetric, giving a total wave function

$$|xx, S=0\rangle = \frac{1}{\sqrt{2}}(|\uparrow\downarrow\rangle - |\downarrow\uparrow\rangle)|p_x p_x\rangle. \tag{8.4}$$

Here, in a ket like $|\uparrow\downarrow\rangle$, the first (second) arrow gives the z component of the spin of the first (second) electron. In the spin part of (8.4), we recognise the spin function of a singlet state, where two spins 1/2 are combined to give a total spin $S = 0$.

Alternatively, the two electrons can be put in two different orbitals, e.g., $|p_x p_y\rangle$. This product clearly has no well defined symmetry properties. We thus consider instead the symmetric and antisymmetric linear combinations $(|p_x p_y\rangle + |p_y p_x\rangle)/\sqrt{2}$ and $(|p_x p_y\rangle - |p_y p_x\rangle)/\sqrt{2}$, respectively. The first also gives rise to an antisymmetric wave function ($S = 0$)

$$|x, y, S=0\rangle = \frac{1}{\sqrt{2}}(|\uparrow\downarrow\rangle - |\downarrow\uparrow\rangle)\frac{1}{\sqrt{2}}(|p_x p_y\rangle + |p_y p_x\rangle). \tag{8.5}$$

The total wave function associated with the antisymmetric spatial wave function is obtained by multiplying by a symmetric spin wave function, e.g.,

$$|xy, S=1, M_z=1\rangle = |\uparrow\uparrow\rangle \frac{1}{\sqrt{2}}(|p_x p_y\rangle - |p_y p_x\rangle). \tag{8.6}$$

Note that the spin part is the $M_z = 1$ component of the triplet wave function, with total spin $S = 1$, of a pair of spin 1/2 particles, which explains the notation used on

8.2 Magnetism of Atoms

the left-hand side of (8.6). For the symmetric spin wave function required here, we could also take the functions $(|\uparrow\downarrow\rangle + |\downarrow\uparrow\rangle)/\sqrt{2}$ or $|\downarrow\downarrow\rangle$, which are the $M_z = 0$ and $M_z = -1$ components of the spin triplet.

The singlet states (8.4) and (8.5) and the triplet state (8.6) constructed here are eigenstates of the two-electron Hamiltonian

$$H_1 = \frac{p_1^2}{2m_0} + V_{at}(\mathbf{r}_1) + \frac{p_2^2}{2m_0} + V_{at}(\mathbf{r}_2).$$

They are degenerate and have energy $2E_{2p}$. However, in order to treat the *repulsive Coulomb interaction between the two p electrons* in a fully rigorous way, we must use the Hamiltonian

$$H = \frac{p_1^2}{2m_0} + V_{at}(\mathbf{r}_1) + \frac{p_2^2}{2m_0} + V_{at}(\mathbf{r}_2) + V_c(\mathbf{r}_2 - \mathbf{r}_1). \qquad (8.7)$$

Given the eigenstates of H_1, the effects of the Coulomb interaction term denoted by $V_{12} = V_c(\mathbf{r}_2 - \mathbf{r}_1)$ can be treated as a perturbation. In this case, the energies of the different eigenstates are given to first order in perturbation theory by

$$E_{xx,S=0} = 2E_{2p} + \langle p_x p_x | V_{12} | p_x p_x \rangle = 2E_{2p} + K_{xx}, \qquad (8.8)$$

$$E_{xy,S=0} = 2E_{2p} + \frac{1}{2}\big(\langle p_x p_y| + \langle p_y p_x|\big) V_{12} \big(|p_x p_y\rangle + |p_y p_x\rangle\big)$$

$$= 2E_{2p} + \langle p_x p_y | V_{12} | p_x p_y \rangle + \langle p_y p_x | V_{12} | p_x p_y \rangle$$

$$= 2E_{2p} + K_{xy} + J_0, \qquad (8.9)$$

$$E_{xy,S=1} = 2E_{2p} + K_{xy} - J_0. \qquad (8.10)$$

In these expressions, the terms K_{xx}, K_{xy}, and J_0 are all positive. Note that, in the integral K_{xx}, the electrons are on the same orbital p_x and have higher probability of being at the same place than in K_{xy}, where they occupy two different orbitals. The Coulomb repulsion is therefore stronger in the first case than in the second, so $K_{xx} > K_{xy}$. The Coulomb repulsion between electrons thus tends to make the two electrons occupy different orbitals. The energy level diagram is as shown in Fig. 8.5.

Note that this level diagram highlights the importance of the integral

$$J_0 = \left\langle p_y(\mathbf{r}_1) p_x(\mathbf{r}_2) \middle| V_c(\mathbf{r}_2 - \mathbf{r}_1) \middle| p_x(\mathbf{r}_1) p_y(\mathbf{r}_2) \right\rangle. \qquad (8.11)$$

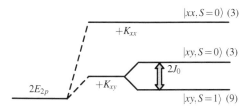

Fig. 8.5 Level diagram taking into account Coulomb repulsion between the two p electrons. The degeneracies of the levels are given in brackets

This integral is called the *intra-atomic exchange integral* because it involves states in which the electrons are exchanged from one side to the other. It requires the ground state to be the spin triplet state $S = 1$ rather than the singlet state $S = 0$ constructed with the same radial wave functions. This effect can be understood qualitatively by examining (8.5) and (8.6). In the first spin singlet case, the symmetry of the spatial wave function allows the electrons to be at the same position, whereas in the case of the triplet, the antisymmetry of the spatial wave function means that the probability of the two electrons being at the same position is zero. The divergence of V_c at short range does not contribute to the corresponding matrix element, and this implies a weaker Coulomb repulsion for the triplet state. The maximal spin state $S = 1$ is thus the state with lowest energy.

This is a first example of one of the key ideas underpinning almost every magnetic phenomenon: *the Coulomb interaction between electrons and the Pauli principle constitute the microscopic explanation for atomic magnetism*. Note that no direct spin-dependent interaction is needed to remove the degeneracy between singlet and triplet states.

8.2.4 Atoms with Partially Filled Shells: Hund Rules

The above considerations for two electrons can be generalised to any situation in which an atomic shell is partially filled, and this leads to the *Hund rules*, which determine the orbital and spin angular momenta of the atomic ground state:

1. Choose the wave function with maximal total spin S.
2. Among the remaining possibilities, choose the wave function with maximal orbital angular momentum L, taking care to respect the Pauli exclusion principle, i.e., to build a totally antisymmetric wave function.
3. The total angular momentum J is given by $|L-S|$ if the shell is less than half-filled, and by $L+S$ if it is more than half-filled (an exactly half-filled shell always gives $L = 0$).

Rules 1 and 2 are explained by a generalisation of the previous argument, appealing to the Coulomb interaction between the electrons and the Pauli principle, while rule 3 is due to a relativistic correction to quantum mechanics, namely the spin–orbit interaction. Finally, an atom of total angular momentum J carries a magnetic moment

$$\boldsymbol{\mu} = -g_J \mu_B \mathbf{J}, \tag{8.12}$$

where the *Landé factor* g_J is given by

$$g_J = \frac{3}{2} + \frac{1}{2}\left[\frac{S(S+1) - L(L+1)}{J(J+1)}\right]. \tag{8.13}$$

8.2 Magnetism of Atoms

Note that, in the case of carbon, rule 3 gives a total angular momentum $J = 0$, and hence zero magnetic moment, but there are clearly going to be many other situations, e.g., three or more electrons in a p shell, two or more electrons in a d shell, etc., where J will not vanish. The most important cases (when the d or f shells are partially filled) are given in Tables 8.1 and 8.2, which specify the order in which the electronic shell must be filled in order to satisfy the first two Hund rules.

For example, for the d shell, we have $l = 2$ and $l_z = 2, 1, 0, -1, -2$, and the corresponding electron can have spin ↑ or ↓, so there are a maximum of 10 electrons in the shell. To satisfy the first Hund rule, we have to put in a maximum of spin ↑ electrons before filling states with spin ↓. Choosing the maximum orbital quantum numbers for each electron then ensures the correct application of the second Hund rule.

Consider, for instance, the case of Mn ($3d^5 4s^2$). The Mn^{3+} ion will thus have a $3d^4$ configuration, i.e., $n = 4$ electrons in the d shell. The occupation of electronic states for $n = 4$ is specified in the fourth line of Table 8.1. The electrons must all have spin ↑ with l decreasing from 2 to -1. The resulting value for the spin is therefore $S = 4(1/2) = 2$, and for the orbital angular momentum $L = 2 + 1 + 0 - 1 = 2$. The value of J results from the third Hund rule and corresponds to $J = |L - S| = 0$ since the shell is less than half-filled. Naturally, the Hund rules imply that there is no magnetic moment for filled shells with $L = S = J = 0$, whereas for a half-filled shell, $L = 0$ and the magnetism is entirely due to the spin, as is the case for the Mn^{2+} ion, for example.

In a similar way, Table 8.2 specifies the occupation of the f shell, which corresponds to $l = 3$ and the rare earth series.

Table 8.1 Minimum energy configurations of the partially filled d shell. n is the total number of electrons and l_z the z component of the orbital angular momentum of an electron. L, S, and J satisfy the Hund rules

n	$l_z = 2$	1	0	-1	-2	S	L	J
1	↑					1/2	2	3/2
2	↑	↑				1	3	2
3	↑	↑	↑			3/2	3	3/2
4	↑	↑	↑	↑		2	2	0
5	↑	↑	↑	↑	↑	5/2	0	5/2
6	↑↓	↑	↑	↑	↑	2	2	4
7	↑↓	↑↓	↑	↑	↑	3/2	3	9/2
8	↑↓	↑↓	↑↓	↑	↑	1	3	4
9	↑↓	↑↓	↑↓	↑↓	↑	1/2	2	5/2
10	↑↓	↑↓	↑↓	↑↓	↑↓	0	0	0

Table 8.2 Minimum energy configurations of the partially filled f shell. n is the total number of electrons and l_z the z component of the orbital angular momentum of an electron. L, S, and J satisfy the Hund rules

n	$l_z = 3$	2	1	0	−1	−2	−3	S	L	J
1	↑							1/2	3	5/2
2	↑	↑						1	5	4
3	↑	↑	↑					3/2	6	9/2
4	↑	↑	↑	↑				2	6	4
5	↑	↑	↑	↑	↑			5/2	5	5/2
6	↑	↑	↑	↑	↑	↑		3	3	0
7	↑	↑	↑	↑	↑	↑	↑	7/2	0	7/2
8	↑↓	↑	↑	↑	↑	↑	↑	3	3	6
9	↑↓	↑↓	↑	↑	↑	↑	↑	5/2	5	15/2
10	↑↓	↑↓	↑↓	↑	↑	↑	↑	2	6	8
11	↑↓	↑↓	↑↓	↑↓	↑	↑	↑	3/2	6	15/2
12	↑↓	↑↓	↑↓	↑↓	↑↓	↑	↑	1	5	6
13	↑↓	↑↓	↑↓	↑↓	↑↓	↑↓	↑	1/2	3	7/2
14	↑↓	↑↓	↑↓	↑↓	↑↓	↑↓	↑↓	0	0	0

Spin–Orbit Interaction

The fact that the nucleus is charged and moving relative to the electron implies that it will create a magnetic field that couples to the electron spin. This is called the spin–orbit coupling because it involves the electron spin and its orbital degrees of freedom, and it takes the form $E_{SO} = \lambda \mathbf{l} \cdot \mathbf{s}$. It leads to a coupling $E_{SO} = \Lambda \mathbf{L} \cdot \mathbf{S}$ between the total spin and the total orbital angular momentum of the multielectron atom. When there is no spin–orbit coupling, the mutually commuting operators $\mathbf{L}^2$, L_z, $\mathbf{S}^2$, and S_z each commute with the Hamiltonian, whence one can find a common basis of eigenfunctions which are also eigenfunctions of the Hamiltonian. In this case, (L, m_L, S, m_S) are good quantum numbers for characterising an eigenstate. On the other hand, the spin–orbit coupling term does not commute with these operators. It can be shown that the operator $\mathbf{J} = \mathbf{L} + \mathbf{S}$ is such that J_z and $\mathbf{J}^2 = (\mathbf{L} + \mathbf{S})^2 = \mathbf{L}^2 + \mathbf{S}^2 + 2\mathbf{L} \cdot \mathbf{S}$ commute with $\mathbf{L} \cdot \mathbf{S}$. It then follows that J_z, $\mathbf{J}^2$, and the Hamiltonian including the spin–orbit coupling term have a common basis of eigenstates. The good quantum numbers for the Hamiltonian with spin–orbit coupling are therefore J and m_J. The value of J corresponding to the ground state is the one giving the lowest value to the spin–orbit coupling term, which is proportional to $2\mathbf{L} \cdot \mathbf{S} = \mathbf{J}^2 - \mathbf{S}^2 - \mathbf{L}^2$.

The expectation value of the total magnetic moment of the ground state, viz.,

$$\boldsymbol{\mu} = \mu_B (\mathbf{L} + 2\mathbf{S}),$$

is $\langle \boldsymbol{\mu} \rangle = -\langle J, m_J | \mathbf{L} + 2\mathbf{S} | J, m_J \rangle \mu_B$. This can be determined using the Wigner–Eckart theorem, which tells us that the expectation value of $\boldsymbol{\mu}/\mu_B$ in the subspace $| J, m_J \rangle$ is proportional to the expectation value of $\mathbf{J}$, with multiplicative coefficient [21]

$$g_J = -\frac{\langle J | \boldsymbol{\mu} \cdot \mathbf{J} | J \rangle / \mu_B}{J(J+1)}.$$

It is easy to check, using the expressions for $\mathbf{J} \cdot \mathbf{S}$ and $\mathbf{J} \cdot \mathbf{L}$ in terms of $\mathbf{J}^2$, $\mathbf{L}^2$, and $\mathbf{S}^2$, that $-2\boldsymbol{\mu} \cdot \mathbf{J}/\mu_B = 3\mathbf{J}^2 - \mathbf{L}^2 + \mathbf{S}^2$, which yields the expression (8.13) for g_J.

8.3 Paramagnetism of an Ensemble of Isolated Ions

Consider an ionic solid containing isolated atoms with magnetic moments, like the one discussed in the last section. In the presence of a magnetic field H_a, the spin degeneracy of the electron wave function is removed, the spin levels being

$$E_{J_z} = +g_J \mu_0 \mu_B J_z H_a, \qquad J_z = -J, -J+1, \ldots, J-1, J.$$

If *the ions are independent*, statistical thermodynamics can be used to determine the statistical average of the component μ_z of the magnetic moment operator $\hat{\mu} = -g_J \mu_B \hat{\mathbf{J}}$ from

$$\langle \hat{\mu}_z \rangle = \frac{\sum_{J_z=-J}^{+J} g_J \mu_B J_z \exp(-\beta E_{J_z})}{\sum_{J_z=-J}^{+J} \exp(-\beta E_{J_z})},$$

where $\beta = 1/k_B T$. If the number of paramagnetic ions per unit volume is N, a simple but tedious modification of this expression leads to the following result for the magnetisation $M_z = N \langle \hat{\mu}_z \rangle$ along the z axis:

$$M_z = N g_J \mu_B J \left[\frac{2J+1}{2J} \coth\left(\frac{2J+1}{2J}\alpha\right) - \frac{1}{2J} \coth\left(\frac{\alpha}{2J}\right) \right]$$

$$= N g_J \mu_B J B_J(\alpha), \qquad (8.14)$$

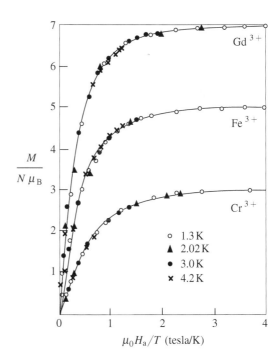

Fig. 8.6 Measurements of the magnetisation per paramagnetic ion for different salts containing the ions indicated in the figure, as a function of the magnetic field and for different temperatures. For a given salt, the magnetisation is indeed a function of the ratio H_a/T. Continuous curves are the Brillouin functions for the ions under consideration. From Henry, W.E.: Phys. Rev. **88**, 559 (1952). With the kind permission of the American Physical Society (APS). http://link.aps.org/doi/10.1103/PhysRev.88.559

where we have set $\alpha = g_J \mu_0 \mu_B J H_a / k_B T$. The function $B_J(\alpha)$ is called the *Brillouin function*. It reduces to $B_{1/2}(\alpha) = \tanh\alpha$ in the special case where $J = 1/2$. At high temperatures, we have $\alpha \ll 1$ and the expansion $\coth u = u^{-1} + u/3$ for small u gives $B_J(\alpha) = \alpha(J+1)/3J$, whence the magnetic susceptibility obeys *Curie's law*:

$$\chi(T) = \lim_{H_a \to 0} \frac{1}{\mu_0} \frac{\partial M}{\partial H_a} = N \frac{(g_J \mu_B)^2}{3} \frac{J(J+1)}{k_B T}. \tag{8.15}$$

This law is found to apply well to many paramagnetic solids, as can be seen from Fig. 8.6.

Question 8.1. From Fig. 8.6, can you determine the values of g_J and J for the various ions considered there? Recall that the configurations of the neutral atoms are Cr ($3d^4\ 4s^2$), Fe ($3d^6\ 4s^2$), and Gd ($4f^7\ 5s^2\ 5p^6\ 5d^1\ 6s^2$). Do the values of g_J and J correspond to what would be expected from Tables 8.1 and 8.2? What conclusion can you draw?

8.4 Ordered Magnetic States

We saw in Sect. 8.1 that in many cases the Curie law will only work at high temperatures, and that an ordered magnetic state appears at low temperatures, e.g., a parallel alignment of the spins in the ferromagnetic state below the Curie temperature T_C. Clearly, such an alignment in zero field at thermodynamic equilibrium is only possible if there are *interactions between the spins* which favour that. (If there were no interactions, the only magnetic effect would be Curie paramagnetism.)

The first type of interaction between spins we can think of would be a dipole interaction. Since each spin carries a magnetic moment **m**, there is a potential energy of interaction between two spins of the form

$$V(\mathbf{r}) = \frac{\mu_0}{4\pi r^3}\left[\mathbf{m}_1 \cdot \mathbf{m}_2 - \frac{3}{r^2}(\mathbf{m}_1 \cdot \mathbf{r})(\mathbf{m}_2 \cdot \mathbf{r})\right], \tag{8.16}$$

where **r** is the displacement vector between the two spins. The order of magnitude of this interaction is easily estimated. The magnetic moment of an atom is of the order of a few times μ_B, so taking a value of μ_B, we have

$$V(\mathbf{r}) \approx \frac{\mu_0}{4\pi} \frac{\mu_B^2}{r^3}. \tag{8.17}$$

With an interatomic distance $r \sim 2$ Å, we then obtain $V(\mathbf{r}) \approx (1/8) \times 10^{-23}$ joule, which corresponds to

$$V(\mathbf{r})/k_B \sim 0.1 \text{ K}. \tag{8.18}$$

This should be compared with typical Curie temperatures, which can be as high as 1,000 K. It thus seems inconceivable that an interaction of the order of 0.1 K should give rise to ferromagnetism at temperatures a thousand times higher. The dipole interactions cannot be the cause of ferromagnetism!

8.4 Ordered Magnetic States

8.4.1 Interatomic Exchange Interaction

It is in fact now well established that the origin of magnetic ordering is actually a purely quantum interaction called the *exchange interaction*, which arises as a consequence of the Coulomb interactions between electrons in combination with the Pauli exclusion principle. We have already seen in the last section how these two effects can give rise to atomic magnetic moments.

In order to understand this interaction, let us consider the simplest case, viz., the hydrogen molecule, comprising two electrons and two protons. The Hamiltonian will then be

$$H = \frac{\hbar^2}{2m_0}(\mathbf{p}_1^2 + \mathbf{p}_2^2) + \frac{e^2}{4\pi\varepsilon_0}\frac{1}{|\mathbf{r}_1 - \mathbf{r}_2|} \quad (8.19)$$

$$-\frac{e^2}{4\pi\varepsilon_0}\left(\frac{1}{|\mathbf{r}_1 - \mathbf{R}_1|} + \frac{1}{|\mathbf{r}_1 - \mathbf{R}_2|} + \frac{1}{|\mathbf{r}_2 - \mathbf{R}_1|} + \frac{1}{|\mathbf{r}_2 - \mathbf{R}_2|}\right),$$

where $\mathbf{r}_i$ and $\mathbf{p}_i$ are the positions and momenta of the two electrons, while the $\mathbf{R}_i$ are the positions of the two protons. The first line of (8.19) represents the kinetic energy of the two electrons and their mutual Coulomb repulsion, and the second line is the Coulomb attraction between electrons and protons. The protons are assumed fixed, so their is no term for their kinetic energy here. Without explicitly solving the Hamiltonian (8.19), we may nevertheless make the following observations: H is invariant under exchange of the two electrons, $\mathbf{r}_1 \leftrightarrow \mathbf{r}_2$, and as a consequence the spatial parts of the wave functions are either even or odd with respect to this transformation:

$$\psi_\pm(\mathbf{r}_1, \mathbf{r}_2) = \pm\psi_\pm(\mathbf{r}_2, \mathbf{r}_1). \quad (8.20)$$

The situation is similar to the one encountered in Sect. 8.2.3 for the two p electrons of the carbon atom. The energies $E_\pm$ of these states will generally differ. The Pauli principle requires the total wave function, including both spatial and spin factors, to be antisymmetric under exchange of the two electrons. We thus have either a spin singlet wave function ($S = 0$), viz.,

$$\psi_{\text{tot},+} = \frac{1}{\sqrt{2}}(|\uparrow\downarrow\rangle - |\downarrow\uparrow\rangle)\psi_+(\mathbf{r}_1, \mathbf{r}_2), \quad (8.21)$$

or a triplet wave function ($S = 1$) with three possible components ($M = -1, 0, 1$):

$$\psi_{\text{tot},-} = \begin{cases} |\downarrow\downarrow\rangle\psi_-(\mathbf{r}_1, \mathbf{r}_2), \\ \frac{1}{\sqrt{2}}(|\uparrow\downarrow\rangle + |\downarrow\uparrow\rangle)\psi_-(\mathbf{r}_1, \mathbf{r}_2), \\ |\uparrow\uparrow\rangle\psi_-(\mathbf{r}_1, \mathbf{r}_2). \end{cases} \quad (8.22)$$

A classic picture of these two situations is one in which the two spins are antiparallel ($S = 0$) or parallel ($S = 1$).

For the hydrogen molecule, the ground state is the + type, i.e., $S = 0$, whereas the first excited state is of − type, i.e., $S = 1$. (This is where the Coulomb interaction between the electrons comes in. Without it, these different states would be degenerate.). If we are interested in relatively low energy phenomena, where only the ground state and lowest excited state are involved, we can then specify these states by their spin, with $S = 0$ for the ground state and $S = 1$ for the first excited state.

In order to represent the system, we can take an effective Hamiltonian that is diagonal in this $S = 0$, $S = 1$ representation. We note that the Hamiltonian $\hat{\mathbf{S}}_1 \cdot \hat{\mathbf{S}}_2$ satisfies this condition. Indeed $\hat{\mathbf{S}}^2 = (\hat{\mathbf{S}}_1 + \hat{\mathbf{S}}_2)^2$ implies that

$$\hat{\mathbf{S}}_1 \cdot \hat{\mathbf{S}}_2 = \frac{1}{2}\left(\hat{\mathbf{S}}^2 - \hat{\mathbf{S}}_1^2 - \hat{\mathbf{S}}_2^2\right), \tag{8.23}$$

so for two spin 1/2 particles,

$$\hat{\mathbf{S}}_1 \cdot \hat{\mathbf{S}}_2 = \frac{1}{2}\hat{\mathbf{S}}^2 - \frac{3}{4}, \tag{8.24}$$

and then

$$\hat{\mathbf{S}}_1 \cdot \hat{\mathbf{S}}_2 |S=0\rangle = -\frac{3}{4}|S=0\rangle, \tag{8.25}$$

$$\hat{\mathbf{S}}_1 \cdot \hat{\mathbf{S}}_2 |S=1\rangle = \frac{1}{4}|S=1\rangle. \tag{8.26}$$

The Hamiltonian can thus be represented in the subspace of the first singlet and triplet states by

$$\boxed{\hat{\mathbf{H}} = -\mathscr{J}\hat{\mathbf{S}}_1 \cdot \hat{\mathbf{S}}_2}, \tag{8.27}$$

up to an additive constant, with

$$\mathscr{J} = E_+ - E_-, \tag{8.28}$$

where $\mathscr{J}$ is called the *exchange constant*. This *exchange Hamiltonian* (8.27) determines the spin of the ground state. For $\mathscr{J} < 0$ (this is satisfied for hydrogen), the ground state is a singlet (with 'antiparallel' spins), while for $\mathscr{J} > 0$, it is a triplet (with 'parallel' spins).

It is important to understand that this is an effective Hamiltonian describing the combined effects of the Coulomb interaction and the Pauli exclusion principle. The original Hamiltonian (8.19) contains no terms depending on the spins of the electrons. The process which governs the appearance of magnetism is then similar to the one that yielded atomic magnetism. There, the repulsive interaction between electrons favoured spatially antisymmetric states for the carbon atom, which must be triplet states due to the Pauli principle. We could indeed have treated the subset of the two states (8.5) and (8.6) of the carbon atom using a Hamiltonian $\hat{\mathbf{H}} = -2J_0 \mathbf{s}_1 \cdot \mathbf{s}_2$ whose ground state is the triplet state, with $2J_0$ representing the *intra-atomic exchange coupling*.

Note that the energies involved here are either kinetic or Coulomb and lead to much higher values of $\mathcal{J}$ than the dipole interaction energies. This kind of coupling can therefore explain the high values found for Curie temperatures. Note that, as a general rule, the values of $\mathcal{J}$ are very difficult to determine theoretically, and are usually obtained experimentally.

Exchange Hamiltonian

The term 'exchange Hamiltonian' demands a little more explanation. Consider the effect of the operator $\hat{\mathbf{S}}_1 \cdot \hat{\mathbf{S}}_2$ on the quantum states of the two spins. Noting that

$$\hat{\mathbf{S}}_1 \cdot \hat{\mathbf{S}}_2 = \hat{S}_{1z}\hat{S}_{2z} + \frac{1}{2}\left(\hat{S}_1^+ \hat{S}_2^- + \hat{S}_1^- \hat{S}_2^+\right),$$

we have

$$\hat{\mathbf{S}}_1 \cdot \hat{\mathbf{S}}_2 |\uparrow\uparrow\rangle = \frac{1}{4}|\uparrow\uparrow\rangle, \quad \hat{\mathbf{S}}_1 \cdot \hat{\mathbf{S}}_2 |\downarrow\downarrow\rangle = \frac{1}{4}|\downarrow\downarrow\rangle,$$
$$\hat{\mathbf{S}}_1 \cdot \hat{\mathbf{S}}_2 |\uparrow\downarrow\rangle = -\frac{1}{4}|\uparrow\downarrow\rangle + \frac{1}{2}|\downarrow\uparrow\rangle, \quad \hat{\mathbf{S}}_1 \cdot \hat{\mathbf{S}}_2 |\downarrow\uparrow\rangle = -\frac{1}{4}|\downarrow\uparrow\rangle + \frac{1}{2}|\uparrow\downarrow\rangle.$$
(8.29)

This can be rewritten in the more compact form

$$\left(\hat{\mathbf{S}}_1 \cdot \hat{\mathbf{S}}_2 + \frac{1}{4}\right)|\alpha\beta\rangle = \frac{1}{2}|\beta\alpha\rangle,$$
(8.30)

where α and β represent one of the two orientations $\uparrow$ or $\downarrow$. The operator $\hat{\mathbf{S}}_1 \cdot \hat{\mathbf{S}}_2 + 1/4$ thus exchanges the two spins. Up to the additive constant $1/4$, it is precisely this operator that appears in the Hamiltonian (8.27), which justifies the given name "Exchange Hamiltonian".

8.4.2 Heisenberg Model

In a magnetic insulating solid, there are a great many spins, one for each magnetic atom. These spins interact through exchange forces like those discussed previously, and the exchange constant $\mathcal{J}$ between two atomic moments depends on the overlap between the electron wave functions of the two atoms. These functions decrease exponentially with distance, so the exchange constant also decreases very fast as the distance between atoms increases. In the vast majority of cases, it is then enough to take into account only the exchange interactions between nearest neighbours. This leads to the *Heisenberg model* for a magnetic insulator, described by the Hamiltonian

$$\hat{H} = -\mathcal{J}\sum_{\langle RR'\rangle} \mathbf{S}_R \cdot \mathbf{S}_{R'} + 2\mu_B \mathbf{B} \cdot \sum_R \mathbf{S}_R,$$
(8.31)

where the $\mathbf{R}$ are the sites occupied by the magnetic atoms (the lattice points of a Bravais lattice in the simplest cases), and $\mathbf{S}_R$ is the spin operator of the atom at the site with position $\mathbf{R}$. The notation $\langle RR'\rangle$ on the sum indicates that it is to be taken over all pairs of nearest neighbours. The first term of $\hat{H}$ thus describes the exchange interactions. The second term represents the interaction between the spins and an external magnetic field $\mathbf{B} = \mu_0 \mathbf{H}_a$.

Consider the case where all the atoms have spin 1/2 (the generalisation to the case of arbitrary spins, rather important in practice, is quite straightforward), and also $\mathscr{J} > 0$, which is usually called the ferromagnetic case for reasons that will soon become clear. Suppose also that the external field is oriented along the z axis and positive. Under these conditions, the ground state of the Heisenberg model (8.31) is simply a state in which all the spins are parallel:

$$|0\rangle = \prod_\mathbf{R} |\downarrow\rangle_\mathbf{R}, \tag{8.32}$$

where $|\downarrow\rangle_\mathbf{R}$ is the spin-down state of the atom at site $\mathbf{R}$. It is easy to check that $|0\rangle$ is effectively an eigenstate of $\hat{H}$. The exchange operator acting on a state in which the two spins are parallel simply reproduces this state up to a prefactor of 1/4 [see (8.29)]. We then obtain

$$\hat{H}|0\rangle = \left(-\frac{Nz\mathscr{J}}{8} - N\mu_B B\right)|0\rangle, \tag{8.33}$$

where N is the number of atoms per unit volume in the solid and z is the number of nearest neighbours. Since $|0\rangle$ minimises each term of $\hat{H}$ individually for $\mathscr{J} > 0$, this state is indeed the ground state. This is still true even if $B = 0$. The spins are all aligned, and this is indeed a ferromagnetic state.

Consider in particular the case where the applied field is zero. Note that the magnetic moment

$$\mathbf{M} = -2\mu_B \langle \mathbf{S} \rangle \tag{8.34}$$

of the state (8.32), where $\mathbf{S} = \sum_\mathbf{R} \mathbf{S}_\mathbf{R}$ is the total spin, is oriented along the positive z axis. However, there are other states with the same energy $-Nz\mathscr{J}/8$. For example, the state

$$|0\rangle' = \prod_\mathbf{R} |\uparrow\rangle_\mathbf{R} \tag{8.35}$$

has all spins pointing in the same direction and opposite magnetisation to $|0\rangle$. More generally, any state obtained by repeated application of the operator

$$S^+ = \sum_\mathbf{R} S^+_\mathbf{R} = \sum_\mathbf{R} (S^x_\mathbf{R} + iS^y_\mathbf{R}) \tag{8.36}$$

to the state $|0\rangle$ is an eigenstate of $\hat{H}$ with the same energy. (This is easily shown by noting that $\hat{H}$ and S^+ commute and that S^+ increases the z component of $\hat{S}$ by 1.) Note in particular that $|0\rangle' = (S^+)^N |0\rangle$ and that there is a total of $N+1$ states of the same energy.

Which state is actually chosen by the system from among these $N+1$ states will depend on how the system is prepared, i.e., the experimental circumstances. However, each of these states is characterised by a nonzero magnetisation oriented in some specific direction in space. This is an example of an important phenomenon known as *symmetry breaking*. The Hamiltonian itself is invariant under simultaneous rotation of all the spins, since it depends only on their scalar products. However, each of its ground states, and more generally, as we shall see in Chap. 10, each of its

8.4 Ordered Magnetic States   249

thermodynamic equilibrium states at low enough temperatures, is characterised by a preferred direction in space. The ground state thus has lower symmetry than the Hamiltonian.[2]

8.4.3 Ferromagnetism: Molecular Field Approximation

The explanation for ferromagnetism arose from a simple idea put forward at the beginning of the 1900s by Weiss and Curie, based on the properties of an ensemble of paramagnetic spins. An applied magnetic field partially aligns the magnetic moments. In the magnetic state, a field thus seems to exist even when there is no applied field. They called this the *molecular field*. Their idea was to treat this as the field created at a vacant site by all the moments. It therefore seemed natural to assume that it would be proportional to the magnetisation itself. If we consider the Heisenberg model, we can understand the basis for this approximation. Indeed, we need only consider all the terms in this Hamiltonian involving a spin at a given site $\mathbf{R}$, viz.,

$$H_\mathbf{R} = \mathbf{S}_\mathbf{R} \cdot \left(-\mathscr{J} \sum_{\mathbf{R}'} \mathbf{S}_{\mathbf{R}'} + 2\mu_B \mu_0 \mathbf{H}_a \right). \qquad (8.37)$$

Here the sum over $\mathbf{R}'$ refers to nearest neighbours of the site $\mathbf{R}$. The molecular field approximation consists in replacing the spin operators at sites $\mathbf{R}'$ in this Hamiltonian by their *mean value*, the underlying assumption being that fluctuations about this mean value are small. If we carry out the transformation

$$\mathbf{S}_{\mathbf{R}'} \longrightarrow \langle \mathbf{S}_{\mathbf{R}'} \rangle = -\frac{\mathbf{M}}{2\mu_B N} = \langle \mathbf{S}_\mathbf{R} \rangle, \qquad (8.38)$$

the Hamiltonian of site $\mathbf{R}$ then takes the simple form of the Hamiltonian of a spin in an effective magnetic field $\mathbf{B}_{\text{eff}}$:

$$H_\mathbf{R} = 2\mu_B \mathbf{B}_{\text{eff}} \cdot \mathbf{S}_\mathbf{R}. \qquad (8.39)$$

The effective magnetic field is given by

$$\mathbf{B}_{\text{eff}} = \mu_0 (\mathbf{H}_a + \lambda^* \mathbf{M}), \qquad \lambda^* = \frac{z\mathscr{J}}{4\mu_B^2 \mu_0 N}, \qquad (8.40)$$

where z is the number of nearest neighbours of the site $\mathbf{R}$. It remains only to determine $\mathbf{M}$. In this approach, all the spins have the same expectation value, as

---

[2] In practice, these considerations need to be treated cautiously. In a solid, there are usually magnetic anisotropies, to be discussed in the next chapter, stemming from the dipole interaction between spins and/or the spin–orbit interaction. These terms tend to orient $\mathbf{M}$ with respect to the crystal axes. However, the operation $\mathbf{S} \to -\mathbf{S}$ remains a symmetry of the Hamiltonian even in this case. A ferromagnetic state will thus always break at least one symmetry.

Fig. 8.7 Graphical solution of (8.42). There is only a solution with $M \neq 0$ if the gradient $T/N\mu_B$ of the straight line representing $M/N\mu_B$ as a function of M/T is less than the gradient at the origin of the function tanh. The value (8.43) is then obtained for T_C

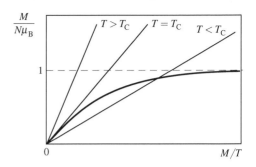

one would expect in a ferromagnetic material. We thus have $\mathbf{M} = -2N\mu_B \langle \mathbf{S_R} \rangle$. However, the thermodynamic average for an ensemble of spins subjected to a field $\mathbf{B}$ is simply given by (8.14). Assuming that $\mathbf{M}$ is parallel to $\mathbf{z}$, and for $J = S = 1/2$, this gives

$$M = N\mu_B \tanh \frac{\mu_B B_{\text{eff}}}{k_B T}, \qquad (8.41)$$

which becomes, for zero applied field,

$$M = N\mu_B \tanh \frac{\lambda^* \mu_0 \mu_B M}{k_B T}. \qquad (8.42)$$

This equation thus determines M in a self-consistent way. Apart from the obvious solution $M = 0$ which yields the paramagnetic state, there is only a solution with $M \neq 0$ if $\lambda^* N \mu_0 \mu_B^2 > k_B T$, as can be seen from the graphical solution shown in Fig. 8.7.

It is easy to check that the solution $M \neq 0$ has lower free energy than the paramagnetic solution $M = 0$, and hence that the ferromagnetic solution is stable for $T < T_C$, where

$$\boxed{T_C = \frac{z \mathscr{J}}{4 k_B}}. \qquad (8.43)$$

Fig. 8.8 The magnetisations of Ni and Fe in Fig. 8.2 are plotted in reduced coordinates. Results are compared with the solutions obtained using the molecular field approximation for different values of the total angular momentum J. From [12], Springer ©1939

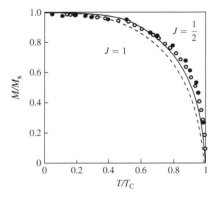

8.4 Ordered Magnetic States

We can then identify T_C with the experimentally measured Curie temperature. The results in Fig. 8.2 for Fe and Ni are plotted in Fig. 8.8 in reduced coordinates, and it can be checked that the agreement is qualitatively quite good.

> **Question 8.2.** Apply the mean field theory to the general case of the Brillouin function for an arbitrary total angular momentum J. Can you deduce the value of the effective field for Fe and Ni? Take $g = 2$ and $J = 1$ and $J = 1/2$, respectively, for Fe and Ni.

Let us point out here that Fe and Ni are ferromagnets which display good metallic properties. They cannot therefore be described correctly by the Heisenberg model developed so far, which is restricted to localised electrons in an insulator. We shall sketch in Sect. 8.6 the fundamental aspects which differentiate the electronic structure of such magnetic insulators from the band insulators considered so far, as well as some model approaches which allow one to describe the coexistence of magnetism and metallicity.

Note, however, that $M(T)$ is observed to deviate from the predictions of the molecular field theory both at low and at high temperatures. In Chap. 12, we shall see in more detail that the mean field approximation is in fact just one of the more elementary approximations for describing magnetic phase transitions, but that it often provides an excellent qualitative description of the phenomena coming into play in a phase transition. Table 8.3 gives some values for T_C and the spontaneous magnetisation at low temperatures in various ferromagnetic solids.

Table 8.3 Curie temperature T_C and spontaneous magnetisation at low temperature ($T \to 0$) for various solids. Elements or compounds that are ferromagnetic at room temperature are all metallic

	T_C (K)	M_0 (kA/m)
Fe	1043	1752
Co	1388	1446
Ni	627	510
Gd	293	1980
Dy	85	3000
CrBr$_3$	37	270
Au$_2$MnAl	200	323
Cu$_2$MnAl	630	726
Cu$_2$MnIn	500	613
EuO	77	1910
EuS	16.5	1184
MnAs	318	870
MnBi	670	675
GdCl$_3$	2.2	550

8.5 Antiferromagnetism and Ferrimagnetism

In nature, the exchange integral $\mathcal{J}$ is negative in many cases. This situation is often encountered in oxides, sulfides, and fluorides, for which the magnetic ions are separated by ions with fully occupied shells like O^{2-}, S^{2-}, or F^-. The interaction $\mathcal{J} < 0$ between magnetic ions then occurs via the filled shells of the anion. This is known as a *superexchange interaction*.

If the spins are considered as classical vectors, it is natural to assume that they may arrange themselves so as to be antiparallel between nearest neighbours. Clearly, such an arrangement is possible for crystal structures which separate simply into equivalent sublattices A and B, as is the case for the square lattice, for instance (see Fig. 8.9), the simple cubic lattice, or the body-centered cubic lattice. (But note that this is not possible for the triangular or fcc lattices.) It seems natural to introduce the *Néel state* (named after the French physicist Louis Néel, who won the Nobel prize in 1970 for demonstrating the existence of antiferromagnetism):

$$|\text{Néel}\rangle = |\uparrow\downarrow\uparrow\downarrow\ldots\rangle . \tag{8.44}$$

In this state, all the spins of sublattice A are up, while all those of sublattice B are down. This clearly minimises the energy *if the spins are treated as classical vectors*. This state has zero magnetisation, since the partial magnetisations $\mathbf{M}_A$ and $\mathbf{M}_B$ cancel one another. However, we may still consider the *alternating magnetisation*

$$\mathbf{M}_{\text{alt}} = \mathbf{M}_A - \mathbf{M}_B . \tag{8.45}$$

This quantity is nonzero in the Néel state, and since this vector introduces a preferred direction in space, as in the ferromagnetic case, the symmetry is broken once again. Compounds with nonzero alternating magnetisation are said to be *antiferromagnetic*, and the temperature below which the alternating magnetisation appears is called the *Néel temperature*. This temperature is given in Table 8.4 for some compounds.

Fig. 8.9 (a) Division of the square lattice into sublattices A (*dark disks*) and B (*clear disks*). (b) Two-dimensional antiferromagnetic square lattice

8.5 Antiferromagnetism and Ferrimagnetism

Table 8.4 Some antiferromagnetic compounds and their Néel temperatures

	T_N (K)		T_N (K)
MnO	122	MnF_2	67.3
FeO	198	FeF_2	78.4
CoO	291	CoF_2	37.7
NiO	600	$MnCl_2$	2
$RbMnF_3$	54.5		
$KFeF_3$	115		
$KMnF_3$	88.3	Sr_2CuO_3	5
$KCoF_3$	125	VS	1040
$KCuF_3$	39	Cr	311

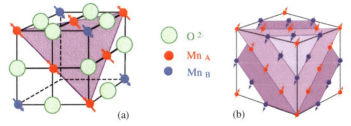

Fig. 8.10 (a) Crystallographic and magnetic structure of MnO. The Mn^{2+} ions occupy the lattice points of a face-centered cubic lattice. The O^{2-} ions are in the middle of the sides and at the center of the cube. In the antiferromagnetic phase, neighbouring moments have alternating orientations pointing along the diagonals of the faces. (b) Primitive cell of the antiferromagnetic lattice of MnO. Only the Mn are shown

The case of MnO is illustrated in Fig. 8.10. The existence of an antiferromagnetic state is not so easy to demonstrate as the existence of a ferromagnetic state, since there is no measurable spontaneous magnetisation.[3] In fact the most direct evidence comes from crystallographic measurements. Owing to the fact that the spins alternate, the unit cell of an antiferromagnetic solid is twice that of the same crystal without antiferromagnetism. The magnetic unit cell of MnO is shown in Fig. 8.10b. The doubled cell then gives rise to new Bragg reflections. These can be observed by neutron diffraction (a neutron has a magnetic moment and thus interacts with the spins of the crystal). In Fig. 8.11, new Bragg reflections are clearly visible for MnO below the Néel temperature. In principle, X-ray diffraction should yield the same result. However, the coupling of light with the atomic magnetic moment is weaker, and it is only just becoming possible to exploit this method with the advent of very intense sources of synchrotron radiation.

---

[3] The magnetic susceptibility of an antiferromagnetic compound has specific features that often allow one to establish the existence of the antiferromagnetic state (see Problem 18: *Properties of an Antiferromagnetic Solid*).

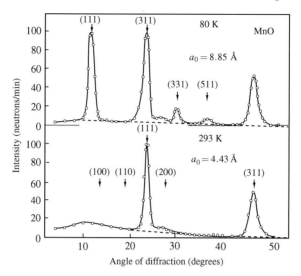

Fig. 8.11 Intensity of neutron diffraction by an MnO powder above ($T = 293$ K) and below ($T = 80$ K) the Néel temperature. Additional Bragg peaks are clearly detected in the antiferromagnetic state. From Shull, C.G., Strauser, W.A., Wollan, E.O.: Phys. Rev. **83**, 393 (1951). With the kind permission of the American Physical Society (APS). http://link.aps.org/doi/10.1103/PhysRev.83.333

Question 8.3. Given the primitive cell of MnO shown in Fig. 8.10, can you see whether the Bragg reflections expected for MnO are indeed observed in the paramagnetic state? What reflections are expected in the antiferromagnetic state if the magnetic structure is indeed the one given? Confirm this using the indexation of Bragg peaks and the lattice parameters given in Fig. 8.11. Are any reflections missing? Why is this?

The Néel state should be treated with some care. Due to the relation (8.29), this state is not an eigenstate of the Heisenberg Hamiltonian. Instead, we have

$$\mathbf{S}_1 \cdot \mathbf{S}_2 | \uparrow\downarrow\uparrow\downarrow\uparrow\downarrow \ldots \rangle = -\frac{1}{4} | \uparrow\downarrow\uparrow\downarrow\uparrow\downarrow \ldots \rangle + \frac{1}{2} | \downarrow\uparrow\uparrow\downarrow\uparrow\downarrow \ldots \rangle , \quad (8.46)$$

and similar relations for the other pairs of nearest neighbours. The second term on the right of this relation shows that |Néel⟩ cannot be an eigenstate. However, systematically taking into account the extra terms generated in (8.46), it can be shown that these effectively reduce the alternating magnetisation, without completely destroying it. The very existence of antiferromagnetism is not therefore in doubt, but rather the exact form of the ground state is more complicated than the simple Néel state.

Due to their zero magnetisation, antiferromagnetic compounds are not of great technological interest. However, compounds exhibiting a spontaneous magnetisation can be obtained from antiferromagnetic exchange interactions (more common in nature than ferromagnetic interactions). It suffices to place atoms of different total

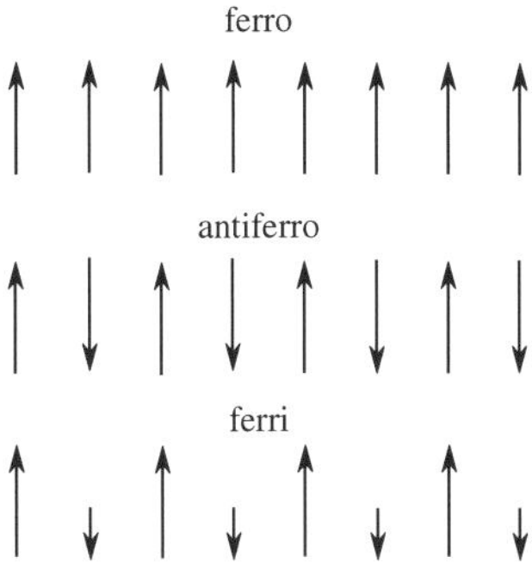

Fig. 8.12 Spin arrangement in ferromagnetic, antiferromagnetic, and ferrimagnetic solids

Table 8.5 Curie temperature (T_C) and spontaneous magnetisation at low temperature ($T \to 0$) for several ferrimagnetic solids

	T_C (K)	M_0 (kA/m)
Fe_3O_4	858	510
$CoFe_2O_4$	793	475
$NiFe_2O_4$	858	300
$CuFe_2O_4$	728	160
$MnFe_2O_4$	573	560
$Y_3Fe_5O_{12}$	560	195

spin on the two sublattices. Even an antiparallel alignment will then give rise to a nonzero magnetisation. Such compounds, called *ferrimagnetic* compounds, are extremely important because they often have high saturation magnetisations, even at room temperature, while being insulators, which is important for alternating current applications, especially at high frequency. (Note that most ferromagnetic solids at room temperature are metals, hence good conductors.) For comparison, Fig. 8.12 shows schematically the spin arrangement in ferromagnetic, antiferromagnetic, and ferrimagnetic solids. The characteristics of some ferrimagnetic solids are given in Table 8.5.

8.6 From Insulator Magnetism to Metallic Magnetism

We have examined so far two completely different limiting descriptions of electronic states in a solid. In the band structure approach used in the four first chapters of this course we have described the case of electrons considered as independent, their interactions being restricted to an averaged potential. The delocalisation of these electrons between the atomic sites driven by the transfer integrals may yield metallic states. In contrast, in the present chapter we are considering the specific situation for which electrons localised on ionic states lead to local atomic magnetic moments. Those arise when the Pauli principle and on site inter-electronic Coulomb repulsion are taken into account properly. We have assumed implicitly that the transfer integrals between electrons on neighbouring ions are small enough in such solids, so that these electrons do not delocalise. This then corresponds to an insulating magnetic state quite different from the band insulating states considered so far in the independant electron band approach.

The actual situation in real materials does indeed sometimes correspond to these limiting cases, but a wide variety of solids correspond to intermediate situations, like the one already mentioned for ferromagnetic metals such as Fe or Ni. These intermediate cases are quite important both for the fundamental questions raised and for the applications of the novel physical effects which come into play.

In this chapter we shall demonstrate first in Sect. 8.6.1 how the local Coulomb interaction on atomic sites and the transfer integrals between atoms compete, which explains the occurence of these intermediate situations. We recover then a comprehensive definition of the magnetic insulator cases named after the approach proposed by Mott and Hubbard. In Sect. 8.6.2 we shall understand under which conditions a Mott–Hubbard insulator can exhibit a transition into a correlated metallic state either by application of a pressure, or by chemical doping. The latter case has been emphasized by the experimental discovery of the strange metallic properties of High Temperature superconducting cuprates, which will be sketched in Sect. 8.6.3. Finally in Sect. 8.6.4 we shall show that magnetism in metals can also be described in a band structure approach, provided that the local Coulomb repulsion is taken into account at least approximately.

8.6.1 Mott–Hubbard Insulator

The very existence of magnetic insulators raises a fundamental problem for the band theory. According to the discussion in Chap. 3, all insulators should be non-magnetic (or more precisely, slightly diamagnetic). In order to explain the magnetic behaviour discussed in this chapter, we must therefore reassess the underlying approximations that led to the band theory, and especially the averaging approach to the Coulomb interactions between electrons.

Let us begin by considering the case of an isolated atom (on the left in Fig. 8.13), which was the starting point for the tight-binding theory in Chap. 1. In this context, it was assumed that the energy brought to the system by an extra electron would be ε_0, and that a second electron on the same atom would also bring ε_0, so that the total energy would be $2\varepsilon_0$ for a doubly negatively charged ion. But this is obviously not very realistic, owing to Coulomb repulsion. Apart from its 'orbital' energy ε_0, the second electron will also be subject to the Coulomb repulsion of the first electron, and its energy will thus be higher than ε_0 by an amount usually denoted by U, which represents the Coulomb repulsion between the first and second electrons added to an initially neutral atom. The total energy of the doubly negative ion is thus $2\varepsilon_0 + U$. Note that U can vary considerably depending on the atom (from about 1 eV to more than 10 eV). This very simple approach goes by the name of the *Hubbard model*.

8.6 From Insulator Magnetism to Metallic Magnetism

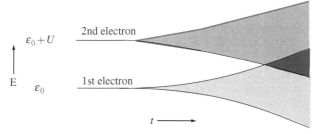

Fig. 8.13 Level diagram for the Mott–Hubbard model. *Left*: Isolated atom. *Center*: Mott–Hubbard insulator obtained for a small hopping integral $t < U$. *Right*: Metallic situation corresponding to $t > U$

If we now consider this ion in a crystal, the hopping integrals between nearest neighbours will broaden the discrete atomic levels into bands. If to begin with we consider the limiting case of small hopping integral compared with U, we find ourselves in a situation corresponding to the middle of Fig. 8.13. There are two allowed energy bands called the upper and lower Hubbard bands, separated by a band gap. This gives the impression that we have a typical insulator (or semiconductor). But this is not in fact correct. There is one additional one-electron state per atom, so that, in a solid comprising N_n atoms, the lower band of the middle column can contain up to N_n electrons, rather than up to $2N_n$ electrons, as was the case in the context of the band theory developed in Chap. 3. In particular, if there is now one electron per atom (or more generally an odd number of electrons per primitive cell), the lower band will be completely filled and the upper band completely empty. We will thus have *an insulator with an odd number of electrons per primitive cell*, in contrast to the predictions of Chap. 3. This is of course a consequence of the interactions U between electrons. The very existence of such an insulator (usually called a Mott–Hubbard insulator in recognition of the two British scientists who first studied them in the 1960s) is thus a consequence of the Coulomb interaction between electrons. Important examples of Mott–Hubbard insulators are undoped cuprates in which the Cu^{2+} ions are in a $3d^9$ state.[4]

So what are the magnetic properties of a Mott–Hubbard insulator? If we begin by considering the limiting case of very small hopping integrals, we end up with isolated atoms. The electron in the level ε_0 can then have spin ↑ or spin ↓, behaving like an isolated spin 1/2. In the solid, these spins taken together will give rise to Curie paramagnetism $\chi \sim 1/T$, and we then have a *paramagnetic insulator*, once again contradicting band theory, which predicted that any insulator must be weakly diamagnetic. If we also take into account the finite value of the hopping integrals, we find that, at low enough temperatures, the spins will arrange themselves antiferromagnetically.

---

[4] See Problem 17: *Electronic Properties of La_2CuO_4*.

We have seen in this chapter that there are not only magnetic insulators of spin 1/2 (in fact these are in the minority), but that in most cases the spin per atom is much higher. This is due to the fact that, in almost all cases, the atomic orbitals in question are not s levels (hence non-degenerate), but d- or f-type (hence five- or seven-fold degenerate). In this situation, Hund's rules determine the angular momentum of each atom in the Mott–Hubbard insulating state.

The Hubbard model, which replaces the true Coulomb potential $V(\mathbf{r}) \sim 1/r$ by a repulsion which only acts if the two electrons are located on the same atom, is clearly a drastic simplification of the actual physical situation. However, it is rather naturally justified in the context of the theory of magnetic phenomena. We have seen in this chapter that in this context the Pauli exclusion principle is of fundamental importance. Pauli exclusion is essentially important for two electrons on the same atom, since two electrons separated by some multiple of the lattice constant will hardly be affected by the Pauli principle, so the interaction between them will depend very little on the relative orientation of their spins, and will have very little influence on their magnetic properties.

The main conclusion to this section is that, going beyond the possibilities offered by band theory (paramagnetic metals and diamagnetic insulators), *the presence of Coulomb interactions between electrons, if they are strong enough, can give rise to an insulating state with a variety of magnetic properties*, such as Curie paramagnetism, ferromagnetism, antiferromagnetism, and so on.

8.6.2 Mott Transition and Doped Mott–Hubbard Insulators

In a Mott–Hubbard insulator, if we increase t (or if we consider compounds with lower values of U), for a certain critical value of t/U, the upper and lower Hubbard bands begin to overlap (see Fig. 8.13 right), causing the band gap to disappear and leading to a metallic state. Such an increase in t can be produced by bringing the atoms closer together.

This was first achieved in the case of doped semiconductors by increasing the donor concentration, e.g., by increasing the concentration of phosphorus in silicon. This causes the hydrogen-like orbitals of P to move much closer together (see Sect. 4.5b) and increases the hopping integrals, while remaining in a configuration corresponding to one electron per donor atom.[5] A simpler way to achieve this situation directly without changing the number of electrons in a material is to apply an external pressure. This increases the hopping integrals t by bringing the atoms closer together, provided that the material is compressible.

In the metallic state thereby induced, one then observes a Pauli paramagnetism and a specific heat that is linear in temperature, but the Wilson ratio discussed in Chap. 3 is generally greatly increased.

---

[5] See Problem 7: *Insulator–Metal Transition*.

Note that a Mott–Hubbard insulator looks at first glance like a band insulator, the only difference being that here each Hubbard band contains only N_n states rather than $2N_n$ states in the case of the band theory described in Chap. 3. Chemical treatment may be envisaged to change the number of electrons in a Mott insulator. For example, it can be doped with holes, reducing the number of electrons in the lower Hubbard band by a number N_e smaller than N_n.

This is exemplified by the high-temperature superconducting compounds, where one typically has $N_e = 0.75$–$0.9 N_n$. It is interesting to ask *how many states there are in the upper Hubbard band*. The answer is simple. A state in the upper band corresponds to a doubly occupied atom. But in order for an atom to be doubly occupied, a first electron must clearly be present. Since there are N_e electrons in the lower Hubbard band, the number of states in the upper Hubbard band will be N_e, i.e., it will depend on the number of electrons present. *The situation is very different in a band insulator.*

Furthermore, the total number of states in the upper and lower Hubbard bands must be $2N_n$. For the lower band, we thus find $2N_n - N_e$ available states. For the so-called doped case, i.e., $N_e < N_n$, there are thus more states than electrons, and a metallic situation is expected, with nonzero conductivity and Pauli paramagnetism. A very similar analysis can be applied to the case $N_e > N_n$, with analogous results. *The 'doped' Mott–Hubbard insulator is therefore a metal.*

8.6.3 Magnetism and Superconductivity

According to the above discussion, cuprates such as $YBa_2Cu_3O_6$ or La_2CuO_4 are antiferromagnetic Mott insulators. At low doping levels, their metallic properties arise from the situation described in the last section, and their metallic behaviour is far from being like that of conventional metals described in Chaps. 3 and 4. Experimental investigations carried out on the cuprates, and also on certain other classes of doped Mott insulators, have shown that doping gradually reduces the Néel temperature of the antiferromagnetic state. This AF state is completely suppressed for a low level of doping, of the order of $n_h \approx 0.05$, as can be seen in the phase diagram for the cuprates in Fig. 8.14. The static magnetism gives way to a metallic and superconducting state for $n_h > 0.05$. However, most of the relevant physical quantities turn out to be very different from those of more 'ordinary' metals. Up to about $n_h \approx 0.15$, the existence of magnetic correlations between the electron moments of Cu tends to suppress low temperature electronic excitations, which justifies the name *pseudogap phase* for the corresponding electronic state. The magnetic susceptibility does not exhibit a Pauli behaviour in this regime, and the existence of the pseudogap can be seen in all the thermodynamic quantities, as in optical absorption, photoemission (ARPES), or tunnel effect experiments.

The temperature T^* at which this pseudogap appears decreases with increasing n_h and eventually meets the superconducting critical temperature when the latter

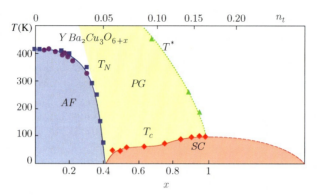

Fig. 8.14 Cuprate phase diagram versus hole doping n_h of the CuO$_2$ planes. The latter can be modified in the YBa$_2$Cu$_3$O$_{6+x}$ compound by changing the oxygen content of the Cu intermediate planes (and formation of CuO chains). The approximate correspondence between n_h and x is given on the upper scale. The insulating Mott antiferromagnetic (AF) state is destroyed for small dopings and opens the way to the 'pseudogap' (PG) phase in the 'underdoped' range. In this compound the 'optimal' doping for which T_c reaches its 93 K maximum value is reached for $x \approx 0.95$ and $n_h \approx 0.15$. For this hole content, the metallic state remains strange with a T linear resistivity. The 'overdoped' regime $n_h > 0.17$ is observed in other cuprate families and exhibits a fast decrease of T_c, with a progressive evolution towards a more classical metallic behaviour. This figure has been composed from experimental results reported by Alloul, H., Bobroff, J., Gabay, M., Hirschfeld, P.: Rev. Mod. Phys. **81**, 45 (2009).

becomes maximal. The pseudogap is said to appear in the *underdoped region* of the phase diagram. When the doping becomes optimal for superconductivity, it is nevertheless observed that the electronic properties of the metallic state still differ from those of normal metals. The pseudogap disappears, but the resistivity varies linearly with temperature, and the measured electron collision rate is given by $\tau^{-1} \approx k_B T/\hbar$, and is much larger than in normal metals. This high level of electron scattering is attributed to the persistence of very short range magnetic correlations, so that this region of the phase diagram is referred to as the *strange metal region*. When the doping is raised above $n_h \approx 0.20$, the physical properties gradually begin to resemble those of a 'normal' metal, or Fermi fluid, whereas the superconductivity disappears rapidly. This region of the phase diagram is called the *overdoped region*.

While this experimental situation gives an impression of consistency, little is really understood on the theoretical level at the present time (apart from the undoped case). The unusual metallic behaviour, the pseudogap, and the origins of superconductivity in the cuprates remain front line research topics.

Note that many families of compounds exhibiting superconductivity suspected of being exotic have remarkably similar phase diagrams to the cuprates, with regard to the proximity of the magnetic and superconducting phases, the system moving from one state to the other for different levels of doping. In many cases, e.g., organic compounds, heavy fermion compounds, and more recently iron pnictides and the fullerene compound Cs$_3$C$_{60}$, the magnetic behaviour can be reduced by applying a pressure, leading to a metallic, and often superconducting state.

8.6.4 Metallicity and Magnetism in a Band Approach

The notion of Mott–Hubbard transition has revealed that it should be possible to treat the electronic structure of a Mott insulator and of a metallic state within a unique framework. Although major steps have been taken in this direction, this is far from being achieved. The developments currently under way are at the forefront of the research on the physics of correlated electron systems and are well beyond the scope of this course. However, in view of the importance of the physical properies of the corresponding materials for applications we shall present here some limiting situations where one can at least reach a significant qualitative description of the relevant physics.

a. Distinct Electronic Bands for Magnetism and Metallicity

In Sects. 8.6.1–8.6.3 we have considered magnetic Mott–Hubbard states for a single orbital model. In many actual compounds with polyatomic unit cells, Mott localised states on some specific atoms of the unit cell may coexist with delocalised electrons on distinct atomic sites. This happens for instance in some compounds involving rare earth atoms and lighter elements. The external f electrons of the rare earths have a limited spatial extension and for particular atomic structures these orbitals can display very little overlap with the other atomic orbitals. An excellent example of this situation can be found again in the $YBa_2Cu_3O_{6+x}$ family of cuprates.

Indeed such compounds can be synthetized with nearly any $4f$ rare earth instead of Y. Let us consider the case of $GdBa_2Cu_3O_{6+x}$. There the Gd atoms, which are (like Y) embedded between two CuO_2 planes, are in a 3^+ state but, while Y^{3+} has no f electron, Gd^{3+} has 7 electrons on its unfilled $4f$ orbitals. The Gd^{3+} ion is then in its well defined $4f^7$ state with a total spin $S = 7/2$, as expected for an isolated Gd^{3+} ion (see Table 8.2). Adding an electron on Gd to bring it into the $4f^8$ electronic state would require a prohibitive energy U which cannot be provided by the weak transfer integrals from the orbitals of the CuO_2 planes. Consequently Gd^{3+} remains a well isolated ion with a local magnetic moment of $7\mu_B$ very weakly coupled to the electronic states of the CuO_2 plane which are responsible for the 2D metallic behaviour and superconductivity. The coupling is so small that the metallic and superconducting properties of the CuO_2 planes are absolutely identical to those of $YBa_2Cu_3O_{6+x}$ with T_c values which differ by at most 1 K. However the small transfer integrals are sufficient to induce a weak coupling between Gd ionic spin states with an antiferromagnetic ordering of the Gd moments below a Néel temperature $T_N = 2.25$ K.

This simple example just demonstrates that magnetism can be sustained by some orbitals (those of Gd^{3+}) which remain partly filled in a Mott–Hubbard state, while a metallic character can occur in independent bands (those of the CuO_2 planes). In cuprates, from what we have seen in Sect. 8.6.3, the problem of strange metallicity in a correlated band, which occurs in the CuO_2 planes themselves, nevertheless remains the same there for the Y and Gd compounds.

Other systems are such that the metallic bands display weak correlations and might be treated in an independent electron picture. This is for instance the case for the family of comounds discovered by Chevrel, an archetype being $HoMo_6S_8$. Here the Mo_6S_8 units give rise to $4d$ metallic bands while the rare earth Ho^{3+} order ferromagnetically below $T_C = 0.69$ K, so that magnetism and metallicity can be considered as independent. In that case a classic BCS superconductivity with $T_c = 1.2$ K is sustained by the $4d$ bands. This T_c being small, superconductivity is nevertheless suppressed below $T_C = 0.69$ K by the large internal field induced on the Mo sites in the magnetic ordered state of Ho^{3+}, contrary to the case of $GdBa_2Cu_3O_{6+x}$ for which the high temperature superconductivity is quite robust to the internal field developed below the AF ordering temperature of the Gd^{3+} moments.

Such specific situations of coexistence between magnetism and metallicity are not so difficult to understand and to model. In those cases we just need to remember that the strong Coulomb interaction pins the occupancy of the rare earth orbitals at a fixed number (7 for Gd), and does not allow electrons to transfer from the other sites to complete the filling of the rare earth levels. In the band theory of Chaps. 1, 2, 3 and 4 we often considered the case of atomic ions with filled electronic shells. Here the difference is that the partial occupancy of some atomic levels which sit below the Fermi level leads to local moment and magnetism somewhat disconnected from the properties of the metallic bands.

b. Metallicity and Magnetism in the Same Bands

If we now consider cases such as that of Fe or Co we address a quite different problem, since in these elemental metals, although s electrons contribute to the metallic transport, magnetism and metallicity also occur for the electrons of the 3d bands. So how can we describe this possibility to get ordered magnetism for delocalised electronic states ? We immediately see that, as in the Mott–Hubbard case, we have to consider that the strong Coulomb interaction U on the Fe (or Co) sites forces us

(i) to relax the rule that every $3d$ orbital should be doubly populated
(ii) to break the symmetry between the possible orientations of the electron spins.

One way to implement that in the band structure theory was suggested initially by Stoner who considered that in a magnetic state one should differentiate the energy bands for up and down spin orientations, the simplest approach being then to shift the energy band of the down spin orientation with respect to that of the up spin orientation of an energy $2\delta E$ which will be controlled by the on site Coulomb interaction U, as displayed in Fig 8.15. This reduces the electronic correlation energy at the expense of an increased kinetic energy.

One can note that the total correlation energy for the $3d$ band can be written $Un_\uparrow n_\downarrow$ where $n_\uparrow$ and $n_\downarrow$ are the total occupancies of the two sub-bands, which are initially equal, so that the variation of correlation energy due to the shift $2\delta E$ of the $\uparrow$ spin energy band with respect to the $\downarrow$ spin band is given by

8.6 From Insulator Magnetism to Metallic Magnetism

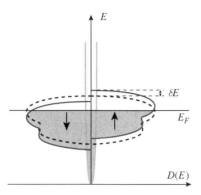

Fig. 8.15 In a purely paramagnetic state the densities of states $D(E)$ of the up and down spin energy bands are degenerate in the absence of any applied field (broad s band and dashed lines for the narrow $3d$ band). Strong enough electronic correlations in the $3d$ band might favor energetically a state where the degeneracy of the two spin directions is lifted (full lines). Such a splitting $2\delta E$ between the two $3d$ electronic bands yields non integer moments per site and differentiates the density of carriers of the two spin states which also acquire distinct mobilities

$$U\left[n_\uparrow - D(E_F)\delta E\right]\left[n_\downarrow + D(E_F)\delta E\right] - Un_\uparrow n_\downarrow = -U[D(E_F)\,\delta E]^2$$

where $D(E)$ is the density of states per spin diection, and we have assumed for simplicity that δE is smaller than the bandwidth and that the bands are rigidly shifted. The total kinetic energy of the band increases by

$$D(E_F)(\delta E)^2,$$

so that the total electronic energy variation

$$[1 - UD(E_F)]D(E_F)(dE)^2$$

can only be negative if

$$UD(E_F) > 1.$$

The latter equation is then the condition for which magnetic moments can appear in the metallic bands and may give rise to magnetic order at low T.

As $D(E_F) \approx 1/W$ (with $W \approx zt$, where z is the number of neighbours), we naturally recover a condition analogous to the Mott–Hubbard condition $U/W > 1$ for the metal to insulator transition in Sect. 8.6.1. This condition can be met for high values of the density of states $D(E_F)$, as found in transition metals. This simple qualitative argument explains furthermore why elements of the $3d$ series like Fe, Co and Ni are magnetic contrary to the $4d$ elements, for which the d orbitals have larger spatial extension, so that t and W are therefore larger. A further important merit of this approach is that, contrary to the Heisenberg model, it explains the non integer measured values of the magnetic moment per atom in the transition metals.

This approach initiated by Stoner has been taken as giving a criterion for a ferromagnetic instability in a metallic band. But as the spin quantization axis is not defined in this approach and could vary from site to site, it only if fact tells us that moments may appear. The ground state of the electronic system will only be ferromagnetic if the effective exchange coupling between electrons mediating this instability is indeed ferromagnetic.

One may notice that the Pauli principle prevents electrons with the same wave vector $\mathbf{k}$ and the same spin from sitting at the same point $\mathbf{r}$, so that even in a free electron band the wave function of a $\mathbf{k} \uparrow$ electron at $\mathbf{r}$ feels a lower density of $\mathbf{k} \uparrow$ than of $\mathbf{k} \downarrow$ electrons. This reduces the total Coulomb repulsion energy for electrons with the same spin direction, and leads then to a reduction of the ground state energy for electrons with the same spin orientation with respect to that of electrons with antiparallel spin orientation, which does indeed correspond to an effective ferromagnetic interaction. So the Pauli principle for delocalised electrons seems to favor ferromagnetism while the short range Coulomb repulsion (like the on site local U) rather favours antiparallel ordering, which yields AF if the lattice is bipartite.

We have therefore sketched here that both ferromagnetism and antiferromagnetism may be sustained in metallic bands if the Pauli principle and Coulomb repulsions are taken into account. Subsequent efforts are being made to generalize the band structure calculations (similar to those introduced in Chaps 1, 2, 3 and 4) by introducing U explicitly. Such spin polarized band structures, computed with different approximation schemes are under intense scrutiny in comparison with experimental data, such as ARPES.

Now such a differentiation of $\uparrow$ and $\downarrow$ spin bands also has implications for the transport properties of ferromagnetic metals since the numbers of carriers differ for the two spin bands (see Fig. 8.15). Then the contribution of the majority and minority spin carriers to the conductivity will be different, as the scattering times of these two types of carriers on magnetic defects will also differ.

This spin dependent transport is at the origin of the so-called GMR (Giant Magneto-Resistance) effect which has led to the attribution in 2007 of the Nobel prize to A. Fert and P. Grümberg. These authors had noticed that the resistance of multilayers of metallic films can be dependent on the relative magnetic ordering of these layers. For instance Fe/Cr multilayers can be synthetized with alternating orientations of the magnetization of the ferromagnetic Fe layers, while the intermediate Cr layers are non magnetic. A small applied field can suffice to align the magnetism of the Fe layers in the same direction so that electrons are more scattered in the antiparallel case than in the parallel case. This variation of resistance is used to detect the magnetic field induced by the magnetic bits recorded in a disk memory, so that the GMR is nowadays efficiently used in read heads for hard disk drives.

More subtle effects induced by spin dependent scattering are being investigated in order to attempt to detect and/or reverse the magnetization of domains in nanostructured materials. The global effort in this field, known as "spintronics", associates fundamental and applied research with many potential applications.

8.7 Summary

The magnetism of multielectron atoms is related to the removal of orbital degeneracy in the electron levels resulting from the Coulomb repulsion between electrons in these orbitals and the Pauli exclusion principle. The existence of atomic magnetic moments then results indirectly from interactions between electrons.

An ensemble of independent magnetic ions with magnetic moment $\boldsymbol{\mu} = g\mu_B \mathbf{J}$ has Curie magnetic susceptibility

$$\chi(T) = N \frac{(g_J \mu_B)^2}{3} \frac{J(J+1)}{k_B T} .$$

The magnetisation of ferromagnetic materials results from the macroscopic alignment of the atomic moments. The magnetic dipole interaction between atomic magnetic moments is too weak to be the cause of ferromagnetism. The Coulomb interaction between electrons on neighbouring atoms removes the degeneracy of their electronic levels. Together with the Pauli exclusion principle, this leads to an effective interaction between the spins of near-neigbouring atoms, as expressed by the exchange Hamiltonian

$$\hat{\mathbf{H}} = -\mathscr{J} \hat{\mathbf{S}}_1 \cdot \hat{\mathbf{S}}_2 ,$$

where $\mathscr{J}$ depends on the distance between neighbouring atoms and their orbital configurations.

The Heisenberg model describes the magnetic properties of an insulator, taking into account only the exchange interactions between nearest neighbours:

$$\hat{H} = -\mathscr{J} \sum_{\langle \mathbf{RR}' \rangle} \mathbf{S_R} \cdot \mathbf{S_{R'}} + 2\mu_B \mathbf{B} \cdot \sum_{\mathbf{R}} \mathbf{S_R} .$$

A ferromagnetic ground state is obtained for $\mathscr{J} > 0$. A molecular field theory, in which each moment is subjected to an effective field due to its nearest neighbours, leads for a moment $\mathscr{J} = 1/2$ to a magnetisation satisfying

$$M = N\mu_B \tanh \frac{\lambda \mu_0 \mu_B M}{k_B T} ,$$

with a Curie temperature

$$T_C = \frac{z \mathscr{J}}{4 k_B} .$$

When $\mathscr{J} < 0$, neighbouring spins prefer to be antiparallel. When the crystallographic structure allows it, a configuration in which each spin is antiparallel to its neighbours is obtained. This antiferromagnetic (Néel) structure has zero macroscopic magnetisation in systems involving only one type of magnetic ion. With an ordered structure of different magnetic ions, a macroscopic magnetic moment can survive (ferrimagnetism).

The existence of such magnetic insulators is shown to result naturally when the energy cost associated with a large on site atomic Coulomb ineraction U prevents double occupancy of the corresponding atomic orbitals. When U overcomes the effect of the transfer integrals t which favours the electron delocalisation, the resulting Mott–Hubbard insulator displays only single occupancy of the electronic states. However if t and U become of comparable magnitude a transition to a metallic state with unusual magnetic properties may occur.

In undoped cuprates the CuO_2 planes are in a Mott–Hubbard insulating state with one hole per copper site. The superconductivity of cuprates is obtained by chemical doping of the CuO_2 planes which induces a metallic state in which strong signatures of the large electronic correlations remain.

In many compounds Mott localised states on some specific atoms of the unit cell may coexist with delocalised electrons on distinct atomic sites. In such situations a conventional non interacting band approach might apply for the delocalised electrons and the metallic and magnetic properties are somewhat independent.

However in other limiting cases for which transfer integrals are comparable to U for the same band, a magnetic metallic state might be described in a band approach as long as one relaxes the rule that each state should be doubly occupied and one then differentiates states with up and down spin electrons. In such a situation the spin polarized bands correspond to two distinct types of carriers and the resulting spin dependent transport properties are prominent and might be quite useful in a variety of "spintronics" devices.

8.8 Answers to Questions

Question 8.1

We obtain the values $g_J J = 7$, 5, and 3 for Gd^{3+}, Fe^{3+}, and Cr^{3+}, respectively. The slope at the origin is $g_J^2 J(J+1)\mu_B/3k_B$, i.e., $0.22 g_J^2 J(J+1)$ in SI units. The experiment can thus be used to determine $(J+1)/J$ and hence also J. However, the accuracy of the experiment is rather poor. The most accurate value obtained is for Cr^{3+}, for which we find $1 + 1/J = 1.66$, so that $J = 3/2$ and hence $g_J \sim 2$.

According to Table 8.2, we expect $J = S = 7/2$ for Gd^{3+}, which corresponds to a $4f^7$ ion, and therefore $g_J = 2$. This would appear to be confirmed experimentally. Likewise, from Table 8.1, we expect $J = S = 5/2$ for Fe^{3+}, which is a $3d^5$ ion, and hence $g_J = 2$. This is also confirmed experimentally.

However, Cr^{3+} is a $3d^3$ ion, for which we expect $J = 3/2$, $S = 3/2$, and $L = 3$, and hence $g_J = 0.4$. We should therefore find $g_J J = 0.4(3/2) = 0.6$. This is not confirmed experimentally. We have instead $J = 3/2$ and $g_J = 2$, as though only the total spin were contributing to the total angular momentum, i.e., $S = 3/2$ and $L = 0$. This is effectively the case. The reason is that the Hund rules apply to free

8.8 Answers to Questions

ions, but must be modified for solids. In an ionic crystal, the sum of the Coulomb potentials due to the charges of the ions surrounding a given ion must be added to the Coulomb potential due to the atomic nucleus. This term, called the crystal field potential, cannot be neglected in the case of $3d$ ions and changes the order of the atomic electronic levels. For ions in the $3d$ series, this means that we obtain $L = 0$ for the lowest level, and the orbital angular momentum is said to be *quenched* by the crystal field.

Question 8.2

The value of M at saturation is $M_{\text{sat}} = g_J \mu_B J$. The relation between T_C and $\mathscr{J}$ becomes

$$k_B T_C = z \mathscr{J} \frac{J(J+1)}{3}, \qquad \lambda^* = \frac{\mathscr{J} z}{N \mu_0 g_J^2 \mu_B^2},$$

which implies, at low temperatures, when the magnetisation is saturated, that

$$B_{\text{eff}} = \frac{3 k_B T_C}{g_J \mu_B (J+1)}.$$

For Ni ($J = 1/2$), this gives

$$B_{\text{eff}} = \frac{k_B T_C}{\mu_B},$$

whereas for Fe, we obtain

$$B_{\text{eff}} = \frac{3}{4} \frac{k_B T_C}{\mu_B}.$$

With the values of T_C in Fig. 8.2, $T_C = 630$ K for Ni and 1,050 K for Fe, we obtain values $B_{\text{eff}} \sim 1,000$ tesla, well above anything that could be achieved in the laboratory.

Question 8.3

The MnO lattice is face-centered cubic with lattice constant a, with one Mn and one O per primitive cell of the Bravais lattice. The reciprocal lattice is body-centered cubic (see Chap. 2). If we consider the reciprocal lattice of the cubic lattice of side $a^* = 2\pi/a$, the lattice points correspond to $(2h, 2k, 2l)$ and $(1+2h, 1+2k, 1+2l)$. We thus expect reflections for (111), (200), and (311), which are indeed observed, although (200) is weak, whereas there is no reflection for (100).

The magnetic lattice is obtained by doubling the lattice spacing in the three space directions, and is also fcc with a basis. We find the same reflections with half the lattice constant $a^*/2$ in the reciprocal lattice. The corresponding reflections have the same indices (111), (200), (311), (331), (511), and so on, with half the spacing.

Note that the Bragg reflection angles given by $\sin\theta = n\lambda/2d_{hkl}$ are very small and expressed in arc minutes, because the wavelength of the neutrons used is very short compared with the size of the primitive cell. In this case, $\sin\theta \simeq \theta$ and the reflection angles are divided by 2 for the magnetic lattice as compared with the atomic lattice, as can be checked by comparing the top and bottom figures in Fig. 8.11. (Reflections with the same indices occur at half the angles.) Note, however, that there is no magnetic reflection at (200) or (220). This corresponds to the fact that there is extinction here. The structure or form factor (2.22) vanishes due to the high level of symmetry of the magnetic basis.

Chapter 9
Magnetic Anisotropy, Domains, and Walls

Contents

9.1 Magnetic Anisotropy 272
    9.1.1 Magnetocrystalline Anisotropy 272
    9.1.2 Effect of Anisotropy: Irreversibility 273
9.2 Dipole Interactions, Demagnetising Fields, and Domains 277
    9.2.1 Demagnetising Fields 277
    9.2.2 Demagnetisation Energy and Magnetic Domains 282
9.3 Bloch Walls 285
9.4 Magnetic Hysteresis Cycles 288
    9.4.1 Bitter Method for Observing Domains 288
    9.4.2 Magnetisation Process and Hysteresis 289
9.5 Summary 291
9.6 Answers to Questions 292

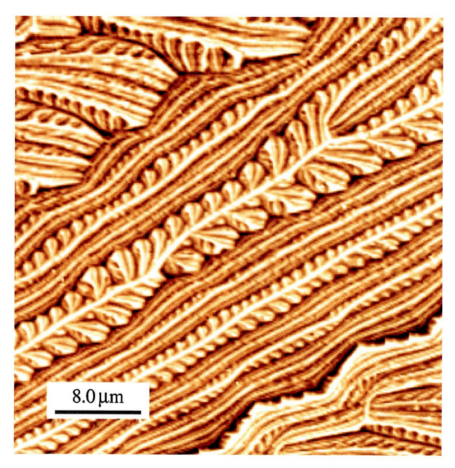

Ferromagnetic materials do not necessarily have a detectable magnetisation in zero magnetic field. This is well known for iron: dressmaker's pins are only magnetised when subjected to the magnetic field of a magnet! When the applied magnetic field is zero, domains of opposite magnetisation appear spontaneously and cancel each other out in the macroscopic magnetisation. These domains have been imaged here by magnetic force microscopy (MFM) measurements of the magnetisation at the surface of an Fe single crystal cleaved in the direction perpendicular to the (111) axis (see Chap. 10 for more on MFM). The domains appearing naturally here display a structure that many artists would have been pleased to produce. Scale 40×40 μm^2. Image courtesy of Miltat, J., Thiaville, A.: Laboratoire de Physique des Solides. Orsay, France

9 Magnetic Anisotropy, Domains, and Walls

We have seen that the exchange couplings are responsible for the appearance of ordered magnetic states, determining for example the Curie temperature T_C of the ferromagnetic state. But everyday experience shows that a piece of iron at room temperature does not generally exhibit a magnetic moment, even though it is well below its Curie temperature $T_C \approx 1,000$ K. In fact, in order to give it a macroscopic magnetic moment, it must be magnetised by placing it in a relatively weak magnetic field, of the order of a few 10^{-3} tesla. If this field is then removed after the application, the piece of iron will retain a significant magnetisation, called *remanent magnetisation*. But how can such a weak applied field, corresponding to such a small magnetic energy compared with $k_B T_C$, produce such a marked effect? The explanation for this phenomenon is to be found in various interactions that we have neglected in the Heisenberg model, to be discussed in the present chapter.

In Sect. 9.1 we shall show that several sources of anisotropy related to the crystal structure introduce energy terms that tend to stabilise the magnetisation along one crystal axis. We can thus understand the existence of magnetic hysteresis, but not the vanishing magnetisation when a material has never been in the presence of a magnetic field. This effect can be attributed to the existence of magnetic dipole interactions between the microscopic moments, which we shall investigate in Sect. 9.2. Since these interactions are long ranged, we shall show that they induce effects that are very sensitive to the geometric shape of the material. This observation is quite specific to the magnetic properties of highly magnetised materials. As a rule, physical properties of materials such as the resistivity, specific heat, and so on, are bulk quantities that are insensitive to surface effects, insofar as the material has macroscopic dimensions. They can be measured without regard for the shape of the sample. Magnetisation is also a bulk quantity, but since there can be no material that is not bounded in space, any measurement of **M** will depend critically on the shape of the outer surfaces of the material. The effects of demagnetising fields due to the dipole interaction, and the existence of a demagnetisation energy, will allow us to understand why, in zero applied field, a magnetic material will spontaneously decompose into magnetic domains. The magnetic configuration of the interfaces between magnetic domains, known as *Bloch walls*, will be discussed in Sect. 9.3, where we will determine their formation energy. Taken together, these phenomena will enable us to understand in Sect. 9.4 the various contributions to magnetic hysteresis, which is very important for technical applications of magnetic materials.

In the last chapter, we stressed the quantum origins of magnetism. But in this one we shall appeal to many semi-classical descriptions, which are adequate for the task of accounting for the main physical effects. A fully quantum description would barely change the result, and would increase the complexity to little advantage given that the relevant quantities are generally hard to predict and must often be deduced from experiment. However, the reader should bear in mind that these semi-classical arguments can only be justified if we accept the existence of magnetic moments and interactions of purely quantum origins. There is no hope here of constructing a classical model of magnetism by letting Planck's constant h tend to zero.

9.1 Magnetic Anisotropy

It is observed experimentally that, for single crystals in which **M** = 0 in the absence of an applied field, the magnetisation is more easily saturated if a field is applied in certain crystallographic directions. For example, for nickel, which has a face-centered cubic structure, **M** grows most rapidly for a field applied along the [111] axis, whereas for iron, which has a body-centered cubic structure, **M** is most easily established in the [100] direction (see Fig. 9.1). These directions are said to correspond to *easy magnetisation axes*. For Co (hexagonal close-packed), the easy magnetisation axis is the *c* axis of the hexagonal structure (see Fig. 9.1). This tells us that certain axes are favoured by the crystal structure, and hence that the rotational symmetry of the Heisenberg Hamiltonian is not preserved. This Hamiltonian must therefore be supplemented by terms that take into account the loss of symmetry.

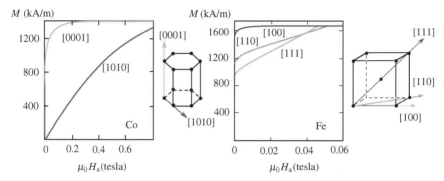

Fig. 9.1 Magnetisation as a function of the applied field for single-crystal samples of cobalt (*left*) and iron (*right*) and for various orientations of the applied field with respect to the crystallographic axes ($T \approx 300$ K). Adapted from Monda, K., Kaya, S.: Sci. Rep. Tohoku Univ. **15**, 721 (1926); **17**, 1157 (1928)

9.1.1 Magnetocrystalline Anisotropy

Magnetic anisotropy is mainly due to the *spin–orbit coupling* and the *crystal field* (see Chap. 8) which introduce correction terms into the energy levels of an ion in the crystal. These levels depend on the orientation of the magnetic moment with respect to the crystal axes. In the simplest case, this leads to the addition of an extra term in the Heisenberg Hamiltonian (8.31):

$$H_A = D \sum_{\mathbf{R}} (S_{\mathbf{R}}^z)^2 . \quad (9.1)$$

Such a term is called the *single-ion magnetocrystalline anisotropy*. It clearly breaks the rotational symmetry of the Heisenberg Hamiltonian. More precisely, if $D < 0$, the energy is minimised when the spins and hence the magnetisation are aligned

9.1 Magnetic Anisotropy

with the $\pm z$ axis, which is the easy axis of magnetisation. Otherwise, when $D > 0$, the spins prefer the (x, y) plane, which is then the easy plane of magnetisation.

In real solids, the magnetic anisotropy terms can take various forms, but generally more complicated than the one indicated above. Note in particular that (9.1) attributes a special role to the z axis, which can be the case for a crystal with tetragonal or hexagonal structure (e.g., cobalt), but which is incompatible with the cubic structure of many magnetic solids. In these structures, which do not distinguish the x, y, and z axes, one might try to remedy the problem by writing

$$H_A = D \sum_{\mathbf{R}} \left[(S_{\mathbf{R}}^x)^2 + (S_{\mathbf{R}}^y)^2 + (S_{\mathbf{R}}^z)^2 \right].$$

However, $(S_{\mathbf{R}}^x)^2 + (S_{\mathbf{R}}^y)^2 + (S_{\mathbf{R}}^z)^2 = S(S+1)$ is constant and this term is isotropic. In a cubic solid, the simplest form for an anisotropy term is rather given by

$$\boxed{H_A = D \sum_{\mathbf{R}} \left[(S_{\mathbf{R}}^x)^4 + (S_{\mathbf{R}}^y)^4 + (S_{\mathbf{R}}^z)^4 \right]}. \tag{9.2}$$

> **Question 9.1.** What will be the easy magnetisation axes or planes for this form of anisotropy?

A further consequence of the spin–orbit interaction is the existence of *exchange anisotropy*: the exchange constant in the Heisenberg model is then replaced by a rank 2 tensor. However, this effect is generally less important than single-ion anisotropy. The above discussion shows that the exact form of the anisotropy terms depends heavily on the structure of the solid under examination (and also on the spin of each atom). It is nevertheless clear that one can expect to find easy magnetisation axes (or even planes).

9.1.2 Effect of Anisotropy: Irreversibility

If the material is magnetised along the easy magnetisation axis z, an applied field $\mathbf{H}_a$ in a direction other than z will tend to rotate the magnetisation, whereas the anisotropy term will tend to oppose this rotation. If the applied field lies along the easy magnetisation axis, it will stabilise this magnetisation if it points the same way along the axis, but if it is then decreased and reversed to point the other way along the easy magnetisation axis, it will then favour the opposite direction. There will therefore be a value of the field for which the magnetisation reverses. We shall see that the existence of the anisotropy term brings about *magnetic hysteresis*.

The behaviour of the magnetisation can be calculated as a function of the applied field by making several simple assumptions. Suppose the exchange interaction is strong enough to ensure that the spins remain parallel and that the macroscopic

Fig. 9.2 Influence of a magnetic field $\mathbf{H}_a$ applied along the OZ axis on the magnetisation $\mathbf{M}$ of a single-domain magnetic crystal. The direction of the magnetisation $\mathbf{M}$ is specified relative to the easy axis of magnetisation Oz

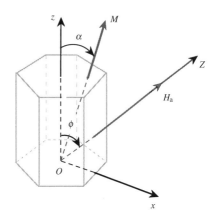

magnetisation behaves like a vector of constant magnitude $M_0 = Ng\mu_B$. In this case, the effect of the field will simply be a *rotation of the macroscopic magnetisation*. One can then carry out a classical treatment of the anisotropy terms and interaction with the field.

Suppose the easy magnetisation axis is aligned with the Oz axis and $\alpha = (\mathbf{Oz}, \mathbf{M})$ as shown in Fig. 9.2. If the anisotropy term is given by (9.1), the associated energy is

$$E_a = NDS^2 \cos^2 \alpha \, . \tag{9.3}$$

As D is negative, we can define a macroscopic anisotropy energy per unit volume by $K = -NDS^2 = -DM_0^2/Ng^2\mu_B^2$. Up to a shift in the energy origin, the energy associated with the anisotropy term can be taken as

$$\boxed{E_{\mathrm{an}} = K \sin^2 \alpha} \, , \tag{9.4}$$

the zero minimum value of E_{an} being obtained when $\mathbf{M}$ lies along the easy axis of magnetisation.

Now apply the magnetic field in a direction $\mathbf{OZ}$, with signed magnitude H_a. The energy of the coupling between the magnetisation and the applied field is

$$E_Z = -\mu_0 H_a M_0 \cos(\phi - \alpha) \, , \tag{9.5}$$

where $\phi = (\mathbf{Oz}, \mathbf{OZ})$, and the total energy is therefore

$$E_T = K \sin^2 \alpha - \mu_0 H_a M_0 \cos(\phi - \alpha) \, . \tag{9.6}$$

The equilibrium positions of the magnetisation are thus given by

$$\frac{\mathrm{d}E}{\mathrm{d}\alpha} = K \sin 2\alpha - \mu_0 H_a M_0 \sin(\phi - \alpha) = 0 \, , \tag{9.7}$$

with $\mathrm{d}^2 E/\mathrm{d}\alpha^2 > 0$.

9.1 Magnetic Anisotropy

There are generally two minima. One corresponds to a metastable magnetisation state. If we begin at a position where the magnetisation is in stable equilibrium, e.g., H_a positive and infinite, hence $\mathbf{M} \parallel \mathbf{H}_a$, and the magnitude of H_a is decreased, the magnetisation remains in the direction corresponding to the same minimum of $dE/d\alpha$ until $d^2E/d\alpha^2 = 0$, the condition for this minimum to become unstable. The magnetisation then flips over to the second minimum for

$$\frac{d^2E}{d\alpha^2} = 2K\cos 2\alpha + \mu_0 H_a M_0 \cos(\phi - \alpha) = 0. \quad (9.8)$$

If we set $\mu_0 H_K = 2K/M_0$, and $h = H_a/H_K$, the simultaneous solution of (9.7) and (9.8) is obtained for

$$\tan(\phi - \alpha) = -\frac{\tan\alpha}{1 - \tan^2\alpha}, \quad (9.9)$$

which means that the magnetisation reverses for $\alpha = \alpha_{\text{rev}}$ and $h = h_{\text{rev}}$ given by

$$\tan^3 \alpha_{\text{rev}} = -\tan\phi, \quad (9.10)$$

and

$$h_{\text{rev}}^2 = 1 - \frac{3}{4}\sin^2 2\alpha_{\text{rev}}. \quad (9.11)$$

In the simple case ($\phi = 0$) where the *field is applied along the z direction*, $\alpha = 0$ for all $h > 0$. If the field is reversed, the reversal occurs for $\alpha_{\text{rev}} = 0$ and $h_{\text{rev}} = -1$. If the field is then increased from $h < -1$ with $\alpha = \pi$, the magnetisation is stable up to $h = 0$. After reversing the field, the reversal occurs for $h_{\text{rev}} = 1$. We thus have a *rectangular hysteresis cycle*, with reversal of the magnetisation for $H_a = \pm H_K$ (see Fig. 9.3).

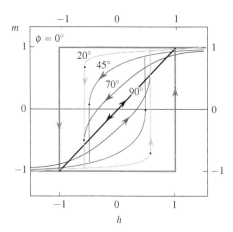

Fig. 9.3 Hysteresis cycles of the magnetisation component of a single-domain magnetic particle in the direction of the applied field as a function of the angle ϕ between the field and the easy axis of magnetisation. Here $m = M_\phi/M_0$ is plotted as a function of $h = H_a/H_K$. Adapted from [14, p. 336] Wiley ©1972

Quite generally, the magnetisation has a component

$$M_\phi = M_0 \cos(\phi - \alpha)$$

along the axis $\mathbf{OZ} \parallel \mathbf{H}_a$. This follows a hysteresis cycle that can be calculated from the above equations. The results are graphed in Fig. 9.3 as a function of the magnitude H_a of this field for different values of the angle ϕ between the direction of $\mathbf{H}_a$ and the easy axis of magnetisation.

In the special case where the field is applied perpendicularly to the easy axis of magnetisation, i.e., for $\phi = \pi/2$, we find $\alpha_{\text{rev}} = -\pi/2$ and $h_{\text{rev}} = -1$, but this does not correspond to $\mathbf{M}$ reversing, because m has then already reached the value -1, the solution of (9.7) being simply $\sin\alpha = h$, i.e., $M_x = M_0 H_a/H_K$. The magnetisation then rotates reversibly in the $(\mathbf{Oz}, \mathbf{Ox})$ plane between $M_x = M_0$ and $M_x = -M_0$ as the applied field varies from H_K to $-H_K$ (see Fig. 9.3).

Apart from this case, it is clear that the introduction of an anisotropy energy term explains why there are hysteresis cycles. The reversal of the magnetisation is an irreversible and sudden process that corresponds to sudden switch to a state of minimum energy. The total energy dissipated during one cycle is given by the area of the cycle, which is $4M_0 H_K$, for example, when $\phi = 0$.

However, the above analysis does not explain why a ferromagnetic sample that has not been subjected to a magnetic field generally has zero magnetisation. One would expect it to magnetise spontaneously along one of its easy magnetisation axes. Moreover, it is often found that a much smaller field than H_K will produce an appreciable magnetisation of a ferromagnetic material.

Question 9.2. Can the results of Fig. 9.1 be explained with the model developed above? Consider instead the case of Co, which has uniaxial anisotropy.

Anisotropies and Broken Symmetries

Note first that all the anisotropic interactions (single ion or exchange anisotropy, dipole interactions) break the perfect rotational symmetry of the initial Heisenberg model. However, the operation $\mathbf{S} \to -\mathbf{S}$ (which implies $\mathbf{M} \to -\mathbf{M}$) is still a symmetry of the full Hamiltonian, and as a consequence, a ferromagnetic state always breaks (at least) this symmetry. In addition, the anisotropy energies per spin are generally very small compared with the exchange energy and therefore play a minor role as long as there is no magnetic order (the spins behaving basically as though they were independent in the paramagnetic phase). On the other hand, in the ordered state, all the spins are aligned, and thus have common dynamics, which multiplies the anisotropy energies by the (considerable) number of spins present and thereby gives rise to quite significant anisotropy effects. More generally, the appearance of a quantity (the spontaneous magnetisation here) which characterises collective behaviour in a reduced symmetry phase at low temperature can bestow importance on parameters that would be negligible in the high symmetry phase at high temperature, e.g., the transition from a liquid to a solid.

9.2 Dipole Interactions, Demagnetising Fields, and Domains

The anisotropies mentioned so far (single-ion or exchange anisotropies) are short-range interactions, between nearest neighbours at the very most. However, we have ignored the *dipole interactions* between spins given by (8.16). We have seen that the dipole interaction between nearest neighbours is negligible compared with the exchange coupling, and quite often even with the magnetocrystalline anisotropy. But the dipole interaction is long ranged, decreasing as $1/r^3$, and it concerns a number of spins that increases in proportion to the volume. The detailed effects of these dipole interactions are often hard to calculate owing to the very large number of spins involved, and the fact that they depend to a large extent on the shape of the sample.

9.2.1 Demagnetising Fields

Up to now we have decoupled the field sources and the materials. We have considered currents in external coils as the sources of the field H_a. But this is no longer valid when we consider a material that is spontaneously magnetic. It is then itself a source of field. It is well known that a magnetised material and a solenoid with a current through it are equivalent, creating fields with similar behaviour at long range (see Fig. 9.4a and b). In the first case, the field is the sum of the fields induced by the different turns in the coil, while in the second it is the sum of the dipole fields produced by the different magnetic dipoles. The identity of the resulting fields at long range arises because the fields produced by a magnetic dipole or by a current loop are equivalent far from the source. When we study the magnetism of a material, the field produced by the material itself, both outside and inside, plays an important role in determining its properties. But this field depends to a large extent *on the shape of the outer surfaces of the material*.

To begin with, we consider a uniformly magnetised material, and from there we shall be able to understand why the thermodynamic equilibrium of a given sample might lead to situations where the magnetisation is non-uniform. For the little

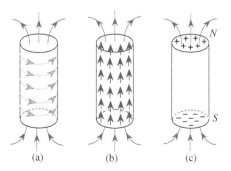

Fig. 9.4 Equivalence of a solenoid (**a**) and a cylindrical magnet (**b**). (**c**) The field produced by the magnet can be determined by considering the + and − magnetic charges at the north and south ends of the bar

magnet considered above, the field results from the existence of north and south poles at its ends (see Fig. 9.4c). This is obtained by analogy with the electric field created by the $+$ and $-$ electric charges at the surface of an electric dipole.

Field Produced by a Magnetisation Distribution

Let us apply Maxwell's equations to a magnetised material when there are no macroscopic currents, which means that we consider only the fields due to the magnetic moments. (Fields due to macroscopic currents are superposed additively, so this restriction does not lead to a loss of generality.) In this case, **curl H** = 0 and **H** is then associated with a magnetic "potential" $U(\mathbf{r})$

$$\mathbf{H} = -\mathbf{grad}\, U(\mathbf{r}). \tag{9.12}$$

Moreover, div $\mathbf{B} = \mu_0 \text{div}\,(\mathbf{H}+\mathbf{M}) = 0$, which implies that

$$\Delta U(\mathbf{r}) = \text{div}\,\mathbf{M}. \tag{9.13}$$

Since **M** is zero outside the material, the magnetic potential thus satisfies the above equation inside the material and $\Delta U(\mathbf{r}) = 0$ outside it. The continuity conditions for $H_\parallel$ and $B_\perp$ at the surface of the material require $U(\mathbf{r})$ to be continuous at the surface, and

$$\left(\frac{\partial U}{\partial n}\right)_{\text{int}} - \left(\frac{\partial U}{\partial n}\right)_{\text{ext}} = \mathbf{M}\cdot\mathbf{n}. \tag{9.14}$$

It can be shown that the solution to the problem summarized by equations (9.13) and (9.14) for $U(\mathbf{r})$ is

$$U(\mathbf{r}) = \frac{1}{4\pi}\left[-\int_V \frac{\text{div}\,\mathbf{M}(\mathbf{r}')}{|\mathbf{r}-\mathbf{r}'|}d^3\mathbf{r}' + \int_S \frac{\mathbf{n}\cdot\mathbf{M}(\mathbf{r}')}{|\mathbf{r}-\mathbf{r}'|}d^2S'\right]. \tag{9.15}$$

We shall not prove this here, but this is similar to the problem for the electrostatic potential produced by a charge distribution which satisfies analogous equations. **M** results from the magnetic dipole moments, which can be considered as made up of fictitious magnetic charges of opposite signs. The magnetic potential goes as $1/r$, like the electrostatic potential.

The **H** field produced by the magnetic dipoles is analogous to the electric field produced by a distribution of electric charges. The first term in (9.15) is the potential associated with the bulk distribution of these magnetic dipoles. It is zero if the magnetisation is uniform, as for the electric field created by a uniform distribution of electric dipoles. The second term is the potential associated with the surface density of magnetic poles $\sigma = \mathbf{M}\cdot\mathbf{n}$, which arises from the discontinuity in the magnetisation at the surface. Its equivalent in electrostatics is a distribution of electrical charges at the surface. The charges are positive (north pole N) when the magnetisation points out of the surface, and negative (south pole S) when it points inwards.

Equation (9.15) can be used to find $U(\mathbf{r})$ and hence $\mathbf{H}(\mathbf{r}) = -\mathbf{grad}\, U(\mathbf{r})$, given $\mathbf{M}(\mathbf{r}')$ at all space points $\mathbf{r}'$. Although these expressions are in principle rather simple, it is quite a different matter to implement them. We shall not go into great detail with such calculations, but instead aim for a qualitative understanding of the physical effects that follow from these dipole fields. For example, for the little uniformly magnetised cylindrical rod, the magnetic potential is simply due to the second term associated with the discontinuities in $\mathbf{M}(\mathbf{r}')$ at the ends of the rod, i.e., the surface densities $+M$ and $-M$ corresponding to the north and south poles N and S of the permanent magnet. Outside the rod, we obtain the standard field lines of the dipole which leave the north pole N and curve round to enter the south pole S.

9.2 Dipole Interactions, Demagnetising Fields, and Domains

a. Influence of Sample Shape on Magnetisation

In order to understand the importance of these magnetostatic effects, consider the small cylindrical rod of ferromagnetic material magnetised along its axis in the absence of any applied field. The field $\hat{\mathbf{H}}_d$ produced inside the rod by the north and south poles N and S is of opposite sign to the field produced outside the rod and along its axis. It opposes the magnetisation of the rod (see Fig. 9.5), which is why it is called the *demagnetising field*.

The magnetisation of the rod is not therefore the one that characterises the material when the applied magnetic field is zero, but rather the one corresponding to the material when it is subjected to its own demagnetising field $\mathbf{H}_d$ with opposite sign to $\mathbf{M}$. To minimise this effect due to the shape of the sample, we must take an infinitely long rod, so that the magnetic poles are pushed out to infinity. In this case, this effect due to the demagnetising field will disappear, and the material will really be subjected to zero field.

The argument is the same if we place a magnetic material in an applied magnetic field $\mathbf{H}_a$ parallel to its axis. The field produced by the dipole adds to the external field and modifies the field lines so that they are no longer parallel (see Fig. 9.6a, b). The magnetisation of the material induces a demagnetising field $\mathbf{H}_d(\mathbf{M})$ inside the rod which opposes the applied field.[1] The field seen by the material is the effective field

$$\mathbf{H}_i = \mathbf{H}_a + \mathbf{H}_d(\mathbf{M}),$$

rather than $\mathbf{H}_a$. If $\mathbf{M}$, whilst being uniform, depends on the applied field, as we saw for example in Sect. 9.1.2 when there is anisotropy, then the magnetisation measured with a magnetometer is not $\mathbf{M}(\mathbf{H}_a)$, but $\mathbf{M}(\mathbf{H}_a + \mathbf{H}_d)$. To make a direct measurement

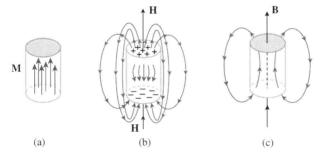

Fig. 9.5 Demagnetising fields of a short cylindrical sample, assumed uniformly magnetised. (**a**) The magnetic poles appearing at the surface induce the magnetic field distribution illustrated in (**b**). Inside the sample, the demagnetising fields oppose magnetisation. (**c**) Spatial distribution of the induction field

---

[1] Note also that, in this example of the short cylindrical sample, (9.15) is such that $\mathbf{H}_d(\mathbf{r})$ depends on the position $\mathbf{r}$ inside the rod, and then the magnetic material is not subjected to a uniform field. Apart from the physical effects which may lead to non-uniformity of $\mathbf{B}$ inside the material, such as vortices in a superconductor, the sample shape can itself lead to a non-uniform distribution of the effective field $\mathbf{H}_a + \mathbf{H}_d$ as seen by the sample.

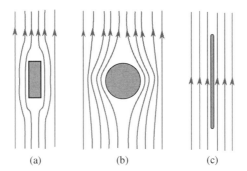

Fig. 9.6 Effect of the demagnetising field in the presence of an applied field. Field distribution outside samples for which the demagnetising field is strong (**a**) and (**b**) or weak (**c**)

of $\mathbf{M}(\mathbf{H_a})$, one must use a long cylindrical sample in order to make the edge effects negligible (see Fig. 9.6c).

b. Demagnetising Fields for Ellipsoidal Samples

To be able to assess the effects of demagnetising fields, we shall consider the special case of an ellipsoidally shaped material, for which it can be shown that the demagnetising field inside the material is uniform if the magnetisation is uniform (see Fig. 9.7). In this case,

$$\boxed{\mathbf{H_d} = -\overline{\overline{\mathbf{N}}}\mathbf{M}}, \qquad (9.16)$$

where $\overline{\overline{\mathbf{N}}}$ is a tensor of unit trace whose eigenvectors are the three principal axes x, y, z of the ellipsoid. In this case, the effective field $\mathbf{H_i}$ seen by the sample is

$$\mathbf{H_i} = \mathbf{H_a} + \mathbf{H_d} = \mathbf{H_a} - \overline{\overline{\mathbf{N}}}\mathbf{M}. \qquad (9.17)$$

c. Expression for the Tensor $\overline{\overline{\mathbf{N}}}$ in Several Special Cases

For an ellipsoid of revolution about the z axis, we have by symmetry

$$N_x = N_y = \frac{1}{2}(1 - N_z). \qquad (9.18)$$

In the special case of an elongated ellipsoid, for which the ratio r of the diameters of the ellipsoid in the directions z and (x, y) is greater than unity, using the formalism developed in (9.15), we find

$$N_z = \frac{1}{r^2 - 1}\left\{\frac{r}{(r^2-1)^{1/2}}\ln\left[r + (r^2-1)^{1/2}\right] - 1\right\}. \qquad (9.19)$$

9.2 Dipole Interactions, Demagnetising Fields, and Domains

Fig. 9.7 Distribution of the magnetic induction in an ellipsoidal sample. **B**, **H**, and **M** are uniform inside the sample

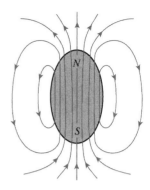

In particular, for an infinitely long cylinder, obtained as the limit of an ellipsoid whose elongation in the z direction tends to infinity, i.e., r infinite, H_d vanishes for $\mathbf{H_a} \parallel \mathbf{Oz}$, as indicated intuitively above [we do indeed obtain $N_z = 0$ from (9.19)], so that

$$N_x = N_y = 1/2 \, .$$

Likewise, an infinite plate parallel to the (x,y) plane corresponds to the limit of an ellipsoid of revolution whose diameter perpendicular to the axis of revolution tends to infinity. The magnetic poles are pushed out to infinity for a field applied parallel to the plane of the plate, implying that

$$N_x = N_y = 0, \quad \text{then} \quad N_z = 1 \, .$$

Finally, note that, in the case of a sphere, we clearly have

$$N_x = N_y = N_z = 1/3 \, .$$

In order to identify the properties of the material in the absence of demagnetising field effects, we must use the value of M to determine H_i given the known value of $\overline{\overline{N}}$. The magnetisation curve in the absence of demagnetising field effects is given by the function $M = f(H_i)$. These effects due to the geometric shape of the sample can seriously affect the value of M, especially in the case of superconductors or ferromagnetic materials, with significant consequences, as we shall see in Chap. 10.

Question 9.3. Consider a spherical sample of a material with magnetisation proportional to the field, i.e., with a susceptibility $\tilde{\chi} = M/H$. Determine the apparent susceptibility, taking the demagnetising field into account. What would be measured for an insulating sample containing no paramagnetic ions and for which $\tilde{\chi} = -10^{-5}$? Or again, for a superconductor exhibiting a perfect Meissner effect?

9.2.2 Demagnetisation Energy and Magnetic Domains

We saw above that the magnetic fields produced by the elementary magnetic dipoles induce demagnetising fields that can influence the magnetisation of a uniformly magnetised sample. But more generally, the energy associated with these interaction terms between spins must be taken into account in order to describe the ground state of the spin system. The problem we have already highlighted above is that these terms depend on the shape of the free surfaces of the magnetic material. We begin by determining the energy associated with these dipole interactions, which we shall call the *demagnetisation energy*.

a. Demagnetisation Energy

If $\mathbf{h}_{ij}$ is the dipole field created by the moment $\boldsymbol{\mu}_j$ at the site i, the energy associated with the dipole couplings is

$$E_\mathrm{d} = -\frac{1}{2}\mu_0 \sum_{ij} \boldsymbol{\mu}_i \cdot \mathbf{h}_{ij}, \tag{9.20}$$

where the factor of 1/2 is introduced to ensure that we only count the coupling energy between $\boldsymbol{\mu}_i$ and $\boldsymbol{\mu}_j$ once. Seperating the sums over i and j

$$E_\mathrm{d} = -\frac{1}{2}\mu_0 \sum_i \boldsymbol{\mu}_i \sum_j \mathbf{h}_{ij} = -\frac{1}{2}\mu_0 \sum_i \boldsymbol{\mu}_i \mathbf{H}_\mathrm{d}(i), \tag{9.21}$$

where $\mathbf{H}_\mathrm{d}(i)$ is nothing other than the value of the demagnetising field at site i. Going to an integral over the volume,[2] the demagnetising energy is thus

$$\boxed{E_\mathrm{d} = -\frac{1}{2}\mu_0 \int_V \mathbf{M}(\mathbf{r})\mathbf{H}_\mathrm{d}(\mathbf{r})\mathrm{d}^3\mathbf{r}}. \tag{9.22}$$

For a uniformly magnetised ellipsoidal sample, this term can be written in the form

$$\boxed{E_\mathrm{d} = \frac{\mu_0}{2}\mathbf{M} \cdot \overline{\overline{\mathbf{N}}} \cdot \mathbf{M}}, \tag{9.23}$$

using (9.16). At thermodynamic equilibrium, a magnetic configuration can be established that minimises the demagnetisation energy.

---

[2] The replacement of the sum by a volume integral is not rigorous in the vicinity of site i. In this case, one must retain a discrete sum in a small region around this site. The correction to (9.22) is found to correspond to a crystalline anisotropy term of dipole origin. This can be included with the other magnetic anisotropy terms that are difficult to obtain in any other way than by experiment.

9.2 Dipole Interactions, Demagnetising Fields, and Domains

> **Question 9.4.** Consider a sample with ellipsoidal external shape and uniform magnetisation. What will be the orientation of its magnetisation when the magnetocrystalline anisotropy energy is zero? What about an extensive thin film? Consider the case of an ellipsoid of revolution whose major axis c lies along the axis of revolution (like a rugby ball). Give the form of the magnetostatic energy as a function of the angle between the magnetisation and the axis c. What conclusion can be drawn?

b. Magnetic Domains

We see immediately that, for zero applied field, the configuration corresponding to uniform magnetisation is not the most favourable. The demagnetisation energy is only zero for certain geometrical shapes of the sample. To minimise E_d, it is intuitively clear that it will be favourable to divide the sample up into ever thinner layers in such a way as to minimise the demagnetising field in each individual layer. Examples of configurations gradually reducing E_d for a parallelepiped of material are shown in Fig. 9.8. This subdivision could be continued indefinitely, but the number of magnetic domains is bounded owing to the energy cost associated with creating interfaces. Indeed, these spatial regions, shaded in Fig. 9.8 and referred to as *Bloch walls*, separate regions with different magnetisation orientations. This problem is reminiscent of the one encountered with type II superconductivity. However, while in the superconductivity case the physical origin of the subdivision was to be sought in the microscopic physical properties, the subdivision into domains in the present case is related to the external shape of the material. In the case of an infinitely long cylindrical wire, there would be only one magnetic domain along the axis of the wire, and zero demagnetisation energy, since the magnetic poles would be at infinity.

Note that the expression (9.22) cannot inform us about all the factors that determine the configurations minimising the demagnetisation energy. The expression for E_d can be expressed in a form implying that E_d will be minimal if both the demagnetising fields and the fields produced outside the material are minimised (see

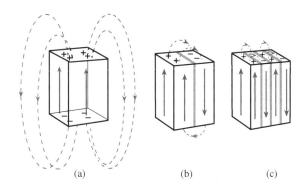

Fig. 9.8 Different arrangements of the magnetisation of a ferromagnetic solid. The demagnetisation energy of the layers decreases as they are made thinner, leading to a simultaneous reduction in the field produced outside the material

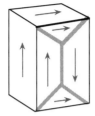 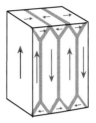

Fig. 9.9 Closure domains allow the magnetisation to be parallel to the surface of the material and cancel the field outside

below). This therefore leads to a subdivision into magnetic domains with **M** parallel to the surface insofar as this is possible. This is indeed achieved by the subdivision into layers as shown in Fig. 9.8. However, the field produced at the surface is not zero, even though it is reduced by these alternating magnetisation configurations. It may be just as interesting to reduce the surface charges to zero by forming domains known as closure domains such that the magnetisation is always parallel to the outer surfaces, as shown in Fig. 9.9.

Question 9.5. Do you think closure domains will be equally favourable for single crystals of Co and Fe?

Another Expression for the Demagnetisation Energy

In the general case, we integrate $\mathbf{H} \cdot \mathbf{B}$ over a volume V enclosing the whole magnetic material. Using the expression (9.12) for $\mathbf{H}$ as a function of the magnetic potential U,

$$\int \mathbf{H} \cdot \mathbf{B} \mathrm{d}^3\mathbf{r} = -\int \mathbf{B} \cdot \mathbf{grad}\, U \mathrm{d}^3\mathbf{r} = -\int \nabla(U\mathbf{B}) \mathrm{d}^3\mathbf{r} = -\int U\mathbf{B} \cdot \mathbf{n} \mathrm{d}S. \quad (9.24)$$

The second equality is obtained using the identity $\nabla \cdot (U\mathbf{B}) = \mathbf{B} \cdot \mathbf{grad}\, U + U \mathrm{div}\, \mathbf{B}$, in which the second term is zero. If the integration volume is extended over the whole of space, we note that U goes as $1/r$ at infinity, and hence $\mathbf{B} = \mu_0 \mathbf{H}$ goes as $1/r^2$, so the surface integral vanishes. The integral of $\mathbf{H} \cdot \mathbf{B}$ over the whole of space is therefore zero, and

$$\int_{\text{space}} \mathbf{H} \cdot \mathbf{B} \mathrm{d}^3\mathbf{r} = \int_{\text{space}} \mu_0 \mathbf{H} \cdot (\mathbf{H}+\mathbf{M}) \mathrm{d}^3\mathbf{r} = \int_{\text{space}} \mu_0 \mathbf{H}^2 \mathrm{d}^3\mathbf{r} + \int_V \mu_0 \mathbf{H}_\mathrm{d} \cdot \mathbf{M} \mathrm{d}^3\mathbf{r} = 0. \quad (9.25)$$

In the second equality, we have decomposed the integral into two terms and used the fact that **M** vanishes outside the magnetic material of volume V, within which the field is none other than the demagnetising field $\mathbf{H}_\mathrm{d}$. This implies another expression for the demagnetisation energy:

$$E_\mathrm{d} = \frac{\mu_0}{2} \int_{\text{space}} \mathbf{H}^2 \mathrm{d}^3\mathbf{r}, \quad (9.26)$$

where the integral is carried out over the whole of space, both inside and outside the sample. This second expression shows that E_d is always positive. In order to minimise E_d, the field must be minimised everywhere, both inside and outside the sample. In particular, this means avoiding surface charges which create a magnetic field outside the sample.

9.3 Bloch Walls

The formation of domains leads to a gain in magnetostatic energy, but it requires the formation of interfaces. As we have already seen in the case of superconductivity, the formation of an interface generally involves an energy cost. It is the balance between these two energy terms that will eventually specify the minimum energy state.

So what is the energy of formation of an interface? It might be thought a priori that there is a sharp boundary on the atomic scale, and that the moments to the right of this wall will all be oriented downwards, while those to the left will all be oriented upwards. In a simple cubic crystal, such a spin arrangement would cost an exchange energy equal to $2S^2|\mathscr{J}|$ per primitive cell of the wall. It is easy to see that this is not the lowest energy arrangement. It is better to rotate the spins progressively through a small angle for each lattice spacing, as shown in Fig. 9.10, because the angle between two neighbouring spins is small, so the cost in exchange energy will be reduced, even if the number of spins for which the exchange energy is not minimal is actually increased.

Exchange tends to favour a 'spreading' of the wall, in that it makes the angle between neighbouring spins as small as possible while extending the region over which the transition between the $\uparrow$ and $\downarrow$ orientations takes place.

On the other hand, the existence of magnetocrystalline anisotropy has the opposite effect. If the wall is sharp between two domains for which the magnetisation lies along an easy axis of magnetisation, the anisotropy term is minimised on either side of the wall, and the cost in anisotropy energy is zero. But if the magnetisation rotates gradually through the wall, there will on the contrary be an anisotropy energy cost which increases as the width of the wall (indeed the number of spins deviating from the easy magnetisation direction increases as the width of the wall). Any magnetic anisotropy thus tends to make the wall 'thinner'. We thus see that the configuration that minimises the wall formation energy will depend on the relative values of the exchange energy and the magnetic anisotropy term in the Hamiltonian of the spin system.

To be more precise, consider a specific case in which the anisotropy term takes the simple form (9.1) with $D < 0$, giving an easy axis of magnetisation parallel

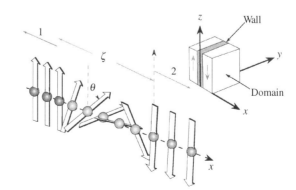

Fig. 9.10 Spin configuration in a Bloch wall. The magnetisation rotates steadily about the x axis in the (y,z) plane. The angle $\theta = (\mathbf{z}, \mathbf{M})$ varies from 0 to π for increasing x. Adapted from [14, p. 289] Wiley ©1972

to the z axis. To obtain the structure of a wall, the spins are treated as classical vectors by writing $\mathbf{S} = S(\sin\theta\cos\phi, \sin\theta\sin\phi, \cos\theta)$. A wall separates regions with $\mathbf{S} = (0,0,+S)$ and $\mathbf{S} = (0,0,-S)$. If we assume that the wall is parallel to the (y,z) plane, the two values of $\mathbf{S}$ are reached for $|x| \to \infty$, and the orientation of a spin $\mathbf{S_R}$ depends only on the x component of its position (see Fig. 9.10).

To simplify, consider also that we are dealing with a simple cubic lattice with cubic unit cell of side a, and assume that the x component of the spin is everywhere zero ($\phi \equiv \pi/2$). Under these conditions the total exchange and anisotropy energy is given by

$$E = \sum_l \left(-\mathscr{J}\mathbf{S}_l \cdot \mathbf{S}_{l+1} + DS_z^2\right) = \sum_l \left[-\mathscr{J}S^2 \cos(\theta_{l+1} - \theta_l) + DS^2 \cos^2\theta_l\right], \quad (9.27)$$

for a row of spins along the x axis, with coordinates $x = la$ indexed by l.

The minimal energy configuration of the wall is obtained by minimising E with respect to any change in the orientation of an arbitrary spin of the wall, that is, by setting $dE/d\theta_l$ equal to zero, which yields the condition

$$\mathscr{J}\left[\sin(\theta_{l+1} - \theta_l) + \sin(\theta_{l-1} - \theta_l)\right] + D\sin 2\theta_l = 0. \quad (9.28)$$

We may also assume that the angle between neighbouring spins is small. We may then:

- replace the first two sine functions in (9.28) by their arguments,
- treat θ_l as a continuous function θ of the variable $x = la$.

Equation (9.28) then becomes the differential equation

$$\mathscr{J}a^2 \frac{d^2\theta}{dx^2} + D\sin 2\theta = 0. \quad (9.29)$$

Setting $x' = x/\varsigma$ with

$$\boxed{\varsigma^2 = \frac{\mathscr{J}a^2}{2|D|}}, \quad (9.30)$$

this can be written

$$\frac{d^2\theta}{dx'^2} - \frac{\sin 2\theta}{2} = 0. \quad (9.31)$$

This is an Euler equation whose solution must in this case satisfy the boundary conditions $\theta \to 0$ for $x' \to -\infty$ and $\theta \to \pi$ for $x' \to +\infty$. The solution satisfying these conditions and for which $\theta = \pi/2$ for $x' = 0$ is

$$\boxed{\theta(x') = 2\arctan\left(e^{x'}\right)}. \quad (9.32)$$

9.3 Bloch Walls

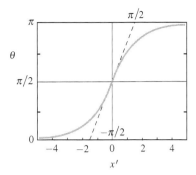

Fig. 9.11 Dependence of the angle θ between the spin and the z axis on the position in the wall

This function is plotted in Fig. 9.11. The region in which $\theta(x')$ deviates significantly from the limiting values 0 and π can be estimated by a linear approximation around $x' = 0$ (dotted line in Fig. 9.11).

Note that the function $\theta(x')$ satisfies the simple relation

$$\frac{d\theta}{dx'} = -\sin\theta, \qquad (9.33)$$

and that the linear extrapolation corresponds to a width $x' = \pi$, or $x = \pi\varsigma$ for the wall. The length ς given by (9.30) thus characterises *the thickness of the wall*.

The *energy of formation* of the wall can be estimated by substituting this solution for $\theta(x')$ into (9.27). The increase in energy compared with the value for a uniform magnetisation for the row of spins defines the energy of the wall E_{wall} per primitive cell of the wall surface. It is given by

$$E_{\text{wall}} = \sum_{l}\left[\frac{\mathscr{J}S^2 a^2}{2}\left(\frac{d\theta_l}{dx}\right)^2 - DS^2\sin^2\theta_l\right]. \qquad (9.34)$$

Replacing the sum over l by an integral over x and hence over θ, and then using $dx' = -d\theta/\sin\theta$, it is found that the two contributions to the energy of the wall due to exchange and anisotropy are actually equal, and that this wall energy per area a^2 of the primitive cell is given by

$$\boxed{E_{\text{wall}} = 2S^2\sqrt{2|D|\mathscr{J}}}. \qquad (9.35)$$

It can be checked that, for fixed $\mathscr{J}$, a zero value of D would correspond to an infinitely thick wall, with zero energy cost for creation of the wall. On the other hand, an increase in the anisotropy energy, i.e., an increase in $|D|$, raises the energy required for the creation of the wall, but reduces its thickness ς. Similar calculations can be carried out for other forms of the anisotropy energy, or for walls in which the orientation of the magnetisation changes by an angle less than π. The results only differ from those obtained here by numerical factors in the expressions for the energy and thickness of the wall. However, it should be noted that this type of calculation is only valid for small anisotropy energies, for which the thickness of the wall remains greater than the lattice constant a. In this case the step from (9.28) to (9.31) is legitimate. In the other case, $|D| \gg \mathscr{J}$, the wall corresponds to a sudden reversal over one primitive cell and its energy is simply $2S^2 \mathscr{J}$. In many materials, the anisotropy energy is small enough for the energy cost of the walls to be too low to prevent the formation of domains. The wall energy will nevertheless limit the number of walls per unit volume and will thus determine the size of the magnetic domains.

Note that this investigation of the Bloch wall shows that, in a magnetic material, the balance between exchange forces, anisotropy, and demagnetising fields can lead to complex spatial configurations for the magnetisation on the microscopic scale.[3]

9.4 Magnetic Hysteresis Cycles

We have set up the tools required to understand how a magnetic material divides up into alternating magnetisation domains. The equilibrium configuration in zero field for a sample of given shape results from minimisation of the sum of the exchange, anisotropy, and demagnetisation energies. It is the minimisation of this demagnetisation term of dipole origins that imposes the absence of macroscopic magnetisation of a material in zero field conditions.

9.4.1 Bitter Method for Observing Domains

It was obviously important to be able to determine the domain configuration of magnetic materials by experimental means. Now we know that a magnetic material subjected to a magnetic field gradient suffers a force. When it became possible to make large amounts of small magnetic particles with well determined sizes, in particular for the fabrication of magnetic recording tapes, it was interesting to see what would happen when such particles were sprinkled on the surface of a magnetic material. The little particles are attracted by regions where the gradient of the magnetic field is high. This is the idea behind the method used by Bitter in 1931 to visualise the Bloch walls in ferromagnetic materials. Consider, for example, two domains with opposite magnetisations. The field induced is tangent to the surface except in the region above a Bloch wall, where the moments in the wall induce a field component perpendicular to the surface. The field gradient is then high at the surface, in the region where the Bloch wall emerges. If a suspension of magnetic particles in a fluid is deposited on the surface, the magnetic particles (usually ferrite Fe_3O_4) naturally tend to move into this region (see Fig. 9.12a). The walls can then be visualised by optical microscope, as shown in Fig. 9.12b and c.

---

[3] A relatively simple case is discussed in Problem 20: *Magnetism of a Thin Film*.

9.4 Magnetic Hysteresis Cycles

Fig. 9.12 (a) Bitter's method for observing domain walls. The magnetic powder tends to accumulate vertically above the Bloch wall. (b) The wall appears lighter when observed optically with transverse illumination. (c) Observation of the walls at the surface of an iron single crystal with excellent crystal quality. The displacement of the walls under the application of a magnetic field can be observed. Adapted from [14, p. 293] Wiley ©1972

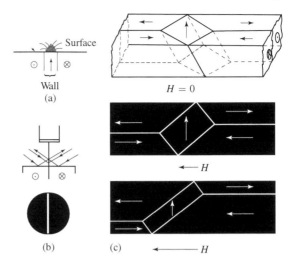

9.4.2 Magnetisation Process and Hysteresis

We may wonder what happens when a material is magnetised by increasing the applied magnetic field H_a? Consider the relatively simple case of a single-crystal material. Initially, the domains oriented parallel to the field will grow, to the detriment of those that are antiparallel to it, as can be seen in Fig. 9.12c.

In a weak applied magnetic field, *the displacements of the walls are reversible*, leading to reversible changes in the magnetisation. This is called the *Rayleigh regime*, shown schematically in Fig. 9.13 ($H = 0 \to H_1$). Above a certain field strength H_a, sudden displacements of the walls can occur and domains can disappear altogether (between H_1 and H_2).

Finally, for sufficiently strong H_a, the magnetisation increases by *rotation of domains* ($H_2 \to H_3$), until it reaches its saturation value, when all the moments are parallel to H_a. If the applied field is now reduced, the magnetisation decreases more slowly. Note in particular that, for $H_a = 0$, the magnetisation has a nonzero value M_r called the *remanent magnetisation*, which can only be reduced by applying a field in the opposite direction. The field $-H_c$ needed to cancel the magnetisation is called the *coercive field*.

If a sample is subjected to a magnetic field cycle of the form $(0,H)$, $(H,-H)$, $(-H,H)$, the change in M follows a *hysteresis cycle*, as shown in Fig. 9.14. The

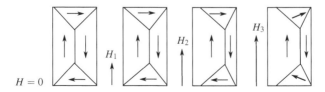

Fig. 9.13 Typical domain configurations in a ferromagnetic material under increasing field

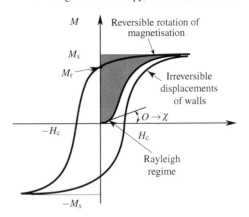

Fig. 9.14 First magnetisation of a ferromagnetic material showing the reversible Rayleigh regime. The hysteresis cycle obtained after applying a high field H_a shows the saturation magnetisation M_s, the remanent magnetisation M_r, and the coercive field H_c

two quantities M_r and H_c characterise the hysteresis cycle qualitatively. This cycle has a more or less rectangular shape, depending on how suddenly the magnetisation reverses, which in turn depends on the system. The fact that a frictional force prevents the Bloch walls from moving around freely leads to a dissipation of energy given by the area enclosed by the cycle. M_r and H_c are key quantities for specifying applications of magnetic materials.

In order to make permanent magnets, we seek materials with high remanence and high coercive field (see Fig. 9.15), referred to as *hard ferromagnetic materials*. In contrast, *soft ferromagnetic materials*, for which H_c is negligible, have small losses and high susceptibility. They are used to build transformers, inductances, and read heads for hard disks. For magnetic recording media, we use materials with square hysteresis cycles and with a coercive field that is not too high, in order to facilitate magnetic inscription.

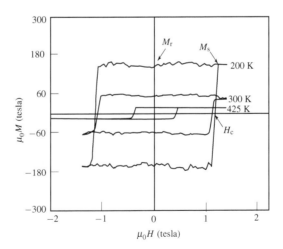

Fig. 9.15 Characteristic hysteresis cycle of a hard ferromagnetic material. Figure courtesy of Pascard, H.: LSI. Ecole Polytechnique, France

9.5 Summary

Magnetic anisotropy terms, e.g., $H_A = D\sum_{\mathbf{R}}(S_{\mathbf{R}}^z)^2$ must be added to the Heisenberg Hamiltonian, whereupon they determine the equilibrium orientations of the magnetic moments relative to the crystallographic axes, but not their directions. For this form of anisotropy, the z axis is the easy axis of magnetisation of the material.

The fact that a reversal of all the spins will not modify the energy for the spin system leads to the existence of magnetic hysteresis and an associated energy dissipation. We can define an anisotropy energy $K = -DM_0^2/N\mu_B^2$ and an anisotropy field $H_K = 2K/\mu_0 M_0$, which corresponds to the reversal of the uniform magnetisation when the magnetic field is applied along the easy axis of magnetisation.

Magnetic dipole interactions generate magnetic fields both outside and inside magnetised materials, and these depend on the external shape of the material. The fields produced can be attributed to fictitious magnetic charges that appear at the surface of the material and whose surface density is given by the component of the magnetisation normal to the surface. The fields $\mathbf{H}_d(\mathbf{r})$ created within the material are demagnetising, i.e., opposed to $\mathbf{M}$. For uniformly magnetised materials with ellipsoidal shape,

$$\mathbf{H}_d = -\overline{\overline{\mathbf{N}}}\mathbf{M},$$

where $\overline{\overline{\mathbf{N}}}$ is a tensor of unit trace with eigenvectors along the three principal axes x, y, z of the ellipsoid. These demagnetising fields have a significant effect on the magnetisation of a sample. This magnetisation $\mathbf{M}(\mathbf{H}_a + \mathbf{H}_d)$ is indeed strongly affected by the external shape of the sample.

The energy associated with the dipole couplings between spins can be written in the form

$$E_d = -\frac{1}{2}\mu_0 \int_V \mathbf{M}(\mathbf{r})\mathbf{H}_d(\mathbf{r}) d^3\mathbf{r}.$$

In zero field, the minimisation of the total energy of the spin system, sum of the exchange energy, the anisotropy energy, and the demagnetisation energy, leads to the formation of magnetic domains magnetised in different directions. This magnetic domain structure depends on the external shape of the sample. The minimisation of the demagnetisation energy implies a domain configuration minimising the external fields, i.e., the surface charges.

Two domains differ through the direction of their magnetisation $\mathbf{M}$. They are separated by a Bloch wall in which $\mathbf{M}$ rotates steadily between its directions in the two domains. In the case of a single-axis anisotropy, the thickness ς of the wall and its energy per primitive cell are given respectively by

$$\varsigma^2 = \frac{\mathscr{J}a^2}{2|D|}, \qquad E_{\text{wall}} = 2S^2\sqrt{2|D|\mathscr{J}}.$$

Images of the distribution of magnetic domains at the surface of a material can be obtained by Bitter's method, which consists in sprinkling a powder of submicron magnetic particles on the material. These tend to accumulate where the Bloch walls reach the surface.

An increasing applied magnetic field gradually magnetises a material through the displacement of the Bloch walls, then by rotation of the magnetisation in the domains. The fact that there are irreversible processes in which the walls are displaced and domains disappear leads to a magnetic hysteresis cycle. It is characterised by the remanent magnetisation M_r obtained in zero field and by the coercive field H_c, which is the field for which the magnetisation is reversed.

9.6 Answers to Questions

Question 9.1

We may write

$$E_a = D(S_x^2 + S_y^2 + S_z^2)^2 - 2D(S_x^2 S_y^2 + S_y^2 S_z^2 + S_z^2 S_x^2).$$

As the first term is isotropic, only the second contributes to anisotropy. It is easy to check that, if $D < 0$, the second term is minimal and zero for (100), i.e., when the magnetisation lies along one of the sides of the cube, as happens for the iron in Fig. 9.1. On the other hand, for $D > 0$, the minimum of E_a is obtained for (111), i.e., when the magnetisation lies along the diagonals of the cube, as happens in nickel.

Question 9.2

We do not expect to obtain $M = 0$ for $H = 0$ as happens experimentally in Fig. 9.1, and this because the magnetisation of the single crystal is made up of domains. The curves shown are first magnetisation curves rather than hysteresis cycles.

However, when the applied field is perpendicular to the easy magnetisation axis, a significant contribution to the magnetisation comes from rotation of the domains. The situation is then close to the one considered for $\phi = \pi/2$ in the case of a single domain. This is valid in particular in the case of Co in Fig. 9.1. The gradient of the curve in the [1010] direction corresponds to

$$\frac{M}{M_0} = \frac{H_a}{H_K},$$

and then

$$\chi = \frac{M}{\mu_0 H_a} = \frac{M_0^2}{2K}.$$

The gradient at the origin gives $\chi = 1.4 \times 10^6/0.6$, and this implies the estimate $K = 4.2 \times 10^5$ A/m^3.

Question 9.3

We have
$$M = \tilde{\chi}(H_a - M/3),$$
implying an apparent susceptibility
$$\tilde{\chi}_{app} = \frac{\tilde{\chi}}{1 + \tilde{\chi}/3}.$$

For a weakly diamagnetic material, the correction is negligible because $\tilde{\chi} \ll 1$. For a superconductor exhibiting a perfect Meissner effect, $\tilde{\chi} = -1$, and the above relation gives $\tilde{\chi}_{app} = -3/2$. Such a sample can be used to carry out an absolute calibration of a magnetometer.

Question 9.4

For a particle in the form of an ellipsoid of revolution about the z axis (see Fig. 9.16),
$$N_x = N_y = \frac{1}{2}(1 - N_z).$$

If M_z and M_p are the components of **M** parallel and perpendicular to the z axis, respectively,
$$\frac{2E_d}{\mu_0} = N_z M_z^2 + \frac{1 - N_z}{2} M_p^2.$$

Since $M_z^2 + M_p^2 = M_s^2$,
$$\frac{2E_d}{\mu_0} = N_z(M_s^2 - M_p^2) + \frac{1 - N_z}{2} M_p^2$$
$$= N_z M_s^2 + \frac{1 - 3N_z}{2} M_p^2.$$

If $N_z < 1/3$, E_d is minimal for $M_p = 0$. If $N_z > 1/3$, E_d is minimal for $M_p = M_s$. So for a sphere, the orientation of **M** is indeterminate, for a flattened ellipsoid, **M** lies in the (x,y) plane, and for an elongated ellipsoid, **M** lies along the z axis.

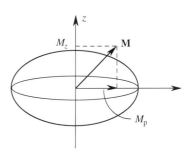

Fig. 9.16 Ellipsoid of revolution for Question 9.4

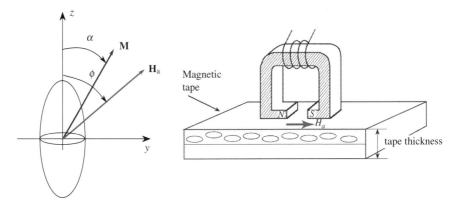

Fig. 9.17 *Left*: Elongated ellipsoid for Question 9.4. *Right*: Elongated particles with the long axis aligned with the feed direction in a magnetic tape

Note that these cases correspond to situations in which the magnetic surface charges are as far away as possible. In the case of an almost infinite plane, the magnetisation must lie in the plane in order to push the surface charges to infinity.

In the case of an ellipsoid that is elongated along the z axis, with the notation in Fig. 9.17 (left), the demagnetisation energy can be expressed in the form

$$\frac{2E_d}{\mu_0 M_s^2} = N_z \cos^2 \alpha + \frac{1-N_z}{2} \sin^2 \alpha = +\frac{1-3N_z}{2} \sin^2 \alpha + \text{constant} .$$

Setting $K = (1 - 3N_z)/2 > 0$, this has the same form as for the anisotropy energy in (9.4). The magnetisation will therefore describe hysteresis cycles as in Fig. 9.3. We thus see that *even when there is no magnetocrystalline anisotropy energy*, a small particle will exhibit hysteresis with high coercive field if it has non-spherical shape.

A ferromagnetic material with weak magnetocrystalline anisotropy can therefore be used to make a magnetic recording tape if the particles have elongated shape, provided that they lie parallel to the tape feed axis on the polymer substrate (see Fig. 9.17 right). Such an arrangement is convenient for inscribing a magnetisation produced parallel to the substrate (see Fig. 10.10).

Question 9.5

In the example, the closure domains are oriented at an angle of $\pi/2$ to the magnetisation in the main domains. In the case of Co, as the anisotropy is single axis, this corresponds to a difficult magnetisation axis, and a high cost in anisotropy energy. On the other hand, in Fe, which is cubic with easy axes lying along the (001) directions, the domains oriented at $\pi/2$, i.e., along (100), are also easy axes. These domains at $\pi/2$ are often observed in Fe (see, for example, Fig. 9.12).

Chapter 10
Measurements in Magnetism: From the Macroscopic to the Microscopic Scale

Contents

10.1	Macroscopic Magnetic Measurements	297
	10.1.1 Susceptibility or Magnetisation	298
	10.1.2 Examples of Measurements on Superconductors	301
	10.1.3 Hysteresimeters: Toroidal Geometry	303
10.2	Magnetic Surface Measurements and Magnetic Imaging	304
	10.2.1 Microscale Measurements and Surface Imaging	305
	10.2.2 Towards Submicron Scales	308
10.3	Summary	312
10.4	Answers to Questions	312

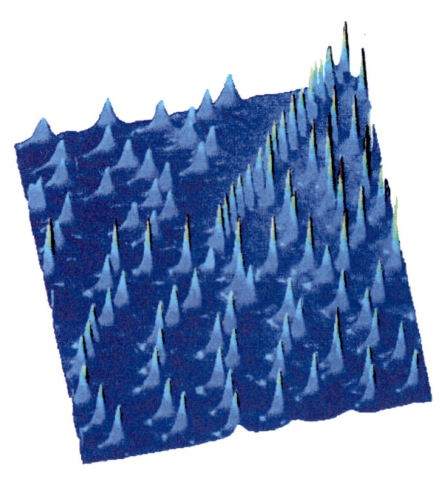

The magnetic field induced by vortices at the surface of a superconductor can be measured by displacing the loop of a micrometric SQUID across the surface. In this way, the penetration of the magnetic flux in a film of the high-T_c superconductor $BiSr_2CaCu_2O_8$ deposited on a single-crystal substrate can be detected directly. In this system, the vortices are not arranged in a lattice, because they are trapped by defects (vortex pinning). Here, the deliberate choice of a substrate comprising three crystals allows one to generate a film which itself comprises three crystals. Note in particular the alignments of vortices along the boundaries between the three crystals. The field pointing from the vortex located at the meeting point of these three crystals has a smaller magnitude than the fields pointing from the other vortices. This particular vortex corresponds to half a flux quantum $\phi_0/2$. This experiment establishes a special feature of the cuprate superconductors as compared with metal superconductors: their band gap vanishes in some directions of the reciprocal lattice (four-leaf clover, or d type symmetry). Image courtesy of Tsuei, C.C. (IBM Yorktown Heights) and Raffy, H. (Physique des Solides, Orsay, France). The corresponding experiment is described in Europhys. Lett. **36**, 707 (1996)

10.1 Macroscopic Magnetic Measurements 297

An unadvertised observer usually gains information on the magnetic properties of a material by testing its interaction with a known magnet, e.g. attraction strength for a ferromagnet or levitation height for a superconductor. These observations involve the field induced by one of the materials on the other, and are very difficult to exploit as the fields depend sensitively on the shape of the materials.

To characterise the magnetic response of a material, it is obviously simpler to subject it to a uniform magnetic field (Sect. 10.1). However, even in this case, we have seen in the preceding chapters that the magnetic response of a material depends strongly on its shape. Unlike many other physical properties of solids, manifestations of magnetism can be extremely non-uniform, both in superconductors and in magnetic materials, even when the material is ideal in the sense that its other properties are uniform. In fact, quantities such as χ and M which are averaged quantities, represent only the first level of magnetic information.

The ideal observation would be one that could visualise the currents and spins in every detail. Such information is clearly inaccessible! However, in order to get a better understanding of the fundamental phenomena, we must try to visualise vortices and walls between magnetic domains, and devise local observational methods. Some such attempts will be discussed briefly in Sect. 10.2. This is an area where scientific perfectionism encounters practical considerations. The aim is to write data in magnetic form on substrates with the highest possible density, hence the need to devise methods for writing and reading data on microscopic or even nanoscopic scales.

10.1 Macroscopic Magnetic Measurements

All methods for making magnetic measurements use magnetic fields. These fields must therefore be known to great accuracy. The field produced by a current in a coil can be accurately calculated if the geometry of the coil is sufficiently well controlled. Moreover, since the field–current relationship is linear, one calibration for a given current value is sometimes enough. However, if very high accuracy is required, there are many physical effects that can be used to establish measurement standards for magnetic fields. Simple but important methods consist of measuring the *Hall voltage* induced in a semiconductor sample or the resistance of a sample with high *magnetoresistance*. A more sophisticated method to implement, but which proves much more accurate, consists of measuring the *Larmor frequency* of the nuclear spins contained in a reference sample, e.g., protons in water (see Chap. 11).

Measurements of magnetisation or magnetic induction appeal to simple electromagnetic techniques. In Sect. 10.1.1, we shall discuss the main methods currently used, with a geometry that minimises sample shape effects. Examples of measurements are presented in Sect. 10.1.2, and these will illustrate the way the sample shape can modify the spatial structure of the magnetism in unexpected ways. Finally, in Sect. 10.1.3, we discuss the most propitious geometry for studying hysteresis cycles in ferromagnetic materials with low coercivity.

10.1.1 Susceptibility or Magnetisation

The basic idea of a magnetometer is to detect the flux of the magnetic induction B in a coil A in which the sample has been inserted, the whole thing being subjected to an applied field H_a produced by a long external solenoid S_0 (see Fig. 10.1a). In this section, we take the sample and coils to have very long cylindrical shapes in order to make the effects of the demagnetising field negligible.

The flux inside the coil is $\Phi = nBS$, where S and n are the cross-sectional area and the number of turns in the coil, respectively, and we assume that this coil perfectly fits the shape of the sample. This method can in principle be used to make a direct measurement of B if the field H_a is applied suddenly at time $t = 0$. The EMF induced in the coil by the change in the flux $\Phi(t)$ is then

$$V(t) = \frac{d\Phi}{dt} = nS\frac{dB(t)}{dt}.$$

If we use a device that integrates $V(t)$, we thus measure $\int V(t)dt = nSB$. This method can be used to measure B directly, but it requires a variation in H_a and hence in M to carry out the measurement.

In order to get round this difficulty, it is more convenient to consider a system comprising two opposite coils. In a fixed magnetic field, the sample is extracted from coil A and inserted in coil B, as shown in Fig. 10.1b. In this case, the total flux changes from

$$(nS)_A B - (nS)_B \mu_0 H_a \quad \text{to} \quad (nS)_A \mu_0 H_a - (nS)_B B.$$

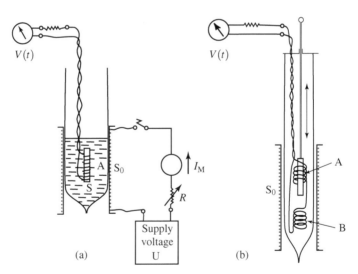

Fig. 10.1 (a) Setup for measuring B by integrating the voltage induced in the coil when the applied field is varied from 0 to H_a. (b) Setup for measuring M in a fixed field H_a by integrating the voltage induced in two opposite coils when the sample is extracted from coil A and inserted in coil B. Adapted from [18, pp. 49–51]

10.1 Macroscopic Magnetic Measurements

If the two coils are identical, $(nS)_A = (nS)_B$ and the flux variation when the cylindrical sample is moved from A to B is therefore

$$\Delta\Phi = -2nS(B - \mu_0 H_a) = -2nS\mu_0 M . \tag{10.1}$$

With this method, the sample can be kept in the field H_a all the time and the magnetisation $\mu_0 M = B - \mu_0 H_a$ can be measured directly.

Note that this measurement technique shows why it was necessary to introduce the notion of magnetisation. We do not know a priori how to measure the spatial dependence of B, i.e., the spatial dependence of the currents in a superconductor, or the domains in a ferromagnet. However, we are able to measure the total flux, or rather the difference in the flux with and without the sample. It is thus natural to consider that the sample is equivalent to a fictitious surface current density that produces the field M which must be added to H_a. This is in fact perfectly accurate in the case of a superconductor exhibiting a perfect Meissner effect. But in all other cases, M only represents the current density that would produce the same variation as the one produced by the sample. The value of B in (10.1) is $S^{-1} \int_{(S)} B(r) \mathrm{d}^2 r$, the spatially averaged value of $B(r)$ over the cross-section of the sample.

> **Question 10.1.** Must the sample and coil have the same cross-sectional area in the setups of Figs. 10.1a and b? What are the advantages and disadvantages of these two methods?

The variation of the flux can be obtained by electronic or digital integration of the temporal variation of the voltage induced in the coil during extraction. In modern magnetometers, *exceptional sensitivity is obtained using a SQUID to make the flux measurement*. This is done by making the coils A and B with superconducting wire. The free ends of the coils, joined in series, are reconnected on a superconducting coil, which is itself inductively coupled to the loop of a SQUID. This forms a flux transformer. The variation of the flux induced on the SQUID is proportional to $\Delta\Phi$ (see Fig. 10.2). It is thus by counting the number of flux quanta in the SQUID loop

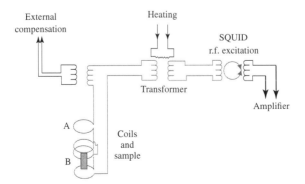

Fig. 10.2 Superconducting flux transformer coupled to a SQUID for measuring the flux variation in the magnetometer of Fig. 10.1b. Adapted from the technical notes for the SQUID Quantum Design, San Diego, CA (1990)

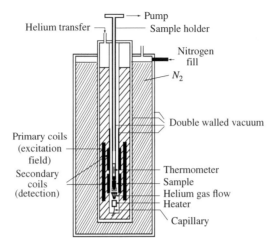

Fig. 10.3 Magnetic susceptometer allowing to measure the voltage induced by mutual inductance between the primary winding and two opposite secondary windings. The detected signal in presence of the sample is proportional to $\nu\chi$. The thermalised cell in which the sample is housed is represented

that the magnetisation of the sample can be determined. These magnetometers are highly versatile, and they are so sensitive that there is no rival method for measurements on weakly magnetic materials or very small samples. It is commonplace to measure the magnetisation of an atomic monolayer of Fe with an area of only a few square millimeters.

The magnetic susceptibility χ of a sample can be measured directly using an alternating exciting field. The latter is produced by an alternating current at frequency ν in a primary coil (see Fig. 10.3). The signal induced across the terminals of the two oppositely wound secondary coils is zero when there is no sample. The difference in the mutual inductance due to the sample gives a signal at the frequency ν, with amplitude proportional to $\nu\chi$. The signal induced in this kind of susceptometer can also be measured using a SQUID, if very high sensitivity is required.

A setup known as the Faraday balance measures the force acting on a sample when it is subjected to a constant magnetic field gradient $\nabla \mathbf{H}_a$. The force exerted on the sample is

$$\mathbf{F} = -\mu_0 \nabla (\mathbf{M} \cdot \mathbf{H}_a). \tag{10.2}$$

$\mathbf{H}_a$ and $\mathbf{M}$ are horizontal in the setup of Fig. 10.4 ($\mathbf{M}$ is induced by the main component of $\mathbf{H}_a$ which is horizontal), while a gradient of $\mathbf{H}_a$ is produced in the z direction by appropriately shaped pole pieces on the magnet. In this case,

$$F_z = -\mu_0 M \frac{dH_a}{dz},$$

and M can be found by measuring F_z with an electronic balance. For ferromagnetic materials, the sensitivity required is not very high and less sophisticated methods

10.1 Macroscopic Magnetic Measurements

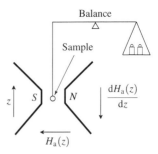

Fig. 10.4 Faraday magnetometer for measuring the force acting on a sample placed in a horizontal field with vertical gradient

are still very convenient. In particular, in Sect. 10.1.3, we shall discuss the usual method for measuring the hysteresis cycles of soft ferromagnetic materials.

Question 10.2. If we needed to carry out magnetisation measurements at different temperatures, would the various measurement setups discussed above be equally well suited to the task?

10.1.2 Examples of Measurements on Superconductors

a. Magnetisation of Type II Superconductors

If the magnetisation of a bulk type I superconductor, e.g., a long cylinder of diameter $d \gg \lambda$, is measured with a magnetometer, a perfect Meissner effect is found for $H_a < H_c$. For a type II superconductor, the perfect Meissner effect occurs up to H_{c_1}, but when the field penetrates in the form of vortices, B gradually increases, and the diamagnetic magnetisation decreases to zero when H_a reaches H_{c_2} (see Fig. 10.5).

Figure 10.6 shows the magnetisation curves for pure lead (A) and Pb–In alloys with increasing concentrations of indium (B, C, D).

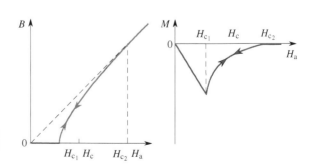

Fig. 10.5 B and M in a type II superconductor for different values of the applied field

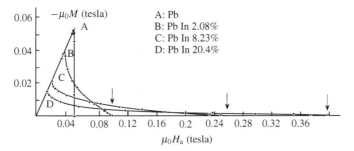

Fig. 10.6 Curves of $M(H_a)$ for Pb–In alloys with different indium concentrations. From Livingston, J.D.: Phys. Rev. **129**, 1943 (1963). With the kind permission of the American Physical Society (APS). http://link.aps.org/doi/10.1103/PhysRev.129.1943

Question 10.3. What type of superconductors do the magnetisation curves in Fig. 10.6 belong to? What does the area under the magnetisation curve represent? How do the thermodynamic quantities characterising superconductivity vary with the indium concentration?

The results shown in Fig. 10.6 correspond to a system in thermodynamic equilibrium, where penetration occurs in the form of a regular vortex lattice. In many experimental cases, the observations are not as beautiful as those shown above. In particular, the measured magnetisation often exhibits irreversibility, and can even change sign (see Fig. 10.7). The existence of irreversibility suggests that *it is difficult to establish a thermodynamic equilibrium situation*. This should not happen in an ideal material. On the other hand, in any real material, *structural defects* are inevitable, and can introduce frictional effects which slow down or inhibit vortex motion. For example, Fig. 10.7 shows the magnetisation curves of a Pb–In 8% alloy after cold-working, then after longer and longer annealing at room temperature. In each case, the sample was cooled in zero field and the magnetisation was measured in increasing then decreasing field.

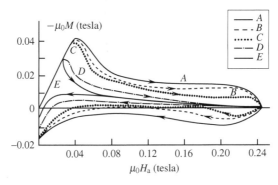

Fig. 10.7 Magnetisation curves of a Pb–In 8% alloy. The sample, initially cold-worked to introduce structural defects, was then annealed for longer and longer periods at room temperature to gradually remove them. From Livingston, J.D.: Phys. Rev. **129**, 1943 (1963). With the kind permission of the American Physical Society (APS). http://link.aps.org/doi/10.1103/PhysRev.129.1943

Question 10.4. Compare these results with those for the sample with the same concentration in Fig. 10.6. What differences do you observe? Explain these observations qualitatively.

Note that the simple fact that there is a free surface can prevent nucleation or elimination of vortices when the magnetic field is changing.[1]

b. Effect of Shape on the Magnetisation of a Type I Superconductor: Intermediate State

We have seen that, as a rule, the existence of demagnetising fields can significantly alter magnetisation measurements in magnetic materials. This will also be true for superconducting materials whose diamagnetic magnetisation is high in the Meissner state. Let us consider the special case of a type I superconducting sphere in an applied field. It is observed (see Fig. 10.8) that its magnetisation curve does not exhibit a perfect Meissner effect, because the diamagnetic magnetisation does not vanish suddenly, but instead decreases gradually when the applied field exceeds a value H'_c.

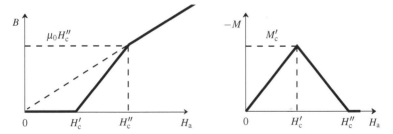

Fig. 10.8 Average magnetic induction B and magnetisation M measured for a sphere of type I superconducting material

Question 10.5. Using the demagnetising fields defined in Chap. 9, show that the absolute value of the apparent susceptibility of this spherical superconductor in weak fields is greater than unity. Determine the values of H'_c, H''_c, and M'_c for the case shown in Fig. 10.8 as a function of the thermodynamic critical field H_c. What happens in the applied field interval $H'_c < H_a < H''_c$?

10.1.3 Hysteresimeters: Toroidal Geometry

For strongly magnetised materials, it is important to reduce the effects of demagnetising fields when carrying out a magnetic measurement. To do this, long cylinders

---

[1] See Problem 13: *Magnetisation of a Type II Superconductor*.

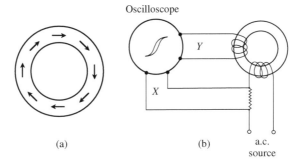

Fig. 10.9 (**a**) Magnetisation configuration in a toroidal sample. (**b**) Hysteresimeter used to observe hysteresis cycles in insulating ferrites with weak coercivity

or thin films can be used, arranged in the direction of the measurement field. However, it is generally preferable to use a closed magnetic circuit, since this completely avoids the appearance of magnetic poles. The geometry usually employed is that of a torus, which is just a cylinder that closes up on itself. In this case, the demagnetisation energy is minimised when the magnetisation is everywhere parallel to the surface. There is then no demagnetising field and no field is created outside (see Fig. 10.9a).

By winding an excitation coil and a detection coil on the torus, B can be measured directly as a function of H_a. Indeed, the current injected into the excitation solenoid determines the applied external field H_a, while the flux detected in the measurement coil can be used to measure B. With an alternating excitation current, the curve $V(I)$ allows a direct determination of the hysteresis cycle in the material. In this configuration the flux is uniform in the torus (see Fig. 10.9b).

This method is particularly well suited to measurements on soft magnetic materials, i.e., those with weak coercive field. In this case the fields H_a required are also weak and can be produced by a small coil. With an alternating current I, one can also find out how the hysteresis cycle varies with frequency. The losses associated with the pinning of Bloch walls by defects may depend significantly on the frequency and amplitude of the exciting field (see Chap. 11). These characteristics are important when ferrites are used in high-frequency electronic devices.

10.2 Magnetic Surface Measurements and Magnetic Imaging

In combination with suitable models, macroscopic magnetisation measurements help to understand certain features of the thermodynamics of magnetic materials. However, the existence of domains, walls, and vortex lattices obviously encourage the development of methods for observing the distribution of B in greater detail. Several techniques were devised in the second half of the twentieth century to

10.2 Magnetic Surface Measurements and Magnetic Imaging

achieve magnetic imaging. They all exploit the fact that the non-uniform distribution of magnetisation within the material induces a non-uniform distribution of the magnetic field at its surface.

Most observations are thus made at the surface of magnetic or superconducting materials. This configuration is also used to record information on magnetic substrates. The developments made to extend our fundamental understanding of magnetism have produced methods and materials that will improve magnetic data storage methods in the coming decades. To begin with, we present methods for carrying out microscale measurements. This is the current scale on which magnetic materials are manipulated by read/write systems in the hard disks of our computers (see Sect. 10.2.1). Methods for attaining the submicron scale will be outlined in Sect. 10.2.2, since they are already opening up new prospects.

10.2.1 Microscale Measurements and Surface Imaging

Here we shall distinguish methods that carry out local quantitative measurements from those that only visualise the distribution of magnetism at the surface either directly or indirectly.

a. Measurement by Microscopic Probes

The first method developed for the read/write heads of magnetic tapes and hard disks used a microscale magnetic field sensor. This could be an air gap several microns across in a closed magnetic circuit (see Fig. 10.10a). When the magnetic substrate is moved across the air gap, the magnetisation component parallel to the surface produces a flux variation in the circuit and this induces a voltage in the windings (see Fig. 10.10b). This method requires a steady movement of the substrate with respect to the measurement head.

To avoid any limitation of sensitivity due to irregularity in the mechanical displacement of the tape, it is also possible to use sensors giving a signal that is directly

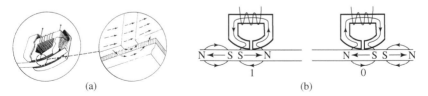

Fig. 10.10 (a) Magnetic writing in a tape recorder. Data is written onto the tape by a soft iron magnet with a very narrow air gap. The signal injected into the coil induces a magnetisation of the magnetic particles embedded in a polymer film. (b) Method for reading the data on a magnetic tape. As the tape passes through, the signal induced by the magnetisation of the particles trapped in the tape is detected and amplified. From [14, pp. 588–589] Wiley ©1972

sensitive to the magnetic field. This is what happens in magnetoresistive probes or probes measuring the Hall effect in a semiconductor. Now that recent progress in microelectronics can significantly reduce component sizes, sensors have been developed for determining the magnetic field on a piece of a surface of size less than a few µm².

Sensors of this kind comprising a rod holding a dozen or so Hall effect probes have been used to study the magnetic induction distribution in superconductors. Positioned on the surface of a single crystal, they can measure B at points only ten microns apart. This method cannot be used to obtain either λ or the vortex structure. In the mixed state, however, it can identify the spatial variation in the average magnetic induction associated with the geometric shape of the sample (see Fig. 10.11a and b). Indeed, we have seen that, when the magnetic field penetrates the superconductor, an irreversibility is observed in the magnetisation curves. This may arise because the vortices produced at the surface are subject to an attractive force that holds them near the surface. The density of the vortex lattice, i.e., the average magnetic induction, thus varies with the distance from the surface.

The measurement made with a given probe is used to study the functional dependence $B(H_a)$ locally in a region of the crystal where the magnetic induction is uniform. In superconducting samples with high T_c, in the mixed phase, between H_{c_1} and H_{c_2}, it was thereby shown that there is a sudden change in M as a function of H_a (see Fig. 10.11c). This measurement reveals a phase transition between a low temperature region with an ordered vortex lattice and a liquid vortex phase at high temperatures. Indeed, the high value of T_c allows thermal fluctuations to appear before superconductivity is destroyed. The (H, T) phase diagram of these systems is

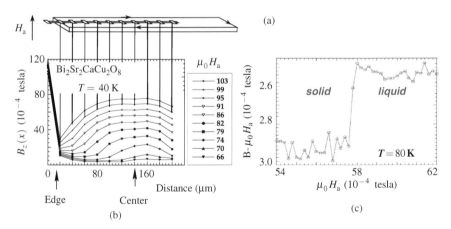

Fig. 10.11 (a) Arrays of Hall effect probes used for local measurement of the magnetic induction at the surface of a superconducting crystal. (b) The value of B depends on the distance to the edge of the sample in the irreversible flux penetration regime. (c) Transition to a liquid vortex phase between H_{c_1} and H_{c_2} in a sample of $Bi_2Sr_2CaCu_2O_8$. Illustrations courtesy of Konzcykowski, M.: Laboratoire des Solides Irradiés. Ecole Polytechnique, France

10.2 Magnetic Surface Measurements and Magnetic Imaging

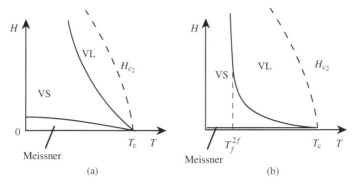

Fig. 10.12 Schematic phase diagrams determined for high-T_c cuprate superconductors. VS = vortex solid, VL = vortex liquid. (**a**) $YBa_2Cu_3O_7$ and (**b**) $Bi_2Sr_2CaCu_2O_8$ which is more anisotropic

thus very different from those observed in conventional superconductors. The extent of the liquid vortex phase increases with the anisotropy of these layered materials (see Fig. 10.12).

b. Magneto-Optical Method: Kerr and Faraday Effects

The method described above makes local measurements, whereas the Bitter method was able to visualise domains. An important method which can combine observation and measurement uses the rotation of the polarisation of a light wave when it is reflected (the Kerr effect, shown in Fig. 10.13a) or transmitted (the Faraday effect) by a magnetic medium. These optical properties of materials should be distinguished from birefringence, which is observed in crystallographically anisotropic materials.

Fig. 10.13 (**a**) Kerr effect. (**b**) Magnetic dots obtained by X-ray lithography on a Au/Co/Au film. The Co layer comprises 5 atomic planes. The dots are squares of side 0.5 μm. The *black squares* are images of dots with $M = +M_s$, while the invisible squares correspond to dots with $M = -M_s$. Image courtesy of Jamet, J.P.: Laboratoire de Physique des Solides. Orsay, France

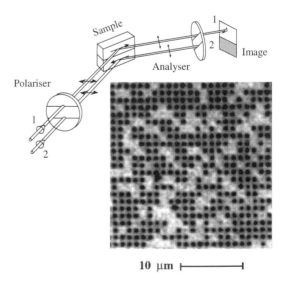

Magneto-optical effects are generally weaker and tend to be used to study cubic materials for which there is no birefringence, in the absence of mechanical stresses. Some amorphous alloys of transition metals and rare earths, or cubic garnets, give rise to significant magneto-optical effects. While these methods can only be used for specific magnetic materials, they are nevertheless of great interest because they allow observations at any temperature and for any magnetic field.[2] Moreover, the dynamics of domain wall displacements can be studied directly.

Future methods of digital magnetic data storage might consist in delimiting domains, or engraving magnetic dots on thin magnetic films (see Fig. 10.13b) with anisotropy perpendicular to the substrate. For such media, a read method using the Kerr effect with a laser has long been envisaged. Writing might be achieved by locally heating the relevant region with a laser beam in the presence of a magnetic field. Selective heating of a dot reduces its coercive field and thus alters its magnetisation. With the development of semiconductor lasers and optical compact disks (CD), such methods might become available in the form of magneto-optical CDs.

10.2.2 Towards Submicron Scales

The methods discussed above can only be extended to nanometric scales with the help of modern methods of microscopy. Electron microscopy was used quite early on, e.g., to visualise vortices in superconductors. The discovery in 1985 of new methods of microscopy called near-field microscopy, able to make observations on the submicron scale, opened the way to a new field of investigation.

a. Bitter Method and Microscopy: Vortices in Superconductors

This method was used by Trauble and Essman in 1967 to visualise for the first time the vortex lattice in a type II superconductor. Indeed, at the surface of a superconducting sample, in the presence of an applied field H_a in the z direction, perpendicular to the surface, the field component parallel to the z axis is only significant at the points where the vortices reach the surface. Very small magnetic particles are thus attracted to points vertically above the vortex lines. It is not easy to transpose the Bitter technique to this situation. Ultrasmall particles measuring about 10 nm across are required, because the distance between vortices is less than the micron. The particles deposited on the sample at low temperatures are fixed by a non-magnetic deposition. A replica of the surface layer was then made and observed by electron microscopy. Figure 10.14 shows quite clearly that the clusters of ferromagnetic particles form a hexagonal lattice characteristic of the vortex lattice.

---

[2] See for example Problem 19: *Magnetism of thin films and Magneto-Optic Applications*.

10.2 Magnetic Surface Measurements and Magnetic Imaging

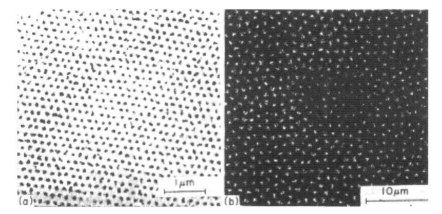

Fig. 10.14 Observation of a vortex lattice using the Bitter method. Images of clusters of very small magnetic particles deposited below the superconducting critical temperature on the surface of superconductors are observed by electron microscopy. (**a**) Note the high level of order in the case of Nb, on a light background, at 1.2 K in a field of 82 mT. (**b**) For $YBa_2Cu_3O_7$, near 15 K in a weak field (2 mT), the significant level of disorder is due to pinning of vortices by defects. From [15, p. 151]

More recently, such observations were carried out on high-T_c superconductors. In this case, one can observe positional irregularities due to defects in the sample.

Question 10.6. In Fig. 10.14, it is observed that, for a sample of niobium, the vortices form an almost perfect hexagonal lattice. But significant defects are nevertheless visible in this 2D crystal lattice, in the top left corner of the figure. Can you locate them? To get a better view of them, you should look at the figure at grazing incidence in various space directions. What kind of defects can you locate?

b. Lorentz Microscopy

The Bitter method played an important role in the first observations of walls and vortices. More recently, electron microscopy has been able to visualise these features directly. Lorentz microscopy exploits the fact that an electron beam is deflected in a direction that depends on the sign of the magnetic induction. This method is able to visualise walls or vortices, and determine the direction of the magnetisation inside each domain. Indeed, the walls show up as white or black lines, depending on the relative orientations of the domains (see Fig. 10.15).

More detailed electron microscopy methods exploit the information contained in changes in the phase of the electron wave function as it crosses the sample. Indeed, for a beam of electrons with the same kinetic energy, the phase change is directly related to the circulation of the vector potential in crossing the sample (see Chap. 5). This difference can be used to determine the magnetisation profile across the sample. For thin films, the method can give quantitative information.

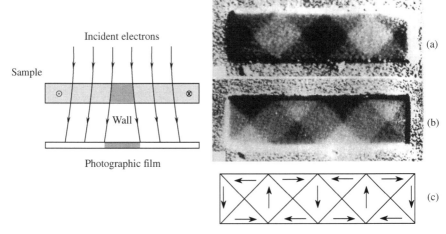

Fig. 10.15 Lorentz microscopy. The electron beam crossing a thin sample (thickness 0.1 µm) is deflected in a direction that depends on the direction of magnetisation of the domain encountered. This contrast can be identified by analysing the electron beam (*left*). Observation by Lorentz microscopy of a sample of permalloy of dimensions 3 µm × 0.75 µm and thickness 60 nm. The intensity of the image is related to the component of the magnetisation perpendicular (**a**) or parallel (**b**) to the long dimension of the sample. The orientation of the magnetisation can thus be determined (**c**). From [15, p. 260] (*right*)

c. Near-Field Microscopy

It is natural to end this chapter on imaging by mentioning recent developments in near-field microscopy. These techniques consist in scanning the surface of a material with a very fine tip of nanometric dimensions. We saw in Chap. 2 how the tunneling effect between a metal tip and a metallic sample could be used to image its surface topography (Scanning Tunneling Microscopy, STM).

This method cannot be used to study insulating samples. However, a tip on a very thin and flexible support called the cantilever, as shown in Fig. 10.16 (left), can be very sensitive to extremely weak forces of attraction between tip and surface. For example, a non-magnetic insulating tip is attracted toward the surface by the Van der Waals forces. Any slight bending of the cantilever toward the surface is then detected by a reflected laser beam, as shown in Fig. 10.16 (right), for different positions of the sample. The height of the tip above the sample is controlled by V_z. This method, known as atomic force microscopy (AFM), can determine the surface topography of many different kinds of material, from insulators, to metals and soft matter.

If the tip and sample are magnetic, a magnetic attractive force adds to the Van der Waals force. Since these forces depend on the height z of the tip as z^α with very different values of the exponent α, the two forces can be distinguished by varying z. As the magnetic forces decrease slowly with z, a magnetic image of the domains at the surface of the sample can be made by scanning the tip across at a distance of

10.2 Magnetic Surface Measurements and Magnetic Imaging

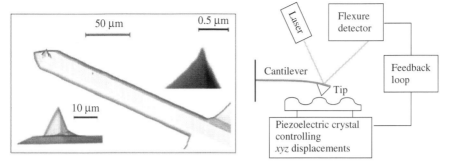

Fig. 10.16 Cantilever and microtip made from silicon for an atomic force microscope. Image courtesy of Miltat, J., Thiaville, A.: Laboratoire de Physique des Solides. Orsay, France (*left*). Setup for an atomic force or magnetic force microscope (AFM or MFM). The feedback loop controls the force acting on the cantilever (*right*)

a few hundred angstrom units from the surface. This method, known as magnetic force microscopy (MFM), is currently used to resolve magnetic structures on scales below 0.1 µm (see Fig. 10.17).

Such methods should benefit from future developments in nanolithography, which will produce ever more precisely controlled tips. It should be possible by nanolithography to make very small Hall effect and magnetoresistive probes on the cantilever, allowing more detailed quantitative measurements of local magnetisation values and eventually leading to imaging on scales of a few tens of nanometers.

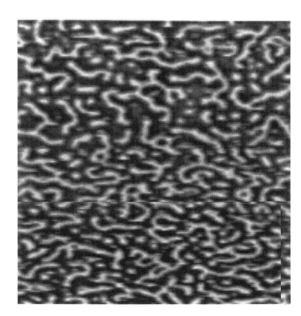

Fig. 10.17 Image of magnetic domains obtained in zero field on a Pt/Co multilayer by magnetic force microscopy (MFM). The scale of the image is 5 µm × 5 µm. The resolution is 50 nm. Image courtesy of Thiaville, A.: Laboratoire de Physique des Solides. Orsay, France

10.3 Summary

The most basic information about the magnetic response of a system is its magnetisation or macroscopic magnetic susceptibility. These quantities are obtained by measuring the change in the magnetic flux inside a coil. The most sensitive methods use a SQUID to detect the flux. A SQUID is a superconducting loop containing a superconductor–insulator–superconductor (SIS) junction. Sensitivities well below the flux quantum Φ_0 are thereby reached. Determination of the magnetisation of type II superconductors and hysteresis cycles in ferromagnets illustrate the importance of such basic measurements.

The existence of demagnetising fields associated with the long range of the magnetic dipole interaction mean that the sample shape can have important consequences for macroscopic measurements. In type I superconductors, the sample shape can induce a non-uniform magnetic penetration for $H < H_c$. In this intermediate state, normal and superconducting regions alternate through the sample.

Detailed measurements of the spatial dependence of the magnetic induction are made at the surface of a material by using microscopic sensors. Magnetic recording and reading use magnets with air gaps that have gradually been reduced in size to reach micron dimensions today in computer hard disk readers. Hall effect sensors are used to study with great sensitivity the phase diagrams of high-T_c superconductors in the mixed state. The vortex liquid phase observed in these materials is due to their anisotropy and the high value of T_c.

Images of the spatial distribution of magnetic induction at the sample surface can be obtained using microscopy techniques. The Bitter method uses submicronic magnetic powders which tend to aggregate where domain walls or vortices (in a superconductor) emerge from the sample surface. Magneto-optical methods exploit the rotation of polarisation of light induced by interaction with the surface of a magnetic medium. Likewise, the interaction of the electron spin with the magnetisation of a thin sample can produce an observable contrast under electron microscopy. Near-field methods using the interaction of a tip with a surface to explore its topography are currently the subject of promising developments. Making a magnetic deposit on the tip, the magnetic interaction force between tip and surface can be ascertained. These methods can simultaneously reveal the surface topography and the spatial dependence of its magnetisation with a resolution better than 0.1 μm.

10.4 Answers to Questions

Question 10.1

With these methods, it is not necessary to use a sample with the same cross-section as the coil. However, in the first case (Fig. 10.1a), the change in B outside the sample due to the change in the applied field H_a will soon become much greater than the change in B inside the sample, and this will quickly reduce the sensitivity of the measurement.

The measurement by the method of Fig. 10.1b is less sensitive to the geometry of the sample compared with that of the coil. A simple calibration with a sample of known susceptibility will suffice, e.g., a ferromagnetic sphere with $\chi_{app} = 3$ (see the solution to Question 9.3).

The first method can measure the instantaneous value of M induced by the change in the field H_a, while the second will be better suited to a measurement of the thermodynamic equilibrium value of the magnetisation in the field H_a.

Question 10.2

We may cool or heat both the coil and sample, or just the sample. In the case of Fig. 10.1a, the first solution is preferable, in order to maintain sensitivity (see the solution to Question 10.1). Otherwise, a calorimetric container must be inserted into the coil or coils, while ensuring that this container does not itself exhibit a magnetic response.

This type of complication is minimal in the case of Fig. 10.1b, because only the magnetisation of the moving part of the sample holder will contribute to the observed signal. In this kind of setup, the coils can be held at fixed temperature, which is absolutely essential if they are to be made from a superconducting material, as in the case of a flux measurement made by a SQUID.

Note that these measurements with a SQUID nevertheless require a cryostat containing liquid helium, as would be necessary to make low temperature measurements. On the other hand, this complicates very high temperature measurements, above 1,000 K, e.g., on materials with high Curie temperature. In such cases, the magnetisation to be measured is generally high and the Faraday balance achieves a good compromise.

By using SQUIDs made with high-T_c superconductors operating at the temperature of liquid nitrogen, it should soon be possible to insert very high temperature ovens inside detection coils, and thus increase the sensitivity of very high temperature measurements.

Question 10.3

We saw in Chap. 7 that pure metals like Pb are usually type I superconductors, as shown by the magnetisation curve (A), which exhibits a perfect Meissner effect. Introducing indium, the material is transformed into a type II superconductor. A gradual reduction of H_{c_1} and an increase in H_{c_2} are observed when the In concentration is increased.

According to Chap. 6, the area of the magnetisation curve, viz., $\int \mu_0 M dH_a$, represents the condensation energy in zero field of the electronic system in the superconducting state. We observe in Fig. 10.6 that this varies very little as a function of the indium concentration, suggesting that the condensation energy and hence T_c are barely modified by introducing indium. The main effect of the indium is therefore to increase the ratio λ/ξ, which explains the transition from type I to type II, then a gradual reduction in H_{c_1} and an increase in H_{c_2}. The effect of impurities and defects in BCS superconductors is effectively to reduce the coherence length ξ, and also to increase the penetration depth λ.

Question 10.4

In every case the Meissner effect is perfect at low fields and even seems to lead to ever higher values of H_{c_1} as the defect concentration increases. On the face of it, this contradicts the observations made with different indium concentrations in Fig. 10.6. However, above 0.24 tesla, the magnetisation disappears and the metal becomes normal. The value of H_{c_2} (and hence of H_{c_1}) does not seem to be modified by cold-working and repeated annealing. This therefore suggests that the defects introduced by cold-working do not alter the thermodynamic properties of the superconductor.

On the other hand, the achievement of thermodynamic equilibrium seems to be strongly inhibited. Indeed, when an applied field H_a is increased above H_{c_1}, the first vortices must appear and the density of the vortex lattice must increase through the nucleation of vortices at the surface and the compression of the vortex lattice. This process requires macroscopic displacements of vortices. The defects induced by cold-working play the role of pinning centers, restricting or even preventing such motions.

Likewise, after reaching the normal state for $H > H_{c_2}$, when H_a is decreased, as soon as the superconducting state is reached, the magnetic field is screened and leads to strong diamagnetism, because a high vortex density cannot suddenly penetrate the medium.

Question 10.5

The initial magnetisation of a spherical sample is given by

$$M = \tilde{\chi}\left(H_a - \frac{M}{3}\right), \quad \text{which yields} \quad M = \frac{\tilde{\chi}}{1+\tilde{\chi}/3}H_a.$$

For a superconductor exhibiting a perfect Meissner effect, $\tilde{\chi} = -1$, so $M = -3H_a/2$, and $\tilde{\chi}_{app} = -3/2$, which corresponds to a higher apparent diamagnetism than for a perfect diamagnetic material.

The effective field $H_i = 3H_a/2$ is equal to the critical field H_c for $H_a = 2H_c/3$. Therefore, beyond this value, the superconductivity should disappear. We would then have a normal metal sphere for $2H_c/3 < H_a < H_c$, which would be paradoxical! It suggests that, for $H_a > 2H_c/3$, owing to the effects of the demagnetising field, the behaviour of the material can no longer be uniform. Indeed, it is easy to understand

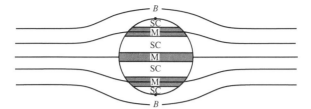

Fig. 10.18 For some values of an applied field, a spherical type I superconductor has an intermediate state comprising normal metal (M) and superconducting (SC) regions of macroscopic size

that in this case there must be some kind of compromise, in which superconducting and normal regions are produced (see Fig. 10.18).

For $2H_c/3 < H_a < H_c$, the associated magnetisation curve varies steadily from $-H_c$ to zero, and resembles the curve for a type II superconductor. In Fig. 10.8, we thus have

$$H'_c = 2H_c/3, \qquad M'_c = -H_c, \qquad H''_c = H_c.$$

Note that this situation, referred to as an *intermediate state*, is in no way related to the *mixed state*. In a type II superconductor, the penetration of the flux in vortex lines corresponds to an intrinsic thermodynamic equilibrium state of the material, whereas the state described here for a type I superconductor is simply induced by surface effects. So for more than ten years, experimenters were tempted to attribute the fact that the magnetisation in superconducting samples does not vanish suddenly above a given field to sample shape effects rather than to the appearance of a mixed state. This misunderstanding was all the more comprehensible in that it was well known that demagnetising fields induce a domain decomposition in ferromagnets.

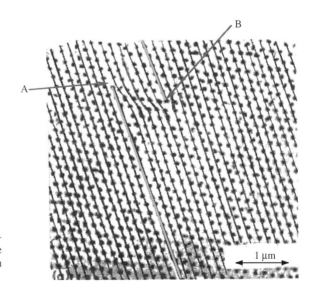

Fig. 10.19 Pair of dislocations A and B observed in the hexagonal vortex lattice in superconducting niobium

Question 10.6

Two defects are visible. Each corresponds to the addition (or suppression) of a half-line of vortices which ends at point A or B, as shown in Fig. 10.19. To accommodate this excess, the lattice is deformed locally. Such defects are called *dislocations* in the crystal lattice. They often occur in pairs to minimise local stresses.

In a 3D crystal, a dislocation is obtained in the crystal lattice by inserting an extra half-plane of atoms in a direction corresponding to an atomic plane. The dislocation is usually represented by the line of dislocation, which is visualised by the row of atoms along the edge of this extra half-plane. For example, two 3D dislocations would be obtained by stacking up planes like the one in Fig. 10.19. The corresponding dislocation lines would then be the vertical lines above A and B. Dislocations play a key role in the mechanical properties of materials.

Chapter 11
Spin Dynamics and Magnetic Resonance

Contents

11.1 Dynamics in Magnetism: General Considerations 319
    11.1.1 Linear Response and Dissipation 320
    11.1.2 Pulse Response and Frequency Response 321
11.2 Dynamics in Ferromagnets 322
    11.2.1 Low Frequency Losses in Ferromagnets 322
    11.2.2 Ferromagnetic Resonance 324
11.3 Resonance in the Paramagnetic Regime 331
    11.3.1 Resonance for a Thermodynamic Ensemble of Spins 331
    11.3.2 Nuclear Magnetic Resonance (NMR) 333
    11.3.3 Spin Echoes: Transverse Relaxation 336
    11.3.4 Applications 339
11.4 Summary 341
11.5 Answers to Questions 342

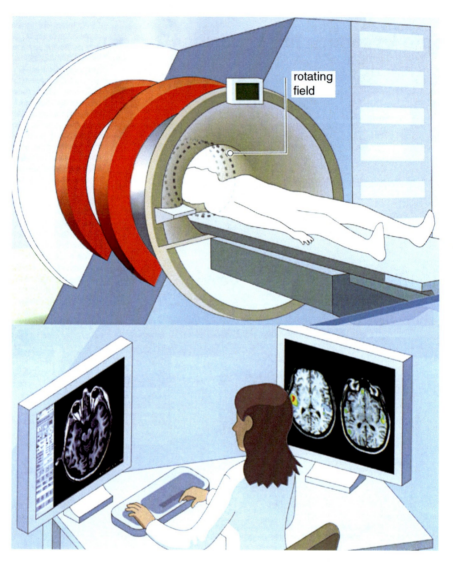

Magnetic resonance imaging (MRI) is a technique developed in the 1980s to visualise the spatial distribution of the nuclear magnetisation in the body of a patient submitted to a magnetic field produced by a superconducting coil. The patient lies along the axis of the superconducting coil. Subsidiary coils specify a straight line along which a uniform field gradient is produced. The Larmor frequencies of the nuclear spins (usually protons in hydrogen) can then be coded as a function of their position along the straight line. Fourier analysis of the precessing nuclear magnetisation by the techniques described in this chapter is then able to determine the proton density as a function of the position on the straight line. A cross-sectional image of the organism is produced by displacing this straight line across a plane. With this imaging technique, the operator can choose the cross-section, and also specify various types of contrast between the protons of the organism, depending on whether they belong to fluid or solid, moving or fixed tissues, and so on. Taken from Les défis du CEA, 119, 17 (2006), edited by the Commissariat à l'Energie Atomique, France

So far we have considered cases where the response of a magnetic or superconducting system can be treated as static. However, it is obvious that in most practical applications the dynamics of the response will be very important. The materials used in transformers, motors operating at low frequencies, or fast electronic components have very different physical characteristics.

Furthermore, we have noted that in many cases the response is not linear in the magnetic field, and may become irreversible, in particular in the case where high fields are applied. This hysteresis of the magnetisation curves reflects a dissipation of energy which is determined by the area of the cycle. Materials must be optimised to reduce such losses, in order to render them usable in setups involving high alternating currents. On the other hand, in weak enough applied fields, the magnetic response does become linear, and under such conditions the dissipation may be extremely weak.

In addition, as with many other physical systems, magnetic systems exhibit resonant eigenmodes whose specific features can lead to novel physical phenomena and a great many applications.

In this chapter, we begin with a simple and very general representation for treating the magnetic response of a physical system in the linear regime and discussing dissipation (see Sect. 11.1). In Sect. 11.2, after illustrating this formalism in the case of a ferromagnet in a weak field, we introduce the idea of magnetic resonance in the case of a saturated ferromagnet. These general ideas are then used in a more detailed way to discuss paramagnetic systems in Sect. 11.3, as exemplified by electron paramagnetic resonance (EPR) and nuclear magnetic resonance (NMR). Finally, we discuss the couplings between electronic and nuclear moments, known as hyperfine couplings. It is through these interactions that the nuclear moments can be used as atomic probes for local measurement of electronic magnetic properties.

11.1 Dynamics in Magnetism: General Considerations

In many physical situations, we wish to study the response of a system to some form of excitation. In a solid, for example, we investigate the response to an electric or magnetic field. If the excitation modifies the energy level spectrum, a thermodynamic quantity conjugate to the excitation will be altered. In every case, *for a weak enough excitation, the response will be a linear function of the excitation*. The magnetisation of an ensemble of paramagnetic ions is a linear function of the applied field for $g\mu_B B \ll k_B T$, while the magnetisation of a type II superconductor is linear for $H < H_{c_1}$, and so on. However, in these different cases, the response is never instantaneous, because a finite time is required for the system to reach thermodynamic equilibrium. If the excitation varies in time, the response will only follow the excitation if the frequency is low compared with the reciprocal of the time required to reach equilibrium. Moreover, the energy exchanges between the system and the energy reservoir constituted by the thermostat will depend on the excitation frequency. *This suggests a relationship between the frequency response and the energy*

11.1.1 Linear Response and Dissipation

Consider an arbitrary system subjected to an alternating magnetic field. Under alternating excitation conditions, a dissipative phenomenon is revealed by a phase difference between response and excitation. This happens with the losses in a capacitor, or with the resistive loss associated with the resistance of an inductance L. For a magnetic material, the response to an excitation by an alternating magnetic field

$$\mu_0 H_a(t) = B_1 \cos \omega t$$

will lead to a magnetisation offset by a phase difference δ:

$$M(t) = M_0 \cos(\omega t - \delta) = M_0(\cos \omega t \cos \delta + \sin \omega t \sin \delta),$$

or

$$M(t) = \frac{M_0 \cos \delta}{B_1} B_1 \cos \omega t + \frac{M_0 \sin \delta}{B_1} B_1 \sin \omega t.$$

We may thus set

$$M(t) = \chi'(\omega) B_1 \cos \omega t + \chi''(\omega) B_1 \sin \omega t. \tag{11.1}$$

Using the complex notation $B(t) = B_1 e^{i\omega t}$, we note that (11.1) corresponds to the real part of

$$\boxed{M(t) = \operatorname{Re}\left\{ [\chi'(\omega) - i\chi''(\omega)] B_1 e^{i\omega t} \right\}.} \tag{11.2}$$

We may thus define a *complex magnetic susceptibility*

$$\chi(\omega) = \chi'(\omega) - i\chi''(\omega) = |\chi(\omega)|(\cos \delta - i \sin \delta).$$

Note that the average energy absorbed by the material per cycle $T = 2\pi/\omega$ and per unit volume is

$$W = -\int_0^T M \frac{dB}{dt} dt,$$

or

$$W = \int_0^T \omega B_1^2 \left[\chi'(\omega) \sin \omega t \cos \omega t + \chi''(\omega) \sin^2 \omega t \right] dt = \pi \chi''(\omega) B_1^2,$$

which corresponds to an absorbed power

$$\boxed{P = \frac{\omega}{2} \chi''(\omega) B_1^2.} \tag{11.3}$$

The component in quadrature $\chi''(\omega)$ thus corresponds to the energy *absorption*, while $\chi'(\omega)$ is called the *dispersion*. The real part of the susceptibility $\chi'(0)$ at zero frequency is the *static susceptibility*.

11.1.2 Pulse Response and Frequency Response

Here we examine the relation between the frequency response and the temporal response for a magnetic system. In a linear response regime, an exciting field $B(t')$ applied at time t' for a time interval dt' induces a magnetisation $m(t-t')B(t')dt'$ at a *later* time t. The magnetisation induced at time t by a time-varying magnetic field $B(t')$ is therefore

$$M(t) = \int_{-\infty}^{t} m(t-t')B(t')dt' . \tag{11.4}$$

In this expression, the function $m(t)$ can be identified as the temporal response function to a pulse excitation produced at time $t' = 0$. [To see this, consider an excitation $b\tau\delta(t'=0)$, which leads to $M(t) = b\tau m(t)$.]

The response to a sinusoidal excitation $B(t') = B_1 \exp(i\omega t')$ is

$$M(t) = \int_{-\infty}^{t} m(t-t') B_1 \exp(i\omega t')dt' = B_1 \exp(i\omega t) \int_{-\infty}^{t} m(t-t')\exp\left[i\omega(t'-t)\right]dt'$$

$$= B_1 \exp(i\omega t) \int_{0}^{\infty} m(t)\exp(-i\omega t)dt ,$$

and comparing with (11.1), we find

$$\boxed{\chi(\omega) = \int_{0}^{\infty} m(t)\exp(-i\omega t)dt} , \tag{11.5}$$

and hence also

$$\boxed{m(t) = \frac{1}{2\pi} \int_{-\infty}^{\infty} \chi(\omega)\exp(i\omega t)d\omega} . \tag{11.6}$$

Quite generally, *in a linear response regime, the frequency response and the pulse response are related by Fourier transform*.

Note that these relations are obtained by assuming that the system has a *linear*, *stationary*, and *causal* response. The first two of these conditions require a response function which depends only on $t-t'$ and also a monochromatic response for a monochromatic excitation. The *causality relation* arising from the fact that the response cannot precede the cause leads to a response function $m(t)$ that is zero for $t < 0$, and real for all t, because it corresponds to the physical response to a real excitation $\delta(0)$. It can be shown completely generally that these conditions require the functions $\chi'(\omega)$ and $\chi''(\omega)$ to be interdependent. Knowing one of them for all ω, one can fully determine the other via the Kramers–Kronig relations [11]:

$$\boxed{\begin{aligned}\chi'(\omega) - \chi'(\infty) &= \pi^{-1} P \int_{-\infty}^{\infty} \frac{\chi''(\omega')}{\omega' - \omega} d\omega', \\ \chi''(\omega) &= -\pi^{-1} P \int_{-\infty}^{\infty} \frac{\chi'(\omega') - \chi'(\infty)}{\omega' - \omega} d\omega',\end{aligned}}$$

(11.7)

where P denotes the principal part, i.e.,

$$P = \lim_{\varepsilon \to 0} \left(\int_{-\infty}^{\omega - \varepsilon} + \int_{\omega + \varepsilon}^{\infty} \right).$$

In these expressions, which apply to many other response functions, e.g., the dielectric constant, the imaginary part χ'' of the response vanishes for $\omega \to \infty$, otherwise the absorbed power would tend to infinity.[1] Note also that χ'' vanishes by definition for $\omega = 0$. In the next section, we shall study some simple cases.

11.2 Dynamics in Ferromagnets

Many applications in magnetism concern materials with high magnetic permeability in transformers, motors, and so on. At low frequencies, metallic magnetic materials, in particular iron-based materials, can be used. In these materials, the resistive losses due to Foucault currents are the main source of dissipation. This explains why transformer plates are made from cold-worked metal structures, as a way of limiting resistive losses. Metallic materials cannot be used in the radiofrequency range ($\simeq 10$ MHz) or ultrahigh frequency (UHF) range (GHz). For these applications ferrimagnetic insulating materials are essential. These ferrites are usually made of agglomerates of grains (materials called ceramics), and their properties depend to a large extent on the grain size and porosity, i.e., on their microstructure. The frequency behaviour of such materials is quite different depending whether they are non-magnetised, with high permeability, or fully magnetised (saturated magnetisation). We shall now examine these two situations in turn.

11.2.1 Low Frequency Losses in Ferromagnets

In ceramic materials, the magnetisation is zero when there is no applied field. This is due to the presence of walls, generally scarce in each grain. The high permeability is due to the high mobility of the walls in a structure without defects. While the

---

[1] However, the real part of the response can generally have a finite value for $\omega \to \infty$. This is not true for the magnetic susceptibility, because there is no physical system for which the magnetisation can keep up with an excitation at infinite frequency. On the other hand, if we consider the magnetic permeability $\mu = 1 + \chi$ as response function, this contains a term that corresponds to the exciting magnetic field, and $\mu' \to 1$ for $\omega \to \infty$. In (11.7), we have kept $\chi'(\infty)$ for the purposes of generality.

11.2 Dynamics in Ferromagnets

applied field is still very weak, in the so-called Rayleigh regime described in Chap. 9 (see Fig. 9.14), the displacement of walls is reversible. Under such conditions, the dissipation is due to friction between the walls and defects in the material. Let us assume that the time required for a wall to reach equilibrium is an exponential function, with a characteristic time τ determined by the mobility of the walls. In this case the magnetisation response function $m(t)$ in (11.4) can be written

$$m(t) = \frac{\chi_0}{\tau} \exp\left(-\frac{t}{\tau}\right). \tag{11.8}$$

Then, from (11.5), we may deduce that

$$\chi(\omega) = \frac{\chi_0}{\tau} \int_0^\infty \exp\left[-\left(\frac{1}{\tau} + i\omega\right)t\right] dt,$$

or

$$\chi(\omega) = \frac{\chi_0}{\tau}\left(\frac{1}{\tau} + i\omega\right)^{-1} = \chi_0 \frac{1 - i\omega\tau}{1 + \omega^2\tau^2}, \tag{11.9}$$

which corresponds to

$$\chi'(\omega) = \frac{\chi_0}{1 + \omega^2\tau^2}, \quad \chi''(\omega) = \frac{\chi_0 \omega\tau}{1 + \omega^2\tau^2}. \tag{11.10}$$

This kind of behaviour is often observed in insulating ferrites (see Fig. 11.1). These ceramic materials are obtained by solid phase synthesis from fine powders. The initial susceptibility and the relaxation time τ depend heavily on the microstructure of the material. To achieve minimum losses, very pure materials are clearly needed, with a minimum of stoichiometric defects and without inclusions. In this case, the wall mobility is limited by friction with intergrain defects. Detailed studies

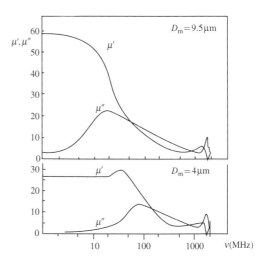

Fig. 11.1 Real and imaginary parts, μ' and μ'', respectively, of the magnetic permeability $\mu = 1 + \chi$ in two samples of the ferrite $NiFe_2O_4$ as a function of the frequency $\nu = \omega/2\pi$ (logarithmic scale). A relaxation behaviour close to the one given by (11.10) is observed. Figure courtesy of Pascard, H.: Laboratoire des Solides Irradiés. Ecole Polytechnique, France

for different mean grain sizes D_m in the ferrites $NiFe_2O_4$ do in fact show that τ and χ_0 are proportional to D_m^2 and D_m, respectively. The choice of microstructure therefore determines the initial permeability and the range of frequencies for which losses are acceptable in a given application. This exemplifies the key importance of microstructure for the physical properties of materials. The microstructure is also a determining factor for critical currents in superconductors, or the mechanical properties of metal alloys.

11.2.2 Ferromagnetic Resonance

Now consider a ferromagnet in which there are no walls. This would happen either in the case of a high magnetic field B_0, or because the anisotropy is sufficient to ensure that the remanent state corresponds to uniform magnetisation. The exchange coupling maintains the coherence of the spin system, and here we shall only study the response to a spatially uniform excitation, for which the magnitude M_s of the macroscopic magnetisation is conserved. These modes correspond to *a collective rotation of all the magnetic moments*. They can be obtained by arguments from classical mechanics.

a. Larmor Precession

To begin with, we consider a situation in which the demagnetising field is zero and neglect anisotropies. The total magnetic moment is associated with an angular momentum $\mathbf{M}/\gamma$, where $\gamma = -g\mu_B$ is the gyromagnetic ratio. The time derivative is equal to the couple $\mathbf{M} \wedge \mathbf{B}$ exerted on the magnetisation. We thus obtain

$$\boxed{\frac{d\mathbf{M}}{dt} = \gamma \mathbf{M} \wedge \mathbf{B}}. \tag{11.11}$$

This relation shows that, if we apply a fixed magnetic field $\mathbf{B}_0$ parallel to the z axis at time $t = 0$, the magnetisation will precess at the *Larmor frequency*

$$\boxed{\omega_L = -\gamma B_0}, \tag{11.12}$$

about the z axis. Indeed, $d\mathbf{M}_z/dt = 0$, and since $|\mathbf{M}|$ is constant, $\mathbf{M}$ moves with constant angular frequency ω_L on a cone with vertex angle $\theta = (\mathbf{Oz}, \mathbf{M}(0))$, i.e., the angle between the applied field and the magnetisation at time $t = 0$ (see Fig. 11.2).

The Larmor frequency is an eigenfrequency of the spin system. It is determined by the gyromagnetic ratio, itself fixed by the Landé factor g (see Chap. 8). Note that, according to (11.11), this motion is perpetual and does not bring $\mathbf{M}$ into line with the z axis. However, this could not be otherwise, because the system we have been considering is *isolated*, and the magnetic energy $-\mathbf{B}_0 \cdot \mathbf{M}$ is a *constant of the motion*. In order for the magnetisation to come back to the minimum energy configuration

11.2 Dynamics in Ferromagnets

Fig. 11.2 Larmor precession

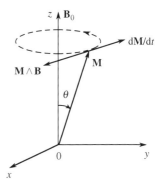

corresponding to thermodynamic equilibrium, there must be mechanisms allowing energy exchanges with a reservoir. The time required to return to equilibrium will be determined by the microscopic nature and magnitude of the couplings between the magnetisation and the degrees of freedom characterising the reservoir.

b. Detection of Ferromagnetic Resonance

A term can be included to allow a return to equilibrium $\mathbf{M}(\mathbf{B})$ depending exponentially on time, with characteristic time T_1 called the *spin lattice relaxation time*:

$$\frac{d\mathbf{M}}{dt} = \gamma \mathbf{M} \wedge \mathbf{B} - \frac{\mathbf{M} - \mathbf{M}(\mathbf{B})}{T_1} . \tag{11.13}$$

Projecting this equation in the z direction for a field $\mathbf{B} = B_0 \mathbf{z}$ for which the equilibrium position is $\mathbf{M}(\mathbf{B}_0) = M_s \mathbf{z}$, we find that

$$\frac{dM_z}{dt} = -\frac{M_z - M_s}{T_1},$$

and then

$$M_s - M_z = (M_s - M_s \cos\theta)\exp\left(-\frac{t}{T_1}\right),$$

for the initial condition considered here. This produces a spiralling return to equilibrium, as shown in Fig. 11.3 (left).

Under these conditions, we can investigate the dynamic susceptibility of our system at equilibrium $\mathbf{M} \parallel \mathbf{B}_0 \parallel \mathbf{Oz}$, i.e., the response to an alternating magnetic field. If the latter is parallel to the z axis, the response is zero, since this field does not couple with the magnetisation. However, we can find the response to a transverse field, e.g., a field applied in the x direction:

$$\mathbf{B}(t) = B_1' \cos\omega t \, \mathbf{x} . \tag{11.14}$$

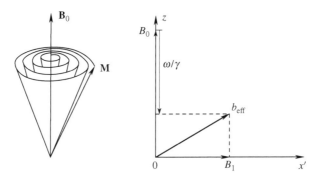

Fig. 11.3 *Left*: Larmor precession with relaxation in a ferromagnet. *Right*: Effective field in the rotating frame

Experimentally, such a field is easily obtained by placing the sample in a coil perpendicular to the z axis. However, since the eigenmode of **M** is circularly polarised, it is preferable to use a circularly polarised excitation field. Note that the linearly polarised field (11.14) is precisely the sum of two circularly polarised fields of magnitude $B_1 = B_1'/2$, rotating in opposite directions, viz.,

$$\mathbf{B}_+(t) = B_1(\cos\omega t\, \mathbf{x} + \sin\omega t\, \mathbf{y}) \quad \text{and} \quad \mathbf{B}_-(t) = B_1(\cos\omega t\, \mathbf{x} - \sin\omega t\, \mathbf{y}). \quad (11.15)$$

To solve (11.13) with the excitation field $B_+(t)$ it is convenient to use a frame $Ox'y'z$ rotating with angular speed ω about **Oz**. The time derivative of a vector **A** in the stationary frame can be expressed relative to its value **a** in the rotating reference frame by

$$\frac{d\mathbf{A}}{dt} = \frac{d\mathbf{a}}{dt} + \mathbf{\Omega} \wedge \mathbf{a},$$

where $\mathbf{\Omega} = \omega\mathbf{z}$. Under these conditions, the equation of motion (11.13) in the laboratory frame is given in a completely analogous way in the rotating frame:

$$\frac{d\mathbf{m}}{dt} = \gamma \mathbf{m} \wedge \mathbf{b}_{\text{eff}} - \frac{\mathbf{m} - M_s\mathbf{z}}{T_1}, \quad (11.16)$$

with a static effective field $\mathbf{b}_{\text{eff}}$ in the rotating frame given by (see Fig. 11.3 right)

$$\boxed{\mathbf{b}_{\text{eff}} = \left(B_0 + \frac{\omega}{\gamma}\right)\mathbf{z} + B_1\mathbf{x}' = \frac{\omega - \omega_{\text{L}}}{\gamma}\mathbf{z} + B_1\mathbf{x}'}. \quad (11.17)$$

Setting

$$\mathbf{m} = m_x\mathbf{x}' + m_y\mathbf{y}' + (M_s - m_z')\mathbf{z}, \quad (11.18)$$

11.2 Dynamics in Ferromagnets

we obtain

$$\frac{dm_x}{dt} = (\omega - \omega_L)m_y - \frac{m_x}{T_1},$$
$$\frac{dm_y}{dt} = \gamma B_1(M_s - m'_z) - (\omega - \omega_L)m_x - \frac{m_y}{T_1}, \quad (11.19)$$
$$-\frac{dm'_z}{dt} = -\gamma B_1 m_y + \frac{m'_z}{T_1}.$$

These equations are not linear, and will lead to special features to be investigated further in the case of nuclear magnetic resonance. But for the moment, let us consider the linear response, i.e., when m_x and m_y are proportional to B_1, and neglect all quadratic terms, in particular $m'_z \sim O(m_x^2 + m_y^2)$. We then have

$$\frac{dm_x}{dt} = (\omega - \omega_L)m_y - \frac{m_x}{T_1},$$
$$\frac{dm_y}{dt} = \gamma B_1 M_s - (\omega - \omega_L)m_x - \frac{m_y}{T_1}. \quad (11.20)$$

Setting $dm_x/dt = dm_y/dt = 0$, we immediately obtain the *stationary solutions* for which m_x and m_y are independent of time in the rotating frame. These are reached after a transient regime. Putting $m_x = \chi'_+(\omega)B_1$ and $m_y = \chi''_+(\omega)B_1$, it follows that

$$\boxed{\begin{aligned}\chi'_+(\omega) &= \gamma M_s \frac{(\omega - \omega_L)T_1^2}{1 + (\omega - \omega_L)^2 T_1^2}, \\ \chi''_+(\omega) &= \gamma M_s \frac{T_1}{1 + (\omega - \omega_L)^2 T_1^2}.\end{aligned}} \quad (11.21)$$

In the laboratory frame, the magnetisation in the x direction is therefore[2]

$$M_x(t) = \chi'_+(\omega)B_1 \cos \omega t + \chi''_+(\omega)B_1 \sin \omega t.$$

---

[2] Since $B_-(t)$ is obtained from $B_+(t)$ by changing the sign of ω in (11.15), we immediately obtain $\chi'_-(\omega) = \chi'_+(-\omega)$ and $\chi''_-(\omega) = \chi''_+(-\omega)$. The susceptibilities $\chi'(\omega)$ and $\chi''(\omega)$ corresponding to the linearly polarised excitation field $B'_1 = 2B_1$ are thus given by

$$\chi'(\omega) = \frac{1}{2}[\chi'_+(\omega) + \chi'_+(-\omega)], \qquad \chi''(\omega) = \frac{1}{2}[\chi''_+(\omega) - \chi''_+(-\omega)].$$

It is easy to check that the solutions given in (11.21) are such that, for narrow resonances, i.e., when $\omega_L \gg 1/T_1$, the $-\omega$ components are negligible for $\omega \approx \omega_L$, and only the component of the field rotating in the right direction gives a response close to ω_L. We can therefore identify $\chi'(\omega)$ and $\chi''(\omega)$ with $\chi'_+(\omega)/2$ and $\chi''_+(\omega)/2$, respectively, in practical cases. Note in particular that the expressions for the real and imaginary parts of the susceptibilities given above satisfy the Kramers–Kronig relations.

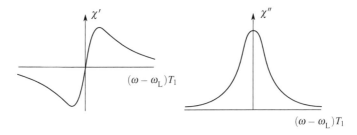

Fig. 11.4 Frequency dependence of the real and imaginary parts, χ' and χ'', respectively, of the magnetic susceptibility. They correspond to the dispersion and absorption signals of the ferromagnetic resonance

The forms of $\chi''_+(\omega)$ and $\chi'_+(\omega)$, which are absorption and dispersion signals, are shown in Fig. 11.4. They are even and odd in $(\omega - \omega_L)$, respectively. The absorption signal has a Lorentzian form, precisely because we assumed an exponential relaxation.

The resonant mode of the magnetisation gives rise to energy absorption at the Larmor frequency. This resonance can be observed by placing the sample in an inductance L, in which there is a current of angular frequency ω. The impedance $iL\omega[1 + \chi(\omega)]$ acquires a resistive term proportional to $\chi''(\omega)$ when ω reaches the Larmor frequency. The resonance will therefore be detected by observing the energy absorption at the frequency of the excitation field when ω crosses the value ω_L. For experimental reasons, it is more convenient to hold ω fixed and vary the applied field. The dispersion signal can also be detected in a coil placed in the y direction, i.e., orthogonal to the coil inducing the excitation field.

c. Ferromagnetic Modes in the Presence of Demagnetising Fields or Magnetic Anisotropy

In practice, the existence of magnetic anisotropy and/or demagnetising fields will alter the resonance conditions. We know that, even when it is uniform, the magnetisation of a ferromagnet is subject to a demagnetising field $\mathbf{H_d}$ determined by the shape of the sample (see Chap. 9). For an ellipsoid of revolution $\mathbf{H_d} = -\overline{\overline{\mathbf{N}}} \cdot \mathbf{M}$, where $\overline{\overline{\mathbf{N}}}$ is the demagnetisation field tensor. Dropping the relaxation term, the equation of motion (11.11) is

$$\boxed{\frac{d\mathbf{M}}{dt} = \gamma \mathbf{M} \wedge \left(\mathbf{B}_0 - \mu_0 \overline{\overline{\mathbf{N}}} \cdot \mathbf{M}\right)}. \tag{11.22}$$

If the field $\mathbf{B}_0$ is oriented along an eigenvector of $\overline{\overline{\mathbf{N}}}$ in the z direction, these equations can be projected in the x and y directions. Terms of second order can be neglected, assuming that $\mathbf{M}$ deviates little from $\mathbf{B}_0$, and we may then write

11.2 Dynamics in Ferromagnets

$$\frac{dM_x}{dt} = \gamma[B_0 + \mu_0 M_s(N_y - N_z)]M_y,$$
$$\frac{dM_y}{dt} = \gamma[-B_0 + \mu_0 M_s(N_z - N_x)]M_x.$$
(11.23)

This implies that

$$\frac{d^2 M_x}{dt^2} = \gamma^2[B_0 + \mu_0 M_s(N_y - N_z)][-B_0 + \mu_0 M_s(N_z - N_x)]M_x.$$

The frequency of the eigenmode is thus given by

$$\boxed{\omega^2 = [\omega_L - \gamma\mu_0 M_s(N_y - N_z)][\omega_L - \gamma\mu_0 M_s(N_x - N_z)]},$$ (11.24)

which only reduces to ω_L in the case of a sphere, since then $N_x = N_y = N_z = 1/3$. We can see from Fig. 11.5 that, for a given material, the ferromagnetic resonance frequency depends on the sample shape.

> **Question 11.1.** 1. For what values of the applied field should one expect to observe the resonance for the three situations in Fig. 11.5. Compare with the experimental results. (Assume that the long cylinder can be treated as infinitely long.)
> 2. Can you estimate the Landé factor and the magnetisation of the ferrite? Use the values $\mu_B = 9.2740 \times 10^{-24}$ J/tesla and $\hbar = 1.05457 \times 10^{-34}$ J s.
> 3. In your opinion, what leads to the shape and width of the observed lines?

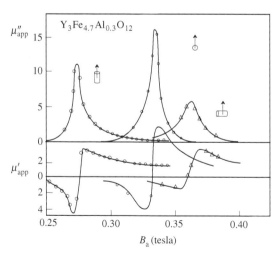

Fig. 11.5 Real and imaginary parts, μ' and μ'', respectively, of the magnetic permeability $\mu = 1 + \chi$ as a function of the applied field at a frequency of 9.3 GHz. Measurements for spherical and cylindrical ferrite samples with two directions of the applied field for the latter. From Berteaud, A.J.: Le Journal de Physique et le Radium **23**, 546 (1962), EDP Sciences copyright

Likewise, when there is magnetic anisotropy, the frequencies of the modes will be perturbed. For example, for a uniaxial anisotropy with easy axis of magnetisation in the z direction, an anisotropy energy term of the form $-DS_z^2$ leads to a macroscopic anisotropy energy of the form $E_a = K \sin^2 \theta$, where $\theta = (\mathbf{Oz}, \mathbf{M})$. When $\mathbf{M}$ does not lie exactly along the easy magnetisation axis, a restoring couple is exerted on it which tends to bring it back along the easy axis. The most direct way to determine the magnetocrystalline anisotropy is to measure this couple for different values of the applied magnetic field. Note that, if $\mathbf{M}$ is close to an easy axis, i.e., if θ is small, the couple is

$$C = -\frac{dE_a}{d\theta} = -\theta \left.\frac{d^2 E_a}{d\theta^2}\right|_{\theta=0}. \tag{11.25}$$

This therefore adds to the couple $B_0 M_s \theta$ due to the interaction with the applied magnetic field. So it is just as though the anisotropy corresponded to a magnetic field B_K which adds to the applied field and modifies the resonance condition (see Fig. 11.6). For the single-axis anisotropy we are considering here, $C = -2K\theta$, the field associated with the anisotropy is $B_K = -2K/M_s$, and the resonance condition

$$\boxed{\omega = -\gamma \left(B_0 - \frac{2K}{M_s} \right).} \tag{11.26}$$

There is therefore a resonance *even when no external magnetic field is applied*, if the coercive field is high enough for the sample to be prepared in a state where the remanent magnetisation is saturated. In practice, in polycrystalline samples, the ferromagnetic resonance will be a superposition of resonances corresponding to the different orientations of the crystallites. Its width will be determined by the distribution of demagnetising fields or anisotropy energy. In general, the shape of the spectrum is not determined by the relaxation time T_1.

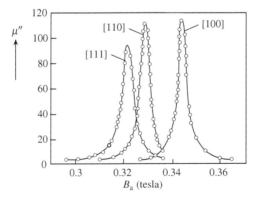

Fig. 11.6 Imaginary part of the magnetic permeability μ'' measured at 9.3 GHz in a garnet single crystal for different orientations of the magnetic field with respect to the crystal axes. From Dillon, J.F., Geshwind, S. Jaccarino, V.: Phys. Rev. **100**, 750 (1955). With the kind permission of the American Physical Society (APS). http://link.aps.org/doi/10.1103/PhysRev.100.750

d. High Frequency Applications of Ferrites

Note that these resonant modes occur at high frequencies, around 30 GHz/tesla. In these frequency ranges, which are very important for telecommunications and correspond to centimeter wavelengths, magnetic materials are very useful in electronic devices. For such electromagnetic waves, the properties of magnetic materials are dominated by the response to the magnetic component of the wave. At higher frequencies, close to optical frequencies, the magnetic susceptibility vanishes, the magnetic permeability tends to 1, and dielectric properties dominate. The insulating ferrites are thus used as filters, tuned to a ferromagnetic resonance. The fact that the resonance only occurs for a wave that is circularly polarised in the right direction allows one to manufacture polarising filters, and consequently isolators and circulators.

11.3 Resonance in the Paramagnetic Regime

The phenomenon of ferromagnetic resonance introduced above leads to a wide range of applications. Larmor precession, directly associated with the idea of angular momentum, was introduced earlier for a single spin [2, Chaps. XI and XIII]. We shall show in Sect. 11.3.1 that this phenomenon also occurs for a thermodynamic ensemble of weakly interacting spins. We shall then be able to introduce electronic paramagnetic resonance (EPR) and nuclear magnetic resonance (NMR) in Sect. 11.3.2. In the radiofrequency range of NMR, specific technical possibilities lead to novel experimental methods (pulse NMR), and the observation of spin echoes (Sect. 11.3.3). This will provide a glimpse of the extended field of applications in magnetometry (Sect. 11.3.4). NMR allows one to perform magnetic measurements on the atomic scale due to the existence of couplings between electronic and nuclear moments (Sect. 11.3.4).

11.3.1 Resonance for a Thermodynamic Ensemble of Spins

We consider localised moments, such as electronic paramagnetic entities diluted in an insulator. The electron magnetic moment $\boldsymbol{\mu}$ is proportional to the angular momentum of the electron, denoted here by $\hbar\mathbf{S}$, so that $\mathbf{S}$ is a dimensionless quantity and $\boldsymbol{\mu} = -g\mu_B\mathbf{S} = \hbar\gamma\mathbf{S}$. The Zeeman interaction between this magnetic moment and an applied field $\mathbf{B}_0$ in the z direction is

$$\hat{H}_{Z0} = -\hbar\gamma\mathbf{S}\cdot\mathbf{B}_0 = -\hbar\gamma S_z B_0 \,. \qquad (11.27)$$

For simplicity, we shall restrict here to the case of spin 1/2 ($S=1/2$), for which the Hamiltonian has two eigenstates (see Fig. 11.7) with energies $E_+ = -\hbar\gamma B_0/2$ and

Fig. 11.7 Quantum levels of the Zeeman Hamiltonian (11.27) for $S = 1/2$

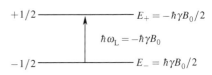

$E_- = \hbar\gamma B_0/2$. The difference between the levels corresponds to the Larmor angular frequency defined by

$$E_+ - E_- = \hbar\omega_\mathrm{L} = -\hbar\gamma B_0 . \tag{11.28}$$

To detect this splitting of the quantum energy levels, we must use a wave of angular frequency ω_L that can induce transitions between states $S_z = 1/2$ and $S_z = -1/2$. If an alternating magnetic field with frequency ω_L is used, the Hamiltonian of the perturbation is

$$\hat{H}_Z^\mathrm{rf} = -\hbar\gamma \mathbf{S} \cdot \mathbf{B}_1 \cos\omega_\mathrm{L} t . \tag{11.29}$$

A transition can only occur between the states $|1/2\rangle$ and $|-1/2\rangle$ if this Hamiltonian has nonzero matrix elements between these two eigenstates of S_z. This in turn will only be possible if $\mathbf{B}_1$ has a component perpendicular to $\mathbf{z}$. Indeed, only the operators S_x and S_y have nonzero matrix elements between these two states. With $\mathbf{B}_1$ in the x direction, we recover the same conditions as those considered to detect the ferromagnetic resonance.

The analogy goes much further than this, because it is easy to show that the time evolution equation for the operator $\hat{\boldsymbol{\mu}}$ in the presence of a time-dependent magnetic field is precisely the same as the one obtained classically in (11.11), viz.,[3]

$$\boxed{\frac{\mathrm{d}\hat{\boldsymbol{\mu}}}{\mathrm{d}t} = \gamma\hat{\boldsymbol{\mu}} \wedge \mathbf{B}} .$$

Likewise, if we now consider a macroscopic ensemble of independent spins placed in a field $\mathbf{B}_0$, the density operator for this system in thermodynamic equilibrium is

$$\hat{\rho} = \frac{\exp(-\hat{H}_{Z0}/k_\mathrm{B}T)}{\mathrm{Tr}\left[\exp(-\hat{H}_{Z0}/k_\mathrm{B}T)\right]} . \tag{11.30}$$

At thermodynamic equilibrium, the magnetisation of the system is

$$\langle\hat{\mu}_z\rangle = \mathrm{Tr}(\hat{\rho}\hat{\mu}_z) = N\mu_\mathrm{B} \tanh\frac{\hbar\gamma B_0}{2k_\mathrm{B}T} , \tag{11.31}$$

and $\langle\hat{\mu}_x\rangle = \langle\hat{\mu}_y\rangle = 0$, where N is the number of paramagnetic centers per unit volume. If the spins only interact with a magnetic field $\mathbf{B}$, possibly time-varying, it can

---

[3] Simply write $\mathrm{d}\mathbf{S}/\mathrm{d}t = (\mathrm{i}/\hbar)[\hat{H}_Z, \mathbf{S}]$, with $H_Z = -\hbar\gamma\mathbf{S}\cdot\mathbf{B}$, and use the commutation relations between S_x, S_y and S_z to obtain this relation.

11.3 Resonance in the Paramagnetic Regime

be shown that the time evolution equation of the density operator leads to a time dependence of $\langle \hat{\boldsymbol{\mu}} \rangle$ satisfying[4]

$$\boxed{\frac{\mathrm{d}\langle \hat{\boldsymbol{\mu}} \rangle}{\mathrm{d}t} = \gamma \langle \hat{\boldsymbol{\mu}} \rangle \wedge \mathbf{B}}.\qquad(11.32)$$

So this equation, used for a ferromagnet in Sect. 11.2.2, also applies to the magnetisation of a statistical ensemble of localised spins when interactions between the spins are neglected. Such interactions can be neglected, for example, when the temperature is well above the Curie or Néel temperatures.

The same experimental setups can be used to detect the ferromagnetic resonance and the paramagnetic resonance of the electronic moments (EPR). Note also that, in the paramagnetic state, the relaxation time T_1 has a simple interpretation: it is the time required to reach thermodynamic equilibrium. So if at time $t=0$ a non-polarised sample is placed in a field B_0, a time T_1 will be required to reach the magnetisation (11.31) corresponding to the difference in population between levels $-1/2$ and $1/2$. In many physical situations, experimental determination of T_1 provides a way of investigating the interactions between the electronic moments and the other degrees of freedom of the solid. It provides useful physical information about the properties of the given solid.

11.3.2 Nuclear Magnetic Resonance (NMR)

The physical properties of solids, and in particular their magnetic properties, are determined by the electronic states. On the other hand, the *nuclear spin moments*, which do not affect these properties, provide an extremely useful probe for the electronic properties. Atomic nuclei are made up of neutrons and protons, which are spin 1/2 particles. They are assembled into quantum states in which the nuclear ground state has a total spin $\mathbf{I}$ that may be integer or half-integer. The associated magnetic moment $\boldsymbol{\mu}_n$ is proportional to the magnetic moment of the proton μ_p, where the multiplicative factor is analogous to the Landé factor for an atomic electronic moment. Each atomic nucleus thus has a specific magnetic moment $\boldsymbol{\mu}_n = \hbar \gamma_n \mathbf{I}$. The gyromagnetic ratio γ_n is known to great accuracy for each of the stable isotopes in the periodic table. Since $\mu_p \simeq 10^{-3} \mu_B$, the nuclear moments are extremely small, as are their mutual interactions. As a consequence, they are almost always in a paramagnetic state. An ordered nuclear magnetic state is only generally accessible for temperatures of the order of 10^{-6} K. According to (11.31), we see immediately that, in a given applied field, the nuclear magnetisation is about 10^6 times smaller than the electronic magnetisation. It is therefore practically impossible to detect using the methods described in Chap. 10.

---

[4] This is established by considering the time evolution equation of the density operator, namely $\mathrm{d}\hat{\rho}/\mathrm{d}t = -(i/\hbar)[\hat{H}_Z, \hat{\rho}]$.

However, everything we have just said for the electron paramagnetic moments can be transposed to the nuclear moments. Although the nuclear magnetic susceptibilities are weak, they can be detected by magnetic resonance. As we shall see below, this technique can significantly increase sensitivity.

a. Technical Principles of NMR

We represent the absorption signal by a function $f(\omega)$ like the one given in (11.21), whose area we assume to have been normalised to unity, viz., $\chi''(\omega) = af(\omega)$. The static nuclear susceptibility χ_0 is found from the Kramers–Kronig relation (11.7) evaluated at $\omega = 0$:

$$\chi_0 = \chi'(0) = \frac{a}{\pi} \int_{-\infty}^{\infty} \frac{f(\omega')}{\omega'} d\omega' \approx \frac{a}{\pi \omega_L},$$

because $f(\omega) \approx 0$ when ω is any distance from ω_L. If $f(\omega)$ has width $\Delta\omega$, then the intensity detected in an NMR experiment is

$$\chi''(\omega_L) \approx af(\omega_L) \approx \pi \omega_L \chi_0 f(\omega_L) \approx \pi \chi_0 \omega_L / \Delta\omega.$$

The susceptibility measured in NMR is therefore increased by a factor of about $\omega_L/\Delta\omega$ as compared with the static susceptibility. Since the line widths can be very small in NMR, this leads to a significant gain in sensitivity, in fact up to 9 orders of magnitude in high-resolution NMR in liquids.

Nuclear Larmor frequencies correspond to radiofrequencies of around 10 MHz per tesla. In this frequency range, exceptionally stable coherent sources are available, with narrower spectral widths than the transitions to be observed. These are obtained by electronic oscillators with frequency stabilised on the resonant mode of a piezoelectric quartz crystal. Moreover, for such frequencies, very powerful amplifiers are also available. With radiofrequency pulses, these can be used to significantly modify the populations of the spin quantum states. Such setups are useful for studying nuclear relaxation.

b. Radiofrequency Pulse Spectroscopy

Consider the action of a very strong radiofrequency field B_1 at a frequency close to the Larmor frequency, applied for a time t_w. In the rotating frame, the effective field felt by the spin system is, according to (11.17),

$$\mathbf{b}_{\text{eff}} = \left(B_0 + \frac{\omega}{\gamma} \right) \mathbf{z} + B_1 \mathbf{x}'. \tag{11.33}$$

If we are trying to detect the NMR of nuclei with a frequency spectrum extended over a width $\Delta\omega/\gamma \ll B_1$, we will obtain $(B_0 + \omega/\gamma) \ll B_1$, and the field b_{eff} will be very close to $B_1 \mathbf{x}'$ for all the nuclear moments to be observed (see Fig. 11.3 right).

11.3 Resonance in the Paramagnetic Regime

In this case the equation governing the time evolution of **m** in the rotating frame is, according to (11.16),

$$\frac{d\mathbf{m}}{dt} = \gamma \mathbf{m} \wedge \mathbf{b}_{\text{eff}} \approx \gamma \mathbf{m} \wedge \mathbf{B}_1 ,$$

if the duration t_w of the radiofrequency pulse is much shorter than the relaxation time T_1. As a consequence, during the pulse, **m** rotates about x' with angular frequency γB_1.

Suppose we start out from thermodynamic equilibrium $M_z = M_0$, and that the field B_1 is applied from time 0 to time t_w. In this case, at time t_w, the magnetisation will have rotated through an angle $\theta = \gamma B_1 t_w$ about the x' axis. If we adjust t_w so that $\theta = \pi/2$, the nuclear magnetisation lies along the y' axis after the pulse, and then remains fixed in the rotating frame. It therefore precesses at the Larmor frequency ω_L in the laboratory frame. This magnetisation rotating in the plane perpendicular to B_0 will induce a voltage oscillating at the frequency ω_L in a coil placed in the (x, y) plane. This signal will fade in time due to the distribution of Larmor frequencies in the sample, and also the relaxation phenomena.

It is easy to check that the pulse experiment just described constitutes a determination in the rotating frame of $m(t)$ as given in (11.5) and (11.6). The free precession signal is in fact the Fourier transform of the NMR spectrum. The two signals

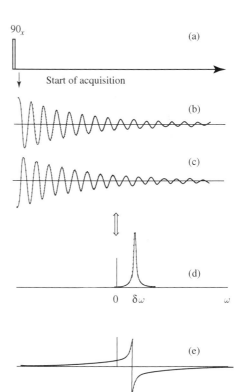

Fig. 11.8 Pulse and Fourier transform NMR. (**a**) Radiofrequency pulse in the x direction. Free precession signal in phase (**b**) and in quadrature (**c**). Spectrum obtained by complex Fourier transform in absorption $\chi''(\omega)$ (**d**) and in dispersion $\chi'(\omega)$ (**e**). From [Callaghan, P.T.: Principles of NMR Microscopy. Oxford University Press, Oxford (1991), p. 43]

$\chi'(\omega)$ and $\chi''(\omega)$ can be reconstructed by Fourier transform in the complex plane, by detecting the two precession signals, in phase and in quadrature with the generator producing the radiofrequency pulses (see Fig. 11.8). This method is currently the main technique for NMR observations. Note that the crucial condition for this to work is $\gamma B_1 \gg \Delta\omega$, so it will only succeed for sufficiently narrow spectra. This is not always the case for magnetic materials.

In order to measure the relaxation time T_1, one could apply a pulse of length $2t_\mathrm{w}$, rotating the magnetisation through π and thereby aligning it with the negative z axis. After a delay of t_D, we observe a free precession signal with a $\pi/2$ pulse. Its magnitude is proportional to the magnetisation reestablished at time t_D along the z axis:

$$M_z(t_\mathrm{D}) = M_0 - 2M_0 \exp\left(-\frac{t_\mathrm{D}}{T_1}\right). \tag{11.34}$$

After a time much longer than T_1, thermodynamic equilibrium is regained. A measurement can be made for a different value of t_D, reproducing the time evolution (11.34) in a step by step manner, whereupon T_1 can be determined (see Fig. 11.9).

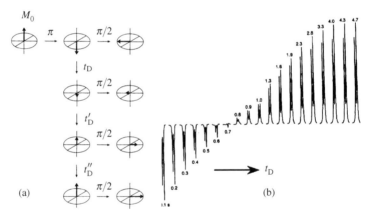

Fig. 11.9 (**a**) Measuring the relaxation time T_1. The equilibrium magnetisation is reversed at time $t = 0$. It is then measured after a delay t_D. Measurements made for various values of t_D can be used to reconstruct the nuclear relaxation. (**b**) The resulting evolution is shown for a spectrum comprising two lines. The time t_D is given in seconds. From [Sanders, J.K.M., Hunter, S.K.: Modern NMR Spectroscopy. Oxford University Press, Oxford (1993), pp. 65–66]

11.3.3 Spin Echoes: Transverse Relaxation

In an experiment in which the magnetisation precesses freely, if there is a magnetic field distribution in the sample that can be related to microscopic physical properties, the nuclear spins can lose their phase coherence by rotating in the (x, y) plane, without exchange of energy with the thermodynamic bath. The free precession signal may disappear well before the magnetisation along the z axis has had

11.3 Resonance in the Paramagnetic Regime

time to reappear. This suggests that the relaxation may be different in the (x,y) plane and in the z direction. Indeed, for a saturated ferromagnet, it may be possible to treat the magnetisation as a classically evolving vector, with conservation of $|\mathbf{M}|$. But for a statistical ensemble of spins, $|\mathbf{M}|$ is not conserved during the evolution to equilibrium. Here it is possible to produce a transverse magnetization as the system is prepared in a state that establishes phase coherence within the nuclear spin ensemble. *But the microscopic couplings required to establish thermodynamic equilibrium can destroy this transverse coherence without necessarily reestablishing global thermodynamic equilibrium.* This will be the case if *dipole interactions* between nuclear spins are taken into account.

Indeed the dipole coupling Hamiltonian for two neighbouring nuclear spins $\mathbf{I}_1$ and $\mathbf{I}_2$ contains terms of type $I_1^+ I_2^- /r^3$ which induce simultaneous reversals of $\mathbf{I}_1$ and $\mathbf{I}_2$, e.g.,

$$\left| I_1^z = -1/2; I_2^z = +1/2 \right\rangle \implies \left| I_1^z = +1/2; I_2^z = -1/2 \right\rangle .$$

Such transitions modify the dipole field felt by neighbouring spins, and hence lead to a loss of transverse coherence without changing the macroscopic magnetisation component M_z, because the change in the total spin is zero.

These considerations led F. Bloch to introduce two different relaxation times T_1 and T_2 for M_z and (M_x, M_y), respectively, so that the equation of motion for $\mathbf{M}$ in the laboratory frame (11.13) becomes

$$\frac{d\mathbf{M}}{dt} = \gamma \mathbf{M} \wedge \mathbf{B} - \frac{(M_z - M_0)\mathbf{z}}{T_1} - \frac{M_x \mathbf{x} + M_y \mathbf{y}}{T_2}, \qquad (11.35)$$

where $\mathbf{z}$ is a unit vector along the z axis. This expression is not strictly accurate, because there is no reason why the loss of coherence in the transverse plane modelled by T_2 should be exponential. Moreover, it may stem from different origins. A width $\Delta \omega_i$ can be associated with a static distribution of the magnetic field of experimental origin (e.g., non-uniformity of the magnet) or of the Larmor frequencies due to microscopic differences within the sample. The local fields of dipole origins between nuclear spins lead to a broadening $\Delta \omega_{dd}$ of the spectrum. The total width $\Delta \omega \approx \Delta \omega_i + \Delta \omega_{dd}$ leads to a decrease in the free precession over a time $T_2^* \approx \Delta \omega^{-1}$.

The physical difference between these two sources of broadening can be demonstrated by a sequence of pulses giving rise to a *spin echo*. After a $\pi/2$ pulse of duration t_w, a pulse of duration $2t_w$ is applied at time $\tau > T_2^*$, which rotates the magnetisation through π about the y axis. This amounts to reversing the magnetisation distribution in the (x,y) plane. It can be seen from Fig. 11.10 that, at time 2τ, the magnetisation refocuses along the y axis, because the time evolution associated with the local field distribution leading to $\Delta \omega_i$ is reversed. Coherence thus reappears at time 2τ in the transverse plane. The so-called *spin echo signal* increases in the opposite direction to the free precession, to go to zero once again like the free precession after a time T_2^*. However, its magnitude will be reduced because only the static field distributions associated with $\Delta \omega_i$ are reversible. Non-deterministic

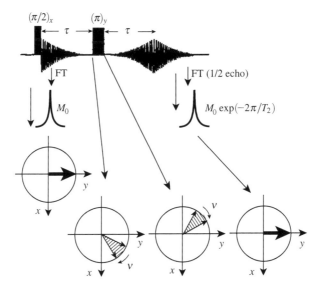

Fig. 11.10 Spin echoes. Sequence of radiofrequency pulses giving rise to a spin echo. The pulse π applied at time τ in the y direction reverses the magnetisation distribution in the (x,y) plane. The magnetisation then refocuses at time 2τ. From [Canet, D.: La RMN, Concepts et méthodes. InterEditions, Paris (1991), p. 170]

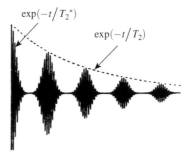

Fig. 11.11 Relaxation time T_2. Spin echoes produced at different values of 2τ can be used to determine the irreversible component of T_2. From [Freeman, R.: A Handbook of Nuclear Magnetic Resonance. Longman Scientific and Technical, Oxford (1997), p. 270] Wiley copyright

processes like those leading to $\Delta\omega_{dd}$ induce an irreversible loss that characterises a true transverse relaxation time T_2. This can be determined by plotting the magnitude of the echo as a function of 2τ (Fig. 11.11).

As a general rule, any evolution under the action of a given Hamiltonian preserves coherence, and it may be possible to retrieve this coherence by methods that reverse the equations of motion. However, non-deterministic terms increase the entropy of the spin system in an irreversible way.

11.3.4 Applications

a. Magnetometry

NMR is clearly an important method for carrying out very accurate field measurements. One only has to select a liquid for which the NMR line is very narrow, because the fast Brownian motions of the atoms tend to average out the dipole interactions between nuclear spins. Measuring the field reduces to measuring the Larmor frequency. Since time and frequency standards are now very precise, the gyromagnetic ratios are determined with great accuracy, allowing absolute measurements of the magnetic field.

b. Distribution of B in the Mixed State of a Superconductor

In superconductors, we have seen that the magnetic induction varies significantly in space in the mixed state. This leads to a distribution of Larmor frequencies for the nuclear spins in the material. The shape of the NMR spectrum reconstitutes the histogram of the magnetic fields. Close to H_{c_2}, singularities appear in the spectrum for values of the magnetic field corresponding to the extrema of the field distribution. In high-T_c superconductors, the line width of the observed resonance can be used to deduce the London penetration depth.

However, for experimental reasons, NMR is not the best method for studying the superconducting state. A related technique uses elementary particles called *muons*. These behave like heavy electrons (or light protons), and have the property of decaying by emission of positrons in the direction of their spin. A muon whose spin is initially polarised perpendicularly to the field B_0 is implanted in the sample at time zero. We then observe the direction of the emitted positrons when it decays. By repeating this experiment for a large number of events, the free precession signal of the muon spin can be reconstructed statistically. This experiment is equivalent to an NMR experiment, and can be used to determine $\lambda(T)$. Since muons can be implanted in almost any sample, it has been possible to make comparative measurements of λ in a wide range of superconducting materials.

Atomic Scale Magnetic Measurements

We have seen above that changes in the magnetic induction in a material can be detected directly by a change in the nuclear Larmor frequency. In weakly magnetic materials, for which the magnetisation is negligible, $B = \mu_0 H_a$ and the nuclear Larmor frequency should be determined by the applied external field alone. It would be difficult to obtain information about the physical properties of materials by this technique.

But we are forgetting here that the nucleus is a kind of atomic scale microscopic probe, coupled to the electrons. Interactions like the dipole interactions between nuclear and electronic spins are such that *the nuclear spin feels a magnetic field associated with the polarisation of the electronic magnetic moments*. This means that the magnetic field felt by the nuclear spins is modified with respect to the applied field. It is the spectroscopy of these fields that provides atomic scale information about the immediate vicinity of the nuclei in the material. Let us examine the different interactions between the nuclear spins and the magnetic moments of electronic origins, known collectively as *hyperfine interactions*.

a. Hyperfine Interactions

The *dipole interaction* between the moments associated with a nuclear spin **I** and an electron spin **S** separated by a displacement **r** is

$$\hat{H}_{dd} = -\frac{\mu_0}{4\pi} \frac{\gamma_e \gamma_n \hbar^2}{r^3} \left[\mathbf{I} \cdot \mathbf{S} - 3\frac{(\mathbf{I} \cdot \mathbf{r})(\mathbf{S} \cdot \mathbf{r})}{r^2} \right],$$

where γ_n and γ_e are the nuclear and electronic gyromagnetic moments, respectively, and **S** and **I** are dimensionless quantities. This dipole interaction diverges when r tends to zero, and is only therefore valid for electrons with zero probability of being at the site of the nucleus. This is the case for electrons in the p, d, or f shells.

On the other hand, the s electrons have nonzero probability of being at the site of the nucleus. The Dirac Hamiltonian can be used to show that the corresponding interaction, called the *contact interaction* $\hat{H}_c$ is scalar, and is given in this case by

$$\hat{H}_c = \frac{\mu_0}{4\pi} \frac{8\pi}{3} \gamma_e \gamma_n \hbar^2 \mathbf{I} \cdot \mathbf{S} \delta(\mathbf{r}).$$

Finally, the interaction with the magnetic field associated with the *orbital angular momentum* $\boldsymbol{\ell}$ of the electron is

$$\hat{H}_{orb} = -\frac{\mu_0}{4\pi} \gamma_e \gamma_n \hbar^2 \mathbf{I} \cdot \boldsymbol{\ell}/r^3.$$

These Hamiltonians can all be written in the form

$$\hat{H}_{eff} = -\gamma_n \hbar \mathbf{I} \cdot \mathbf{B}_{eff},$$

and we may consider that each electron induces a magnetic field $\mathbf{B}_{eff}$ at the nuclear site. As the temporal fluctuations of the electronic moments are very fast compared with the nuclear Larmor frequency, the static component of $\mathbf{B}_{eff}$ is its time average. The position of the NMR for a given nucleus is thus determined by the time average $\langle \mathbf{B}_{eff} \rangle$ of the resultant of the fields due to the different electrons in the material.

It is easy to see that the hyperfine interaction will vanish for filled electronic shells, because they have zero total spin and total orbital angular momentum. When there is no applied field, $\langle \mathbf{B}_{eff} \rangle$ can only be nonzero for materials in which there is a static spin or orbital magnetic moment. This will be the case for ferromagnetic materials.

b. Chemical Shift

In substances where the electrons are paired in atomic or molecular levels, the static part of the hyperfine coupling is only nonzero in the presence of an applied field B_0, and is proportional to B_0, like the magnetisation. The *resonance is shifted* with respect to that of the free atom in a gas. The relative shift $\langle B_{eff} \rangle / B_0$ may be due to the orbital part of the hyperfine coupling. This is the case, for example, for the displacement due to the orbital currents induced by the external magnetic field in electronic or molecular shells close to the nucleus. Since this shift depends on the electronic charge distributions, it is highly sensitive to the chemical environment of the given atom, hence the name *chemical shift*.

These effects are generally small, but can be used to distinguish the nuclear spin resonances of the different atoms depending on their environment. This has become a very powerful tool, used universally in chemistry and biology. Routine chemical analyses are carried out by NMR. It can also determine the 3D structures of biological molecules, with the help of multipulse methods, which have reached an exceedingly high level of refinement.

c. NMR in Metals and Magnetic Materials

When the electron states are not paired, a component of $\langle \mathbf{B}_{\text{eff}} \rangle$ due to the electronic atomic moment may arise via the contact hyperfine term. The *Pauli spin susceptibility* of a metal induces a frequency shift given in %, whereas the chemical shifts are expressed in parts per million (ppm). This comes about because the contact coupling is much stronger than the dipole or orbital couplings, and the Pauli susceptibility is high compared with the susceptibilities of orbital origins in insulating compounds.

In *magnetic materials*, the electronic moments are static at low temperatures, as compared with their behaviour at the Curie or Néel temperatures. It follows that the static effective fields are nonzero even when there is no applied field. For atomic nuclei carrying an electronic moment, this field will be very large (several tesla in general), and will give rise to a resonance at the Larmor frequency $\omega_L = \gamma_n B_{\text{eff}}$, *even for zero applied field*. The fields induced on the nuclei distinct from those of magnetic atoms are generally weak.

Note that, in *antiferromagnetic compounds*, the magnetic susceptibilities are low, and it is not always possible to determine whether a material is in fact antiferromagnetic by macroscopic susceptibility measurements. In contrast, NMR measurements generally reveal the existence of magnetic moments below the Néel transition. NMR often provides a way of characterising the existence of antiferromagnetic phases.

Other experimental techniques are sensitive to the existence of a static hyperfine field. An example is provided by the *Mössbauer effect*, which detects the existence of a hyperfine field from the removal of degeneracy of the electronic energy levels of some radioactive ions, including ^{57}Fe. Detection is achieved by monitoring the change in energy of a radioactive nuclear decay transition.

11.4 Summary

The response of a magnetic system to a time-varying excitation is easy to study in the linear response regime. For excitation by an alternating field $B_1 \cos \omega t$, the response $M(t)$ is used to define a complex magnetic susceptibility

$$M(t) = \text{Re}\left\{ \left[\chi'(\omega) - i\chi''(\omega) \right] B_1 e^{i\omega t} \right\}.$$

In this expression,

$$\chi(\omega) = \chi'(\omega) - i\chi''(\omega) = \int_0^\infty m(t) \exp(-i\omega t) dt$$

is the Fourier transform of the time response $m(t)$ to an impulsive excitation proportional to $\delta(t' = 0)$, and its real and imaginary components χ' and χ'' are related by the Kramers–Kronig relations. The energy absorbed by the material is $P = (\omega/2)\chi'' B_1^2$, which is why the term χ'' is called the absorption.

Low frequency losses in ferromagnets can be explained by a response function $m(\tau) = (\chi_0/\tau) \exp(-t/\tau)$, defined by a relaxation time τ which depends on the mobility of the Bloch walls.

In contrast, a saturated ferromagnet has a resonant mode at the eigenfrequency $\omega_L = -\gamma B_0$ which corresponds to the Larmor precession of the magnetisation around the applied field. This precession is damped by the existence of a relaxation time T_1 which brings the magnetisation back into line with $\mathbf{B}_0$. The time dependence of $\mathbf{M}(t)$ is found by solving

$$\frac{d\mathbf{M}}{dt} = \gamma \mathbf{M} \wedge \mathbf{B} - \frac{\mathbf{M} - \mathbf{M}(\mathbf{B})}{T_1}.$$

In ferromagnetic and ferrimagnetic materials, the frequency of the resonant mode is shifted by the existence of demagnetising fields and magnetic anisotropy terms. These modes lead to many applications in ultrahigh frequency electronic devices (several GHz).

In paramagnetic spin systems, Larmor precession of the macroscopic magnetisation $\mathbf{M}_0$ aligned with the field in the z direction can also be observed. In the case of the NMR, it is technically possible to study the pulse response of $\mathbf{M}_0$. An intense radiofrequency pulse $\mathbf{B}_1$ in the x direction, of duration t_w and at the Larmor frequency, rotates $\mathbf{M}_0$ through an angle $\theta = \gamma B_1 t_w$ about the x axis. By choosing $\theta = \pi/2$, the magnetisation is brought into the (x, y) plane. By detecting its precession after the pulse, one obtains the NMR spectrum by Fourier transform. Pulse methods also provide a way of determining the longitudinal relaxation time T_1 of the nuclear magnetisation. A spin echo sequence is used to distinguish the decrease in the transverse magnetisation due to the non-uniformity of the magnetic field from that due to the transverse relaxation time T_2 of the nuclear magnetisation.

The existence of hyperfine couplings between electronic and nuclear spins shifts the NMR and provides a way of distinguishing the sites within a chemical structure, or measuring magnetic properties on the atomic scale.

11.5 Answers to Questions

Question 11.1

1. According to (11.24), the ferromagnetic resonance frequency is given by

$$\omega_0^2 = \left[\omega_L - \gamma \mu_0 M_s (N_y - N_z)\right]\left[\omega_L - \gamma \mu_0 M_s (N_x - N_z)\right], \quad (11.36)$$

when the field is applied along the z axis.

Sphere. In this case, $N_x = N_y = N_z = 1/3$ and $\omega_0^2 = \omega_L^2$. The resonance condition is thus obtained for $B_a = B_s$ such that

$$\omega_0 = -\gamma B_s. \quad (11.37)$$

Infinite cylinder with axis in the z direction. The cylinder axis is parallel to $\mathbf{B}_a$, which is in the z direction. In this case, we have $N_z = 0$, $N_x = N_y = 1/2$, and hence

11.5 Answers to Questions

$$\omega_0^2 = \left(\omega_L - \frac{1}{2}\gamma\mu_0 M_s\right)^2 .$$

The resonance condition is thus obtained for the field $B_{c\|}$ such that

$$\omega_0 = -\gamma\left(B_{c\|} + \frac{1}{2}\mu_0 M_s\right) . \tag{11.38}$$

Infinite cylinder with axis in the x direction. The cylinder axis is now perpendicular to $\mathbf{B}_a$, which is in the z direction. We thus have $N_x = 0$, $N_y = N_z = 1/2$, and then

$$\omega_0^2 = \omega_L\left(\omega_L + \frac{1}{2}\gamma\mu_0 M_s\right) .$$

The resonance condition is therefore obtained for a field $B_{c\perp}$ such that

$$\omega_0^2 = -\gamma B_{c\perp}\left(-\gamma B_{c\perp} + \frac{1}{2}\gamma\mu_0 M_s\right) . \tag{11.39}$$

From (11.37) and (11.38), it follows that

$$\boxed{B_{c\|} = B_s - \frac{1}{2}\mu_0 M_s} . \tag{11.40}$$

Equations (11.39) and (11.37) yield

$$B_{c\perp}^2 - \frac{\mu_0 M_s}{2}B_{c\perp} - B_s^2 = 0 ,$$

or

$$B_{c\perp} = \frac{\mu_0 M_s}{4} + B_s\sqrt{1 + \frac{(\mu_0 M_s)^2}{16 B_s^2}} ,$$

whereupon

$$\boxed{B_{c\perp} \simeq B_s + \frac{\mu_0 M_s}{4}} . \tag{11.41}$$

It can be seen from Fig. 11.12 that the configurations (cylinder $\perp$) and (cylinder $\|$) give resonance fields on either side of the resonance for the sphere. The difference is indeed twice as large for (c$\|$) as for (c$\perp$).

2. The resonance for the spherical ferrite is obtained for

$$B_s = 0.3338 \pm 0.0009 \text{ tesla} ,$$

or $g = h\nu_0/\mu_B B_s = 1.990 \pm 0.005$. Likewise, from (11.40), $\mu_0 M_s/2 = B_s - B_{c\|}$. Measurement gives $\mu_0 M_s = 0.12 \pm 0.002$ tesla.

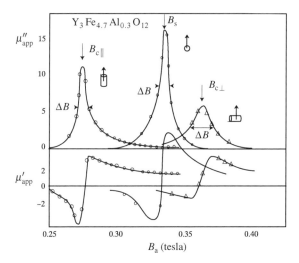

Fig. 11.12 Determinations of the resonance conditions and widths for the three geometrical shapes of samples considered

3. Note that the the width of the resonance at half-maximum decreases from the (c⊥) configuration to the (c∥) configuration to the (s) configuration. (This is also in increasing order of resonance intensity.) This suggests that the width of the resonance can be associated with the fact that the demagnetising fields are not uniform. This is clearly the case for the cylinder which is not infinite. The resonance is narrower in the (c∥) configuration than in the (c⊥) configuration, because the first configuration is more symmetrical. The resonance also has noticeable width for the spherical configuration. It might be thought that the line is Lorentzian and that the width corresponds to a relaxation time T_1. In actual fact, T_1 is very long and even in this case it is the demagnetising field distribution associated with imperfections in the sample that determines the line width (deviation from sphericity, porosity of the ferrite).

Chapter 12
The Thermodynamics of Ferromagnets

Contents

12.1 Excited States and Low Temperature Properties 347
    12.1.1 Magnons 347
    12.1.2 Low Temperature Thermodynamic Properties 349
    12.1.3 Experimental Detection of Magnons 351
12.2 The Magnetic Phase Transition 352
    12.2.1 Mean Field Theory 352
    12.2.2 Landau and Ginzburg–Landau Theories 354
    12.2.3 Critical Behaviour 357
12.3 Summary 359
12.4 Answers to Questions 360

The temperature dependence of the thermodynamic properties of solids is determined by the excited energy levels above the ground state. In order to detect these excited states directly, a choice method is to observe the scattering of neutrons produced by a nuclear reactor. A low temperature finger is used to cool the neutrons emitted by the reactor, thereby reducing their energy. The directions of their velocity vectors are selected by means of a monochromator, in which a specific single crystal selects a Bragg diffraction of the neutrons. When these neutrons are scattered by the sample, their wave vectors change and there is a loss or gain in energy that can be determined by diffraction on a crystal analyzer. The photograph shows a three-axis spectrometer (monochromator output axis, orientation of the single crystal sample, output analyser axis). In magnetic materials, the magnon excitation spectra can be determined by spectroscopic analysis of this inelastic neutron scattering mediated by the interaction between the spin of the neutron and the electronic atomic moments. Likewise, neutron scattering by interaction with the atomic nuclear core is used to determine the phonon excitation spectra in solids (see Problem 9: *Phonons in Solids*). Photograph of the TRISP spectrometer at the FRM-II reactor in Garching (Germany). Image courtesy of Keller, T., Keimer, B. who run this instrument on behalf of the Max Planck Institute for Solid State Research

The discussion in Chap. 8 showed that the Heisenberg model provides a good description of the ordering of magnetic moments in a magnetic material, especially in the case of insulating compounds. However, this discussion considered only the properties of the ground state, so it cannot inform as to the thermodynamic properties of magnetic materials. This will be the task of the present chapter. We begin by discussing the low energy excited states in the Heisenberg model and deduce some low temperature properties. We shall then examine the vicinity of the critical temperature where an alternative approach will be needed, based on the molecular field theory.

12.1 Excited States and Low Temperature Properties

12.1.1 Magnons

The excited states of the Heisenberg Hamiltonian (8.31) are obtained from the ferromagnetic ground state (8.32) by reversing one or more spins. To begin with, consider the case in which a single spin is reversed. The state for which the spin at site $\mathbf{R}$ is reversed is

$$|\mathbf{R}\rangle = S_{\mathbf{R}}^{+}|0\rangle = |\uparrow\rangle_{\mathbf{R}} \prod_{\mathbf{R}'\neq \mathbf{R}} |\downarrow\rangle_{\mathbf{R}'} . \tag{12.1}$$

But this state is not an eigenstate of the Hamiltonian. Indeed, for $B=0$, the identity (8.30) implies that

$$H|\mathbf{R}\rangle = -\frac{1}{4}\left(\frac{N}{2}-1\right)\mathscr{J}z|\mathbf{R}\rangle - \frac{1}{2}\mathscr{J}\sum_{\mathbf{R}'}{}' \left(|\mathbf{R}'\rangle - \frac{1}{2}|\mathbf{R}\rangle\right) . \tag{12.2}$$

The first term on the right-hand side comes from those terms in H not involving the site $\mathbf{R}$, while the second term comes from terms coupling this site with its z neighbours. The sum over $\mathbf{R}'$ is thus restricted to nearest neighbours of $\mathbf{R}$, as indicated by the prime over the summation symbol.

Introducing the energy $E_0 = -Nz\mathscr{J}/8$ of the ground state, (12.2) can be rewritten in the form

$$H|\mathbf{R}\rangle = E_0|\mathbf{R}\rangle - \frac{1}{2}\mathscr{J}\sum_{\mathbf{R}'}{}' \left(|\mathbf{R}'\rangle - |\mathbf{R}\rangle\right) . \tag{12.3}$$

Since $\hat{H}$ couples the states $|\mathbf{R}\rangle$ and $|\mathbf{R}'\rangle$, we can obtain the eigenstates of $\hat{H}$ as linear combinations of the states $|\mathbf{R}\rangle$. It is easy to see that, given the periodicity of the lattice, these can be constructed as linear combinations of the states $|\mathbf{R}\rangle$ in the following way:

$$|\mathbf{k}\rangle = \frac{1}{\sqrt{N}}\sum_{\mathbf{R}} e^{i\mathbf{k}\cdot\mathbf{R}}|\mathbf{R}\rangle . \tag{12.4}$$

If as usual we now impose periodic boundary conditions on a cube of volume L^3, the components of **k** are quantised in the reciprocal lattice. It is straightforward to obtain

$$H|\mathbf{k}\rangle = E_0|\mathbf{k}\rangle + \frac{\mathscr{J}}{2}\sum_{\mathbf{R}}\sum_{\mathbf{R}'}{}' \frac{1}{\sqrt{N}}e^{i\mathbf{k}\cdot\mathbf{R}}(|\mathbf{R}\rangle - |\mathbf{R}'\rangle). \quad (12.5)$$

Setting $\mathbf{R}' = \mathbf{R} + \mathbf{R}''$, the $\mathbf{R}''$ are the lattice vectors from the origin to nearest neighbours. The change of variables $(\mathbf{R}, \mathbf{R}') \to (\mathbf{R}', \mathbf{R}'')$ makes the sums over $\mathbf{R}'$ and $\mathbf{R}''$ independent and the order of summation can then be changed, which gives

$$H|\mathbf{k}\rangle = E_0|\mathbf{k}\rangle + \frac{\mathscr{J}}{2}\sum_{\mathbf{R}''}\sum_{\mathbf{R}'}{}' \frac{1}{\sqrt{N}}e^{i\mathbf{k}\cdot(\mathbf{R}'-\mathbf{R}'')}(|\mathbf{R}' - \mathbf{R}''\rangle - |\mathbf{R}'\rangle)$$

$$= E_0|\mathbf{k}\rangle + \frac{\mathscr{J}}{2}\sum_{\mathbf{R}''}{}' \left(|\mathbf{k}\rangle - e^{i\mathbf{k}\cdot\mathbf{R}''}|\mathbf{k}\rangle\right).$$

This in turn implies that

$$H|\mathbf{k}\rangle = [E_0 + \varepsilon(\mathbf{k})]|\mathbf{k}\rangle. \quad (12.6)$$

We thus find that $|\mathbf{k}\rangle$ is an energy eigenstate with excitation energy (taken above the ground state)

$$\varepsilon(\mathbf{k}) = \mathscr{J}\sum_{\mathbf{R}''}{}' \frac{1 - e^{i\mathbf{k}\cdot\mathbf{R}''}}{2} = \mathscr{J}\sum_{\mathbf{R}''}{}' \sin^2\left(\frac{1}{2}\mathbf{k}\cdot\mathbf{R}''\right). \quad (12.7)$$

The second equality here results from the symmetry of the lattice with respect to the origin, which clearly implies a real value for $\varepsilon(\mathbf{k})$.

These states $|\mathbf{k}\rangle$, usually called *spin waves* or *magnons*, represent precession of the spins around the z axis with an angular frequency $\varepsilon(\mathbf{k})/\hbar$ and spatial period $2\pi/|\mathbf{k}|$ (Fig. 12.1). In particular, for $|\mathbf{k}| \to 0$, the precession frequency tends to zero. This is explained by the fact that, for $\mathbf{k} = 0$, a global rotation of all the spins does not alter their energies, owing to the invariance of the Hamiltonian under spin rotations.

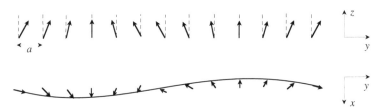

Fig. 12.1 Precession of spins about the z axis in a spin wave excitation. Note that each spin is out of phase with respect to its neighbours

12.1 Excited States and Low Temperature Properties

No restoring force is therefore generated. For nonzero values of **k** the restoring force comes from the phase difference between neighbouring spins in their precessional motion.

12.1.2 Low Temperature Thermodynamic Properties

It is important to note that the states $|\mathbf{k}\rangle$ are eigenstates of the operator S^z with a lowered absolute eigenvalue with respect to that of the ground state:

$$S^z|\mathbf{k}\rangle = \sum_{\mathbf{R}} S^z_{\mathbf{R}}|\mathbf{k}\rangle = -\left(\frac{N}{2} - 1\right)|\mathbf{k}\rangle. \tag{12.8}$$

As a consequence, the magnetic moment is also lowered. This has an important consequence when we consider the finite temperature equilibrium properties, as the thermally excited spin waves will lead to a reduction in the magnetisation.

To estimate this effect quantitatively, we must establish whether spin waves satisfy boson or fermion statistics. Note that we obtain the same quantum state if we first reverse the spin at site $\mathbf{R}_1$ and then reverse the spin at another site $\mathbf{R}_2$, or vice versa. More formally, we say that the spin operators associated with two different sites commute, expressed mathematically by $[\mathbf{S}_{\mathbf{R}_i}, \mathbf{S}_{\mathbf{R}_j}] = 0$ if $\mathbf{R}_i \neq \mathbf{R}_j$. The wave function does not therefore change sign when two spin reversals are exchanged, and these reversals must therefore be treated as bosons.

Note that, for $S = 1/2$, the same spin cannot be reversed twice $[(S^+)^2 = 0]$, so these are bosons with a short-range interaction that prevents one site from being occupied by two bosons (sometimes called hard-core bosons). At low temperatures, when not many spin waves are excited and the number of bosons is thus small, this interaction will have no significant consequences. In a dilute gas, interactions between particles play little role.

The number of spin waves with a given vector **k** is thus determined by Bose–Einstein statistics:

$$n(\mathbf{k}) = \frac{1}{e^{\beta \varepsilon(\mathbf{k})} - 1}, \tag{12.9}$$

in which the chemical potential does not appear because the number of spin waves excited is not fixed. We may now calculate the effect of the excited spin waves on the magnetisation:

$$\begin{aligned} M(T) &= -2\mu_B \langle S^z \rangle \\ &= -2\mu_B \left[-\frac{1}{2}N + \sum_{\mathbf{k}} n(\mathbf{k}) \right] \\ &= M(T=0)\left[1 - \frac{2\Omega}{N} \int \frac{d^3k}{(2\pi)^3} \frac{1}{e^{\beta \varepsilon(\mathbf{k})} - 1} \right]. \end{aligned} \tag{12.10}$$

In the last line, the sum over **k** has been replaced as usual by an integral. The general case of this integral is not easy to evaluate, but at low enough temperatures, it can be simplified. In that case, only states with low excitation energies, i.e., with low values of $|\mathbf{k}|$, contribute significantly, so (12.7) can be replaced by its quadratic expansion in **k** about $\mathbf{k} = 0$, i.e., $\varepsilon(\mathbf{k}) = \alpha k^2$. Setting $\mathbf{q} = \beta^{1/2}\mathbf{k}$, we thus obtain

$$M(T) = M(T=0)\left[1 - \frac{2\Omega}{N}(k_\mathrm{B}T)^{3/2}\int \frac{\mathrm{d}^3 q}{(2\pi)^3}\frac{1}{e^{\alpha q^2}-1}\right]. \quad (12.11)$$

The integral, which cannot be evaluated analytically, is independent of temperature. We obtain

$$M(T) = M(T=0)\left(1 - \mathrm{const.} \times T^{3/2}\right). \quad (12.12)$$

This is *Bloch's law*. It has been confirmed experimentally in many cases (see, for example, Fig. 12.2). A detailed investigation of the integral in (12.11) determines the value of the constant in Bloch's law. It also shows that the terms omitted in the truncated expansion of (12.7) give contributions going as higher powers of T than $T^{3/2}$, which can thus be neglected at low temperatures.

Note that in 1D and 2D the integral in (12.11) diverges. This suggests, at least formally, that the first thermal correction to the magnetisation will be larger than the order zero term (the correction at zero temperature). The natural interpretation of this, as confirmed by more rigorous arguments, is that a ferromagnetic state at finite temperature is not possible in 2D or 1D. This point is important when we come to investigate magnetic effects in certain substances and structures that are currently receiving much attention.

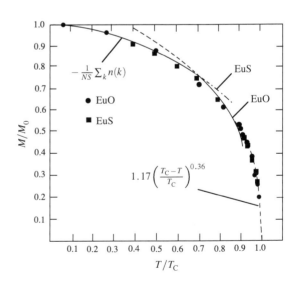

Fig. 12.2 Spontaneous magnetisation of EuO and EuS at different temperatures. The temperature is normalised by the Curie temperature and the spontaneous magnetisation by its value at zero temperature. Bloch's law is satisfied at low temperatures. From Als-Nielsen, J., et al.: Phys. Rev. B. **14**, 4908 (1976). With the kind permission of the American Physical Society. http://link.aps.org/doi/10.1103/PhysRevB.14.4908

Note finally that spin waves also occur in antiferromagnets and ferrimagnets. However, an important difference in antiferromagnets is that the energy is linear rather than quadratic in **k** for small values of $|\mathbf{k}|$.

12.1.3 Experimental Detection of Magnons

Spin wave modes can be studied directly by neutron scattering. Indeed, a neutron with incident wave vector **k** can be scattered by a magnetic material by emitting or absorbing a *magnon*. This scattering process is inelastic, because it changes the energy and wave vector of the neutron. Indeed, the scattered neutron has wave vector $\mathbf{k}+\mathbf{q}$, where **q** is the wave vector of the absorbed magnon. Spectrometers with three axes of rotation, like the one shown on p. 346, are used to carry out such experiments. The first axis corresponds to that of the monochromator which specifies the orientation of the incident neutron beam, while the second fixes the orientation of the sample with respect to this beam, and the third selects the neutrons scattered in a given direction. Writing down the conservation of energy and wave vector, the magnon dispersion relation $\varepsilon(\mathbf{k})$ can then be determined. Figure 12.3 shows the dispersion relation observed for a ferromagnetic cobalt compound ($Co_{0.92}Fe_{0.08}$).

Question 12.1. Can you explain the shape of the experimental curve in Fig. 12.3? What experimental quantities characterising ferromagnetism can be deduced from this experiment? The magnetisation of this compound with face-centered cubic structure is $1.84\mu_B$/atom (neglecting the difference between Co and Fe). It has a Curie temperature of 1,300 K. Can this information be used to check the applicability of mean field theory?

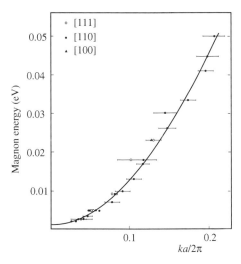

Fig. 12.3 Magnon dispersion curves obtained by inelastic neutron scattering in $Co_{0.92}Fe_{0.08}$. The wave vector is renormalised in the three directions indicated. From Sinclair, R.N., Brockhouse, B.N.: Phys. Rev. **120**, 1639 (1960). With the kind permission of the American Physical Society. http://link.aps.org/doi/10.1103/PhysRev.120.1638

12.2 The Magnetic Phase Transition

The Curie temperature separates two different magnetic states or phases, one with and one without spontaneous magnetisation ($T \leq T_C$ and $T \geq T_C$, respectively). This temperature thus characterises a *phase transition*. The theory of spin waves described in the last section is adequate at low temperatures because the total number of spin wave states remains small. However, in order to understand what is happening close to the phase transition, where the spontaneous magnetisation is small, we would need a theory of dense spin wave gases. On the other hand, this is a difficult undertaking and not particularly instructive. In Chap. 8, we pursued another approach called the *mean field theory*, based on the molecular field of Curie and Weiss.

12.2.1 Mean Field Theory

In the framework of the mean field theory, we considered the part of the Heisenberg Hamiltonian involving a spin at a given site $\mathbf{R}$:

$$H_{\mathbf{R}} = \mathbf{S}_{\mathbf{R}} \cdot \left(-\mathscr{J} \sum_{\langle \mathbf{R}' \rangle} \mathbf{S}_{\mathbf{R}'} + 2\mu_B \mathbf{B} \right). \tag{12.13}$$

The *mean field approximation* replaces the spin operators at neighbouring sites $\mathbf{R}'$ of $\mathbf{R}$ by their average value. The underlying hypothesis is that fluctuations about this average value are small. Setting

$$\mathbf{S}_{\mathbf{R}'} \longrightarrow \langle \mathbf{S}_{\mathbf{R}'} \rangle = -\frac{\mathbf{M}}{2\mu_B N}, \tag{12.14}$$

the Hamiltonian at site $\mathbf{R}$ then takes the simple form

$$H_{\mathbf{R}} = 2\mu_B \mathbf{B}_{\text{eff}} \cdot \mathbf{S}_{\mathbf{R}}, \tag{12.15}$$

where the spin at $\mathbf{R}$ is subjected to an effective field

$$\mathbf{B}_{\text{eff}} = \mathbf{B} + \lambda^* \mathbf{M}, \qquad \lambda^* = \frac{z\mathscr{J}}{4\mu_B^2 N}, \tag{12.16}$$

with z the number of nearest neighbours of the site $\mathbf{R}$. For a spin $S = 1/2$, this led simply to

$$M = N\mu_B \tanh\left(\frac{\mu_B B_{\text{eff}}}{k_B T}\right). \tag{12.17}$$

12.2 The Magnetic Phase Transition

A ferromagnetic solution was obtained in zero field below a critical temperature

$$T_C = \frac{z\mathscr{J}}{4k_B}, \qquad (12.18)$$

which can be identified with the experimentally observed Curie temperature. For an arbitrary value of the spin, the result (12.18) must be multiplied by a factor of $4S(S+1)/3$.

Note that this theory does indeed explain the self-stabilising nature of the ferromagnetic state. On the one hand, the nonzero magnetisation of site **R** is due, via the effective field, to the nonzero magnetisation of the neighbouring sites **R'**, while on the other hand, site **R** contributes to the creation of a magnetisation at the sites **R'** via the very same effective field. This effect will propagate from site to site. Starting from interactions only between nearest neighbours, we thus obtain a coherent magnetic state over the whole crystal.

Note also that any nonzero value of the spontaneous magnetisation breaks the spin rotational symmetry of the Heisenberg model. The Curie temperature thus corresponds to the symmetry breaking already discussed.

The mean field theory makes quite detailed predictions concerning certain physical quantities. Consider first the spontaneous magnetisation. From (12.17), the magnetic moment per atom for temperatures just below T_C is given by

$$m = \frac{M}{N} = \mu_B \sqrt{3} \left(1 - \frac{T}{T_C}\right)^{1/2}. \qquad (12.19)$$

Note in particular the square root growth, i.e., with infinite gradient just below T_C.

For temperatures much lower than T_C, the solution of (12.17) obviously tends to unity, agreeing with the magnetic moment of the ground state in the Heisenberg model. However, it is easy to show that the solution of (12.17) tends to unity exponentially when $T \to 0$, whereas spin wave theory predicts a very different law, going as $T^{3/2}$ [see (12.12)]. This disagreement is due to the fact that spin waves are not correctly taken into account in the mean field theory.

Above T_C, (12.17) can be used to find the behaviour in a weak applied field B. A small magnetic moment is then expected, so the equation can be linearised to obtain

$$m = \frac{\mu_B^2}{k_B T} B_{\text{eff}} = \frac{\mu_B^2}{k_B T} B + \frac{T_C}{T} m. \qquad (12.20)$$

This yields a linear relation between m and B, whence the temperature dependence of the magnetic susceptibility is

$$\chi(T) = N \frac{\mu_B^2}{k_B(T - T_C)}. \qquad (12.21)$$

The magnetic susceptibility thus diverges as the Curie temperature is approached. This is the *Curie–Weiss law*. This increase in the susceptibility compared with the situation for spins without exchange interactions can be explained by the fact that a given spin 'sees' not only the applied field, but also, through the exchange interaction, the magnetisation of the neighbouring spins. It is therefore exposed to a higher field in a given applied field, and so has a higher magnetisation than a free spin.

We may wonder how to interpret the infinite susceptibility at the Curie temperature. It could not of course be an infinite magnetic moment for a weak applied field. Indeed, at the temperature $T = T_C = z\mathscr{J}/4k_B$, expanding the hyperbolic tangent function to third order in (12.17), we find that

$$B - \frac{z\mathscr{J}}{12\mu_B^4}m^3 = 0, \tag{12.22}$$

that is

$$m = \mu_B \left(\frac{12\mu_B B}{z\mathscr{J}}\right)^{1/3}. \tag{12.23}$$

Consequently, the magnetisation begins by growing faster than linearly in B (which does indeed correspond to an infinite susceptibility), but clearly remains less than or equal to $N\mu_B$.

Finally, by (12.15), we see that the internal energy $U(T)$, i.e., the expectation value of the Hamiltonian, is proportional to the square of the spontaneous magnetisation. As a function of temperature, it thus has a different gradient above and below the Curie temperature. As a consequence, the specific heat $C(T) = \Omega^{-1}\partial U/\partial T$ has a discontinuity at T_C.

12.2.2 Landau and Ginzburg–Landau Theories

We thus find that the mean field approximation gives a reasonable explanation for the behaviour of the thermodynamic properties at the paramagnetic–ferromagnetic transition. We shall show that this approach can be generalised in a simple way to many other phase transitions, following Landau.

a. Landau Theory of Phase transitions

As in any problem of themodynamic equilibrium, we consider the partition function

$$Z = \mathrm{Tr}[e^{-\beta H}] = \sum_n e^{-\beta E_n}, \tag{12.24}$$

where $\beta = 1/k_B T$ and the sum over n includes all the eigenstates of the Hamiltonian (the Heisenberg Hamiltonian in this case). We now observe that the Heisenberg Hamiltonian commutes with the z component of the spin, and hence also with the

12.2 The Magnetic Phase Transition

z component of the magnetisation. Consequently, the sum in (12.24) can be rearranged to give

$$Z = \sum_{M_z} \left({\sum_n}' e^{-\beta E_n} \right) = \sum_{M_z} e^{-\beta L(T, M_z)} . \tag{12.25}$$

The prime on the sum over n implies that the sum only includes states for the given value of M_z. The *Landau function* L is then defined as

$$L(T, M_z) = -k_B T \ln {\sum_n}' e^{-\beta E_n} . \tag{12.26}$$

Naturally, L will be as hard to calculate as the full partition function. However, there are two points to note here:

1. The Landau function can be considered as the free energy associated with a given magnetisation. Like any free energy, L must therefore be extensive, i.e., proportional to the volume Ω of the system.
2. L must have the same symmetry properties as the Hamiltonian. In particular, when there is no magnetic field, we must have $L(T, -M_z) = L(T, M_z)$.

In addition to these two simple observations, Landau assumed that L could be expanded in a Taylor series for small values of M_z

$$L(T, M_z) = \Omega \left[\ell(T) + a(T) M_z^2 + b(T) M_z^4 - H M_z + \cdots \right] . \tag{12.27}$$

We assume here that M_z remains small, as happens near the Curie temperature, and hence that terms of higher order than 4 can be neglected. The form of the coefficients ℓ, a, and b can only be ascertained by detailed microscopic calculation. However, the coupling with the external magnetic field H, assumed to be oriented in the z direction, is determined by the fact that M_z commutes with the Hamiltonian. It follows that the energy of a magnetisation M_z depends on the field according to

$$E_n(M_z, H) - E_n(M_z, H = 0) = -\Omega H M_z . \tag{12.28}$$

We now observe that, in the thermodynamic limit $\Omega \to \infty$, which applies for any macroscopic system, the sum over M_z in (12.25) is dominated by the largest term, i.e., the term minimising L. Up to corrections that tend to zero for $\Omega \to \infty$, the true free energy of the system is then given by

$$F(T, H) = L(T, H, M_{\min}) , \tag{12.29}$$

where $M_{\min}$ is the value of M_z minimising L. Furthermore, the equilibrium value of the magnetisation is simply $M_{\min}$.

In order to obtain a stable minimum of L, the constant b in (12.27) must be positive. For zero applied field, there are then two situations (see Fig. 12.4): either a is also positive, giving rise to a unique minimum at $M_z = 0$ (paramagnetic state), or a

Fig. 12.4 Generic shape of $L(T, M_z)$ for $a > 0$ and $a < 0$

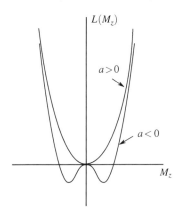

is negative, in which case there are two degenerate minima at $M_z \neq 0$, corresponding to a ferromagnetic state.

Without knowing more about the temperature dependence of the coefficients, the most natural assumption for describing a paramagnetic to ferromagnetic phase transition is a linear variation:

$$a(T) = a'(T - T_c), \qquad (12.30)$$

with $a' > 0$. It is then easy to check that the qualitative description at the end of the last section is confirmed by Landau's theory. We may add that this theory contains two phenomenological parameters, a' and b, but makes predictions about (at least) four physical quantities, leading to non-trivial experimental tests.

The Landau theory exploits very little microscopic information about the given system. It is based primarily on considerations of symmetry and internal consistency. For this reason, it applies to a very broad range of phase transitions: magnetic phase transitions, structural transitions in solids, order–disorder transitions in alloys, phase transitions in liquid crystals, and so on. Naturally, in each case one must identify a physical variable characterising the ordered state, usually called the *order parameter*. In ferromagnetic systems, we took the magnetisation, while in antiferromagnetic systems it would be the alternating magnetisation, in structural transitions the atomic displacements, and so on. In every case, an expansion of type (12.27) can be made, and similar conclusions are drawn: the order parameter varies as $(T_c - T)^{1/2}$, the specific heat has a discontinuity, and there are divergent susceptibilities. In short, Landau theory predicts a *universal* behaviour for thermodynamic quantities near a phase transition.

b. Ginzburg–Landau Functional

In Landau's theory, the magnetisation (or more generally, the order parameter) is treated as a global variable characterising the system. In particular, in a uniform

12.2 The Magnetic Phase Transition

body and far from any surfaces, this implies a constant magnetisation throughout space. However, one often has to study non-uniform systems, as exemplified by the magnetic domains and walls discussed in Chaps. 6 and 9. Ginzburg and Landau generalised the theory to cover this situation. They observed that in general the homogeneous configurations are the most stable, while inhomogeneous configurations, when possible, have a certain energy cost. The simplest case, and realistic in most situations, is simply an energy associated with the magnetisation gradient, of the kind already discussed in Chap. 9. Since the magnetisation is now space dependent, the Landau function must be written as an integral over the whole volume of the sample:

$$L(T, \mathbf{M}(\mathbf{r})) = \Omega \ell(T) + \int d^3 r \left[a(T)\mathbf{M}^2 + b(T)(\mathbf{M}^2)^2 + c(T)(\nabla \mathbf{M})^2 - \mathbf{H} \cdot \mathbf{M} \right]. \tag{12.31}$$

For reasons of stability, the constant c must be positive. Note also that we have reinstated the vectorial nature of the magnetisation, neglected in the above description of the Landau theory to simplify the discussion.

With the simplest boundary conditions, the *Ginzburg–Landau functional L* introduced above is once again minimised by a constant magnetisation state, as in the Landau theory. However, in non-uniform situations, the full description by (12.31) must be used. This is of particular importance when studying superconductivity, for which Ginzburg and Landau originally developed their theory. Indeed, in this case the general principle of gauge invariance imposes a well-determined form for the coupling between the order parameter and the electromagnetic field. A consequence of this coupling is the quantisation of the magnetic flux through a superconducting loop. This effect in turn allows unambiguous determination of the charge of the superconducting carriers. However, it should be noted that these experiments were only carried out after the BCS theory had been developed, at which point the existence of pairs was already well accepted.

Another consequence of the Ginzburg–Landau model (12.31) is that there are now deviations from the minimal L state. These deviations cost a finite energy, no longer proportional to the volume of the system. As a result, in the expansion in a sum of terms with fixed M_z, similar to the one in (12.25), we can no longer restrict to a single dominant term. We must actually sum over all possible configurations, which is clearly a much more difficult task. This is of crucial importance in a deeper understanding of phase transitions, as we shall discuss briefly in the next section.

12.2.3 Critical Behaviour

We have just seen that physical quantities vary in a singular way near phase transitions. All such singularities are commonly referred to as *critical phenomena*, and such behaviour is characterised by a set of *critical exponents*.

It is interesting to compare the theory with experimental results. The temperature dependence of the spontaneous magnetisation of EuO and EuS is shown in Fig. 12.2. The two compounds have the same crystal structure, with magnetic moments carried by the Eu^{2+} ions, with spin $S = 7/2$. Note the almost vertical slope of the magnetisation close to T_C, as predicted by the mean field theory. However, more careful analysis of the experimental results shows that the exponent of the singularity is not $1/2$, but 0.36. Note on the other hand that the behaviour for $T \ll T_C$ agrees extremely well with the spin wave theory.

The magnetic susceptibility of nickel is shown in Fig. 12.5. There is overall agreement with the Curie–Weiss law, but we find similar deviations to those observed for the magnetisation: the critical behaviour $\chi \propto (T - T_C)^{-\gamma}$ is not given by $\gamma = 1$, but rather by $\gamma = 4/3$. Similar discrepancies with respect to the mean field theory are also observed in other physical quantities, and this in a quite general way, independently of the substance under investigation. The experimental observations are summed up in Table 12.1, where the critical exponents are defined.

The differences between the critical behaviour observed experimentally and the predictions of the mean field and Landau theories have been the subject of a great deal of theoretical physics over several decades. The main problem was the lack

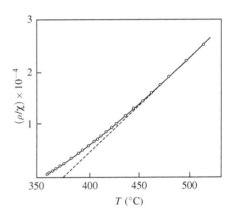

Fig. 12.5 Inverse magnetic susceptibility of nickel as a function of temperature. χ^{-1} is not found to vary linearly with temperature, thus defining a critical exponent γ greater than unity. From Weiss, P., Forrer, R.: Ann. Phys. **5**, 153 (1926), EDP Sciences copyright

Table 12.1 Critical phenomena. The *first column* defines the critical exponents γ, β, α, and δ specifying the behaviour of physical quantities close to T_C. The values predicted by the mean field theory and the experimental data are given in the *second* and *third columns*, respectively. The other columns give the exact results for specific models, viz., the 2D and 3D Ising models, and the 3D Heisenberg model

	Mean field	Experiment	2D Ising	3D Ising	3D Heisenberg
$\chi(T) \propto (T - T_C)^{-\gamma}$	1	1.3–1.4	7/4	1.24	1.39
$M(T) \propto (T_C - T)^{\beta}$	1/2	$\approx 1/3$	1/8	0.324	0.362
$C(T) \propto |T - T_C|^{-\alpha}$	0	−0.1–0.1	log	0.110	−0.115
$M(B, T = T_C) \propto |B|^{1/\delta}$	3	≈ 5	15	4.82	4.82

12.3 Summary   359

of exact or approximate theoretical methods for understanding the physics and making reliable predictions. A variety of methods were thus implemented. One was to simplify as far as possible the original model (the Heisenberg model in our case) to obtain an exactly soluble model.

The most spectacular success here was the solution of the *Ising model* by Onsager in 1944. This is obtained by replacing the spin operators in the Heisenberg model by their z components, simply dropping the x and y components. The eigenstates then become very simple, although the partition function remains extremely difficult to calculate. Even today, it can only be done in the one- and two-dimensional cases. In one dimension, there is no phase transition, while in two dimensions, Onsager's solution leads to the exponents in Table 12.1. Note the considerable difference between the exact results and those of the mean field theory. Apart from a few specific applications, e.g., to surface phase transitions, the importance of the Onsager solution is mainly conceptual: it clearly demonstrated that the critical exponents could be very different from those predicted by the mean field theory.

> Exactly soluble models are scarce and mainly concern 1D and 2D systems. In many cases, a realistic alternative is a (low or high temperature) series expansion of the partition function and other physical quantities. Very precise results can now be obtained in this way. We can accurately obtain the critical temperatures of various models (the mean field theory typically overestimates them by 20–40%), together with their critical exponents. From a quantitative point of view, this method is often satisfactory, but the underlying physical mechanisms of the critical behaviour remain obscure.
>
> The general theory of critical phenomena was finally developed by Wilson at the beginning of the 1970s (Nobel prize in 1982). He introduced the idea of the *renormalisation group* and applied it to the Ginzburg–Landau theory of (12.31). This approach provides a general physical picture of critical phenomena and at the same time a quantitative calculational tool. In particular, it explained the approximate nature of the Landau theory. In fact, the Landau function in the form (12.27) is not correct: assuming uniform magnetisation, L contains non-integer powers of M_z. The idea of the renormalisation group subsequently proved fruitful in many other areas and has become one of the most general tools of theoretical physics today.

12.3 Summary

In the Heisenberg model, the first excited states of a ferromagnetic material are not obtained by reversing a spin, but rather by producing a rotation of each spin with respect to its neighbours, leading to a total variation $\Delta S_z = 1$. The end of each vector **S** thus describes a wave called a spin wave for these excitations. They have an energy spectrum given by

$$\varepsilon(\mathbf{k}) = \mathscr{J} \sum_{\mathbf{R}}{}' \sin^2\left(\frac{1}{2}\mathbf{k}\cdot\mathbf{R}\right),$$

where the sum is taken over nearest neighbours of the origin. Their occupation number is given by Bose–Einstein statistics. At nonzero temperature, spin waves are excited and correspond to a reduction in the equilibrium magnetisation given by

$$M(T) = M(T=0)\bigl(1 - \text{const.} \times T^{3/2}\bigr).$$

This is confirmed by experiment.

The paramagnetic–ferromagnetic phase transition can be explained by a mean field theory. Here the effect of the spins neighbouring a given site is replaced by a uniform mean field $\mathbf{B}_{\text{eff}} = \mathbf{B} + \lambda \mathbf{M}$. The self-consistent solution for the magnetisation provides a determination of the Curie temperature and the temperature dependence of M in the ferromagnetic and paramagnetic states. It also shows that there is a discontinuity in the specific heat.

The mean field approximation is an example of a more general methodology initiated by Landau which starts with a simplified expansion of the free energy as a function of the magnetisation:

$$L(T, M_z) = \Omega\Bigl[\ell(T) + a(T)M_z^2 + b(T)M_z^4 - HM_z + \cdots\Bigr].$$

The magnetisation M is then determined as the variable minimising the Landau functional L. This method is quite general and applies to many phase transitions, for which it predicts universal behaviour of thermodynamic quantities at the transition. Deviations from the predictions of the Landau theory are observed experimentally. The general theory of critical phenomena explains these differences and gives a better account of fluctuations in the vicinity of the phase transition.

12.4 Answers to Questions

Question 12.1

At the center of the Brillouin zone, for small values of $\mathbf{k}$, the dispersion relation of the spin waves takes the form $\varepsilon(\mathbf{k}) = A \mathscr{J} k^2$ when $\mathbf{k}$ is taken in a given direction of the reciprocal space. The experimentally observed relation agrees with this expectation, but with a constant term:

$$\varepsilon(\mathbf{k}) = C + A \mathscr{J} k^2.$$

There is no difficulty in determining $\mathscr{J}$ from exact knowledge of the crystal structure and the normalisation factor in the experiment.

Note that this calculation corresponds to a Heisenberg model, for which the orientation of the magnetisation is indeterminate in space. The spin wave mode for $\mathbf{k} = 0$ is the one for which the overall magnetisation rotates in space without phase shifts between neighbouring moments. This justifies the fact that $\varepsilon(\mathbf{k}) = 0$ for $\mathbf{k} = 0$ in the Heisenberg model ($\mathbf{M}$ can be rotated in space at no energy cost). However, in

12.4 Answers to Questions

the presence of a magnetocrystalline anisotropy, for example, a rotation of **M** with respect to the easy magnetisation axis has an energy cost related to the anisotropy energy, and this explains the constant term C. There is a gap in the excitation energies of the spin system. It can then be shown that the energy of the mode at $\mathbf{k} = 0$ is the energy associated with the ferromagnetic resonance in zero applied field $C = \hbar \gamma B_K$ (see Chap. 11).

Note that the magnetisation does not correspond to $1\mu_B$/atom, indicating that the spin S is not in fact $1/2$. The spin wave calculation must be generalised to $S \neq 1/2$ and $g \neq 2$. This is perfectly feasible, and $\mathscr{J}$ could then be determined experimentally. In a mean field model, the Curie temperature is itself related to $\mathscr{J}$ and S. It thus provides another experimental determination of $\mathscr{J}$. We may then be able to check whether the mean field theory is quantitatively correct by comparing these two experimental values for $\mathscr{J}$.

Chapter 13
Problem Set

Contents

Problem 1	Debye–Waller Factor	367
Problem 2	Reflectance of Aluminium	373
Problem 3	Band Structure of $YBa_2Cu_3O_7$	377
	3.1 Isolated Copper–Oxygen Chain	378
	3.2 Isolated Copper–Oxygen Plane	379
	3.3 Chain and Plane	380
	3.4 Realistic Models of $YBa_2Cu_3O_7$	381
Problem 4	Electronic Energy and Stability of Alloys	391
Problem 5	Optical Response of Monovalent Metals	399
Problem 6	One-Dimensional TTF-TCNQ Compounds	405
	6.1 Isolated Chains	405
	6.2 Experimental Observations	406
	6.3 Dimerised Chain	407
	6.4 Peierls Transition	409
Problem 7	Insulator–Metal Transition	419
	7.1 Tight-Binding Method for Hydrogen-Like Orbitals	419
	7.2 Interactions Between Electrons	420
	7.3 Alkali Elements and Hydrogen	423
	7.4 Insulator–Metal Transition in Si–P	423
Problem 8	Cyclotron Resonance	441
	8.1 Real and Reciprocal Space Paths of an Electron State	441
	8.2 A Semiconductor: Silicon	442
	8.3 Metals	444
Problem 9	Phonons in Solids	453
	9.1 Einstein Model	453
	9.2 Debye Model	453
	9.3 Experimental Detection of Phonons	455
	9.4 Thermodynamic Properties	456
	9.5 Resistivity	457
Problem 10	Thermodynamics of a Thin Superconducting Cylinder: Little–Parks Experiment	469
Problem 11	Direct and Alternating Josephson Effects in Zero Magnetic Field	481
	11.1 Model Josephson Junction	481
	11.2 Realistic Josephson Junction	481
	11.3 Josephson Junction in a Microwave Field	483

H. Alloul, *Introduction to the Physics of Electrons in Solids*, Graduate Texts in Physics,
DOI 10.1007/978-3-642-13565-1_13, © Springer-Verlag Berlin Heidelberg 2011

Problem 12	Josephson Junction in a Magnetic Field	491
	12.1 Current Distribution	492
	12.2 Screening of the Magnetic Field	493
	12.3 Josephson Plasma Resonance	495
Problem 13	Magnetisation of a Type II Superconductor	511
Problem 14	Electronic Structure and Superconductivity of V_3Si	523
Problem 15	Superconductivity of $NbSe_2$	531
Problem 16	Magnesium Diboride: A New Superconductor?	541
	16.1 Atomic and Electronic Structure of MgB_2	541
	16.2 Superconductivity of MgB_2	543
Problem 17	Electronic Properties of La_2CuO_4	557
Problem 18	Properties of an Antiferromagnetic Solid	563
	18.1 Preliminaries: The Ferromagnetic Case	563
	18.2 Antiferromagnetic Transition	564
	18.3 Susceptibility in the Antiferromagnetic State	565
Problem 19	Magnetism of Thin Films and Magneto-Optic Applications	573
Problem 20	Magnetism of a Thin Film	579
	20.1 Uniform Magnetisation	580
	20.2 Non-uniform Situations	581
	20.3 Detailed Investigation of Non-uniform Situations	583

While physics is often considered from a formal point of view, and this is necessary at a certain level of understanding, the basis of the scientific method is observation and experiment. For this reason, experimental observations underpin the many problems presented hereafter. In the picture, Claire and David are now quite clear that their qualitative understanding of the experimental phenomena must be set in the form of models that can account for the observations. Peter, a keen experimenter, prefers to make a careful record of the facts before moving on to this stage

Problem 1: Debye–Waller Factor[1]

Experiments using X-ray diffraction by crystals show that the intensity of Bragg spots depends on temperature. Figure P1.1 shows the decrease in intensity of some of these peaks with temperature for aluminium, which has a face-centered cubic structure. This decrease is due to thermal vibrations of the atoms around their equilibrium positions.

The aim here is to calculate the effect of these thermal vibrations on the structure factor $S(\mathbf{K})$, bearing in mind that this is given by

$$S(\mathbf{K}) = \sum_j f_j(\mathbf{K}) e^{-i\mathbf{K}\cdot\mathbf{r}_j},$$

where $\mathbf{r}_j$ and $f_j(\mathbf{K})$ are the position in the primitive cell and the form factor of each atom belonging to the basis, respectively.

We thus seek to evaluate the thermal average $\langle S(\mathbf{K})\rangle(T)$. Let $\mathbf{u}_j(t)$ be the displacement of each atom j from its equilibrium position $\mathbf{r}_j$:

$$\mathbf{r}_j(t) = \mathbf{r}_j + \mathbf{u}_j(t).$$

It is assumed that each atom undergoes independent fluctuations about its equilibrium position (Einstein model) and that it can be described by an isotropic 3D harmonic oscillator of mass M_j and angular frequency ω_j.

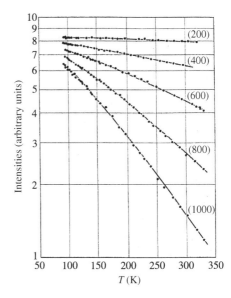

Fig. P1.1 Effect of temperature on the intensity of $(h00)$ reflections for aluminium. $(h00)$ reflections with odd h are forbidden in an fcc structure. From Nicklow, R.M. Young, R.A.: Phys. Rev. **152**, 591 (1966). With the permission of the American Physical Society (© 1966 APS). http://link.aps.org/doi/10.1103/PhysRev152.591

---

[1] This problem has been designed with G. Montambaux.

1. *Express the structure factor $\langle S(\mathbf{K})\rangle(T)$ in terms of $\langle u_j^2\rangle = \langle |\mathbf{u}_j|^2\rangle$, the mean squared displacement of each atom in the material basis. Hint: Use the property of the harmonic oscillator that $\langle e^{iAx}\rangle = e^{-A^2\langle x^2\rangle/2}$, where x is the displacement and A is a scalar.*

2. *Calculate the mean squared displacement $\langle u_j^2\rangle$ of a harmonic oscillator as a function of its mass M_j and angular frequency ω_j and the temperature. Recall that the average potential energy of the harmonic oscillator is equal to half the average total energy.*

In the following, it is assumed that the crystal is monatomic. The mass of the atoms is M and the angular frequency of their oscillations about their equilibrium position is ω.

3. *When $k_B T \gg \hbar\omega$, calculate the intensity $I(\mathbf{K},T)$ diffracted by the crystal in the direction $\mathbf{K}$. Give an expression for $I(\mathbf{K},T)/I_0(\mathbf{K})$, where $I_0(\mathbf{K})$ is the intensity diffracted by the atomic lattice at rest.*

4. *At zero temperature, show that the mean squared displacement of an atom $\langle u^2\rangle$ does not vanish, and that the factor by which the intensity is reduced is given by*

$$I(\mathbf{K}, T=0) = I_0(\mathbf{K})\exp\left(-\frac{\hbar K^2}{2M\omega}\right).$$

5. *Comment on the experimental results. In particular, discuss (a) the temperature dependence, (b) the dependence on the diffraction vector, and (c) explain the nonlinearity of the curves at low temperature.*

Problem 1: Debye–Waller Factor

Solution

1. For atoms at positions $\mathbf{r}_j(t) = \mathbf{r}_j + \mathbf{u}_j(t)$, the structure factor is

$$S(\mathbf{K}) = \sum_j f_j \exp\left[-i\mathbf{K}\cdot\mathbf{r}_j - i\mathbf{K}\cdot\mathbf{u}_j(t)\right].$$

At finite temperature T, the displacements $\mathbf{u}_j(t)$ are finite and the thermal average of the structure factor is given by

$$\langle S(\mathbf{K})\rangle(T) = \sum_j f_j \exp(-i\mathbf{K}\cdot\mathbf{r}_j) W_j(T),$$

where

$$W_j(T) = \left\langle \exp\left[-i\mathbf{K}\mathbf{u}_j(t)\right]\right\rangle.$$

The reduction factor $W_j(T)$ is called the Debye–Waller factor. $\mathbf{u}_j$ is the displacement of a 3D harmonic oscillator described by the energy

$$E = \frac{|\mathbf{p}_j|^2}{2M_j} + \frac{1}{2}M_j\omega_j^2|\mathbf{u}_j|^2.$$

We have

$$\left\langle e^{-i\mathbf{K}\cdot\mathbf{u}_j}\right\rangle = \left\langle \exp\left(-iK_x u_{jx} - iK_y u_{jy} - iK_z u_{jz}\right)\right\rangle$$

$$= \exp\left[-\frac{1}{2}\left(K_x^2\langle u_{jx}^2\rangle + K_y^2\langle u_{jy}^2\rangle + K_z^2\langle u_{jz}^2\rangle\right)\right].$$

Furthermore, $\langle u_{jx}^2\rangle = \langle u_{jy}^2\rangle = \langle u_{jz}^2\rangle = \langle|\mathbf{u}_j|^2\rangle/3$, and hence,

$$\boxed{W_j(T) = \exp\left(-\frac{K^2}{6}\langle|\mathbf{u}_j|^2\rangle\right)}.$$

2. For the harmonic oscillator, the average value of the kinetic energy is equal to the average value of the potential energy and half the average total energy $\langle E_j\rangle$. The average energy of an oscillator is given by

$$\langle E_j\rangle = \frac{\hbar\omega_j}{2} + \frac{\hbar\omega_j}{e^{\beta\hbar\omega_j}-1}.$$

An atom oscillates in three directions, and this corresponds to three independent oscillators. Hence,

$$\boxed{\frac{1}{2}M_j\omega_j^2\langle|\mathbf{u}_j|^2\rangle = \frac{3}{2}\left(\frac{\hbar\omega_j}{2} + \frac{\hbar\omega_j}{e^{\beta\hbar\omega_j}-1}\right)}. \tag{13.1}$$

3. At high temperatures such that $k_B T \gg \hbar\omega_j$, this average value tends to

$$\frac{1}{2}M_j\omega_j^2 \langle |\mathbf{u}_j|^2 \rangle = \frac{3}{2}k_B T .$$

This classic result can be obtained directly using the energy equipartition theorem. There is a corresponding Debye–Waller coefficient

$$W_j(T) = \exp\left(-\frac{K^2}{2}\frac{k_B T}{M_j\omega_j^2}\right) .$$

For a monatomic crystal, all the atoms have the same form factors $f(\mathbf{K})$ and Debye–Waller coefficient $W(T)$, and

$$\langle S(\mathbf{K}) \rangle(T) = f(\mathbf{K}) W(T) \sum_j \exp(-i\mathbf{K}\cdot\mathbf{r}_j) .$$

The intensity of the diffraction spots is the square of the structure factor:

$$\boxed{I(\mathbf{K},T) = I_0(\mathbf{K})\exp\left(-K^2\frac{k_B T}{M\omega^2}\right)} , \qquad (13.2)$$

where

$$I_0(\mathbf{K}) = \left| f(\mathbf{K}) \sum_j e^{-i\mathbf{K}\cdot\mathbf{r}_j} \right|^2$$

is the intensity diffracted by the atomic lattice at rest.

4. At zero temperature, the zero point motion of the atoms is such that $\langle u^2 \rangle$ does not vanish, but is given by $(1/2)M\omega^2\langle |\mathbf{u}|^2 \rangle = 3\hbar\omega/4$. Therefore,

$$\boxed{I(\mathbf{K},T=0) = I_0(\mathbf{K})\exp\left(-\frac{\hbar K^2}{2M\omega}\right)} .$$

5. Figure P1.1 shows the logarithm of the intensity diffracted in a direction $\mathbf{K}$ as a function of temperature. (a) At high temperature, we do indeed observe a limiting linear behaviour described by (13.2) (see Fig. P1.2 left). (b) The gradient of these straight lines is proportional to $K^2 \propto h^2$ (see Fig. P1.2 right). (c) At low temperatures, we observe a deviation due to the zero point motion. The mean squared displacement $\langle u^2 \rangle$ is no longer linear in T. At any temperature, it is given by (13.1) and the above ratio is therefore given by

$$\ln\frac{I(\mathbf{K},T)}{I_0(\mathbf{K})} = -\frac{\hbar K^2}{M\omega}\left(\frac{1}{e^{\beta\hbar\omega}-1}+\frac{1}{2}\right) ,$$

Problem 1: Debye–Waller Factor

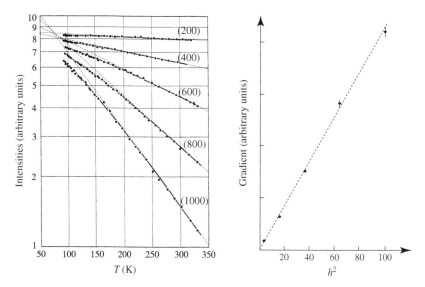

Fig. P1.2 *Left*: Exponential decay of the intensity at high temperatures. *Right*: Slopes of the straight lines on the left are proportional to h^2

which explains the shape of the curves obtained. Experiments at low temperatures would be required to make a better assessment of the applicability of the model.

Problem 2: Reflectance of Aluminium

Aluminium ($1s^2 2s^2 2p^6 3s^2 3p^1$) crystallises into a face-centered cubic Bravais lattice. The side of the conventional unit cell is $a = 4.05$ Å. The overlaps of the orbitals of the inner shells $1s^2$, $2s^2$, and $2p^6$ between nearest neighbours are very small, leading to narrow bands of tightly bound electrons in the solid. In contrast, the three valence electrons are weakly bound and can be treated as nearly free electrons.

1. *Find the density n of nearly free electrons in aluminium.*

2. *Assuming these electrons to be completely free, find the radius of the Fermi sphere and its position with respect to the first Brillouin zone of aluminium. Find the value of the Fermi energy E_F in eV.*

Now consider the Bragg plane bounding the first Brillouin zone in the k_x direction.

3. *Give an expression for the energy of the free electrons for the electron states belonging to this Bragg plane as a function of k_y and k_z. Graph the surface $E(k_y, k_z)$ corresponding to the vectors **k** of this Bragg plane and locate the Fermi level.*

4. *When there is a periodic potential, how will the eigenenergies of these states be modified in the nearly-free electron approximation? Hint: Assume that the matrix element of the periodic potential $V_\mathbf{K}$ associated with the vector **K** of the reciprocal lattice corresponding to the Bragg plane is such that $|V_\mathbf{K}| \ll E_F$.*

A photon incident upon an aluminium surface can only be absorbed by a Bloch electron of the metal with quasi-momentum **k** if the electron can be excited into an unoccupied state with the same quasi-momentum **k**. The absorption of photons by electrons can be determined by measuring the reflectance of aluminium, i.e., the reflected intensity from a polished aluminium surface, as a function of the incident photon energy.

5. *Can you explain the experimentally measured reflectance of aluminium shown in Fig. P2.1? Deduce the value of $|V_\mathbf{K}|$.*

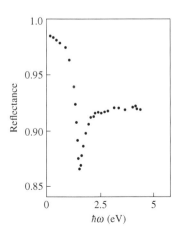

Fig. P2.1 Reflectance of aluminium for different values of the incident photon energy. From Bennet, H.E., Silver, M., Ashley, A.J.: J. Opt. Soc. Am. **53**, 1089 (1963)

Solution

1. There are three nearly-free electrons per aluminium atom. In the fcc lattice, the volume of the primitive cell, which contains one aluminium atom, is $a^3/4$, so the number of nearly-free electrons per unit volume is

$$n = 3/(a^3/4) = 12/a^3 = 0.18 \text{ Å}^{-3} = 1.8 \times 10^{-31} \text{ m}^{-3}.$$

2. The radius of the Fermi sphere is

$$k_\text{F} = (3\pi^2 n)^{1/3} = (36/\pi)^{1/3}(\pi/a) = 1.1272(2\pi/a).$$

The Brillouin zone is the polyhedron in Fig. P2.2. The point furthest from the origin Γ is one of the vertices W of the square faces which belong to two Bragg planes: the face of the cube through the point $(0, 2\pi/a, 0)$ and the plane bisecting and perpendicular to the vector ΓA $(2\pi/a, 2\pi/a, 2\pi/a)$. W is therefore equidistant from Γ and the vertex A of the cubic cell, with coordinates $(0, 2\pi/a, \pi/a)$. Hence $|\Gamma W|^2 = 1.25(2\pi/a)^2$. It follows that

$$\boxed{k_\text{F} \approx 1.008|\Gamma W|}.$$

This shows that *the first Brillouin zone is completely contained in the free-electron Fermi sphere*. The Fermi energy is $E_\text{F} = \hbar^2 k_\text{F}^2/2m_0 = 11.65$ eV.

3. The energy of states corresponding to a vector **k** of the Bragg plane $(2\pi/a, k_y, k_z)$ is given by

$$E(k_y, k_z) = (\hbar^2/2m_0)\left[(2\pi/a)^2 + k_y^2 + k_z^2\right].$$

This is the paraboloid of revolution in Fig. P2.3 (left), with minimum

$$E_0 = (\hbar^2/2m_0)(2\pi/a)^2 = 9.16 \text{ eV}.$$

The intersection with the Fermi sphere is the circle of radius

$$k_\text{i} = \sqrt{(11.65 - 9.16)/9.16}(2\pi/a) = 0.52(2\pi/a).$$

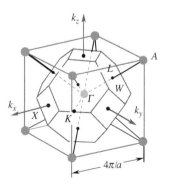

Fig. P2.2 First Brillouin zone of aluminium

Problem 2: Reflectance of Aluminium

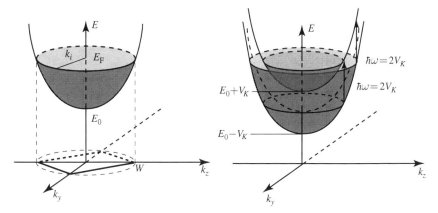

Fig. P2.3 Energy of states corresponding to a vector **k** of the Bragg plane $(2\pi/a, k_y, k_z)$ for a free electron approximation (*left*). In presence of a periodic potential the degeneracy of these states is lifted (*right*)

The square of side $\pi\sqrt{2}/a$ bounding the first Brillouin zone in this plane lies entirely within the projection of this circle. This is to be expected since all states in the first Brillouin zone are occupied.

4. In a periodic potential, states corresponding to the Bragg plane must be handled using degenerate perturbation theory. For all **k** corresponding to the Bragg plane, the dominant matrix element $V_\mathbf{K}$ of the potential is the same, and the degeneracy is removed in the same way: $E(k_y, k_z) \pm V_\mathbf{K}$. We obtain two paraboloids by vertical translations through $\pm V_\mathbf{K}$, as shown in Fig. P2.3 (right). The Fermi energy is barely changed in a nearly-free electron approximation, insofar as $|V_\mathbf{K}| \ll E_F$. Occupied states are shown in the same figure.

5. Degeneracy is removed in such a way that incident photons of energy $\hbar\omega = 2|V_\mathbf{K}|$ can be absorbed by a transition of an electron from an occupied state **k** of the lower paraboloid in Fig. P2.3 (right) to a free state of the upper paraboloid. As there are many possibilities for the same energy, absorption will be significant. This is precisely what is observed for the experimental results in Fig. P2.4. High

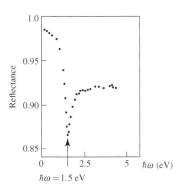

Fig. P2.4 Observed reflectance of aluminium

absorption leads to a marked drop in reflectance, occurring for $\hbar\omega \approx 1.5$ eV, and then

$$\boxed{|V_{\mathbf{K}}| \approx 0.75 \text{ eV}}.$$

Since this is much less than E_F, the nearly-free electron approximation is vindicated.

Problem 3: Band Structure of $YBa_2Cu_3O_7$[2]

In 1986, a family of oxides was found that proved to be superconducting with much higher critical temperatures than metals and alloys. The compounds in this oxide family all contain planes of copper and oxygen with Cu atoms at the nodes of a square lattice (see Fig. P3.1). Some members of this family also contain copper–oxygen chains.

An example is $YBa_2Cu_3O_7$. The real structure is shown in Fig. P3.2 (left). The simplification used in this problem is shown in Fig. P3.2 (right). The primitive cell of this compound contains two copper–oxygen planes (levels 2 and 3) and one family of chains, e.g., level 1.

The aim here is to investigate the band structure of $YBa_2Cu_3O_7$ in a simplified way using the tight-binding approximation (LCAO). 3.1 examines an isolated copper–oxygen chain. 3.2 then looks at an isolated copper–oxygen plane and two

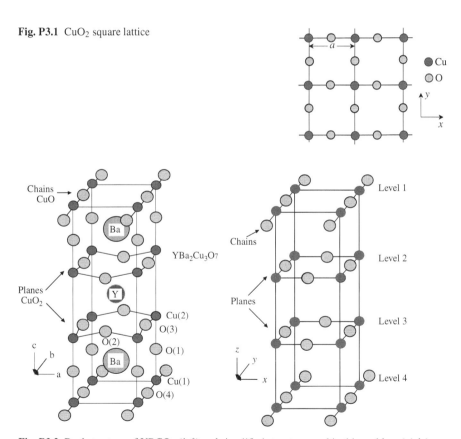

Fig. P3.1 CuO_2 square lattice

Fig. P3.2 Real structure of YBCO$_7$ (*left*) and simplified structure used in this problem (*right*)

---

[2] This problem has been designed with C. Hermann and T. Jolicœur.

coupled planes, while 3.3 deals with the coupling between a plane and the nearest chains. Finally, 3.4 compares the results obtained in this way with a more detailed calculation and experimental data concerning $YBa_2Cu_3O_7$.

Throughout this exercise, we only take into account those orbitals with energy levels close to the Fermi level of the solid, i.e., the $3d$ orbitals of copper and the $2p$ orbitals of oxygen. To simplify, we assume that, within the basis, these atomic orbitals combine to form one atomic orbital per Bravais lattice point. This orbital will be denoted by $\phi_1(\mathbf{r})$ for the chains (3.1) and $\phi_2(\mathbf{r})$ for the planes (3.2 and 3.3). These orbitals are assumed to be isotropic, real, and normalised. We apply the LCAO method to these orbitals. We also assume that the orbitals of two neighbouring sites barely overlap, and take them to be orthogonal for simplicity. There is no need to write down the relevant Hamiltonians explicitly. All the necessary matrix elements will be given.

Note: Questions 4 and 7 in 3.2 are not essential for tackling the neighbouring questions.

3.1: Isolated Copper–Oxygen Chain

Consider an isolated chain of copper and oxygen atoms. Let a be the Cu–Cu distance and $\mathbf{y}$ the unit vector along the chain.

1. *Using Fig. P3.3, specify the Bravais lattice and basis of a copper–oxygen chain.*

2. *What is the associated reciprocal lattice? Specify the corresponding first Brillouin zone.*

3. *Give the form of the Bloch functions in the tight-binding approximation.* Let k be the corresponding wave vector.

4. *Considering only nearest neighbours, calculate the dispersion relation $E_C(k)$. Express the result in terms of the matrix elements*

$$E_C^0 = \int d^3\mathbf{r}\, \phi_1(\mathbf{r}) \hat{H}_C \phi_1(\mathbf{r}), \qquad V_{2\text{chain}} = V = -\int d^3\mathbf{r}\, \phi_1(\mathbf{r}) \hat{H}_C \phi_1(\mathbf{r}+a\mathbf{y}),$$

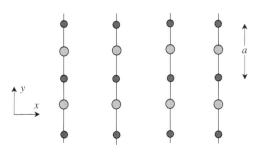

Fig. P3.3 Copper–oxygen chains

Problem 3: Band Structure of YBa$_2$Cu$_3$O$_7$

where **y** is the unit vector along the chain and $\hat{H}_C$ is the Hamiltonian of an electron in the chain.

5. *Plot the dispersion relation, assuming V > 0. If the electron occupation of this band is one electron per unit cell, what is the shape of the Fermi 'surface'? What happens if the electron occupation is very low? Give the effective mass m^* of the electrons in that case.*

3.2: Isolated Copper–Oxygen Plane

Consider an isolated copper–oxygen plane:

1. *From Fig. P3.1, specify the Bravais lattice and basis of a copper–oxygen plane. What is the associated reciprocal lattice. Specify also the corresponding first Brillouin zone.*

As in 3.1: consider one orbital per basis, denoted $\phi_2(\mathbf{r})$. Let $\hat{H}_P$ be the Hamiltonian of an electron in the plane.

2. *Give the form of the Bloch functions in the tight-binding approximation. Let* **k** *be the corresponding wave vector. Calculate the dispersion relation $E_P(\mathbf{k})$, considering only nearest neighbours. Use the matrix elements*

$$E_P^0 = \int d^3\mathbf{r}\, \phi_2(\mathbf{r}) \hat{H}_P \phi_2(\mathbf{r})$$

and

$$V_P = -\int d^3\mathbf{r}\, \phi_2(\mathbf{r}) \hat{H}_P \phi_2(\mathbf{r}+a\mathbf{x}) = -\int d^3\mathbf{r}\, \phi_2(\mathbf{r}) \hat{H}_P \phi_2(\mathbf{r}+a\mathbf{y}),$$

where **x, y** *are unit vectors along the x and y axes.*

3. *If the electron occupation of this band is one electron per unit cell, what is the shape of the Fermi 'surface'? If the number of electrons is $1 \pm \delta$ with $\delta \ll 1$, sketch the Fermi surface in the Brillouin zone. What happens if the band is almost empty?*

Now, and only for this question, consider second nearest neighbours in the plane using the matrix element

$$V_P' = \int d^3\mathbf{r}\, \phi_2(\mathbf{r}) \hat{H}_P \phi_2(\mathbf{r} \pm a\mathbf{x} \pm a\mathbf{y}),$$

where the notation $\pm$ indicates that the four matrix elements are equal by symmetry. Assume that $V_P' > 0$.

4. *For one electron per unit cell, what is the new value of the energy for those values of* **k** *that corresponded to the Fermi surface in the last question? Deduce the approximate position of the new Fermi surface in the region $0 \leq k_x, k_y \leq \pi/a$.*

In the compound $YBa_2Cu_3O_7$, the planes are in fact coupled into pairs [levels 2 and 3 are coupled in Fig. P3.2 (right)]. We treat the case of an isolated double plane. This complicates the square lattice basis considered above. We now consider that there are two orbitals per site of this basis, one for each plane. These are $\phi_2(\mathbf{r})$ and $\phi_2(\mathbf{r}+c\mathbf{z})$, where **z** is the vector joining the two planes. Use the following LCAO function:

$$\psi_\mathbf{k}(\mathbf{r}) = \sum_j \exp(i\mathbf{k} \cdot \mathbf{R}_j) \left[A_k \phi_2(\mathbf{r}-\mathbf{R}_j) + B_k \phi_2(\mathbf{r}+c\mathbf{z}-\mathbf{R}_j) \right],$$

where the sites $\mathbf{R}_j$ run over the points of the Bravais lattice of the plane $z=0$ with j as index. The coefficients A_k and B_k are adjustable parameters.

5. *Show that the LCAO function satisfies Bloch's theorem. The Hamiltonian of an electron in the double plane is denoted by $\hat{H}_{DP}$.*

Project Schrödinger's equation $\hat{H}_{DP}|\Psi_\mathbf{k}\rangle = E(\mathbf{k})|\Psi_\mathbf{k}\rangle$ onto the functions $\phi_2(\mathbf{r})$ and $\phi_2(\mathbf{r}+c\mathbf{z})$. Simplify the problem by neglecting overlaps between distinct sites, i.e., assume that the orbitals $\phi_2(\mathbf{r})$ and $\phi_2(\mathbf{r}+c\mathbf{z})$ are orthogonal.

6. *Derive a homogeneous linear system of equations in which the unknowns are the coefficients A_k and B_k. Specify the coefficients of the system in the form of matrix elements of $\hat{H}_{DP}$. Assume that the matrix elements involving only the orbitals of a given plane are the same as those of $\hat{H}_P$ introduced in question 2. For matrix elements involving orbitals from both planes, keep only the one involving the same site, viz.,*

$$\int d^3\mathbf{r} \phi_2(\mathbf{r}-\mathbf{R}_n)\hat{H}_{DP}\phi_2(\mathbf{r}+c\mathbf{z}-\mathbf{R}_n) = T.$$

7. *How many bands are there? Give their dispersion $E(\mathbf{k})$ as a function of $E_P(\mathbf{k})$ and T. Plot the result for $0 \leq k_x = k_y \leq \pi/a$. Assuming T small, and when there are two electrons per unit cell, one from each plane, locate the Fermi energy on the band diagram $E(\mathbf{k})$ for the given direction. Deduce the shape of the Fermi surface in the region $0 \leq k_x, k_y \leq \pi/a$, with the help of question 3.*

3.3: Chain and Plane

In $YBa_2Cu_3O_7$, there are also copper–oxygen chains, as discussed in 3.1 and shown in Fig. P3.2 (right). Consider now a plane coupled with a lattice of chains. To describe the combined CuO_2 plane in level 2 and Cu–O chains in level 1, use the same Bravais lattice as in 3.2, but consider two orbitals per primitive cell, viz., $\phi_1(\mathbf{r})$ of the chain and $\phi_2(\mathbf{r})$ of the associated plane.

Problem 3: Band Structure of YBa$_2$Cu$_3$O$_7$

1. Start by examining the lattice of Cu–O chains in level 1 using the tight-binding approximation. These chains are barely coupled together in YBa$_2$Cu$_3$O$_7$ and we may completely neglect matrix elements involving different chains. *Using 3.1, give the dispersion $E_{CL}(\mathbf{k})$. Plot this function for $\mathbf{k} = (0, k_y)$ and $-\pi/a \le k_y \le +\pi/a$. Plot the constant energy curves of $E_{CL}(\mathbf{k})$ in the region $-\pi/a \le k_x, k_y \le +\pi/a$. With one electron per unit cell in this band, sketch the Fermi 'surface'.*

2. Now investigate the coupled problem of the plane and the chains using the wave function

$$\Psi_{\mathbf{k}}(\mathbf{r}) = \sum_j \exp(i\mathbf{k} \cdot \mathbf{R}_j) \left[C_{\mathbf{k}} \phi_1(\mathbf{r} - \mathbf{R}_j) + D_{\mathbf{k}} \phi_2(\mathbf{r} + c\mathbf{z} - \mathbf{R}_j) \right].$$

The one-electron Hamiltonian is now $\hat{H}_{PC}$. Assume that the matrix elements within a given plane are the same as those of $\hat{H}_P$ (see question 2 of 3.2: *Isolated Copper–Oxygen Plane*) and that, in the chain lattice, they are the same as those of $\hat{H}_C$ (see question 4 of 3.1 and question 1 of 3.3). *Using the arguments in 3.2, show that there are two bands $E_{\pm}(\mathbf{k})$ and express them in terms of $E_P(\mathbf{k})$, $E_{CL}(\mathbf{k})$, and the matrix element*

$$T' = \int d^3\mathbf{r} \phi_1(\mathbf{r} - \mathbf{R}_n) \hat{H}_{PC} \phi_2(\mathbf{r} + c\mathbf{z} - \mathbf{R}_n).$$

3. Assume that $T' = 0$ and that there are two electrons per unit cell, one per unit cell of the plane and one per unit cell of the chain lattice. Assume also that $E_P^0 = E_C^0$. *Plot the Fermi surface of the ensemble in the square $0 \le k_x, k_y \le +\pi/a$.*

4. Now consider the case $T' \ne 0$, still with $E_P^0 = E_C^0$. This coupling is only important where the Fermi surfaces of the plane and the chains used to intersect. *By examining the neighbourhood of the point $\mathbf{k} = (\pi/2a, \pi/2a)$, make a qualitative sketch of the Fermi surface for the electron occupation of question 3 in 3.3: Chain and Plane, then for an occupation number close to this. What simple remark can be made about the wave functions at the edge of the region $0 \le k_x, k_y \le +\pi/a$?* This may be important for explaining the Josephson effect in YBa$_2$Cu$_3$O$_7$ [see Combescot, R., Leyronas, X.: Phys. Rev. Lett. **75**, 3732 (1995)].

3.4: Realistic Models of YBa$_2$Cu$_3$O$_7$

1. In fact the structure of YBa$_2$Cu$_3$O$_7$ comprises two weakly coupled planes (levels 2 and 3) and a chain lattice (level 1 coupled to 2) in the primitive cell. Each isolated CuO$_2$ plane is described by the dispersion obtained in question 7 of 3.2. This system is more weakly coupled to the chains by $T' \ll T$ in question 4 of 3.3. Figure P3.4 (left) shows the results of a more detailed calculation. *Give a qualitative interpretation of the different parts of the Fermi surface.*

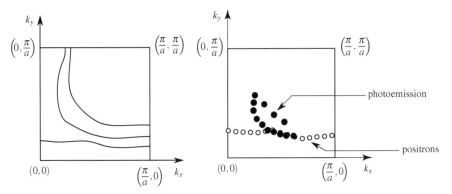

Fig. P3.4 *Left:* Fermi surface obtained from a detailed calculation, adapted from results of Yu, J., et al.: Phys. Lett. A **122**, 203 (1987) *Right:* Data taken by photoemission and positron annihilation adapted from Pickett, W.E., et al.: Science **255**, 46 (1992)

2. In Fig. P3.4 (right), black points represent photoemission measurements of the band structure and white points correspond to results obtained by another technique using positron annihilation. *What parts of the band structure are these techniques able to reveal?*

Problem 3: Band Structure of YBa$_2$Cu$_3$O$_7$

Solution

Isolated Copper–Oxygen Chain

1. The Bravais lattice of a chain is the set of points $\mathbf{R}_n = na\mathbf{y}$, where $n \in \mathbb{Z}$ and $\mathbf{y}$ is a unit vector along the chain axis. The basis then comprises one copper atom at $\mathbf{R}_0 = \mathbf{0}$ and one oxygen atom at $a\mathbf{y}/2$.

2. The reciprocal lattice is then $\mathbf{K}_p = (2\pi/a)p\mathbf{y}$, where $p \in \mathbb{Z}$. The first Brillouin zone is the interval $[-\pi/a, +\pi/a]$.

3. Using the notation adopted in this book, we have

$$\psi_k(\mathbf{r}) = \frac{1}{\sqrt{N_n}} \sum_\ell e^{ik\ell a} \phi_1(\mathbf{r} - \ell a\mathbf{y}),$$

where N_n is the number of unit cells.

4. We obtain

$$E_C(k) = E_C^0 - 2V\cos ka.$$

5. Restricting to the first Brillouin zone, this relation is shown in Fig. P3.5. The function $E_C(k)$ is symmetric under

$$k \longrightarrow \frac{\pi}{a} - k, \qquad E_C(k) \longrightarrow 2E_C^0 - E_C(k).$$

So half the states are in the sub-interval $[-\pi/2a, +\pi/2a]$. One electron per unit cell corresponds to a half-filled band, and the Fermi energy is then E_C^0. The Fermi 'surface' reduces to the two points $k = +\pi/2a$ and $k = -\pi/2a$. If the occupation is very low, only the bottom of the band close to $k \approx 0$ is occupied. Then

$$E_C(k) \simeq E_C^0 - 2V\left(1 - \frac{1}{2}k^2 a^2\right), \qquad \text{whence } m^* = \frac{\hbar^2}{2Va^2}.$$

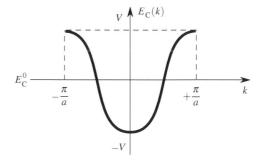

Fig. P3.5 Dispersion relation of the chain obtained in question 4, identical to the one given in Chap. 1

Isolated Copper–Oxygen Plane

1. The Bravais lattice is square: $\mathbf{R}_{n,m} = na\mathbf{x} + ma\mathbf{y}$, $n,m \in \mathbb{Z}$. The basis can be taken as CuO_2 with Cu at $(0,0)$ and two oxygens at $a\mathbf{x}/2$ and $a\mathbf{y}/2$. The reciprocal lattice is square: $\mathbf{K}_{p,q} = (2\pi/a)p\mathbf{x} + (2\pi/a)q\mathbf{y}$, $p,q \in \mathbb{Z}$. The first Brillouin zone is then the square

$$\left(\left[-\frac{\pi}{a}, +\frac{\pi}{a}\right] \text{ along } k_x\right) \times \left(\left[-\frac{\pi}{a}, +\frac{\pi}{a}\right] \text{ along } k_y\right).$$

2. We have

$$\psi_{\mathbf{k}}^{\text{plane}}(\mathbf{r}) = \frac{1}{\sqrt{N_n}} \sum_j e^{i\mathbf{k}\cdot\mathbf{R}_j} \phi_2(\mathbf{r} - \mathbf{R}_j),$$

where N_n is the number of unit cells and j indexes the sites of the Bravais lattice. Further,

$$E_P(\mathbf{k}) \langle \psi_{\mathbf{k}}^{\text{plane}} | \hat{H}_P | \psi_{\mathbf{k}}^{\text{plane}} \rangle = \frac{1}{N_n} \sum_{i,j} e^{i\mathbf{k}\cdot(\mathbf{R}_j - \mathbf{R}_i)} \int d^3r \phi_2(\mathbf{r} - \mathbf{R}_i) \hat{H}_P \phi_2(\mathbf{r} - \mathbf{R}_j).$$

The diagonal terms are all equal to E_P^0. A given site $\mathbf{R}_i$ has four nearest neighbours: $\mathbf{R}_i + a\mathbf{x}$, $\mathbf{R}_i - a\mathbf{x}$, $\mathbf{R}_i + a\mathbf{y}$, and $\mathbf{R}_i - a\mathbf{y}$. For just these cases, we have a nonzero matrix element equal to $-V_P$. Hence,

$$E_P(\mathbf{k}) = E_P^0 - V_P \left(e^{i\mathbf{k}\cdot a\mathbf{x}} + e^{-i\mathbf{k}\cdot a\mathbf{x}} + e^{i\mathbf{k}\cdot a\mathbf{y}} + e^{-i\mathbf{k}\cdot a\mathbf{y}} \right),$$

and

$$\boxed{E_P(\mathbf{k}) = E_P^0 - 2V_P(\cos k_x a + \cos k_y a)}.$$

3. The constant energy terms have the form

$$\boxed{\cos k_x a + \cos k_y a = C}.$$

We have the symmetry

$$k_x \longrightarrow \frac{\pi}{a} - k_x, \quad k_y \longrightarrow \frac{\pi}{a} - k_y, \quad C \longrightarrow -C.$$

The half-filled state thus corresponds to $C = 0$, which reduces to

$$\boxed{|k_x| + |k_y| = \frac{\pi}{a}}.$$

The Fermi 'surface' comprises four straight-line segments [see Fig. P3.6 (center)]. The general shape of the constant energy curves has been given in Chap. 3, Fig. 3.7. If the number of electrons is $1 + \delta$, the immediately adjacent states

Problem 3: Band Structure of YBa$_2$Cu$_3$O$_7$

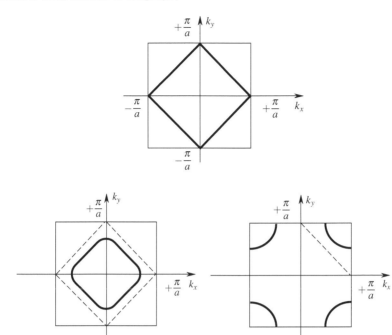

Fig. P3.6 Fermi surface calculations for a 2D square lattice with nearest neighbour hopping, for one electron per unit cell (*center*), $(1+\delta)$ electron per unit cell (*right*) and for $(1-\delta)$ electron per unit cell (*left*)

are filled [see Fig. P3.6 (right)] and if the number is $1-\delta$ the surface becomes connected [see Fig. P3.6 (left)]. If the band is almost empty, $\mathbf{k} \approx 0$ and $E_P(\mathbf{k})$ has almost circular level curves as seen in Fig. 3.7.

4. In the calculation for question 2, for each site $\mathbf{R}_i$, we take into account second neighbours, viz., $\mathbf{R}_i \pm a\mathbf{x} \pm a\mathbf{y}$. This produces an extra term in $\langle \psi_\mathbf{k}^{\text{plane}} | \hat{H}_P | \psi_\mathbf{k}^{\text{plane}} \rangle$ equal to

$$+V_P' \left[e^{i\mathbf{k}\cdot(a\mathbf{x}+a\mathbf{y})} + e^{i\mathbf{k}\cdot(a\mathbf{x}-a\mathbf{y})} + e^{-i\mathbf{k}\cdot(a\mathbf{x}-a\mathbf{y})} + e^{-i\mathbf{k}\cdot(a\mathbf{x}+a\mathbf{y})} \right],$$

and then

$$E_P'(\mathbf{k}) = E_P^0 + 2V_P(\cos k_x a + \cos k_y a) + 4V_P' \cos k_x a \cos k_y a .$$

In the region $0 \leq k_x, k_y \leq \pi/a$, the Fermi surface of question 3 is $k_x + k_y = \pi/a$. With second nearest neighbours, the energy on this straight line is equal to

$$E_P' = E_P^0 - 4V_P' \cos^2 k_x a .$$

Fig. P3.7 Fermi surface taking into account second nearest neighbour contributions

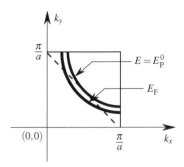

When $V'_P > 0$, this energy remains unchanged at the center of the square, but is always reduced for the other values of k_x. The constant energy curve E_P^0 thus moves toward the corner of the Brillouin zone. For the half-filled band, the new Fermi surface takes the form shown in Fig. P3.7. (Note that it divides the Brillouin zone into two equal areas.)

5. If $\mathbf{r}$ is translated to $\mathbf{r} + \mathbf{R}_0$ for $\mathbf{R}_0$ in the Bravais lattice, the dummy sum changes by

$$\mathbf{R}_j \to \mathbf{R}_j + \mathbf{R}_0 , \qquad \psi_\mathbf{k}(\mathbf{r} + \mathbf{R}_0) = e^{i\mathbf{k}\cdot\mathbf{R}_0} \psi_\mathbf{k}(\mathbf{r}) .$$

6. First project onto $\phi_2(\mathbf{r})$ to obtain

$$\langle \phi_2 | E(\mathbf{k}) | \psi_\mathbf{k} \rangle = E(\mathbf{k}) A_k ,$$

since we neglect non-local overlaps. Then we have

$$\langle \phi_2 | \hat{H}_{DP} | \psi_k \rangle = A_k \sum_j e^{i\mathbf{k}\cdot\mathbf{R}_j} \int d^3 r \phi_2(\mathbf{r}) \hat{H}_{DP} \phi_2(\mathbf{r} - \mathbf{R}_j)$$

$$+ B_k \sum_j e^{i\mathbf{k}\cdot\mathbf{R}_j} \int d^3 r \phi_2(\mathbf{r}) \hat{H}_{DP} \phi_2(\mathbf{r} + c\mathbf{z} - \mathbf{R}_j) .$$

The first sum only involves matrix elements within a plane. These are equal to the matrix elements of H_P. This first sum is thus equal to $E_P(\mathbf{k})$ as obtained in question 2. In the second sum, only the term $\mathbf{R}_j = \mathbf{0}$ is nonzero and equal to T. Therefore,

$$\langle \phi_2 | \hat{H}_{DP} | \psi_k \rangle = E_P(\mathbf{k}) A_k + T B_k = E(\mathbf{k}) A_k .$$

As the two planes enter the expression for $\psi_k(r)$ in a symmetric way, the projection onto $\phi_2(\mathbf{r} + c\mathbf{z})$ leads in a similar manner to

$$\langle \phi_2(\mathbf{r} + c\mathbf{z}) | E(\mathbf{k}) | \psi_\mathbf{k} \rangle = T A_k + E_P(\mathbf{k}) B_k = E(\mathbf{k}) B_k .$$

Problem 3: Band Structure of YBa$_2$Cu$_3$O$_7$

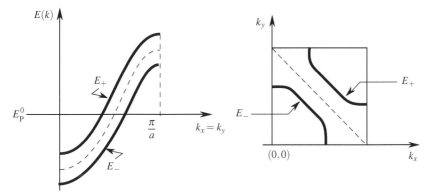

Fig. P3.8 Two bands obtained in the tight-binding approximation for a bilayer of CuO$_2$ planes. Dispersion relation of the two bands obtained in the direction $k_x = k_y$ (*left*). Fermi surface obtained for one electron per CuO$_2$ unit cell per plane (*right*)

The coefficients A_k and B_k thus satisfy the system

$$\begin{cases} A_k E_P(\mathbf{k}) + B_k T = A_k E(\mathbf{k}) \\ A_k T + B_k E_P(\mathbf{k}) = B_k E(\mathbf{k}) \end{cases}.$$

7. There are eigenstates if the determinant of this system is zero:

$$\begin{vmatrix} E_P(\mathbf{k}) - E(\mathbf{k}) & T \\ T & E_P(\mathbf{k}) - E(\mathbf{k}) \end{vmatrix} = 0.$$

There are two solutions, hence two bands of dispersion

$$\boxed{E_\pm(\mathbf{k}) = E_P(\mathbf{k}) \pm T}.$$

Along $k_x = k_y$, $E_P(\mathbf{k}) = E_P^0 - 4V_P \cos k_x$. The two bands $E_\pm$ are related to one another by translation [see Fig. P3.8 (left)]. By symmetry, the Fermi energy for one electron per plane remains equal to E_P^0. This Fermi energy cuts the surface E_+ at a level curve of question 3 with occupation $1 - \delta$, but cuts the surface E_- at a level curve with occupation $1 + \delta$. The Fermi surface thus comprises two arcs [Fig. P3.8 (right)].

Chain and Plane

1. Use the LCAO function

$$\psi_\mathbf{k}^{CL}(\mathbf{r}) = \frac{1}{\sqrt{N_{CL}}} \sum_j e^{i\mathbf{k} \cdot \mathbf{R}_j} \phi_1(\mathbf{r} - \mathbf{R}_j).$$

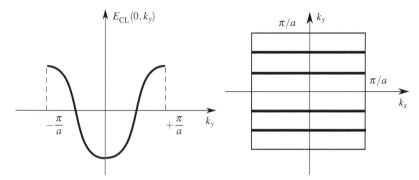

Fig. P3.9 Band structure for a lattice of chains: Dispersion relation $E_{CL}(\mathbf{k})$ *(left)* and constant energy level curves in the region $-\pi/a \leq k_x, k_y \leq +\pi/a$ *(right)*

Only the matrix elements of $\hat{H}_C$ come in when calculating $E_{CL}(\mathbf{k})$ (see question 3 of 3.1), because there is no coupling between chains. As a consequence,

$$E_{CL}(\mathbf{k}) = E_C(k_y),$$

as shown in Fig. P3.9 (left). The constant energy curves of $E_{CL}(k_x, k_y)$ are thus straight lines at fixed k_y [see Fig. P3.9 (right)]. According to question 5 of 3.1, with one electron per unit cell, the Fermi surface thus comprises two straight lines $k_y = +\pi/2a$ and $k_y = -\pi/2a$, as shown in Fig. P3.10.

2. Project $\hat{H}_{PC}|\Psi_\mathbf{k}\rangle = E(\mathbf{k})|\Psi_\mathbf{k}\rangle$ onto $\phi_1(\mathbf{r})$ and $\phi_2(\mathbf{r}+c\mathbf{z})$. The calculation is similar to the one in question 6 of 3.2. For $\phi_1(\mathbf{r})$, we obtain

$$C_k E_{CL}(\mathbf{k}) + D_k T' = C_k E(\mathbf{k}),$$

and with $\phi_2(\mathbf{r}+c\mathbf{z})$, we obtain

$$C_k T' + D_k E_P(\mathbf{k}) = D_k E(\mathbf{k}).$$

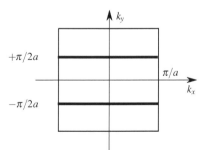

Fig. P3.10 Fermi surface of the chains for one electron per unit cell

Problem 3: Band Structure of $YBa_2Cu_3O_7$

Setting the determinant equal to zero, this yields two bands:

$$E_{\pm}(\mathbf{k}) = \frac{1}{2}\left\{E_P(\mathbf{k}) + E_{CL}(\mathbf{k}) \pm \sqrt{\left[E_P(\mathbf{k}) - E_{CL}(\mathbf{k})\right]^2 + 4T'^2}\right\}.$$

3. Since $E_P^0 = E_C^0$, the plane and chain bands are filled equally. The Fermi surface then comprises the surfaces of these two ensembles [see Fig. P3.11 (left)].

4. The surfaces intersect in $(\pi/2a, \pi/2a)$ when $T' = 0$. If $T' \neq 0$, the result of question 2 above shows that the equality $E_P = E_{CL}$ no longer holds for E_+ and E_-: the levels repel one another. Level crossing disappears and we obtain the qualitative result shown in Fig. P3.11 (right). At the edge of the Brillouin zone, the Bloch states correspond to states completely within the plane or completely within the chains.

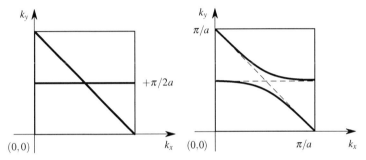

Fig. P3.11 Chain and plane Fermi surface for different hopping T' between chain and plane (*left*) $T' = 0$, (*right*) $T' \neq 0$

Realistic Models of $YBa_2Cu_3O_7$

1. A system of two planes with $V_P' \neq 0$ will have a Fermi surface made up of two segments in $[0, \pi/a]^2$, as shown in Fig. P3.12 (left). If there is a chain lattice as

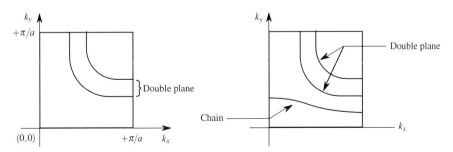

Fig. P3.12 Fermi surfaces for the bilayer of CuO_2 planes (*left*) and for the bilayer coupled to a CuO chain (*right*)

well, it will also give a branch that will avoid crossing the branches of the double plane by the hybridisation phenomenon of questions 3 and 4 of 3.3. This gives Fig. P3.12 (right), which agrees with Fig. P3.4 (left).

2. The dispersion points due to photoemission [see Fig. P3.4 (right)] coincide with contributions coming from double planes. In contrast, positron annihilation sees the chain contribution. A single technique was not enough initially to investigate the whole Fermi surface. It has since been viewed by higher resolution ARPES experiments, and matches the results of a full calculation illustrated in Fig. P3.13.

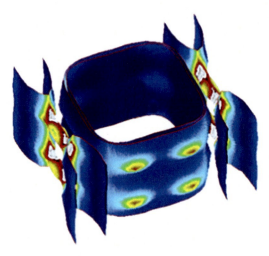

Fig. P3.13 Full calculation of the 3D band structure. Surfaces due to double planes are very close together, while those due to chains are widely spaced. The Fermi surface is almost cylindrical, because the hopping integrals between cells are very small in the c direction. Image courtesy of O. Andersen and I. Mazin from results published in Andersen, O.K., Liechtenstein, A.I., Rodriguez, O., Mazin, I.I., Jepsen, O., Antropov, V.P., Gunnarsson, O., Gopalan, S.: Physica C **185–189**, 147–155 (1991)

Problem 4: Electronic Energy and Stability of Alloys[3]

It is important for industrial applications of metals to investigate alloys in order to obtain as broad a range of properties as possible. For some elements with very similar ionic radii, alloys can be obtained in which the atoms of the two metals are randomly distributed at the lattice points of a Bravais lattice. This happens in particular for many copper-based alloys, denoted here by Cu–B, where the element B could be zinc, for example (giving brass), or aluminium. Let $Cu_{1-x}B_x$ denote the composition of such an alloy, where $0 < x < 1$ is the concentration of B atoms. These crystals can be considered as perfect monatomic crystals, where the Bravais lattice points are occupied by identical 'average' atoms, each with probability $(1-x)$ of being a Cu atom and probability x of being a B atom.

It is observed experimentally that these alloys can adopt different stable thermodynamic structural forms depending on the concentration x. For example, phase diagrams like those in Fig. P4.1 for Cu–Zn and Cu–Al are observed. Note in particular that, at low concentrations x, the stable phase at low temperature is the α phase. This is face-centered cubic (fcc) like pure copper. Above a certain concentration x_c, the stable phase is the β phase, which is body-centered cubic (bcc).

The British physicists Hume and Rothery tried to understand the origin of this transition $\alpha \to \beta$, i.e., from fcc to bcc, along with other structural changes at higher concentrations x. X-ray crystallography observations showed that, in these alloys, the average volume occupied by each atom is the same in the two structures. Let a_f and a_b be the sides of the cubic conventional cells in each structure.

1. *Determine the value of a_f/a_b corresponding to the same volume per atom in these two structures.*

In the remainder of this problem, we assume that this condition is always satisfied, and in particular during the transition fcc $\to$ bcc, with $a_b = 3$ Å.

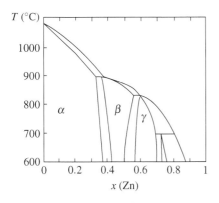
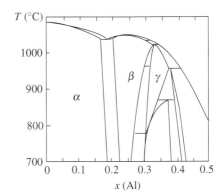

Fig. P4.1 Phase diagrams for Cu–Zn and Cu–Al alloys

---

[3] This problem has been designed with Y. Quéré.

To represent the electronic band structures of these alloys, we assume that the electrons in the filled atomic shells of the Cu^+, Zn^{2+}, and Al^{3+} ions form bound states that fill the deep electronic bands arising from the corresponding atoms. Electrons in the outer valence shells of the atoms will be treated as nearly free and subject to a periodic potential $V_{at}(r)$ which may be considered as an average potential. Hume and Rothery observed that the concentration values x_c at which the fcc → bcc transitions occur in the phase diagrams generally correspond to a well defined value v_c of the number of conduction electrons per atom.

2. Find the average number $v(x)$ of conduction electrons per atom as a function of x for the alloy $Cu_{1-x}B_x$, in the cases $B = Zn$ and $B = Al$. What is the range of this function? Determine the value of v_c for Cu–Zn and Cu–Al using Fig. P4.1.

Consider now that the conduction electrons behave as independent electrons in a constant potential (free electrons), and the temperature of the alloy is $T = 0$ K.

3. Calculate the radius k_F of the Fermi sphere for electrons in the alloy with fcc structure, as a function of v and $2\pi/a_f$.

4. What can you say about k_F for the bcc alloy?

5. Sketch the energy dependence of the density of states $D(E)$ and the position of the Fermi level for a given value of v, in the case of the fcc and bcc structures.

6. State the type and primitive cell of the reciprocal lattices associated with the fcc and bcc structures. For each of these structures, what is the shortest distance between the origin and a point of the reciprocal lattice?

7. Show that, for $x = 0$, the Fermi sphere is entirely contained within the first Brillouin zone (BZ) of the fcc and bcc structures.

When the number v of electrons per atom increases, k_F also increases.

8. Find the 'critical' values of v which correspond to a Fermi sphere that is tangential to the surface of the first Brillouin zone closest to the origin. Denote these by $v(fcc)$ and $v(bcc)$.

Now assume that the conduction electrons behave as nearly free electrons in a periodic potential $V(\mathbf{r})$.

9. Sketch the shape of the curve $E(k)$ in the direction of the vector $\mathbf{k}_0$ from the surface of the BZ closest to the origin. In what way does this differ from the one for free electrons?

10. Using the $E(k)$ of question 9, explain the form of the density of states for nearly free conduction electrons shown in Fig. P4.2.

Assume now that the Fermi surface can always be treated as spherical.

11. What radius k_F does the point G of the curve in Fig. P4.2 correspond to?

Fig. P4.2 Density of states of conduction electrons in a nearly free electron approximation

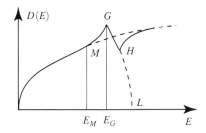

The Fourier component of the periodic potential for the vector **K** of the reciprocal lattice corresponding to the Bragg plane is $V_K = 0.5$ eV for both the fcc and the bcc structure.

12. Give the approximate value of $E_G - E_M$?

13. For different values of v, compare qualitatively the positions of the Fermi level as given by the free electron and nearly-free electron hypotheses.

14. On the same graph, sketch the energy dependences of the densities of states $D_{fcc}(E)$ and $D_{bcc}(E)$, indicating the points M and G corresponding to the two structures. For different values of v, compare the position of the Fermi level in the fcc and bcc structures.

Assume that, at $T = 0$ K, the stable structure of the alloy corresponds to the minimum of its total electronic energy.

15. What is this structure when $v = 1.2$ and $v = 1.45$?

16. What values of v would the stability limits move to if the Fermi surface were no longer taken to be spherical and if the shape determined by the periodic potential $V(\mathbf{r})$ were taken into acount?

Solution

1. For a monatomic crystal, the average volume Ω per atom is the volume of the primitive cell, i.e., $\Omega_f = a_f^3/4$ for an fcc structure and $\Omega_b = a_b^3/2$ for a bcc structure. Equating these, we have

$$\boxed{a_f/a_b = 2^{1/3} \simeq 1.26}.$$

2. An atom in the alloy contributes on average $(1-x)$ electrons in the case of Cu and $2x$ electrons in the case of Zn (or $3x$ electrons in the case of Al). We thus have $v(x) = 1+x$ for B = Zn and $v(x) = 1+2x$ for B = Al. v thus takes values between 1 and 2 in the first case and 1 and 3 in the second. The transition $\alpha \to \beta$ occurs for $x \simeq 0.4$ in the case of Zn and $x \simeq 0.20$ in the case of Al, so that $v_c \simeq 1.4$.

3. We have

$$k_F = (3\pi^2 n)^{1/3} = \left(3\pi^2 \frac{v}{\Omega}\right)^{1/3},$$

implying

$$\boxed{k_F = \frac{2\pi}{a_f} \left(\frac{3v}{2\pi}\right)^{1/3}}.$$

4. k_F is the same for the bcc structure, since we assumed that the primitive cell has the same volume Ω.

5. The Fermi level is thus the same for each structure. For free electrons, the density of states per unit volume is (see Sect. 3.3)

$$D(E) = \frac{m_0^{3/2}}{\pi^2 \sqrt{2}\hbar^3} E^{1/2}.$$

This parabolic form is shown in Fig. P4.3.

6. The reciprocal lattices of the fcc and bcc structures are of types bcc and fcc, respectively, with cubic conventional cells of side $4\pi/a_f$ and $4\pi/a_b$. Their first Brillouin zones (BZ) are shown in Fig. P4.4. In the first case, the shortest distance between reciprocal lattice points is the one between a vertex and the center of the conventional cell. In the second case, it is the distance between a vertex and the

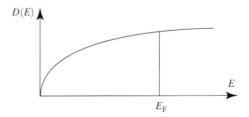

Fig. P4.3 Density of states per unit volume for free electrons

Problem 4: Electronic Energy and Stability of Alloys

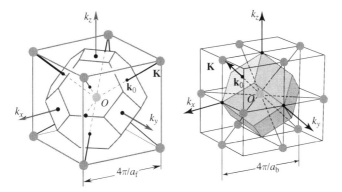

Fig. P4.4 First BZ of the fcc structure (*left*) and of the bcc structure (*right*)

center of an adjacent face. Hence,

$$d_{\text{fcc}} = \frac{4\pi}{a_f} \frac{\sqrt{3}}{2}, \qquad d_{\text{bcc}} = \frac{4\pi}{a_b} \frac{\sqrt{2}}{2}.$$

7. The distances from the origin of the reciprocal lattice to the plane of the first Brillouin zone closest to the origin are half the distances calculated in question 6 for each lattice. For the Fermi sphere to lie within the first BZ, we must have $k_F < d/2$. For $\nu = 1$, we find

$$k_F = 0.781 \times \frac{2\pi}{a_f}, \qquad \frac{d_{\text{fcc}}}{2} = 0.866 \times \frac{2\pi}{a_f}, \qquad \frac{d_{\text{bcc}}}{2} = 0.89 \times \frac{2\pi}{a_f}.$$

We thus find that the condition for inclusion is satisfied in each case.

8. The critical values of ν are determined by the condition $k_F(\nu) = d/2$. We find

$$\boxed{\nu(\text{fcc}) = 1.363, \qquad \nu(\text{bcc}) = 1.48}.$$

9. The curve has the same parabolic shape as for free electrons, except for **k** close to $\mathbf{k}_0$, where it changes curvature and joins with a horizontal tangent onto $E(\mathbf{k}_0) = E_0 - |V_\mathbf{K}|$. Here $E_0 = (\hbar^2/2m_0)k_0^2$ and $V_\mathbf{K}$ is the Fourier component of the periodic potential for the vector **K** of the reciprocal lattice corresponding to the Bragg plane. At $\mathbf{k}_0$, there is a gap $2|V_\mathbf{K}|$. Figure P4.5 shows $E(\mathbf{k})$ in an unfolded band structure diagram.

10. The curve in Fig. P4.5 comprises 4 parts that can be related to the points in Fig. P4.6:

    (i) Between the origin and M, the energy is less than $E_{M'}$ and the corresponding states are close to plane waves, for which the density of states has the parabolic shape for free electrons.

Fig. P4.5 Dispersion $E(k)$ in the direction $\mathbf{k}_0$

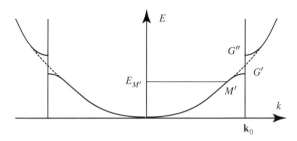

Fig. P4.6 Density of states of nearly free conduction electrons

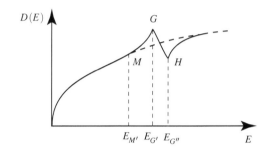

(ii) Between M and G, the curve $E(\mathbf{k})$ is distorted in the direction $\mathbf{k}_0$, and has more states per energy interval due to the curvature. In the other directions, $E(\mathbf{k})$ remains close in behaviour to the function for free electrons. As a consequence, the density of states $D(E)$ increases.

(iii) Between G and H, the energy is less than the gap $E_{G''} - E_{G'}$ in the direction $\mathbf{k}_0$ and neighbouring directions, and there are no states corresponding to these vectors $\mathbf{k}$. Only those states associated with the other wave vectors contribute to the density of states, which therefore decreases.

(iv) Above H, the energy lies above $E_{G''}$. There are once again states in the direction $\mathbf{k}_0$ and neighbouring directions, and $D(E)$ begins to increase again. (We are assuming that $|V_\mathbf{K}|$ is small enough to ensure that the Fermi surface always lies within the first BZ for directions well separated from $\mathbf{k}_0$. This is the case for metals like Cu–Zn or Cu–Al.)

11. Assuming that the Fermi surface remains spherical, the point G corresponds to a situation in which this Fermi surface touches the BZ at $\mathbf{k}_0$, so k_F is given by the condition in question 8, i.e., $k_F = d/2$. It follows that

$$\boxed{k_F(\text{fcc}) = 0.866 \frac{2\pi}{a_f}} \;, \qquad \boxed{k_F(\text{bcc}) = 0.89 \frac{2\pi}{a_f}} \;.$$

12. Still assuming that the Fermi surface is spherical, $E_G = \hbar^2 k_0^2 / 2m_0$. The point M corresponds to the point M' where the curve $E(\mathbf{k})$ in the direction $\mathbf{k}_0$ begins to move significantly away from the free electron curve. The position of M' cannot be estimated from the second order perturbation calculation carried out earlier in this book, because it was assumed there that $E_\mathbf{k} - E_{\mathbf{k}-\mathbf{K}}$ is small compared with

Problem 4: Electronic Energy and Stability of Alloys   397

$|V_\mathbf{K}|$. Now, close to the Bragg plane, and in particular at M', the last two terms are of the same order of magnitude. We must therefore revert to the original formula found by diagonalising $\hat{H} = (p^2/2m_0) + V(r)$ in the two-state basis $|\mathbf{k}\rangle$, $|\mathbf{k} - \mathbf{K}\rangle$. It is easy to show that

$$E(\mathbf{k}) = \frac{\varepsilon_{\mathbf{k}-\mathbf{K}} + \varepsilon_\mathbf{k}}{2} - \sqrt{\left[\frac{(\varepsilon_{\mathbf{k}-\mathbf{K}} - \varepsilon_\mathbf{k})^2}{2}\right]^2 + V_\mathbf{K}^2},$$

with $\varepsilon_\mathbf{k} = \hbar^2 k^2/2m_0$ and $\mathbf{k}_0 = \mathbf{K}/2$. For $\mathbf{k} = \mathbf{K}/2 - \mathbf{q}$, the expansion to order q^2 of the above expression is

$$E(\mathbf{k}) = \varepsilon_{\mathbf{K}/2} + \frac{\hbar^2 q^2}{2m_0} - V_\mathbf{K}\left(1 + \frac{\hbar^2}{4m_0}\frac{\varepsilon_{\mathbf{K}/2}}{V_\mathbf{K}^2}q^2\right)$$

$$= \varepsilon_{\mathbf{K}/2} - V_\mathbf{K} + \frac{\hbar^2}{2m_0}q^2\left(1 - 2\frac{\varepsilon_{\mathbf{K}/2}}{V_\mathbf{K}}\right).$$

This shows that $E(\mathbf{k})$ has parabolic shape near the BZ. We may estimate that the point M' corresponds to the vector $\mathbf{q}$ such that the second term above is of order $-V_\mathbf{K}$. Since $\varepsilon_{\mathbf{K}/2} > V_\mathbf{K}$, we obtain

$$q/k_0 \simeq V_\mathbf{K}/\varepsilon_{\mathbf{k}_0}\sqrt{2}, \qquad E_{M'} \simeq \varepsilon_{\mathbf{k}_0} - 2V_\mathbf{K}.$$

With the above numerical values, $E_G(\text{fcc}) \sim 8$ eV, $E_G(\text{bcc}) \sim 8.45$ eV, and

$$\boxed{E_G - E_M \sim 1 \text{ eV}}.$$

13. If we take ν such that k_F lies below the point M (depending on the structure), the density of states curve is the same as for free electrons and the Fermi level is also similar. When ν is such that k_F (calculated for the spherical Fermi surface) lies between M and G, the density of states is higher than for free electrons, and for a given total number of electrons, the last occupied state will have lower energy than in the case of free electrons. The Fermi level will thus be lower. Beyond G, the Fermi level gradually climbs to reach the level corresponding to free electrons.

14. Given the numerical values found for the positions of the points M and G for the fcc and bcc structures, the density of states curves have the shapes shown in Fig. P4.7. In the same way, the Fermi level $E_F(\text{fcc})$ will be lower than the level $E_F(\text{bcc})$ for ν corresponding to a k_F between M_1 and D, where D is the point such that the integral of the density of states between the origin and D is the same for the two structures. Beyond D and up to the point D', the Fermi level of the bcc structure is lower than that of the fcc structure.

15. The total electronic energy varies in the same way as the Fermi energy. The most stable structure will thus be the one giving the lowest Fermi energy. For $\nu = 1.2$, we are slightly above the point M_1 and the stable structure is fcc. But

Fig. P4.7 Densities of states. *Continuous curve*: fcc structure. *Dashed curve*: bcc structure. *Dotted curve*: free electrons

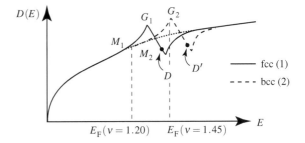

when $v = 1.45$, it is the bcc structure that is the most stable, corresponding to a point close to G_2 on the curve of $D(E)$. *There is thus a limit to the stability of the fcc structure for $v_c = 1.40$.*

16. Considering a Fermi surface for nearly free electrons, there is a distortion of the sphere in the direction $\mathbf{k}_0$ and equivalent directions. The surface tends to protrude in these directions. The Fermi surface will thus touch the surface of the Brillouin zone for lower values of k_F than those calculated above. The energy E_G is shifted to the left, but the energy E_M is not significantly altered. The density of states begins to drop for a lower energy. More detailed numerical calculations must be carried out because the point G no longer corresponds exactly to contact between the two surfaces. The density of states can continue to increase through an increase in the number of states for directions close to $\mathbf{k}_0$.

Problem 5: Optical Response of Monovalent Metals

The alkali elements have an almost spherical Fermi surface and a free electron model can explain most of their electronic properties. We thus expect the Drude model to account correctly for electronic transport. The conductivity predicted by this model was examined in Question 4.1. The real part $\sigma_1(\omega)$ of the conductivity is obtained directly by experiment from the reflectivity of a light wave on a metal film. For potassium, measurements of $\sigma_1(\omega)$ give the result shown in Fig. P5.1.

1. *Does the Drude model apply to this experimental result? How could one determine experimentally the limiting value of σ_1 as $\omega \to 0$?*

To understand the result in Fig. P5.1, we examine the influence of the band structure on the optical reflection coefficient of the metal in the case of the alkali elements. Recall that the alkali elements are monovalent and crystallise into body-centered cubic crystals. Let a be the side of the cube.

2. *Characterise the reciprocal lattice and first Brillouin zone. Specify the reciprocal space directions corresponding to points N of the first Brillouin zone closest to its center O. Find $k_0 = |ON|$.*

For the valence electrons, the band structure of the alkali elements is very well accounted for in the nearly-free electron approximation.

3. *Sketch the first two energy bands in a direction Ok_0: (a) in an extended zone scheme (b) in a restricted zone scheme.*

We may assume that the matrix element of the periodic potential $V(\mathbf{r})$ of the lattice, which removes the degeneracy at the edge of the Bragg zone, is small enough to ensure that the Fermi surface is well separated from the edges of the Brillouin zone. In this case it is easy to find the equations $E_n(\mathbf{k})$ for the energy bands far from the Bragg planes by treating $V(\mathbf{k})$ as a perturbation (to second order).

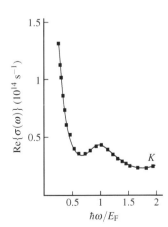

Fig. P5.1 Real part of the conductivity of potassium

Recall that, if $|\mathbf{k}\rangle$ are plane wave functions with eigenenergies $\hbar^2 k^2/2m_0$, the matrix elements of a periodic potential satisfy

$$\langle \mathbf{k}'|V(\mathbf{r})|\mathbf{k}\rangle = V_\mathbf{K}\delta(\mathbf{k}-\mathbf{k}'-\mathbf{K}),$$

where $\mathbf{K}$ is a vector of the reciprocal lattice. For simplicity, we may assume $V_\mathbf{0} = 0$.

4. *Give the equations $E_n(\mathbf{k})$ for the first two bands $n = 0$ and $n = 1$, far from the Bragg planes, in terms of k_0 and the $V_\mathbf{K}$. Consider only those terms in $\mathbf{K}$ giving dominant contributions to second order.*

Apart from the processes whereby a free electron gas can absorb electromagnetic waves that were discussed in Question 4.1, a photon can also be absorbed if it has enough energy to excite an electron of given wave vector $\mathbf{k}$ into an unoccupied electron state of the same wave vector in a folded band structure diagram.

5. *On the band structure diagram sketched for question 3, indicate the interband transition with lowest possible energy $\hbar\omega_s$.*

6. *Assuming that the perturbation terms found in question 4 are totally negligible for $k \leq k_F$, find $\hbar\omega_s$ as a function of the Fermi energy E_F.*

7. *Can you explain how $\sigma_1(\omega)$ differs from the prediction of the Drude model, as obtained for potassium (see Fig. P5.1).*

Solution

1. The Drude model gives

$$\sigma_1(\omega) = \frac{\sigma_0}{1+\omega^2\tau_e^2}.$$

The conductivity $\sigma_1(\omega)$ does indeed fall off steeply with frequency. *The hump on the right of Fig. P5.1 is clearly not described by the Drude model.* The experimental results correspond to optical frequencies and cannot be used to reach low frequencies. We could obtain $\sigma(0)$ by measuring the resistivity of the metal film. Note that the steep decline as a function of ω suggests that $\omega\tau_e \gg 1$ for the frequencies considered experimentally. This therefore corresponds in the Drude model to $\sigma_1(\omega) \simeq [\sigma(0)/\tau_e]^2 \omega^{-2}$. We may thus plot the experimental points obtained for σ as a function of ω^{-2}.

Figure P5.2 shows that the points are aligned and extrapolate to $\sigma \sim 10^{13}$ s^{-1} for $\omega \to \infty$. This implies that the Drude model gives a fair account up to $\hbar\omega/E_F \sim 0.6$. The Drude contribution to $\sigma_1(\omega)$ is plotted in Fig. P5.3. A further

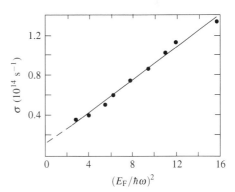

Fig. P5.2 Conductivity as a function of $(E_F/\hbar\omega)^2$ with extrapolation for $\omega \to \infty$

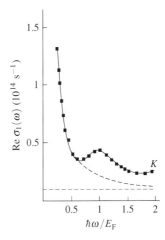

Fig. P5.3 Conductivity as a function of $\hbar\omega/E_F$. *Dashed curve*: Drude contribution

Fig. P5.4 Reciprocal lattice and first Brillouin zone

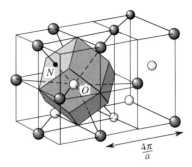

contribution is observed for $\hbar\omega/E_F > 0.6$. Hence τ_e and $\sigma(0) = ne^2\tau_e/m$ can be obtained directly. (But note that the normalisation of σ_1 and the units are not shown correctly in Fig. P5.1.)

2. The reciprocal lattice is fcc with lattice constant $4\pi/a$. The Bragg planes are the orthogonally bisecting planes of the vectors joining O to its nearest neighbours. They bound the first Brillouin zone, which is a regular dodecahedron (see Fig. P5.4). The closest points N are the centers of the faces

$$|k_0| = ON = \frac{2\pi}{a}\sqrt{2}\frac{1}{2} = \frac{\pi\sqrt{2}}{a} = \frac{4.44}{a} = |k_0|.$$

3. Figure P5.5a, b show the changes to the free electron parabola in unfolded and folded band structure diagrams, respectively.

4. We have

$$\hat{H} = \frac{\hat{p}^2}{2m} + V(\mathbf{r}).$$

In the unfolded band structure,

$$E(\mathbf{k}) = \varepsilon_{\mathbf{k}} + \langle \mathbf{k}|V(\mathbf{r})|\mathbf{k}\rangle + \sum_{\mathbf{k}'\neq\mathbf{k}} \frac{|\langle \mathbf{k}'|V(\mathbf{r})|\mathbf{k}\rangle|^2}{\varepsilon_{\mathbf{k}} - \varepsilon_{\mathbf{k}'}}, \quad (13.3)$$

where

$$\varepsilon_{\mathbf{k}} = \hbar^2 k^2/2m_0, \quad \langle \mathbf{k}|V(\mathbf{r})|\mathbf{k}\rangle = 0, \quad \langle \mathbf{k}|V(\mathbf{r})|\mathbf{k}'\rangle = V_{\mathbf{K}}\delta(\mathbf{k}-\mathbf{k}'-\mathbf{K}).$$

Consequently, only the terms $\mathbf{k}' = \mathbf{k} - \mathbf{K}$ contribute to changing $E(\mathbf{k})$ from the free electron parabola, where $\mathbf{K}$ is a vector of the reciprocal lattice.
In the folded band structure, $\mathbf{k}' = \mathbf{k} - \mathbf{K}$ corresponds to $\mathbf{k}$, so the third term of (13.3) only couples states of the same vector $\mathbf{k}$ in the folded band structure. Therefore, for the first band, far from $\mathbf{k} = \mathbf{k}_0$, we have

$$\boxed{E_1(\mathbf{k}) = \frac{\hbar k^2}{2m_0} - \frac{2m_0}{\hbar^2}\frac{V_{2\mathbf{k}_0}^2}{(2\mathbf{k}_0 - \mathbf{k})^2 - k^2} + \cdots},$$

Problem 5: Optical Response of Monovalent Metals

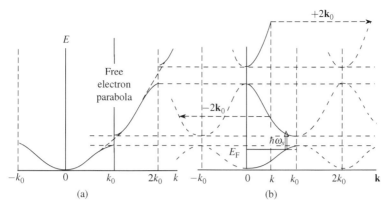

Fig. P5.5 Changes to the free electron parabola in (**a**) unfolded and (**b**) folded band structure diagrams

because the unperturbed state with vector **k** in the second band corresponds to $\mathbf{k} - 2\mathbf{k}_0$. Likewise,

$$E_2(\mathbf{k}) = \frac{\hbar^2(k-2k_0)^2}{2m_0} + \frac{2m_0}{\hbar^2} \frac{V_{2k_0}^2}{(2\mathbf{k}_0 - \mathbf{k})^2 - k^2} + \cdots ,$$

far from $\mathbf{k} = \mathbf{k}_0$ and $\mathbf{k} = 0$. Here we neglect terms due to higher bands for which the energy denominators are large.

5. The lowest energy interband transition is the one involving electrons at E_F. It is indicated as $\hbar\omega_s$ in Fig. P5.5.

6. Neglecting the perturbation terms, the bands are purely parabolic. Then,

$$\hbar\omega_s = \frac{\hbar^2}{2m_0}\left[(k_F - 2k_0)^2 - k_F^2\right] = E_F\left[\left(1 - \frac{2k_0}{k_F}\right)^2 - 1\right].$$

k_F is determined for free electrons, i.e.,

$$\frac{\Omega}{8\pi^3}\frac{4}{3}\pi k_F^3 = \frac{N}{2},$$

with $N = 1$ electron per atom, $\Omega = a^3/2$ (atomic volume), and then

$$k_F = (6\pi^2)^{1/3}/a = 3.90/a,$$

and

$$\boxed{\hbar\omega_s = 0.64 E_F}.$$

7. The absorption induced between bands $n = 0$ and $n = 1$ explains the increase in $\sigma_1(\omega)$ for $\hbar\omega > \hbar\omega_s = 0.64 E_F$.

Colour of Transition Metals

The interband transition $\hbar\omega_s$ is observed for different alkali elements, as shown in Fig. P5.6. It contributes to an absorption term that covers almost the whole visible spectrum. Indeed, E_F is small (1–3 eV in the series Li, K, Na) and the threshold below 2 eV. In addition, the absorption is not very selective, since the empty band is broad. The corresponding metals have no 'colour'.

On the other hand, the noble metals Cu, Ag, and Au are monovalent transition metals ($3d^{10}4s$, $4d^{10}5s$, $5d^{10}6s$). The d and s electron bands overlap, which leads to the situation shown schematically in Fig. P5.7. In this case, optical transitions between the d band and the Fermi level occur in the visible part of the spectrum, at energies below $\hbar\omega_s$ (here $E_F \sim 7$–8 eV). As the d electrons have a very high density of states (a narrow band containing 10 electrons), the absorption $\hbar\omega_d$ is intense. It appears just above 2 eV for Cu and Au, which explains their 'red' and 'yellow' colours, respectively. For silver, the d band is further from the Fermi level and $\hbar\omega_d \sim 4$ eV, so the whole spectrum is reflected.

Fig. P5.6 Interband transitions for three alkali elements. Adapted from [4, p. 296]

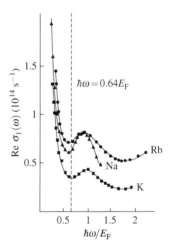

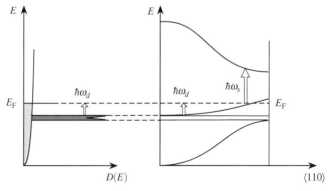

Fig. P5.7 Interband transitions for noble metals

Problem 6: One-Dimensional TTF-TCNQ Compounds[4]

Some organic compounds like TTF-TCNQ are made up of weakly coupled chains. In TTF-TCNQ, tetrathiafulvalene (TTF) and tetracyanoquinodimethane (TCNQ) chains are weakly bound. This gives these compounds a 1D aspect. The aim in this problem is to study some characteristic features of these 1D compounds.

6.1: Isolated Chains

In a very simple approximation, we shall describe these compounds as single chains of N identical molecules, each the same distance a from the next, as shown in Fig. P6.1. Let R_ℓ be the coordinate of the ℓth molecule, so $R_\ell = \ell a$. We assume the usual periodic boundary conditions. We calculate the band structure of each chain in the tight-binding approximation. The quantum states of the electrons in the last occupied orbital of molecule ℓ are modelled by a single non-degenerate orbital state $|R_\ell\rangle$ of energy E_0 with $\langle R_\ell | R_m \rangle = \delta_{\ell m}$. We assume that the Hamiltonian $\hat{H}$ of one electron in the chain can be written in the form

$$\hat{H} = \frac{p^2}{2m} + \sum_{\ell=1}^{N} V(r - R_\ell),$$

where $V(r - R_\ell)$ is the potential at site ℓ. We only take into account hopping integrals involving the two nearest neighbours. These are denoted as they were in Chap. 1 by $-t_0$ and $-t_1$ (with t_0 and t_1 positive).

1. Recall the general form $|\Psi_k\rangle$ for an eigenstate of this Hamiltonian together with the associated energy eigenvalue E_k. Sketch the shape of the dispersion curve in the first Brillouin zone.

2. Sketch the wave functions $|R_\ell\rangle$ with s symmetry, together with the potential contributing to the various hopping integrals entering the expression for E_k.

3. Assume that each molecule gives a conduction electron to the band. What can be deduced about the electrical transport properties of this chain? Find the Fermi wave vector k_F and the Fermi energy E_F.

4. Calculate the total electronic energy of the chain. Hint: It simplifies to integrate over k.

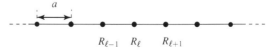

Fig. P6.1 Simple model for a molecular chain

---

[4] This problem has been designed with F. Rullier-Albenque.

6.2: Experimental Observations

The structure of TTF-TCNQ is shown in Fig. P6.2. The TTF and TCNQ molecules are stacked in the direction **b**, aligned with the chains. (The parameter b is identified with a in 6.1 and 6.3). The temperature dependence of the conductivity parallel to the TTF-TCNQ chains is shown in Fig. P6.3.

1. *What can be said about the electrical resistivity of this compound?*

Figure P6.4 shows topographical images obtained by scanning tunnelling microscopy of the surface of a TTF-TCNQ crystal in the plane ab at two different temperatures (70 and 36 K).

2. *On Fig. P6.4a, indicate the primitive cell at the surface. Do you recognise the material basis of Fig. P6.2?*

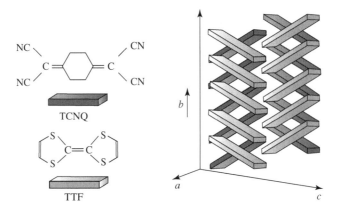

Fig. P6.2 Chemical structure of TTF-TCNQ (*left*) and three dimensional stacking of the molecules (*right*)

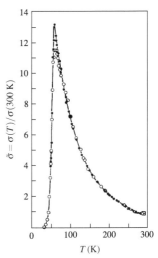

Fig. P6.3 Temperature dependence of the conductivity parallel to TTF-TCNQ chains. From Cohen, M.J., Coleman, L.B., Garito, A.F., Heeger, A.J.: Phys. Rev. B **10**, 1298 (1974). With the permission of the American Physical Society (© 1974 APS). http://link.aps.org/doi/10.1103/PhysRevB.10.1298

Problem 6: One-Dimensional TTF-TCNQ Compounds

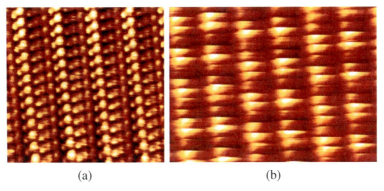

Fig. P6.4 Topographical images obtained by scanning tunnelling microscopy of the surface of a TTF-TCNQ crystal in the plane *ab* at two different temperatures: (**a**) 70 K and (**b**) 36 K. Images courtesy of Wang, Z.Z.: LPN, Marcoussis

3. The structure is radically different at 36 K. How would you characterise it along the chains? Across the chains?

6.3: Dimerised Chain

Although the experimental situation in Fig. P6.4b is more complicated, we shall use a simple model of structural deformation to describe changes in electronic properties. We assume that the chains dimerise at low temperatures, with each molecule moving a distance $\pm u$ as shown in Fig. P6.5.

1. What is the primitive cell for this structure and the size of the first Brillouin zone?

The aim now is to calculate the new band structure of this compound, still using the tight-binding approximation. The Hamiltonian for an electron in the chain can now be written

$$\hat{H} = \frac{p^2}{2m} + \sum_{\ell=1}^{N/2} \left[V(r - R_\ell^+) + V(r - R_\ell^-) \right].$$

2. Explain why solutions of the Hamiltonian will have the form

$$|\Psi_k\rangle = \sqrt{\frac{2}{N}} \sum_{\ell=1}^{N/2} e^{2ik\ell a} \left(a_k |R_\ell^+\rangle + b_k |R_\ell^-\rangle \right),$$

with $|a_k|^2 + |b_k|^2 = 1$.

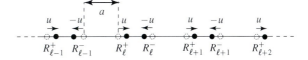

Fig. P6.5 Dimerisation of the chain

In order to find the functions $|\Psi_k\rangle$ and their associated energies E_k, we write down the eigenvalue equation $\hat{H}|\Psi_k\rangle = E_k|\Psi_k\rangle$ for the energies and project onto each of the states $|R_\ell^+\rangle$ and $|R_\ell^-\rangle$ in turn. We still assume that the only nonzero hopping integrals are those involving nearest neighbours, and use the notation

$$-t_0' = \langle R_\ell^+|V_\ell^+(r)|R_\ell^+\rangle = \langle R_\ell^-|V_\ell^-(r)|R_\ell^-\rangle,$$

$$-t_1' = \langle R_\ell^+|V_\ell^+(r)|R_\ell^-\rangle = \langle R_\ell^+|V_\ell^-(r)|R_\ell^-\rangle,$$

$$-t_1'' = \langle R_{\ell-1}^-|V_\ell^+(r)|R_\ell^+\rangle = \langle R_{\ell-1}^-|V_{\ell-1}^-(r)|R_\ell^+\rangle,$$

where $V_\ell^+ = V(r - R_\ell^+) + \sum_{m \neq l} V(r - R_m)$ and similarly for V_ℓ^-.

3. Show that we obtain a system of two equations in a_k and b_k, one of which is

$$a_k(E_0 - E_k - t_0') + b_k\left(-t_1' - t_1'' e^{-2ika}\right) = 0,$$

and derive an expression for the other.

4. Calculate the two energy eigenvalues E_k^1 and E_k^2 in terms of E_0, t_0', t_1', and t_1''. Sketch the band structure, specifying the values at the center and the edge of the Brillouin zone. What happens to the band structure when $u = 0$?

5. Would you say this result is compatible with the observations of Fig. P6.3?

For a more accurate analysis of the temperature dependence of the conductivity σ of TTF-TCNQ, Fig. P6.6 shows the logarithm of σ as a function of $1/T$ down to the lowest temperatures.

6. What can you say about the temperature dependence of the conductivity at the lowest temperatures?

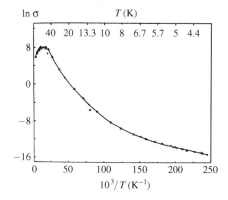

Fig. P6.6 Temperature dependence of the conductivity σ of TTF-TCNQ down to the lowest temperatures. From Ferraris, J., Cowan, D.O., Valatka, N., Jr., Perlstein, J.H.: J. Am. Chem. Soc. **95**, 948 (1973). With permission of the American Chemical Society, © 1973

6.4: Peierls Transition

In this section the aim is to investigate the stability of the dimerised state at $T = 0$. From now on, we consider only the nearest-neighbour contribution to the potentials V_ℓ.

1. *For small displacements u compared with a, justify the following approximations:*
   $t'_0 \approx t_0$, $-t'_1 \approx -t_1 - \alpha u$, and $-t''_1 \approx -t_1 + \alpha u$ with $\alpha > 0$.

2. *Calculate the change in the total electronic energy resulting from dimerisation and show that the gain in electron energy compared with the situation $u = 0$ can be written in the form*

$$\Delta E = -ANu^2 \left(\ln \frac{B}{u} - \frac{1}{2} \right),$$

specifying the values of A and B in terms of t_1 and α.

It is given that

$$\int_0^{\pi/2} \left[1 - (1-z^2)\sin^2 x \right]^{1/2} dx = 1 + \frac{1}{2}z^2 \left(\ln \frac{4}{|z|} - \frac{1}{2} \right),$$

for small z.

3. *Assuming that the atoms are coupled harmonically by an elastic force with stiffness constant K, calculate the static deformation energy of the chain as a function of N, K, and u.*

4. *Discuss the instability of the ion–electron system with regard to dimerisation of the chain. Find the static displacement u_{eq} at equilibrium as a function of K, t_1, and α.*

In TTF-TCNQ, charge transfer between chains is such that a chain carries 0.50 electrons per molecule rather than 1 as assumed above.

5. *Can you explain the structure actually observed in Fig. P6.4b?*

Solution

Isolated Chain

1. According to Chap. 1, we have

$$|\psi_k\rangle = \frac{1}{\sqrt{N}} \sum_{\ell=1}^{N} e^{ik\ell a} |R_\ell\rangle ,$$

and for all ℓ,

$$\left[\frac{p^2}{2m} + V(r - R_\ell)\right] |R_\ell\rangle = \hat{H}_0 |R_\ell\rangle = E_0 |R_\ell\rangle ,$$

$$-t_0 = \langle R_\ell | V_\ell(r) | R_\ell \rangle ,$$

$$-t_1 = \langle R_{\ell-1} | V_\ell(r) | R_\ell \rangle = \langle R_\ell | V_\ell(r) | R_{\ell+1} \rangle ,$$

where

$$V_\ell(r) = \sum_{m=1, m \neq \ell} V(r - R_m) .$$

As shown in Chap. 1, the dispersion curve in the first Brillouin zone has the form

$$E_k = E_0 - t_0 - 2t_1 \cos ka ,$$

shown in Fig. P6.7.

2. From the definition of $-t_0$, the wave function $\phi(r - R_\ell)$ and the potential $V_\ell(r)$ come into the calculation of the transfer integral, where $\langle r | R_\ell \rangle = \phi(r - R_\ell)$. The representation in Fig. P6.8 is similar to the one in Fig. 1.3c. For the integral $-t_1$, the wave functions at two nearest neighbour sites enter as shown in Fig. P6.9.

3. In k space, each state occupies a 'volume' $2\pi/Na$, and taking the spin degeneracy into account, the band contains $2N$ states. If each molecule gives 1 electron to the band, it is thus *half-filled*. We then find $k_F = \pi/2a$ and $E_F = E_0 - t_0$ (see Fig. P6.7). We therefore expect *metallic behaviour*.

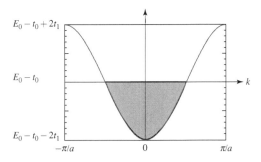

Fig. P6.7 Dispersion curve and occupied states in first Brillouin zone

Problem 6: One-Dimensional TTF-TCNQ Compounds

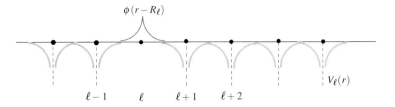

Fig. P6.8 Elements used to calculate $-t_0$

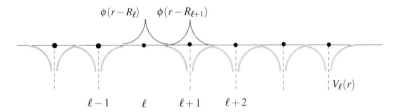

Fig. P6.9 Elements used to calculate $-t_1$

4. The total electronic energy of the system is

$$E = 2 \int_{-\pi/2a}^{\pi/2a} E(k) \times \frac{dk}{2\pi/Na} = N\left(E_0 - t_0 - 4\frac{t_1}{\pi}\right).$$

Experimental Observations

1. In Fig. P6.11, for the temperature range 60–300 K, the conductivity goes down as the temperature goes up (or the resistivity increases with temperature). This is characteristic of metallic behaviour. Below 60 K, the opposite kind of behaviour is observed. The system becomes an insulator.

2. Figure P6.12 is a vertical cross-section in the plane $\|ab$ of the structure shown in Fig. P6.10. The TTF or TCNQ chains are stacked in the b direction. One protrudes from the surface, while the other is indented. Several possible primitive cells are drawn on Fig. P6.12a. Unit cell 1 is also reproduced in Fig. P6.10.

3. At low temperatures, the surface topography observed by STM changes radically. The surface A corresponding to the high-temperature primitive cell is easy to make out. However, the periodicity of the structure is totally different. Along the chains, the period is much longer, apparently multiplied by a factor of 4 (example B). In the direction transverse to the chains, the period appears to be doubled, but in fact more careful observation shows that it is in fact multiplied by 4 (example C) (see the comments at the end of the problem).

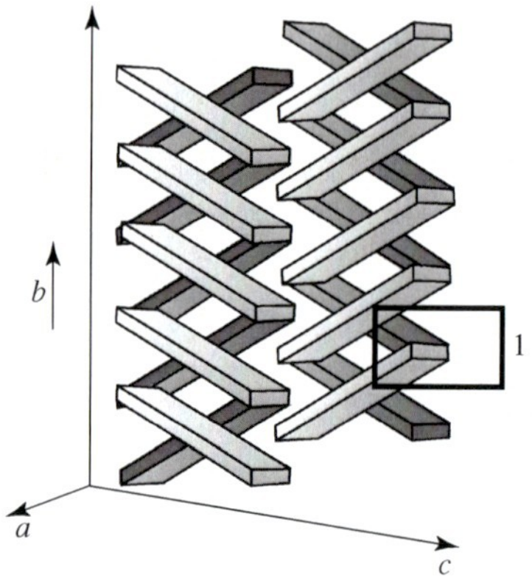

Fig. P6.10 Structure of TTF-TCNQ, showing a primitive cell on the surface area corresponding to Fig. P6.4a

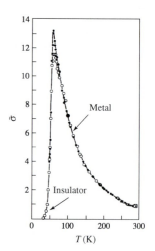

Fig. P6.11 Temperature dependence of conductivity, displaying the metallic and insulating temperature regimes

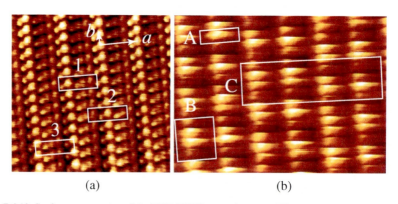

Fig. P6.12 Surface topography of the TTF-TCNQ crystal at two different temperatures: (**a**) 70 K and (**b**) 36 K, showing several possible primitive cells (see text)

Dimerised Chain

1. The new unit cell in Fig. P6.13 has length $2a$. The Brillouin zone is the line segment $[-\pi/2a, +\pi/2a]$.

2. The most general form of the wave function is

$$|\psi_k\rangle = \sqrt{\frac{2}{N}} \sum_{\ell=1}^{N/2} \left(a_{k\ell} |R_\ell^+\rangle + b_{k\ell} |R_\ell^-\rangle \right),$$

with $|a_{k\ell}|^2 + |b_{k\ell}|^2 = 1$. Applying Bloch's theorem with period $2a$, the two coefficients $a_{k\ell}$ and $b_{k\ell}$ can be written in the form

$$\begin{cases} a_{k\ell} = a_k e^{2ik\ell a}, \\ b_{k\ell} = b_k e^{2ik\ell a}, \end{cases} \quad \text{with} \quad |a_k|^2 + |b_k|^2 = 1,$$

whence the proposed form of $|\psi_k\rangle$.

3. The eigenvalue equation is

$$\hat{H}_k |\psi_k\rangle = \hat{E}_k |\psi_k\rangle.$$

Picking out the site R_ℓ^+, we may write $\hat{H}_k$ in the form

$$\hat{H}_k = \frac{\hat{p}^2}{2m} + V(r - R_\ell^+) + V_\ell^+(r) = \hat{H}_0 + V_\ell^+(r).$$

The projection onto the ket $|R_\ell^+\rangle$ then yields

$$\langle R_\ell^+ | \hat{H}_0 + V_\ell^+(r) | \sum_{m=1}^{N/2} e^{2ikma} \left(a_k |R_m^+\rangle + b_k |R_m^-\rangle \right)$$

$$= E_k \langle R_\ell^+ | \sum_{m=1}^{N/2} e^{2ikma} \left(a_k |R_m^+\rangle + b_k |R_m^-\rangle \right),$$

or

$$a_k \left[e^{2ikla} E_0 + e^{2ikla}(-t_0') \right] + b_k \left[e^{2ikla}(-t_1') + e^{2ik(\ell-1)a}(-t_1'') \right] = a_k E_k e^{2ikla}.$$

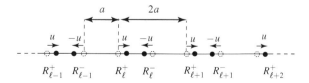

Fig. P6.13 Dimerised chain

Therefore,

$$\boxed{a_k(E_0 - t'_0 - E_k) + b_k(-t'_1 - t''_1 e^{-2ika}) = 0}.$$

Likewise, the projection onto $|R^-_\ell\rangle$ yields

$$\boxed{a_k(-t'_1 - t''_1 e^{+2ika}) + b_k(E_0 - t'_0 - E_k) = 0}.$$

4. The system of two equations in two unknowns only has nonzero solution if its determinant vanishes:

$$(E_0 - t'_0 - E_k)^2 - (+t'_1 + t''_1 e^{-2ika})(t'_1 + t''_1 e^{-2ika}) = 0,$$

that is, if

$$E_k^{1,2} = E_0 - t'_0 \pm \sqrt{t'^2_1 + t''^2_1 + 2t'_1 t''_1 \cos 2ka},$$

or

$$E_k^{1,2} = E_0 - t'_0 \pm \sqrt{(t'_1 + t''_1)^2 - 4t'_1 t''_1 \sin^2 ka}.$$

We conclude that

$$\begin{cases} E_k^{1,2} = E_0 - t'_0 \pm (t'_1 + t''_1), & \text{for } k = 0, \\ E_k^{1,2} = E_0 - t'_0 \pm |t'_1 - t''_1|, & \text{for } k = \pm \pi/2a. \end{cases}$$

A band gap thus appears at the edge of the Brillouin zone (see Fig. P6.14). When $u = 0$, we have $t'_0 = t_0$ and $t'_1 = t''_1 = t_1$. Then $E_k^{1,2} = E_0 - t_0 \pm 2t_1|\cos ka|$, and the band structure is as shown in Fig. P6.15 in the Brillouin zone $[-\pi/2a, +\pi/2a]$. This is indeed the band structure found in question 5 of 6.3 in the case of a non-dimerised chain, where the portions corresponding to line segments $[\pi/2a, \pi/a]$ and $[-\pi/2a, -\pi/a]$ have been translated by $-\pi/a$ and $+\pi/a$, respectively.

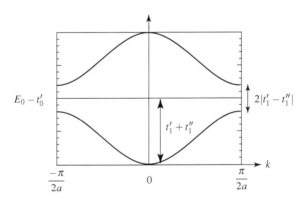

Fig. P6.14 Band gap at the edge of the Brillouin zone

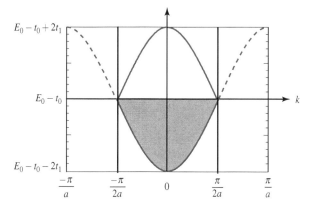

Fig. P6.15 Band structure when $u = 0$

5. When each molecule contributes one electron to the band, the lower band is fully occupied and the upper band is empty. We thus expect insulating behaviour in the case of the dimerised chain. This agrees qualitatively with Fig. P6.11, where a *metal–insulator transition* occurs at 60 K, associated with the structural deformation.

6. For an insulator or semiconductor, the conductivity is always given by

$$\sigma = n_e \frac{e^2 \tau_e}{m_e} + n_h \frac{e^2 \tau_h}{m_h},$$

where n_e and n_h are the electron density in the conduction band and the hole density in the valence band, respectively. The temperature dependence of σ is governed at low T by those of n_e and n_h, which are determined by statistical thermodynamics. This generally leads to an activated temperature dependence for σ, i.e., going as $\exp(-\delta E/k_B T)$, where δE is related to the width of the band gap.

Note: The electron–hole symmetry of the band structure requires a Fermi level in the middle of the band gap E_g. According to (3.56), for $k_B T \ll E_F$,

$$n_e = n_h = N_c(T) \exp(-E_g/2k_B T),$$

where $N_c(T)$ is determined by the effective electron mass, the same in this case as the effective hole mass. [Replace m_0 by m_e in (3.57).] This therefore leads to

$$\sigma = \sigma_0 \exp(-E_g/2k_B T).$$

At low temperatures, the slope of $\ln \sigma$ as a function of $1/T$ can thus be used to estimate $E_g \sim 60$ K (see Fig. P6.16). In fact, since there are many defects in compounds like TTF-TCNQ, and these cause extrinsic doping, the value of E_g obtained in this way is lower than the actual band gap.

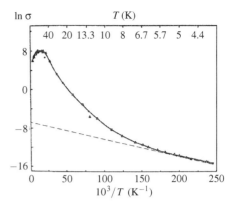

Fig. P6.16 Estimating the band gap from the graph of $\ln\sigma$ vs. $1/T$

Peierls Transition

1. When only nearest neighbours are considered in the definition of V_ℓ, the hopping integrals $-t'_0$, $-t'_1$, and $-t''_1$ can be written in the form

$$-t'_0 = \langle R_\ell^+ | V(r - R_{\ell-1}^-) + V(r - R_\ell^-) | R_\ell^+ \rangle,$$

$$-t'_1 = \langle R_\ell^- | V(r - R_{\ell-1}^-) + V(r - R_\ell^-) | R_\ell^+ \rangle,$$

$$-t''_1 = \langle R_{\ell-1}^- | V(r - R_{\ell-1}^-) + V(r - R_\ell^-) | R_\ell^+ \rangle.$$

Compared with the non-dimerised situation ($u = 0$, distance between atoms a), the overlap integrals now involve either smaller distances ($a - u$) or larger distances ($a + u$). If u is small, the overlap integrals can be estimated by a truncated expansion to first order in u about the non-dimerised situation.

For t'_0, to first order in u, the contributions of the two potentials centered at ℓ^- and $(\ell - 1)^-$ will cancel, and we obtain $t'_0 = t_0$. For t'_1 and t''_1, it is the contributions from the wave functions centered either at ℓ^+ and ℓ^- (for $-t'_1$), or at ℓ^+ and $(\ell - 1)^-$ (for $-t''_1$), which really matter, and we may write

$$\begin{cases} -t'_1 = -t_1 - \alpha u, \\ -t''_1 = -t_1 + \alpha u. \end{cases}$$

For the energies $E_k^{1,2}$, we then obtain

$$\boxed{E_k^{1,2} = E_0 - t_0 \pm 2t_1 \sqrt{1 - \left(1 - \frac{\alpha^2 u^2}{t_1^2}\right) \sin^2 ka}\,.}$$

Problem 6: One-Dimensional TTF-TCNQ Compounds

2. The total electronic energy of the system is written as in question 4 of 6.1:

$$E = 4 \int_0^{\pi/2a} E^1(k) \frac{dk}{2\pi/Na}.$$

Therefore,

$$E = N(E_0 - t_0) - \frac{4Na}{\pi} t_1 \int_0^{\pi/2a} \sqrt{1 - \left(1 - \frac{\alpha^2 u^2}{t_1^2}\right) \sin^2 ka}\, dk.$$

Setting $z = \alpha u/t_1$ and making the change of variable $x = ka$, we obtain

$$E = N(E_0 - t_0) - \frac{4N}{\pi} t_1 \int_0^{\pi/2} \sqrt{1 - (1 - z^2)\sin^2 x}\, dx.$$

For small z (small u), this is equal to

$$E = N(E_0 - t_0) - \frac{4N}{\pi} t_1 \left[1 + \frac{1}{2}\frac{\alpha^2 u^2}{t_1^2}\left(\ln\frac{4t_1}{\alpha u} - \frac{1}{2}\right)\right].$$

The gain ΔE in electronic energy compared with the non-dimerised situation is thus equal to

$$\Delta E = -\frac{2N}{\pi}\frac{\alpha^2 u^2}{t_1}\left(\ln\frac{4t_1}{\alpha u} - \frac{1}{2}\right).$$

Setting $A = 2\alpha^2/\pi t_1$ and $B = 4t_1/\alpha$, the gain in energy ΔE is given by

$$\boxed{\Delta E = -ANu^2 \left(\ln\frac{B}{u} - \frac{1}{2}\right).}$$

3. If the atoms are harmonically coupled by an elastic force, the deformation energy due to dimerisation is

$$E_d = \frac{1}{2}K\left[\sum_{\ell=1}^{N/2}(R_\ell^+ - R_\ell^- - a)^2 + \sum_{\ell=1}^{N/2}(R_{\ell+1}^+ - R_\ell^- - a)^2\right],$$

that is

$$\boxed{E_d = 2NKu^2.}$$

4. The change in the total energy of the system compared with the non-dimerised situation is therefore equal to

$$\Delta E_T = -ANu^2 \left(\ln\frac{B}{u} - \frac{1}{2}\right) + 2NKu^2.$$

Stability is achieved for a displacement u_{eq} such that $\partial \Delta E_T/\partial u = 0$, i.e., for

$$u_{eq} = \frac{B}{e}\exp\left(-\frac{2K}{A}\right).$$

or

$$\boxed{u_{eq} = \frac{4t_1}{\alpha e}\exp\left(-\frac{\pi K t_1}{\alpha^2}\right).}$$

5. We showed above that, for a chain with one electron per molecule, the electron–ion system could be unstable in its ground state with respect to a distortion that produces a band gap at the Fermi level. In this particular case, the distortion results from dimerisation.

Polyacetylene provides a simple example of an insulating polymer due to dimerisation of the polymer chain. It is only by doping that it becomes a conducting compound.

If the occupation of the chain is x electrons per molecule, the corresponding Fermi wave vector is $k_F \neq \pi/2a$. A gap can appear at k_F, allowing a gain in electronic energy, and it is then associated with a distortion of period π/k_F. When $x < 1$, $\pi/k_F > 2a$, and this explains the period of about $4a$ observed in Fig. P6.12b.

Note: This is the general case of the so-called Peierls transition, named after the physicist who first described it. Note that, for an occupation $x \neq 1$, $\pi/k_F a$ is not necessarily an integer. A structure can be obtained whose period is not commensurate with the period of the original lattice, which corresponds to the structure observed in Fig. P6.12b. In the 3D compound, the distortions of the various chains can accommodate a quasi-ordering in 3D. This explains the periodicity in the direction transverse to the chains in Fig. P6.12b.

Problem 7: Insulator–Metal Transition

The aim here is to identify the parameters determining whether a material is conducting or insulating. In 7.1, we apply the tight-binding (LCAO) method to a crystal of hydrogen-like atoms to determine the band width of the electronic states arising from the localised atomic state. In 7.2, the important role played by interactions between electrons will be examined for the simple example of the hydrogen molecule. A condition for the solid described in 7.1 to be metallic is then obtained. It will be compared with experimental observations of hydrogen and the alkali elements in 7.3. A full analogy with doped semiconductors will be made in 7.4. We can then understand how the physical properties of these materials can change when the donor concentration is increased.

The calculations in questions 4 and 5 of 7.1 are useful for understanding the rest of the problem, but not absolutely essential. In 7.2, the problem can be resumed at question 5 if the reader has not fully processed the preceding questions. 7.4 is independent of 7.2. The reader need only understand the physical meaning of 7.1 and 7.2 to be able to answer.

7.1: Tight-Binding Method for Hydrogen-Like Orbitals

We consider a crystalline solid comprising a stack of monovalent atoms arranged in a body-centered cubic structure with conventional unit cell of side b. It is assumed that the electronic energy bands obtained from inner electron shells are extremely narrow and well separated from the band arising from electrons of the outermost s shell of the atom. The latter can be described by the tight-binding approximation. If ε is the eigenenergy and $|\phi(\mathbf{r})\rangle$ the wave function of the s state corresponding to the isolated atom, we look for a crystal wave function of the form

$$|\psi_{\mathbf{k}}(\mathbf{r})\rangle = \sum_j A_{j\mathbf{k}} |\phi(\mathbf{r} - \mathbf{R}_j)\rangle ,$$

where the sum over j runs over all sites of the crystal.

1. *Specify the primitive cell of the lattice. What is its volume?*

It is assumed that:

(a) The potential seen by an electron in the crystal is the sum of the atomic potentials $V(\mathbf{r} - \mathbf{R}_j)$.
(b) $\langle \varphi(\mathbf{r} - \mathbf{R}_i) | \varphi(\mathbf{r} - \mathbf{R}_j) \rangle = \delta_{ij}$.
(c) The only terms $\langle \varphi(\mathbf{r} - \mathbf{R}_j) | V(\mathbf{r} - \mathbf{R}_\ell) | \varphi(\mathbf{r} - \mathbf{R}_m) \rangle$ that are nonzero are those involving two nearest-neighbour sites j and m:

$$-\alpha = \langle \varphi(\mathbf{r} - \mathbf{R}_j) | V(\mathbf{r} - \mathbf{R}_m) | \varphi(\mathbf{r} - \mathbf{R}_j) \rangle$$
$$-\gamma = \langle \varphi(\mathbf{r} - \mathbf{R}_j) | V(\mathbf{r} - \mathbf{R}_j) | \varphi(\mathbf{r} - \mathbf{R}_m) \rangle .$$

2. *Specify the eigenenergy E_k associated with the wave function $|\psi_k(\mathbf{r})\rangle$.*

In the case of hydrogen atoms, the only ones considered up to the end of 7.2: *Interactions Between Electrons*, the corresponding values of $V(\mathbf{r})$, $\varphi(\mathbf{r})$, ε are

$$V(\mathbf{r}) = -\frac{q^2}{4\pi\varepsilon_0 r} = -\frac{e^2}{r},$$

$$\varphi(\mathbf{r}) = \frac{1}{(\pi a_0^3)^{1/2}} \exp\left(-\frac{r}{a_0}\right),$$

$$\varepsilon = -e^2/2a_0,$$

where a_0 is the Bohr atomic radius and q the elementary charge.

3. *Express α and γ as integrals involving a_0 and distances r_A and r_B between electrons at two nearest neighbour sites A and B.*

The quantity γ is found analytically to be

$$\gamma = \frac{e^2}{a_0}\left(1 + \frac{R}{a_0}\right)\exp(-R/a_0), \qquad R = |\mathbf{AB}|.$$

The quantity α is obtained analytically by noting that it corresponds to the potential energy of a point charge q placed at B, subjected to the effects of a spherical charge distribution with electrostatic density $q\varphi(r_A)^2$ centered at A, for which the electrostatic potential is easily found. This can be done by decomposing the charge distribution into two judiciously chosen parts.

4. *Find α in terms of R and a_0.*

5. *Write $-\alpha$ as the sum of two terms, one of which exactly cancels the Coulomb interaction between the nuclear charges. Check that the second term of α can be neglected in comparison to γ in the limit $R/a_0 \gg 1$, and the assumption (b) above only actually applies in this limit.*

6. *Find the width W of the resulting band of electronic states as a function of R and a_0.*

7. *What is the Fermi level at $T = 0$ K? Is the resulting state metallic or insulating? Does it depend on b?*

7.2: Interactions Between Electrons

The result obtained above does not accord with physical reality. This is because we completely neglected the Coulomb interactions between the electrons. We shall illustrate this effect in the simple case of a hydrogen-like molecule, comprising two

Problem 7: Insulator–Metal Transition

atoms in which the protons are located at $\mathbf{r}_A$ and $\mathbf{r}_B$ and the electrons at $\mathbf{r}_1$ and $\mathbf{r}_2$. The total electronic Hamiltonian for this system is

$$\hat{H} = \frac{p_1^2}{2m} + V_{1A} + V_{1B} + \frac{p_2^2}{2m} + V_{2B} + V_{2A} + V_{12},$$

where

$$V_{1A} = -\frac{e^2}{|\mathbf{r}_1 - \mathbf{r}_A|}, \quad V_{1B} = -\frac{e^2}{|\mathbf{r}_1 - \mathbf{r}_B|}, \quad V_{2A} = -\frac{e^2}{|\mathbf{r}_2 - \mathbf{r}_A|}, \quad V_{2B} = -\frac{e^2}{|\mathbf{r}_2 - \mathbf{r}_B|},$$

and

$$V_{12} = +\frac{e^2}{|\mathbf{r}_1 - \mathbf{r}_2|}.$$

Since $\hat{H}$ is independent of the electron spins, the eigenfunctions can be written as a product of space and spin wave functions. For the two-electron spatial wave function, we could adopt the basis

$$|\varphi_{1A}\rangle \otimes |\varphi_{2B}\rangle, \quad |\varphi_{1B}\rangle \otimes |\varphi_{2A}\rangle, \quad |\varphi_{1A}\rangle \otimes |\varphi_{2A}\rangle, \quad |\varphi_{1B}\rangle \otimes |\varphi_{2B}\rangle,$$

with the obvious notation $|\varphi_{1B}\rangle = |\varphi(\mathbf{r}_1 - \mathbf{r}_B)\rangle$. However, it makes more sense to note that $\hat{H}$ is invariant under exchange of the protons A and B, effected by the operator $\hat{T}_{AB}$, which implies that $\hat{H}$ and $\hat{T}_{AB}$ have a common basis of eigenfunctions. Therefore, the eigenfunctions of $\hat{H}$ are either symmetric or antisymmetric under exchange of A and B, and we may choose the following basis of spatial wave functions:

$$|\Phi_1\rangle = \frac{1}{\sqrt{2}}\left(|\varphi_{1A}\rangle \otimes |\varphi_{2B}\rangle + |\varphi_{1B}\rangle \otimes |\varphi_{2A}\rangle\right),$$

$$|\Phi_2\rangle = \frac{1}{\sqrt{2}}\left(|\varphi_{1A}\rangle \otimes |\varphi_{2B}\rangle - |\varphi_{1B}\rangle \otimes |\varphi_{2A}\rangle\right),$$

$$|\Phi_3\rangle = \frac{1}{\sqrt{2}}\left(|\varphi_{1A}\rangle \otimes |\varphi_{1A}\rangle + |\varphi_{1B}\rangle \otimes |\varphi_{2B}\rangle\right),$$

$$|\Phi_4\rangle = \frac{1}{\sqrt{2}}\left(|\varphi_{1A}\rangle \otimes |\varphi_{2A}\rangle - |\varphi_{1B}\rangle \otimes |\varphi_{2B}\rangle\right).$$

Furthermore, since electrons are fermions, the Pauli exclusion principle requires the total wave function, including space and spin components, to be antisymmetric under exchange of electrons 1 and 2.

1. Which of the functions $|\Phi_1\rangle$, $|\Phi_2\rangle$, $|\Phi_3\rangle$, and $|\Phi_4\rangle$ are associated with a spin singlet state, and which with a spin triplet state? Let $|\Phi_n^M\rangle$ be the total wave functions thus obtained, with $M = S$ for a singlet and $M = T$ for a triplet.

2. From the properties given in 7.2: Interactions Between Electrons, explain without calculation why the states obtained in question 1 are eigenstates of $\hat{H}$.

The aim now is to find the eigenvalues and eigenstates of $\hat{H}$ when $R \gg a_0$ (with $\mathbf{R} = \mathbf{r}_B - \mathbf{r}_A$). As we have seen in 7.1, we will then be able to neglect integrals of type α and retain only those of type γ. Moreover, among all the matrix elements of V_{12}, we shall retain only the biggest, involving two electrons on the same atom. The corresponding energy, called the intra-atomic Coulomb energy, is

$$U = \langle \varphi_{1A}| \otimes \langle \varphi_{2A}|V_{12}|\varphi_{1A}\rangle \otimes |\varphi_{2A}\rangle.$$

3. *Show in this limit that the eigenstates found in question 2 have eigenenergies 2ε and $2\varepsilon + U$, and that the 2×2 matrix representing $\hat{H}$ in the subspace spanned by the other two states is*

$$\begin{pmatrix} 2\varepsilon + U & -2\gamma \\ -2\gamma & 2\varepsilon \end{pmatrix}.$$

4. *Specify the energies of the two-electron states. Note that, in the limit $R \to \infty$, only two levels remain. What eigenstates do they correspond to?*

Including the Coulomb interaction V_{12} between the electrons therefore disallows double occupation of an atomic orbital in the ground state. We may thus apply an independent electron description of the kind discussed in 7.1, provided that the spin degeneracy is removed in the one-electron state of energy ε, giving rise to a state of energy ε and an excited state of energy $\varepsilon + U$. In the hydrogen molecule, the effect of the γ terms is, as in 7.1, to remove the degeneracy in each of these states. Extending this remark to a crystal, it can be shown that the electron levels at ε and $\varepsilon + U$ give rise to energy bands with the width W determined in question 6 of 7.1.

5. *Assuming this result, under what conditions is the solid insulating? When does it become metallic?*

In order to make a quantitative estimate of when the insulator–metal transition will occur, we must estimate U.

6. *Express U in its integral form for hydrogen. Use the result for α in question 4 of 7.1 to show that $U = 5e^2/8a_0$.*

7. *Check that the metal–insulator transition occurs when the body-centered cubic solid has a number n_{IM} of atoms per unit volume given by*

$$n_{IM}^{1/3} a_0 = C,$$

where C is a constant of order unity to be determined.

8. *Show that C depends little on the chosen crystal structure by considering the case of a simple cubic crystal.*

7.3: Alkali Elements and Hydrogen

The method of 7.1 and 7.2 applies to monovalent atoms, such as the alkali elements, but the numerical result found in question 7 of 7.2 does not apply directly. Indeed, the valence electrons of the alkali elements are in ns states where, e.g., $n = 2$ for Li, $n = 3$ for Na. The corresponding radial wave functions φ_n do decrease exponentially at long range, but become small for r close to zero, because the s electron barely penetrates the inner electron shells. One can use a pseudopotential approximation, with an infinite potential for $r < R_c$, where the ionic radius R_c increases with the atomic number. It is not difficult to see that, if the wave function is approximated by $\varphi_n(r) = \text{const.} \times \exp\left[-(r - R_c)/a_0\right]$ for $r > R_c$, the condition of question 7 of 7.2 is altered to a first approximation by replacing R by $R - 2R_c$.

1. *Use the values of R_c and the lattice constants of the body-centered crystal lattices of the various alkali elements given in Table 13.1 to decide whether this model suggests the metallic behaviour observed for these crystals.*

Hydrogen solidifies at low temperatures ($T \sim 15$ K) into an insulating *molecular* crystal. The hydrogen molecules are located at the lattice points of a face-centered cubic lattice with lattice constant 5.4 Å.

2. *Calculate the number of hydrogen atoms n_0 per unit volume in the solid state. Can one explain why this phase should be insulating?*

3. *Can you imagine how one could obtain metallic solid hydrogen? In your opinion, under what conditions could certain alkali elements become insulating?*

Table 13.1 Ionic radius R_c and lattice constant for various alkali elements

	Li	Na	K	Rb	Cs
R_c (Å)	1.72	2.08	2.57	2.75	2.98
a (Å)	3.49	4.23	5.23	5.59	6.05

7.4: Insulator–Metal Transition in Si–P

The model exposed in 7.1 and 7.2 is difficult to test in detail for hydrogen or the alkali elements. We shall see, however, that it applies to doped semiconductors with a few minor modifications. Indeed, if we consider silicon doped with phosphorus, at zero temperature the phosphorus atom (valence 5) substituting a silicon atom (valence 4) traps its excess electron in a hydrogen-like orbit with an energy level below the conduction band. The ionisation energy of the donor is the energy of the electron in its enormous Bohr orbit. If a large number n_d of P atoms are introduced per unit volume, we obtain a disordered system of hydrogen-like orbits. Assuming that this disordered state can be modelled by a cubic crystal with lattice constant equal to the average distance between donors, we expect the model of 7.1 and 7.2 to be applicable.

1. *Adapt the condition obtained in question 8 of 7.2 to the case of silicon doped with phosphorus.*

2. *What band structure would you expect for different concentrations n_d of phosphorus? Sketch the energy dependence of the electron densities of states for Si–P in the three cases $n_d \ll n_{IM}$, $n_d \sim n_{IM}$, and $n_d \gg n_{IM}$. Indicate the position of the Fermi level at zero temperature.*

The resistivity of Si–P has been carefully measured for various phosphorus concentrations. It can be seen from Fig. P7.1 that the behaviour changes significantly for different values of n_d.

3. *Find the concentration n_{IM} from the experimental results. Given that the dielectric constant of Si is $\varepsilon = 11.75$ and the effective mass is $m_e/m_0 = 0.33$, compare the experimental value of C with the one obtained in question 8 of 7.2.*

In the following, the aim will be to understand the physical behaviour observed in the two extreme cases $n_d \ll n_{IM}$ and $n_d \gg n_{IM}$.

4. *From Fig. P7.1, deduce the order of magnitude of the electron collision time τ at low temperatures in the most metallic case. What could be the physical origin of the observed temperature dependence? Why is it less than in pure metals?*

5. *What characteristic energy of the band structure in question 2 can be deduced by analysing ρ in the insulating case? Estimate the order of magnitude and compare with the ionisation energy of phosphorus.*

6. *How would you expect this energy to depend on the concentration n_d?*

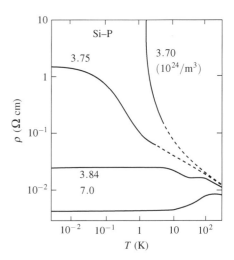

Fig. P7.1 Temperature dependence of the resistivity (logarithmic scales) for four samples of Si–P. Here n_d is given in units of 10^{24} atoms/m^3. From Rosenbaum et al.: Phys. Rev. B **27**, 7609 (1983). With the permission of the American Physical Society © 1983. http:/link.aps.org/doi/10.1103/PhysRevB.27.7509

Fig. P7.2 Magnetic susceptibility (per kg of Si) measured at three different temperatures as a function of the concentration n_d (in atoms per m^3). Logarithmic scales. Adapted from Alloul, H., Dellouve, P.: J. de Physique **C8-49**, 1185 (1988)

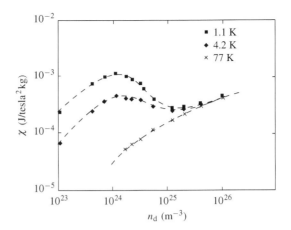

The insulator–metal transition is also accompanied by a major change in the electronic magnetic properties. This is shown in Fig. P7.2, which gives the temperature dependence of the electronic magnetic susceptibility χ. Very different behaviour is observed in the cases $n_d \gg n_{IM}$ and $n_d \ll n_{IM}$.

7. *In the case $n_d \ll n_{IM}$, what magnetic behaviour is expected? Give the expected expression for χ. In Fig. P7.2, plot the calculated expression for the three temperatures $T = 1.1$ K, $T = 4.2$ K, and $T = 77$ K. For what limiting case do the results accord with observation?* The density of Si is 2.33×10^3 kg/m^3.

In the limiting case $n_d \gg n_{IM}$, we may assume that the impurity and conduction bands merge into a single parabolic band.

8. *Give the expected expression for the magnetic susceptibility in this case. How does it depend on n_d and T? From what concentration n_d is this assumption valid?*

Solution

Tight-Binding Method for Hydrogen-Like Orbitals

1. The cubic conventional cell of the body-centered cubic lattice contains two lattice nodes. The Bravais lattice can be constructed in infinitely many different ways, e.g., specifying the three basis vectors (see Fig. P7.3)

$$\mathbf{b}_1 = b\mathbf{x}, \qquad \mathbf{b}_2 = b\mathbf{y}, \qquad \mathbf{b}_3 = \frac{b}{2}(\mathbf{x}+\mathbf{y}+\mathbf{z}).$$

The volume of the primitive cell is $b^3/2$. A more symmetrical cell is given by

$$\mathbf{b}'_1 = \frac{b}{2}(\mathbf{y}+\mathbf{z}-\mathbf{x}), \qquad \mathbf{b}'_2 = \frac{b}{2}(\mathbf{z}+\mathbf{x}-\mathbf{y}), \qquad \mathbf{b}'_3 = \frac{b}{2}(\mathbf{x}+\mathbf{y}-\mathbf{z}).$$

2. E_k is found exactly as done in Chap. 3. For a crystal with N atoms, this gives

$$|\psi_\mathbf{k}(\mathbf{r})\rangle = N^{-1/2} \sum_j e^{i\mathbf{k}\cdot\mathbf{R}_j} |\varphi(\mathbf{r}-\mathbf{R}_j)\rangle,$$

$$E_k = \langle\psi_k(\mathbf{r})|\hat{H}|\psi_k(\mathbf{r})\rangle = \varepsilon - \sum_{\substack{j' \text{ neighbour} \\ \text{of } j}} \alpha - \gamma \sum_{\substack{j' \text{ neighbour} \\ \text{of } j}} e^{i\mathbf{k}\cdot(\mathbf{R}_{j'}-\mathbf{R}_j)}.$$

In the body-centered cubic lattice, an atom has 8 neighbours (the centres of the cubes for which this atom occupies a vertex). The coefficient of γ is then

$$\sum_{\varepsilon_x=\pm\varepsilon_y=\pm\varepsilon_z=\pm 1} \exp\left[i\frac{b}{2}(\varepsilon_x k_x + \varepsilon_y k_y + \varepsilon_z k_z)\right] = 8\cos\left(k_x\frac{b}{2}\right)\cos\left(k_y\frac{b}{2}\right)\cos\left(k_z\frac{b}{2}\right),$$

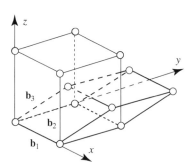

Fig. P7.3 Constructing the Bravais lattice of the bcc crystal

and then

$$E_k = \varepsilon - 8\alpha - 8\gamma \cos\left(k_x \frac{b}{2}\right)\cos\left(k_y \frac{b}{2}\right)\cos\left(k_z \frac{b}{2}\right), \quad (13.4)$$

where **k** belongs to the first Brillouin zone. Here γ and 8α correspond to t_1 and t_0, respectively, in Sect. 3.2.3. The reciprocal lattice of the body-centered cubic lattice is face-centered cubic with conventional cell of side $4\pi/b$, and the first Brillouin zone is shown in Fig. P4.4 (right) (see Problem 4: *Electronic Energy and Stability of Alloys*).

3. $R = b\sqrt{3}/2$ is the distance between nearest neighbours. Let A and B be two atoms at this distance R. With the notation in Fig. P7.4,

$$\alpha = \frac{1}{\pi a_0^3} \int \exp\left(-\frac{r_A}{a_0}\right) \frac{e^2}{r_B} \exp\left(-\frac{r_A}{a_0}\right) d^3 r_A$$

$$= \frac{e^2}{\pi a_0^3} \int \exp\left(-\frac{2r_A}{a_0}\right) \frac{1}{r_B} d^3 r_A . \quad (13.5)$$

$$\gamma = \frac{e^2}{\pi a_0^3} \int \exp\left(-\frac{r_A}{a_0}\right) \frac{e^2}{r_A} \exp\left(-\frac{r_B}{a_0}\right) d^3 r_A . \quad (13.6)$$

4. α/q is the electrostatic potential created at B by a charge distribution

$$q\langle \varphi(\mathbf{r}_A)|\varphi(\mathbf{r}_A)\rangle .$$

The integral giving α can be decomposed into an integral over the sphere of radius R and an integral over the region outside this sphere:

$$\alpha = \frac{e^2}{\pi a_0^3} \int_{r_A < R} \exp\left(-\frac{2r_A}{a_0}\right) \frac{1}{r_B} d^3 r_A + \int_{r_A > R} \exp\left(-\frac{2r_A}{a_0}\right) \frac{1}{r_B} d^3 r_A .$$

To determine the first integral, the charge can be assumed to be concentrated at A, by Gauss' theorem. The second integral corresponds to the potential produced at B by the charge distribution outside the sphere. Since the resulting electric field within the sphere is zero, by Gauss' theorem again, the electric potential is

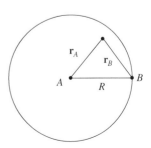

Fig. P7.4 Notation for question 3

constant. It thus suffices to work it out at A, so that

$$\alpha = \frac{e^2}{\pi a_0^3}\left[\frac{1}{R}\int_0^R \exp\left(-\frac{2r_A}{a_0}\right)4\pi r_A^2 dr_A + \int_R^\infty \exp\left(-\frac{2r_A}{a_0}\right)\frac{1}{r_A}4\pi r_A^2 dr_A\right]$$

$$= \frac{e^2}{\pi a_0^3}\left[\frac{1}{R}\int_0^\infty \exp\left(-\frac{2r_A}{a_0}\right)4\pi r_A^2 dr_A + \int_R^\infty \exp\left(-\frac{2r_A}{a_0}\right)\left(\frac{1}{r_A}-\frac{1}{R}\right)4\pi r_A^2 dr_A\right]$$

$$= \frac{e^2}{R} + \frac{4e^2}{a_0^3}\int_R^\infty \exp\left(-\frac{2r_A}{a_0}\right)\left(\frac{1}{r_A}-\frac{1}{R}\right)r_A^2 dr_A .$$

Hence

$$\alpha = \frac{e^2}{R} + \frac{4e^2}{R}\int_{R/a_0}^\infty x\left(\frac{R}{a_0}-x\right)\exp(-2x)dx ,$$

where we have set $x = r_A/a_0$. Integrating by parts with

$$u = \left(\frac{R}{a_0}-x\right)x, \qquad dv = \exp(-2x)dx ,$$

we obtain

$$\alpha = \frac{e^2}{R} + \frac{2e^2}{R}\int_{R/a_0}^\infty \left(\frac{R}{a_0}-2x\right)\exp(-2x)dx .$$

A second integration by parts gives

$$\alpha = \frac{e^2}{R} - \frac{e^2}{R}\left[\left(1+\frac{R}{a_0}\right)\exp\left(-\frac{2R}{a_0}\right)\right] = \frac{e^2}{R} - \frac{e^2}{a_0}\left[\left(1+\frac{a_0}{R}\right)\exp\left(-\frac{2R}{a_0}\right)\right]. \tag{13.7}$$

5. The first term of $-\alpha$ is precisely the Coulomb interaction energy between two charges $+q$ and $-q$ at A and B. It thus balances the Coulomb interaction energy between the protons in the total energy of the system of electrons plus nuclei. For the change in the electronic levels, we thus retain only the second term of α in the expression (13.4) for E_k. According to the statement of the problem,

$$\gamma = \frac{e^2}{a_0}\left(1+\frac{R}{a_0}\right)\exp\left(-\frac{R}{a_0}\right). \tag{13.8}$$

For $R/a_0 \gg 1$,

$$\gamma \simeq \frac{e^2}{a_0}\frac{R}{a_0}\exp\left(-\frac{R}{a_0}\right), \qquad \alpha - \frac{e^2}{R} \simeq -\frac{e^2}{a_0}\exp\left(-\frac{2R}{a_0}\right),$$

which yields

$$\boxed{\left|\alpha - \frac{e^2}{R}\right| \ll \gamma}.$$

Note: From the expression (13.6) for γ, it is clear that, when $R/a_0 \gg 1$, the charge distributions are concentrated around A and B (see Fig. P7.5). The dominant contribution to γ comes from $r_A \sim r_B \sim R/2$, so that the factor $\exp(-R/a_0)$ dominates the expression (13.8) for γ. The same remark leads to

$$\langle \varphi(\mathbf{r} - \mathbf{r}_A) | \varphi(\mathbf{r} - \mathbf{r}_B) \rangle \propto \exp(-R/a_0) \ll 1 ,$$

and this justifies neglecting the overlap integrals $\langle \varphi(\mathbf{r} - \mathbf{R}_j) | \varphi(\mathbf{r} - \mathbf{R}_i) \rangle$ in question 2.

6. Since the term 8α is negligible in (13.4), the atomic level ε gives rise in the solid to an unshifted energy band on average (see Fig. P7.6). The lowest level is $\varepsilon - 8\gamma$, obtained for $\mathbf{k} = 0$, and the highest level is $\varepsilon + 8\gamma$, obtained for $\mathbf{k} = 2\pi\mathbf{x}/b$, i.e., for six vertices of the polyhedron representing the first Brillouin zone, viz., those with 4 adjacent faces [see Fig. P4.4 (right) of Problem 4: *Electronic Energy and Stability of Alloys*]. The band width is $W = 16\gamma$.

7. The resulting band has N spin degenerate states and can therefore accommodate $2N$ electrons. There are N electrons to attribute. The Fermi level at $T = 0$ is in the middle of the band for all R. Although the band narrows very quickly when R increases, it is always half filled and thus corresponds to a metallic state. This clearly contradicts what is observed physically because, when the atoms are widely separated, the electrons are localised on atomic sites and no conduction can occur.

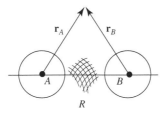

Fig. P7.5 Charge distributions for $R \gg a_0$

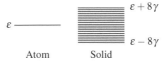

Fig. P7.6 Atomic level ε and resulting energy band

Interactions Between Electrons

1. We have

$$|\Phi_1\rangle = \frac{1}{\sqrt{2}} \left(|\varphi_{1A}\rangle \otimes |\varphi_{2B}\rangle + |\varphi_{1B}\rangle \otimes |\varphi_{2A}\rangle \right) ,$$

$$|\Phi_2\rangle = \frac{1}{\sqrt{2}}\Big(|\varphi_{1A}\rangle \otimes |\varphi_{2B}\rangle - |\varphi_{1B}\rangle \otimes |\varphi_{2A}\rangle\Big),$$

$$|\Phi_3\rangle = \frac{1}{\sqrt{2}}\Big(|\varphi_{1A}\rangle \otimes |\varphi_{2A}\rangle + |\varphi_{1B}\rangle \otimes |\varphi_{2B}\rangle\Big),$$

$$|\Phi_4\rangle = \frac{1}{\sqrt{2}}\Big(|\varphi_{1A}\rangle \otimes |\varphi_{2A}\rangle - |\varphi_{1B}\rangle \otimes |\varphi_{2B}\rangle\Big).$$

The Hamiltonian is spin-independent, so the eigenfunctions separate into a product of a spatial wave function and a spin wave function. For fermions, the Pauli exclusion principle requires the total wave function to be antisymmetric with respect to interchange of the electrons $(1,2)$. Therefore, for a symmetric spatial wave function under the exchange $(1,2)$ there corresponds an antisymmetric wave function under the exchange $(1,2)$, i.e., a spin singlet state. Likewise, for an antisymmetric spatial wave function, there corresponds a triplet spin state. This means that only $|\Phi_2\rangle$ is associated with a spin triplet, while $|\Phi_1\rangle$, $|\Phi_3\rangle$, and $|\Phi_4\rangle$ are all associated with spin singlets. The basis is thus denoted by

$$\boxed{|\Phi_1^S\rangle, \quad |\Phi_2^T\rangle, \quad |\Phi_3^S\rangle, \quad |\Phi_4^S\rangle}.$$

Note that the Pauli exclusion principle has had the effect of reducing the number of possible states from 16 to 6.

2. The Hamiltonian $\hat{H}$ commutes with $\hat{T}_{AB}$ and also with the total spin operator $S^2 = (\mathbf{s}_1 + \mathbf{s}_2)^2$. A common eigenbasis can thus be found for these three operators. Since $|\Phi_2^T\rangle$ is an eigenstate of S^2 orthogonal to the three other states, which are spin singlets, $|\Phi_2^T\rangle$ is also an eigenstate of $\hat{H}$. Furthermore, $|\Phi_1^S\rangle$ and $|\Phi_3^S\rangle$ are symmetric under exchange of A and B, while $|\Phi_2^T\rangle$ and $|\Phi_4^S\rangle$ are antisymmetric under exchange of A of B. The eigenstates of $\hat{H}$ are thus either a linear combination of $|\Phi_1^S\rangle$ and $|\Phi_3^S\rangle$ or a linear combination of $|\Phi_2^T\rangle$ and $|\Phi_4^S\rangle$. Since $|\Phi_2^T\rangle$ is an eigenstate of $\hat{H}$, $|\Phi_4^S\rangle$ is thus also an eigenstate of $\hat{H}$.

3. We seek the energy eigenvalues of the states $|\Phi_2^T\rangle$ and $|\Phi_4^S\rangle$. To do this, we set $\hat{H} = \hat{H}_0 + \hat{V}_{12}$ with

$$\hat{H}_0 = \frac{p_1^2}{2m} + V_{1A} + V_{1B} + \frac{p_2^2}{2m} + V_{2A} + V_{2B}.$$

Consider, for example, $|\Phi_2^T\rangle$, which has eigenenergy given by

$$E_2^T = \langle \Phi_2^T | \hat{H}_0 + \hat{V}_{12} | \Phi_2^T \rangle = \langle \Phi_2^T | \hat{H}_0 | \Phi_2^T \rangle + \langle \Phi_2^T | \hat{V}_{12} | \Phi_2^T \rangle. \quad (13.9)$$

With the assumption concerning the matrix elements of $\hat{V}_{12}$, the second term is zero because

$$|\Phi_2^T\rangle = \frac{1}{\sqrt{2}}\Big(|\varphi_{1A}\rangle \otimes |\varphi_{2B}\rangle - |\varphi_{1B}\rangle \otimes |\varphi_{2A}\rangle\Big)$$

Problem 7: Insulator–Metal Transition

does not contain any terms where the two electrons are at the same site. To calculate the first term of (13.9), note that

$$\hat{H}_0|\varphi_{1A}\rangle \otimes |\varphi_{2B}\rangle = \left[\left(\frac{p_1^2}{2m} + V_{1A} + V_{1B}\right)|\varphi_{1A}\rangle\right] \otimes |\varphi_{2B}\rangle$$

$$+ |\varphi_{1A}\rangle \otimes \left[\left(\frac{p_2^2}{2m} + V_{2A} + V_{2B}\right)|\varphi_{2B}\rangle\right]$$

$$= \left[(\varepsilon + V_{1B})|\varphi_{1A}\rangle\right] \otimes |\varphi_{2B}\rangle + |\varphi_{1A}\rangle \otimes \left[(\varepsilon + V_{2A})|\varphi_{2B}\rangle\right]. \quad (13.10)$$

We obtain a nonzero matrix element in (13.9) by taking the bra containing $\langle\varphi_{2B}|$ for the first term of (13.10), and $\langle\varphi_{1A}|$ for the second. This shows that, in the calculation of E_2^T, the cross terms disappear and

$$E_2^T = 2 \times \frac{1}{2}\langle\varphi_{1A}| \otimes \langle\varphi_{2B}|\hat{H}_0|\varphi_{1A}\rangle \otimes |\varphi_{2B}\rangle$$

$$= 2\varepsilon + \langle\varphi_{1A}|V_{1B}|\varphi_{1A}\rangle + \langle\varphi_{2B}|V_{2A}|\varphi_{2B}\rangle$$

$$= 2\varepsilon - 2\alpha \simeq 2\varepsilon \ .$$

Likewise for

$$|\Phi_4^S\rangle = \frac{1}{\sqrt{2}}\Big(|\varphi_{1A}\rangle \otimes |\varphi_{2A}\rangle - |\varphi_{1B}\rangle \otimes |\varphi_{2B}\rangle\Big) ,$$

we have

$$E_4^S = \langle\Phi_4^S|\hat{H}_0|\Phi_4^S\rangle + \langle\Phi_4^S|V_{12}|\Phi_4^S\rangle = 2\varepsilon - 2\alpha + U \simeq 2\varepsilon + U \ .$$

Applying the same argument to $|\Phi_1^S\rangle$ and $|\Phi_3^S\rangle$, we immediately find

$$\langle\Phi_1^S|\hat{H}|\Phi_1^S\rangle = 2\varepsilon - 2\alpha \simeq 2\varepsilon \ ,$$

$$\langle\Phi_3^S|\hat{H}|\Phi_3^S\rangle = 2\varepsilon - 2\alpha + U \simeq 2\varepsilon + U \ .$$

Regarding the off-diagonal matrix element, $\langle\Phi_3^S|\hat{H}|\Phi_1^S\rangle = \langle\Phi_3^S|\hat{H}_0|\Phi_1^S\rangle$. Taking the term $|\varphi_{1A}\rangle \otimes |\varphi_{2B}\rangle$ in $|\Phi_1^S\rangle$, and considering (13.10) and

$$\langle\Phi_3^S| = \frac{1}{\sqrt{2}}\Big(\langle\varphi_{1A}| \otimes \langle\varphi_{2A}| + \langle\varphi_{1B}| \otimes \langle\varphi_{2B}|\Big) ,$$

it is clear that we must associate the second term of $\langle\Phi_3^S|$ with the first term of (13.10) and vice versa. This yields

$$\langle\Phi_3^S|\hat{H}|\Phi_1^S\rangle = \left(\frac{1}{2}\langle\varphi_{1B}|V_{1B}|\varphi_{1A}\rangle + \frac{1}{2}\langle\varphi_{2A}|V_{2A}|\varphi_{2B}\rangle\right) \times 2 = -2\gamma \ .$$

The matrix representing $\hat{H}$ in the subspace spanned by $|\Phi_1^S\rangle$ and $|\Phi_3^S\rangle$ is thus

$$\hat{H} = \begin{pmatrix} 2\varepsilon & -2\gamma \\ -2\gamma & 2\varepsilon + U \end{pmatrix}.$$

4. The associated eigenenergies are

$$(2\varepsilon - E)(2\varepsilon - E + U) - 4\gamma^2 = 0,$$

or

$$E_\pm = 2\varepsilon + \frac{1}{2}\left(U \pm \sqrt{U^2 + 16\gamma^2}\right).$$

Note that $E_+ > 2\varepsilon + U$ and $E_- < 2\varepsilon$, leading to the level structure in Fig. P7.7. U is an intra-atomic quantity that is independent of the distance between the two atoms. However, γ decreases quickly when the atoms are far apart. In the limit $\mathbf{r}_B - \mathbf{r}_A \to \infty$, we have $\gamma \to 0$, $E_+ \to 2\varepsilon + U$, and $E_- \to 2\varepsilon$. The corresponding eigenstates are then $|\Phi_1^S\rangle$ for E_- and $|\Phi_3^S\rangle$ for E_+.

The energy 2ε corresponds to the states $|\Phi_2^T\rangle$ and $|\Phi_1^S\rangle$ for which only terms of the type $|\varphi_{1A}\rangle \otimes |\varphi_{2B}\rangle$ arise, i.e., in which one electron is localised on each orbital. The energy $2\varepsilon + U$ corresponds to the states $|\Phi_4^S\rangle$ and $|\Phi_3^S\rangle$ for which only terms of the type $|\varphi_{1A}\rangle \otimes |\varphi_{2A}\rangle$ arise, i.e., with two electrons on the same orbital.

Consequently, in the ground state of the system, the two electrons are not simultaneously in the same orbital. In the excited state, two electrons coexist in the same atom, which is thus an H$^-$ ion, and U is the extra energy to be supplied to get two electrons into the same atom.

5. In the solid, as in 7.1, hybridisation due to the γ term broadens the electron states at ε and $\varepsilon + U$ by W. The band due to ε can only accommodate N electrons. It is thus fully occupied for $W < U$, *and the material is an insulator.* On the other hand, for $W > U$, the bands overlap and the system becomes metallic. This happens if the lattice parameter b is varied. *The insulator–metal transition occurs for $U = W$* (see Fig. P7.8).

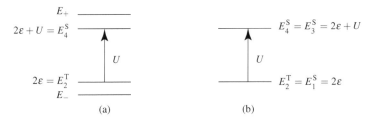

Fig. P7.7 Level structure obtained for the two electron states when (**a**) $\gamma \neq 0$, (**b**) $\gamma = 0$

Problem 7: Insulator–Metal Transition

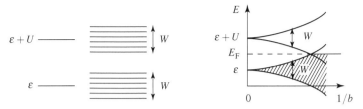

Fig. P7.8 *Left*: Broadening of the states at ε and $\varepsilon + U$. *Right*: Overlap of the bands for $W > U$

Fig. P7.9 Notation for question 6

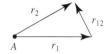

6. By definition, with the notation in Fig. P7.9,

$$U = \left(\frac{1}{\pi a_0^3}\right)^2 \iint \exp\left(-\frac{2r_1}{a_0}\right) \frac{e^2}{|\mathbf{r}_2 - \mathbf{r}_1|} \exp\left(-\frac{2r_2}{a_0}\right) d^3r_1 d^3r_2$$

$$= \frac{1}{\pi a_0^3} \int d^3r_1 \left[\exp\left(-\frac{2r_1}{a_0}\right) \int \exp\left(-\frac{2r_2}{a_0}\right) \frac{e^2}{r_{12}} d^3r_2\right].$$

The second integral over $\mathbf{r}_2$ corresponds exactly to the one in (13.5) for the calculation of α. So according to (13.7), with $r_1 = R$,

$$U = \frac{e^2}{\pi a_0^3} \int \left[\frac{1}{r_1} - \frac{1}{a_0}\left(1 + \frac{a_0}{r_1}\right) \exp\left(-\frac{2r_1}{a_0}\right)\right] \exp\left(-\frac{2r_1}{a_0}\right) d^3r_1 .$$

Hence,

$$U = \frac{4e^2}{a_0} \int_0^\infty \left[\frac{a_0}{r_1} - \left(1 + \frac{a_0}{r_1}\right) \exp\left(-\frac{2r_1}{a_0}\right)\right] \exp\left(-\frac{2r_1}{a_0}\right) \left(\frac{r_1}{a_0}\right)^2 \frac{dr_1}{a_0}$$

$$= \frac{4e^2}{a_0} \int_0^\infty \left[x \exp(-2x) - (x^2 + x)\exp(-4x)\right] dx .$$

After several integrations by parts,

$$\boxed{U = \frac{5e^2}{8a_0}}. \tag{13.11}$$

7. The metal–insulator transition occurs for

$$\boxed{U = W = 16\gamma},$$

Fig. P7.10 Finding the condition for the metal–insulator transition

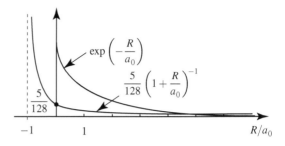

so using (13.8) for γ and (13.11) for U (see Fig. P7.10),

$$\exp\left(-\frac{R}{a_0}\right) = \frac{5}{128(1+R/a_0)}.$$

Graphically, the solution is seen to correspond to $R/a_0 \gg 1$. Numerically, we obtain $R/a_0 = 5.5$. For the body-centered cubic lattice, $R = b\sqrt{3}/2$, and $n = 2/b^3$, so $R = (\sqrt{3}/2^{2/3})n^{-1/3}$ and $R/a_0 = 5.5$ corresponds to the condition for the insulator–metal transition

$$\boxed{n_{\text{IM}}^{1/3} a_0 = 0.218 = C_{\text{bcc}}}.$$

8. In the simple cubic crystal, there are only six nearest neighbours, and the condition is $W = 12\gamma$ or

$$\exp\left(-\frac{R}{a_0}\right) = \frac{5}{96(1+R/a_0)},$$

which yields $R/a_0 = 4.7$. Since $n = R^{-3}$, we then obtain

$$\boxed{n_{\text{IM}}^{1/3} a_0 = 0.213 = C_{\text{sc}}}.$$

For a face-centered cubic lattice (12 neighbours), we find

$$\boxed{n_{\text{IM}}^{1/3} a_0 = 0.204 = C_{\text{fcc}}}.$$

The constant C depends little on the crystal structure.

Alkali Elements and Hydrogen

1. The condition $n_{\text{IM}}^{1/3} a_0 = 0.218$ or $R/a_0 = 5.5$ for a body-centered cubic lattice becomes $(R_{\text{IM}} - 2R_{\text{c}})/a_0 = 5.5$, or $R_{\text{IM}}/a_0 = 5.5 + 2R_{\text{c}}/a_0$ with $R = b\sqrt{3}/2$. We obtain the values shown in Table 13.2. We always have $R < R_{\text{IM}}$ and the density is well above the density required to obtain a metallic state.

Problem 7: Insulator–Metal Transition

Table 13.2 Key parameters for alkali elements

	Li	Na	K	Rb	Cs
b (Å)	3.49	4.23	5.23	5.59	6.05
R_c/a_0	1.13	1.79	2.51	2.79	3.19
R_{IM}/a_0	7.76	9.08	10.52	11.08	11.88
R/a_0	5.70	6.91	8.55	9.13	9.89

The approximation here is somewhat rough. Indeed, although the pseudopotential approximation takes into account the existence of the ionic core (see Fig. P7.11), the radial wave functions do not have the simple form used here. However, for a given distance R, the integrals γ are much larger than for hydrogen atoms, and the approximation is not disastrous, since we did not expect to determine the exact density at which the alkali metals would become insulating.

The above results nevertheless suggest something which is confirmed by experimental observations: it is easier to make Cs or Rb insulating than Li or Na. When $R < R_{IM}$, the integrals are such that $W \gg U$, and in this case U no longer plays the dominating role in the electronic properties. In this case, the free electron model is applicable and the interactions between electrons introduce minor modifications.

Fig. P7.11 Pseudopotential approximation

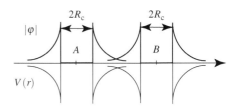

2. The atomic volume is $v = n^{-1} = a^3/8 = 19.7$ Å^3. This gives $n^{1/3}a_0 = 0.196$, just less than the density required for atomic hydrogen to give a metallic solid.

In fact, in the case of molecular hydrogen, the electronic bands arise from energy levels corresponding to molecular orbitals, and an insulator–metal transition can occur when the fully occupied band resulting from the binding orbital overlaps with a higher energy band. We must therefore compare the hopping integral γ' (the analog of γ) between molecular orbitals, which decreases exponentially with the distance between hydrogen molecules, with the energy separation of the molecular levels, which has the same order of magnitude as U. It is not hard to show that γ' for the molecular solid is likely to be less than γ for the monatomic solid of the same density. One thus expects the density required to obtain a molecular conducting state to be well above the value given by the above condition.

3. It is clear that hydrogen must be compressed to obtain a metallic state.

Pressures of the order of 200 GPa (2 Mbar) would nevertheless be required to obtain a metallic state. In the laboratory, such pressures can be only yet be obtained by transient methods, viz., a shock wave produced by a violent explosion. Very high density hydrogen must certainly exist in some planets of the Solar System. The magnetic field of Jupiter detected by Voyager I in 1975 can only be explained by the existence of metallic hydrogen in this planet, at pressures of several hundred GPa.

As far as the alkali elements are concerned, to make them insulating, their density must be reduced. (In the liquid phase, close to the critical point, Cs and Rb become insulating.) Another way to obtain a metal–insulator transition is to produce an alloy of the alkali element with a noble gas. This would tend to distance the alkali atoms from one another.

> This effectively produces a metal–insulator transition, but one has a disordered system of hydrogen-like atoms rather than a crystal, and the transition is in this case more closely related to the existence of disorder (Anderson transition) than to a consequence of interactions between electrons (Mott transition).

Insulator–Metal Transition in Si–P

1. The phosphorus nucleus and the extra electron it contributes to the silicon lattice constitute a large hydrogen-like atom. The interaction potential between the electron and nucleus is reduced by the high dielectric constant ε, and the effective electron mass is reduced ($m_e < m_0$). The main effect is to transform a_1 to $a_p = \varepsilon a_1 m_0 / m_e$. If the phosphorus atoms are distributed over a regular lattice, the arguments of 7.2 can be applied, and we conclude that

$$\boxed{n_{\mathrm{IM}}^{1/3} a_p \simeq 0.20}.$$

2. The energies of states related to the impurity are $E_d = E_c - E_0$, where $E_0 = e^2/2a_p$ is the ionisation energy of the donor, and $E_{P^-} = E_d + U$ corresponding to a phosphorus that traps two electrons (P^-). The latter is slightly higher than E_c because $U = 5e^2/8a_p = 5E_0/4$. It thus lies in the conduction band of silicon. The case $n_d \ll n_{\mathrm{IM}}$ corresponds to the one described in Fig. 4.12a in Chap. 4 of the lecture notes, except that the P^- band is superposed on the conduction band. E_F lies between E_d and E_c at $T = 0$, and the impurity band is fully occupied (see Fig. P7.12). As n_d increases, the widths of the two impurity electronic bands also increase. Close to the conduction band, the electronic band structure is as shown in Fig. P7.13. For $n_d > n_{\mathrm{IM}}$, the bands intersect and a metal is obtained (see Fig. P7.14).

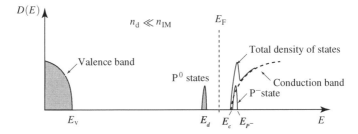

Fig. P7.12 Electronic band structure for the case $n_d \ll n_{\mathrm{IM}}$

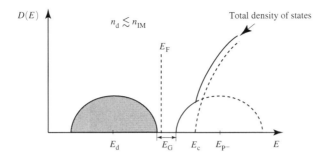

Fig. P7.13 Electronic band structure for a P concentration just below that of the metal–insulator transition

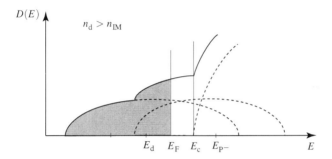

Fig. P7.14 Electronic band structure for the case $n_d > n_{IM}$

3. The experiment identifies *metallic* behaviour for $n_d \geq 3.75 \times 10^{24}/\text{m}^3$ (ρ is constant when $T \to 0$) and *insulating* behaviour for $n_d \leq 3.70 \times 10^{24}/\text{m}^3$ (ρ diverges when $T \to 0$). Note that measurements have to be made at very low temperatures! Taking $n_{IM} \sim 3.74 \times 10^{24}/\text{m}^3$, we obtain

$$C = n_{IM}^{1/3} \frac{\varepsilon}{m_e/m_0} a_1 = (3.74 \times 10^{24})^{1/3} \frac{11.75}{0.33} 0.53 \times 10^{-10} = 0.29 \,.$$

This is close to the calculated value $C \sim 0.20\text{–}0.22$.

4. The electron density in the metallic band is n_d. The conductivity is

$$\sigma = \frac{n_d q^2 \tau}{m_e} = \rho^{-1} \,.$$

For $n_d = 7 \times 10^{24}/\text{m}^3$, we have $\rho \simeq 5 \times 10^{-5}\ \Omega\,\text{m}$, so

$$\tau = \frac{m_e}{n_d q^2 \rho} = \frac{0.33 \times 9.1 \times 10^{-31}}{7 \times 10^{24} \times (1.6 \times 10^{-19})^2 \times 5 \times 10^{-5}} \sim 3.3 \times 10^{-14}\ \text{s}\,.$$

ρ increases slightly with increasing temperature, as would be expected in a metal. This happens because $1/\tau$ is a sum of two contributions:

$$\frac{1}{\tau} = \frac{1}{\tau}\bigg|_{\text{defects}} + \frac{1}{\tau}\bigg|_{\text{vibrations}}.$$

In a pure metal, the second term dominates the first at high temperatures, but here the donor atoms in disordered positions constitute scattering centers for the electrons, and the first term dominates the second across almost the whole temperature range, which explains the almost negligible dependence of τ^{-1} on T. At $T = 0$ the collision time is comparable with $\tau \sim 10^{-14}$ s obtained in a pure metal like aluminium at $T \simeq 300$ K.

5. In the insulating case, the resistivity is due to thermally excited carriers in the conduction band (or the P$^-$ band). For $n_d \ll n_{\text{IM}}$, the number n_e of such carriers is given by

$$n_e \propto \exp(-E_G/2k_BT), \quad \text{with} \quad E_G = E_c - E_d = E_0.$$

When n_d approaches n_{IM}, the distance E_G between the filled band and the empty band decreases (see Fig. P7.15). The resistivity is given by $\rho = (n_e r \mu)^{-1}$, and since $\mu = e\tau/m_e$ is practically independent of temperature, the variation of ρ is dominated by that of n_e:

$$\rho \propto \exp(E_G/2k_BT).$$

In Fig. P7.1, for $n_d = 3.7 \times 10^{24}/\text{m}^3$, ρ does indeed *diverge exponentially as T decreases*. We have $\log \rho = E_G/2k_BT + \text{constant}$.

If $\log \rho$ is plotted as a function of T^{-1} for $2K < T < 10$ K, we therefore obtain a straight line, as in Fig. P7.16. For $\rho_1/\rho_2 = 10$, we obtain $T_1^{-1} - T_2^{-1} = 0.33$ K^{-1}. This implies that $E_G/k_B = 2\log 10/0.33 = 14$ K.

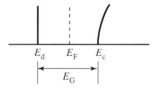

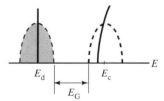

Fig. P7.15 Insulating case. As n_d approaches n_{IM}, the gap E_G decreases

Problem 7: Insulator–Metal Transition

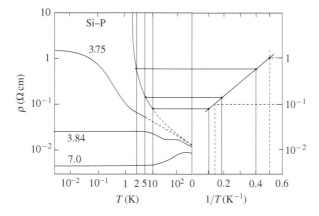

Fig. P7.16 Determination of E_G from $\log \rho = E_G/2k_B T + $ const.

This should be compared with $E_0 = (E_1/\varepsilon^2)(m_e/m_0)$, where $E_1 = 13.6$ eV is the energy of the Bohr atom. This gives $E_0/k_B = 400$ K. *We observe that $E_G \ll E_0$, since the concentration considered is very close to the insulator–metal transition.*

6. E_G must decrease from the value E_0 when n_d is very low, and *vanish* for $n_d \sim n_{IM}$ (see Fig. P7.17).

7. In the limit $n_d \ll n_{IM}$, the electrons are localised on phosphorus sites. This is a classic case of a paramagnetic solid. For electron spins $S = 1/2$, Curie paramagnetism is obtained with

$$\boxed{\chi_C = n_d \frac{\mu_B^2}{k_B T}}.$$

The susceptibility grows linearly with n_d and is inversely proportional to T. This does seem to be the case for $n_d < 10^{24}/\text{m}^3$. The susceptibility is given per kg of silicon. In this case, for example, for $n_d = 10^{23}$ atoms/m³, we find

$$\chi_C = \frac{10^{23}}{2.33 \times 10^3} \frac{(0.927 \times 10^{-23})^2}{1.38 \times 10^{-23} T} = \frac{2.67}{T} 10^{-4} \text{ SI}.$$

We can now plot the three straight lines in Fig. P7.18 for χ_C (1.1 K), χ_C (4.2 K), and χ_C (77 K).

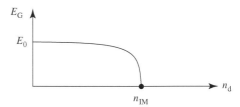

Fig. P7.17 E_G as a function of n_d

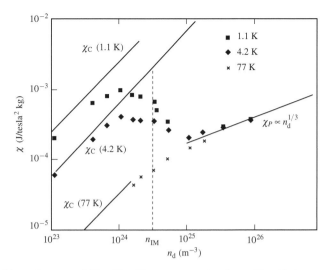

Fig. P7.18 Magnetic susceptibility. Fit of low concentration data with a Curie susceptibility χ_C and high concentration data with a Pauli susceptibility χ_P (see text)

8. For $n_d \gg n_{IM}$, we have a metallic state. In this case, the susceptibility is a temperature-independent Pauli susceptibility. This is indeed observed for $n_d > 10^{25}/\text{m}^3$. The susceptibility is determined by the electronic density of states at the Fermi level:

$$\chi_P = 2\mu_0 \mu_B^2 D_\Omega(E_F).$$

If for high enough n_d the metallic band can be treated as a simple parabolic band, then taking the zero energy at the bottom of this band, we can use (3.47) and (3.49) obtained in Chap. 3 for a free electron gas, replacing m_0 by the effective mass m_e:

$$D_\Omega(E) = \frac{1}{4\pi^2}\left(\frac{2m_e}{\hbar^2}\right)^{3/2} E_F^{1/2}, \qquad E_F = \frac{\hbar^2}{2m_e}\left(3\pi^2 n_d\right)^{2/3}.$$

This yields

$$\chi_P = \left(\frac{3}{\pi^4}\right)^{1/3} \mu_0 \mu_B^2 \frac{m_e}{\hbar^2} n_d^{1/3}.$$

We thus find a susceptibility that approaches this behaviour asymptotically for $n_d > 10^{25}/\text{m}^3$ (see Fig. P7.18).

Note: The insulator–metal transition observed in Si–P cannot be explained as simply as suggested here. The effects of disorder must be included along with the interactions between electrons. A great deal of experimental and theoretical work is currently under way to elucidate the respective roles of these two effects in various insulator–metal transitions.

Problem 8: Cyclotron Resonance[5]

The aim here is to investigate the frequency dependence of the conductivity of a conducting material in the presence of an applied magnetic field $\mathbf{B} = \mu_0 \mathbf{H}$. For a given value of B, the electrons can engage in an oscillatory motion with a characteristic frequency. This is the *cyclotron resonance mode*. The energy associated with a small alternating applied electric field can be absorbed by the electrons if its frequency is tuned to the frequency of the cyclotron mode. In experimental setups, one generally observes the increased absorption at fixed frequency when the magnitude of the magnetic field is modified.

8.1: Real and Reciprocal Space Paths of an Electron State

Consider to begin with a free electron subjected to a magnetic field $\mathbf{B}$. The plane wave packet representing the electron has momentum $\langle \mathbf{p} \rangle$ and group velocity $\mathbf{v}_g = \hbar \mathbf{k}/m_0$, and the Lorentz force acting on the electron is

$$\mathbf{F} = -e\mathbf{v}_g \wedge \mathbf{B}.$$

1. *What are the paths of the electron in real space and its wave vector $\mathbf{k}$ in reciprocal space? What is the period of the motion? Does it depend on the initial value of $\mathbf{k}$?*

Now consider an electron in a solid. It is represented by a Bloch wave packet of quasi-momenta centered on $\mathbf{k}$ and of energy $E_n(\mathbf{k})$. In the presence of a magnetic field, the electron is subjected to a force

$$\mathbf{F} = -e\langle \mathbf{v} \rangle_{n,\mathbf{k}} \wedge \mathbf{B},$$

where $\langle \mathbf{v} \rangle_{n,\mathbf{k}}$ is the group velocity of the electron in the corresponding Bloch state.

2. *Prove that the path C of the point $\mathbf{k}$ representing the electronic state in reciprocal space lies on a surface of constant energy E.*

3. *Show that its motion is planar and periodic.*

4. *Show that the total force on the electrons of a fully occupied band is zero.*

Consider a Bloch state belonging to constant energy surface in the form of an ellipsoid of revolution about the z axis, centered on a point $\mathbf{k}_0$ of the Brillouin zone:

$$E = \hbar^2 \left[\frac{(k_x - k_{0x})^2}{2m_\perp^*} + \frac{(k_y - k_{0y})^2}{2m_\perp^*} + \frac{(k_z - k_{0z})^2}{2m_\parallel^*} \right],$$

where $m_\perp^*$ and $m_\parallel^*$ are the transverse and longitudinal effective masses.

---

[5] This problem has been designed with C. Hermann and B. Sapoval.

5. *Show that the equation of motion of the electrons in real space is*

$$\frac{d\mathbf{v}}{dt} = -\left\|\frac{1}{m^*}\right\| e\mathbf{v} \wedge \mathbf{B},$$

where we have set $\mathbf{v} = \langle \mathbf{v} \rangle_{n,\mathbf{k}}$ *to simplify.*

Let $\theta = (O\mathbf{z}, \mathbf{B})$ be the angle between the axis of revolution of the ellipsoid and the magnetic field, which lies in the (x, z) plane.

6. *Show that the equation of motion has an oscillating solution with* $\mathbf{v} = \mathbf{v}_0 e^{i\omega t}$ *for*

$$\omega = \omega_c = eB/m_c,$$

where m_c *does not depend on the initial Bloch state and is a function of* θ, $m_\perp^*$, *and* $m_\parallel^*$, *to be determined.*

8.2: A Semiconductor: Silicon

Silicon with atomic configuration (Ne) $3s^2\ 3p^2$ crystallises into a diamond cubic structure with lattice parameter a, illustrated in Fig. P8.1. It has a face-centered cubic Bravais lattice.

1. *What is the basis of the primitive cell? What is the reciprocal lattice?*

The dispersion curves corresponding to electrons in the $3s^2\ 3p^6$ shells are shown in Fig. P8.2 in the two directions (100) and (111) of the reciprocal lattice.

2. *Count the states in the bands of Fig. P8.2 and indicate the occupied electron states. Check that silicon is an insulator at zero temperature.*

3. *On Fig. P8.2, indicate the band gap and specify its value. What are the vectors* $\mathbf{k}_0$ *and energies corresponding to the conduction electrons and holes excited at nonzero temperatures?* (1 rydberg = 13.6 eV.)

4. *Do you think cyclotron resonances can be observed in silicon at all temperatures?*

5. *How many cyclotron resonance frequencies would one generally expect to observe for the band structure of silicon given in Fig. P8.2?*

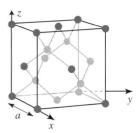

Fig. P8.1 Crystal structure of Si, similar to diamond

Problem 8: Cyclotron Resonance

Fig. P8.2 Band structure of Si, represented in two high symmetry directions of the Brillouin zone

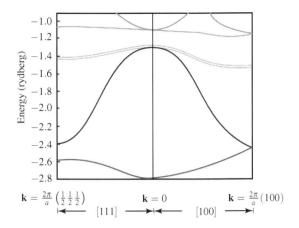

Fig. P8.3 Defining the plane $x'Oz$ in which the field lies

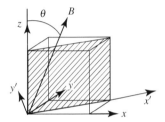

We now assume that the field **B** is applied in a plane $x'Oz$ which bisects the unit cell cube (see Fig. P8.3). For $\theta = \pi/6$ and a frequency $\nu = \omega/2\pi = 2.4 \times 10^{10}$ Hz, we observe the signal shown in Fig. P8.4 when the magnitude of the magnetic field is varied. If **B** is rotated in the plane $x'Oz$, peaks 2 and 3 are found to move, while peaks 1 and 4 do not.

6. Which electrons are responsible for peaks 2 and 3? Which one of them splits into two if the field is shifted outside the plane $x'Oz$?

7. Which electrons are responsible for peaks 1 and 4?

Fig. P8.4 Absorption as a function of the field applied in the direction $\theta = \pi/6$ in the plane $x'Oz$. From Dresselhaus, G., Kip, A.F., Kittel, C.: Phys. Rev. **98**, 368 (1955). With the permission of the American Physical Society © 1955 APS). http://link.aps.org/doi/10.1103/PhysRev.98.368

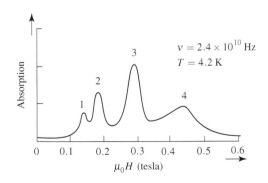

8. *From the experimental results, deduce the effective masses of the electrons and holes in silicon. (For free electrons, the cyclotron resonance occurs at $B = 0.86$ tesla for the given frequency.)*

In the above calculation, we neglected collisions. They can be taken into account by introducing a friction term into the equation of motion, in the presence of the oscillating electric field.

9. *Without calculation, indicate the effect it should have. Is this visible in the experimental curve? What should one expect to see if the temperature is raised or lowered in the above experiment?*

10. *By reconsidering your answer to question 4, can you guess how the experiment in Fig. P8.4 was carried out?*

8.3: Metals

The dominant contribution to the conductivity in a metal comes from electrons at the Fermi level which can absorb energy from the electric field excitation $\mathcal{E}e^{i\omega t}$ associated with the incident electromagnetic wave.

1. *Check that, if the Fermi surface is spherical, the cyclotron angular frequency is the same for all electrons at the Fermi level.*

2. *For an arbitrary Fermi surface, show that the angular frequency ω_c of the motion is related to the area S of the path (C) in $\mathbf{k}$ space by*

$$\omega_c = \frac{2\pi eB}{\hbar^2} \frac{dE}{dS}.$$

Use the fact that the period $T = 2\pi/\omega_c$ of the motion is $T = \oint_{(C)} dt$.

3. *The Fermi surface of copper is shown in Fig. P8.5. What electronic orbits do you expect to dominate the frequency response of the conductivity when the field is applied in the direction (111)?*

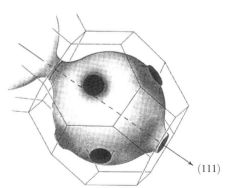

Fig. P8.5 Fermi surface of copper

Solution

Real and Reciprocal Space Paths of an Electron State

1. We have $m_0 \mathbf{v}_g = \hbar \mathbf{k}$, so

$$m_0 \frac{d\mathbf{v}_g}{dt} = -e\mathbf{v}_g \wedge \mathbf{B}.$$

For $\mathbf{B} \parallel z$, $v_{gz} = $ constant and $\mathbf{v}_{g\perp} = (v_{gx}, v_{gy})$ satisfies

$$m_0 \frac{d^2 \mathbf{v}_{g\perp}}{dt^2} + \frac{e^2}{m_0} B^2 \mathbf{v}_{g\perp} = 0.$$

Hence, $\mathbf{k}$ undergoes a circular precession with an angular frequency

$$\boxed{\omega = \omega_c = \frac{eB}{m_0}},$$

which is independent of the initial momentum of the electron. The electron has a spiral motion in real space with the same angular frequency.

2. We have $\langle \mathbf{v} \rangle_{n,\mathbf{k}} = (1/\hbar) \nabla_\mathbf{k} E$ and

$$\frac{d\mathbf{k}}{dt} = -\frac{e}{\hbar^2} \nabla_\mathbf{k} E \wedge \mathbf{B}.$$

Therefore, $\mathbf{k} \cdot \mathbf{B} = $ constant, and since $\nabla_\mathbf{k} E$ is normal to the surface of constant energy E, the point $\mathbf{k}$ moves tangentially to the constant energy surface, following an orbit (C) in the plane perpendicular to B (see Fig. P8.6).

3. If this section of the constant energy surface is a closed curve, as is generally the case, the motion is periodic.

4. Electrons in a fully occupied band are such that the average value over the band of the velocity of the Bloch states is identically zero. The Bloch states

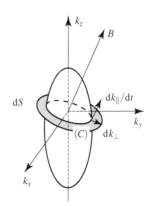

Fig. P8.6 Orbit C of point $\mathbf{k}$ in reciprocal space

corresponding to states **k** on the same orbit (C) have representative states that rotate on this orbit. This means that all the states of the Brillouin zone are conserved in the motion induced by the field. The average velocity of the electrons thus remains identically zero.

5. With

$$E = \hbar^2 \left[\frac{(k_x - k_{0x})^2}{2m_\perp^*} + \frac{(k_y - k_{0y})^2}{2m_\perp^*} + \frac{(k_z - k_{0z})^2}{2m_\parallel^*} \right],$$

we obtain

$$\begin{pmatrix} v_x \\ v_y \\ v_z \end{pmatrix} = \begin{pmatrix} \hbar(k_x - k_{0x})/m_\perp^* \\ \hbar(k_y - k_{0y})/m_\perp^* \\ \hbar(k_z - k_{0z})/m_\parallel^* \end{pmatrix},$$

and hence,

$$\begin{pmatrix} dv_x/dt \\ dv_y/dt \\ dv_z/dt \end{pmatrix} = \begin{pmatrix} 1/m_\perp^* & 0 & 0 \\ 0 & 1/m_\perp^* & 0 \\ 0 & 0 & 1/m_\parallel^* \end{pmatrix} \begin{pmatrix} \hbar dk_x/dt \\ \hbar dk_y/dt \\ \hbar dk_z/dt \end{pmatrix},$$

so finally,

$$\boxed{\frac{d\mathbf{v}}{dt} = -\left\| \frac{1}{m^*} \right\| e\mathbf{v} \wedge \mathbf{B}}.$$

6. With $\mathbf{v} = \mathbf{v}_0 e^{i\omega t}$, this equation becomes

$$0 = i\omega v_{0x} + \frac{eB}{m_\perp^*} \cos\theta\, v_{0y},$$

$$0 = -\frac{eB}{m_\perp^*} \cos\theta\, v_{0x} + i\omega v_{0y} + \frac{eB}{m_\perp^*} \sin\theta\, v_{0z},$$

$$0 = -\frac{eB}{m_\parallel^*} \sin\theta\, v_{0y} + i\omega v_{0z}.$$

There are solutions other than $\mathbf{v}_0 = 0$ for

$$\omega^2 = e^2 B^2 \left(\frac{\cos^2\theta}{m_\perp^{*2}} + \frac{\sin^2\theta}{m_\perp^* m_\parallel^*} \right).$$

The cyclotron frequency is thus

$$\boxed{\omega = \frac{eB}{m_c}}, \quad \text{with} \quad \boxed{\frac{1}{m_c^2} = \left(\frac{\cos^2\theta}{m_\perp^{*2}} + \frac{\sin^2\theta}{m_\perp^* m_\parallel^*} \right)}.$$

Problem 8: Cyclotron Resonance

It is independent of initial conditions and is the same for all orbits perpendicular to **B**. In fact it depends only on the orientation of the magnetic field with respect to the axis of revolution of the ellipsoid.

A Semiconductor: Silicon

1. The basis of the rhombohedral primitive cell comprises two silicon atoms [see Fig. P8.7 (left)]. The reciprocal lattice is body-centered cubic. The first Brillouin zone is shown in Fig. P8.7 (right).
2. There are two silicon atoms ($3s^2\, 3p^2$) per primitive cell. For bands arising from the $3s^2\, 3p^6$ levels, this therefore represents 8 electrons per primitive cell. These will thus fill the states of the first four bands in the first Brillouin zone. The occupied levels are shown in Fig. P8.8. Higher bands are empty and silicon is an insulator at zero temperature.
3. The band gap is shown in Fig. P8.8. It is a little less than 0.1 rydberg across, i.e., about 1 eV. At $T > 0$, electrons will be excited to the bottom of the conduction band, which corresponds to vectors $\mathbf{k}_0 \approx 0.8(2\pi/a, 0, 0)$ and $-\mathbf{k}_0$. With those corresponding to the equivalent directions k_y and k_z, there are therefore *six equivalent electronic pockets* [see Fig. P8.7 (right)]. The states freed at the top of the valence band are close to $\mathbf{k} = 0$ and correspond to *hole pockets in the three last valence bands, of which two are degenerate*.
4. To observe the cyclotron resonances, a significant number of electrons and holes must be excited. In principle, this means that the experiment must be carried out at a high enough temperature to populate electron and hole pockets.
5. For each of these electron and hole pockets, the first term in the Taylor expansion of the energy has the general form

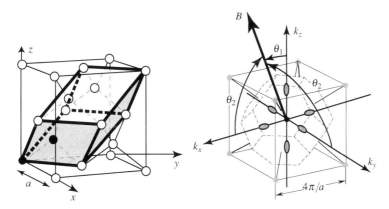

Fig. P8.7 Si real space and reciprocal lattice primitive cell within a basis comprising two Si atoms (*full spheres*) (*left*), reciprocal lattice and first Brillouin zone, showing the six thermally populated ellipsoidal electronic pockets (*right*)

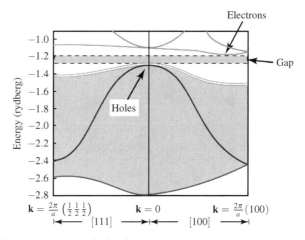

Fig. P8.8 Band structure showing the band gap

$$E = \hbar^2 \left[\frac{(k_x - k_{0x})^2}{2m_x^*} + \frac{(k_y - k_{0y})^2}{2m_y^*} + \frac{(k_z - k_{0z})^2}{2m_z^*} \right].$$

For the holes, $k_{0x} = k_{0y} = k_{0z} = 0$. Given the symmetries of the cubic crystal, the x, y, and z axes are equivalent, and the constant energy surfaces are therefore *spheres with different effective masses m_{h1}^* and m_{h2}^**, as can be seen on Fig. P8.8. For the electrons, we have for example $k_{0x} = k_{0y} = 0$, and $k_{0z} \approx 0.8 \times 2\pi/a$. The symmetry of the cubic cell under rotation through $\pi/2$ about the z axis implies equivalence of the x and y axes, and hence $m_x^* = m_y^*$. The constant energy surfaces are thus *six paraboloids of revolution about the $\mathbf{k}_x$, $\mathbf{k}_y$, and $\mathbf{k}_z$ axes*, characterised by two longitudinal and transverse effective masses, viz., $m_\parallel^*$ and $m_\perp^* = m_x^* = m_y^*$, respectively [see Fig. P8.7 (right)].

A cyclotron frequency is expected for each of the two pockets of holes and for each pocket of conduction electrons. Since ω depends only on the orientation of the magnetic field with respect to the axis of revolution of the ellipsoid, the paraboloids $\mathbf{k}_0$ and $-\mathbf{k}_0$ correspond to the same cyclotron frequencies. There will thus be three cyclotron frequencies for the conduction electrons for an arbitrary direction of the magnetic field. We therefore expect a total of *five cyclotron frequencies* for an arbitrary orientation of the field.

6. Four cyclotron frequencies can be seen in Fig. P8.9. Peaks 2 and 3 are shifted if the field orientation is altered. They cannot therefore correspond to the holes, for which the rotational symmetry implies a cyclotron frequency that is independent of θ. They thus correspond to the *electron ellipsoids*. With the chosen orientation of $\mathbf{B}$ in the plane bisecting xOy, the angles θ between the field $\mathbf{B}$ and the x and y axes are equal and the corresponding cyclotron frequencies are equal. Resonance 3, the most intense, therefore corresponds to the four paraboloids with axes the x and y axes. Cyclotron resonance 2 corresponds to the two z axis

Fig. P8.9 Assignment of the four resonance frequencies (see text)

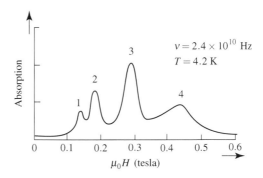

paraboloids. If the field is rotated out of the plane bisecting **xOy**, resonance 3 splits into two distinct resonances, because the x and y axis paraboloids are now distinguished.

7. Resonances 1 and 4 correspond to the *holes in the two valence bands*. Resonance 4 corresponds to the highest resonance field and thus also the highest effective mass. From the bending of the energy bands shown in Fig. P8.8, it is therefore the flattest *doubly degenerate* hole band with the highest energy, and therefore the most intense signal.

8. Resonance 1 occurs for $B \simeq 0.13$ tesla, with $\omega/B = e/m^*_{h1}$. Since $\omega/0.86\,\text{T} = e/m_0$, it follows that

$$\boxed{m^*_{h1} \simeq 0.15 m_0}\,.$$

Resonance 4 occurs for $B \simeq 0.44$ tesla, and corresponds to

$$\boxed{m^*_{h2} \simeq 0.5 m_0}\,.$$

Resonance 2 occurs for $B \simeq 0.18$ tesla, so $m_{c_1} \simeq 0.20 m_0$. It corresponds to $\theta_1 = \pi/6$, and

$$\frac{1}{m^2_{c_1}} = \left(\frac{3/4}{m^{*2}_\perp} + \frac{1/4}{m^*_\perp m^*_\parallel} \right).$$

Resonance 3 occurs for $B \simeq 0.29$ tesla, so $m_{c_2} \simeq 0.34 m_0$. It corresponds to θ_2, shown on Fig. P8.7 (right), and

$$\frac{1}{m^2_{c_2}} = \left(\frac{1/8}{m^{*2}_\perp} + \frac{7/8}{m^*_\perp m^*_\parallel} \right).$$

The above equalities imply that

$$\boxed{m^*_\perp \simeq 0.18 m_0}\,, \qquad \boxed{m^*_\parallel \simeq 1.1\, m_0}\,.$$

9. A cyclotron periodic motion can only be well defined if electrons undergo several cyclotron precessions without suffering collisions. The resonance will not be observed if the collision time τ is shorter than the cyclotron period $2\pi/\omega_c$. Collisions will broaden the resonance by an amount that depends on $1/\tau$, and which will be greater for higher values of $1/\omega_c\tau$. Under the given experimental conditions, ω_c is fixed and it is the field B that varies. The widths at half maximum ΔB of the resonances increase with B_c, and it can be seen that $\Delta B/B_c \simeq 0.2$. This therefore suggests that *the collision time determines these widths* with $\omega_c\tau \approx 10$. If the temperature is lowered, the collision time will increase and the cyclotron resonances should narrow. On the other hand, raising the temperature should broaden the cyclotron resonances.

Note: In principle, collision times differ for electrons and the various hole pockets. Indeed, the values of $\Delta B/B_c$ are $\simeq 0.12$ and $\simeq 0.30$ for the holes (1 and 4, respectively). For resonances 2 and 3, which correspond to electrons, the collision times should be the same. It can be checked that in this case $\Delta B/B_c$ has the same value $\simeq 0.15$ for the two resonances.

10. We observe two contrasting effects of temperature. We would like to increase the number of excited carriers by raising the temperature, but then the resonances broaden and can no longer be detected. But if we lower the temperature to narrow the resonances, the number of thermally excited carriers is reduced. Since the number of carriers goes as $\exp(-E_g/k_B T)$, where E_g is the band gap energy, we lose far more carriers when T is decreased than we gain in spectral resolution, so there is no particular advantage in lowering the temperature. In fact, with $E_g \simeq 1$ eV, the number of thermally excited carriers in silicon at $T = 4.2$ K is too low to be able to detect cyclotron resonances. The experiment shown in Fig. P8.9 was thus carried out by *illuminating the sample with a laser* in which the photons have an energy slightly greater than E_g. In this way, *electron–hole pairs can be excited* optically and enough free carriers are created to be able to observe the cyclotron resonances *with all the benefits of a long collision time at low temperatures*.

Metals

1. For a metal with spherical Fermi surface, question 6 of 8.1 implies that the cyclotron frequency does not depend on the angle θ. Therefore, as for holes in silicon, all electrons at the Fermi level correspond to the same cyclotron frequency, with effective mass determined by the curvature of $E(k)$ at the Fermi level.

2. By question 2 of 8.1,

$$\frac{d\mathbf{k}}{dt} = -\frac{e}{\hbar^2}\nabla_\mathbf{k} E \wedge \mathbf{B}.$$

Problem 8: Cyclotron Resonance

The vector $d\mathbf{k} = dk_\parallel$ is tangential to (C) in the plane perpendicular to $\mathbf{B}$ (see Fig. P8.6). The period of the motion is given by

$$T = \oint_{(C)} dt = \oint_{(C)} \frac{dk_\parallel}{|\nabla_\mathbf{k} E|} \frac{\hbar^2}{eB}.$$

Now, $dE = |\nabla_\mathbf{k} E| dk_\perp$, so

$$T = \frac{\hbar^2}{eB} \frac{1}{dE} \oint_{(C)} dk_\perp dk_\parallel = \frac{\hbar^2}{eB} \frac{dS}{dE}.$$

and finally,

$$\boxed{\omega_c = \frac{2\pi eB}{\hbar^2} \frac{dE}{dS}}.$$

3. The Fermi surface of copper is almost spherical, but the contacts of the Fermi surface with the first Brillouin zone correspond to parts of the Fermi surface which deviate markedly from sphericity. When the field is applied parallel to one of the cube diagonals, some cyclotron orbits at the Fermi level are very different from orbits corresponding to a sphere. Note that, if we consider a cross-section of the Fermi surface parallel to the (111) axis, we find two regions in which the density of cyclotron orbits with similar characteristics is very high. They correspond to orbits for which the Fermi surface can be approximated by a tangential cylinder with axis parallel to B. Figure P8.10 shows orbits at the bellies (B) and at the necks (N) of the Fermi surface, which lead to the *two families of orbits*, called extremal orbits (belly orbits and neck orbits), which dominate the cyclotron resonance spectrum. The belly orbits (B) are not significantly modified if $\mathbf{B}$ moves away from the (111) axis, while the neck orbits (N) only exist for $\mathbf{B} \parallel (111)$.

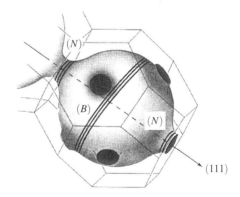

Fig. P8.10 Belly and neck orbits on the Fermi surface of copper

Problem 9: Phonons in Solids[6]

The aim in this problem is to describe the atomic vibrational modes in a crystal, along with their influence on the thermodynamic properties and resistivity of metals.

9.1: Einstein Model

We begin by considering the simplest quantum model that can account for atomic vibrations in a crystal containing N atoms (or ions) with equilibrium positions at the lattice points. This model due to Einstein treats the small-amplitude oscillations of the crystal as harmonic vibrations of its atoms about their equilibrium positions. These atomic vibrations, with angular frequency ω, are assumed to be mutually independent. The Hamiltonian for the ensemble of atoms is then a sum of Hamiltonians of independent 3D harmonic oscillators.

Recall the quantisation of the harmonic oscillator energies:

$$E_n = \left(n + \frac{1}{2}\right)\hbar\omega,$$

where $n = 0, 1, 2, 3, \ldots$. An energy quantum $\hbar\omega$ in the system is called a phonon. Thermal excitation of vibrations in the solid thus corresponds to an increase in the number of phonons with temperature.

1. *Write down the partition function for this ensemble of harmonic oscillators and deduce the temperature dependence of the internal energy.*

2. *Determine the contribution of these vibrational modes to the specific heat of the crystal, and check that, at high temperatures, it does indeed correspond to the classically expected Dulong–Petit law.*

In non-magnetic insulating solids, the specific heat measured at low temperatures varies as a power law in T, usually T^3.

3. *Explain why the specific heat of electronic origins is negligible in such solids. Show that the Einstein model does not apply for low temperatures.*

9.2: Debye Model

A weak point of the Einstein model is that it treats the atomic motions as completely uncorrelated. In reality, whenever an atom moves away from its equilibrium position, this leads to forces on its neighbours. Atomic motions are therefore coupled.

---

[6] This problem has been designed with S. Biermann and G. Montambaux.

The Debye model is the simplest one taking this fact into account. We begin by considering a linear chain of N atoms, and vibrations that can occur along the chain.

We assume that, between two neighbouring atoms of mass M a distance a apart, there is an elastic force $f_{n,n+1} = -K(u_n - u_{n+1})$, where $u_n \ll a$ is the displacement of atom n from its equilibrium position. The Hamiltonian for this system is thus

$$\mathcal{H} = \sum_n \left[\frac{p_n^2}{2M} + \frac{1}{2} M \omega_0^2 (u_{n+1} - u_n)^2 \right],$$

where the eigenfrequency ω_0 is related to the stiffness constant K by $\omega_0 = \sqrt{K/M}$.

a. Eigenmodes

The motion of an atom is thus coupled with that of its neighbours. We start by determining the eigenfrequencies of this chain of atoms. We impose periodic boundary conditions, identifying atom $N+1$ with the first atom of the chain.

4. *Show that, if the motions are decomposed into modes,*

$$u_n(t) = \frac{1}{\sqrt{N}} \sum_q \tilde{u}_q(t) e^{iqna},$$

$$p_n(t) = \frac{1}{\sqrt{N}} \sum_q \tilde{p}_q(t) e^{iqna},$$

the total Hamiltonian can be written as a sum of independent harmonic oscillator terms, viz.,

$$\mathcal{H} = \sum_q \left(\frac{|\tilde{p}_q|^2}{2M} + \frac{1}{2} M \omega_q^2 |\tilde{u}_q|^2 \right),$$

where the frequencies ω_q of these new oscillators are given by

$$\omega_q = 2\omega_0 \left| \sin \frac{qa}{2} \right|.$$

This relation can be used to determine the speed of sound c in the solid in the long wavelength limit $\omega_q = cq$.

5. *Why is the speed of sound greater in a solid than in a gas?*

b. Quantisation: Phonons

This Hamiltonian thus describes an ensemble of independent harmonic oscillators corresponding to the various eigenmodes. The energy of each mode involves a different quantum number n_q through the relation

$$E_q = (n_q + 1/2)\hbar\omega_q .$$

6. *What is the average energy of a harmonic oscillator at finite temperature? Show that the average energy of the atomic chain is*

$$\langle E(T) \rangle = \sum_q \left(\langle n_q \rangle + \frac{1}{2} \right) \hbar\omega_q , \qquad \langle n_q \rangle = \frac{1}{e^{\beta\hbar\omega_q} - 1} .$$

Deduce by analogy with photons that each classical mode can be represented quantum mechanically by bosonic excitations, viz., phonons.

c. Linear Diatomic Chain

We now consider more realistic crystalline solids in which the primitive cell contains several atoms. A simple illustration of the resulting changes in the eigenmodes can be obtained by considering a linear chain comprising two types of atom with masses M_1 and M_2, a distance a apart. To simplify, we assume that nearest neighbours are always coupled via the same stiffness constant K.

7. *Write down the equations of motion in a semiclassical approximation and find the eigenfrequencies.*

8. *Show that, in the limit $q \to 0$, these eigenfrequencies become*

$$\omega_O^2 = 2K \left(\frac{1}{M_1} + \frac{1}{M_2} \right) , \qquad \omega_A^2 = \frac{2K}{M_1 + M_2} q^2 a^2 .$$

9. *Represent the eigenfrequency spectrum in the first Brillouin zone. Why do you think the two branches are called the acoustic branch and the optical branch? What happens if $M_1 \to M_2$?*

9.3: Experimental Detection of Phonons

Phonon modes can be detected by inelastic neutron or X-ray scattering, in which the incident particle absorbs or emits a phonon (see Chap. 12). Energy and momentum conservation are used to determine the dispersion curves $\omega(\mathbf{q})$ of the phonons involved in the scattering process. Figure P9.1 shows just such an experimental determination for graphite. This is a hexagonal crystal obtained by superposing 2D graphene sheets in which the atoms are arranged in a honeycomb pattern, with two carbon atoms per unit cell. The dispersion curves $\omega(\mathbf{q})$ are represented in the directions of highest symmetry of the first Brillouin zone for this structure.

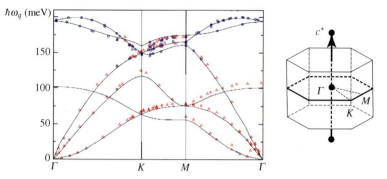

Fig. P9.1 Phonon modes detected in graphite by inelastic X-ray scattering, plotted in directions ΓM, MK, and ΓK of the hexagonal first Brillouin zone of graphite shown *on the right*. Figure courtesy of M. Mohr, from Mohr, M., et al.: Phys. Rev. B **76**, 035439 (2007)

Note that the atomic vibrations are not generally restricted to longitudinal vibrations, and the atoms also vibrate in directions transverse to the wave propagation vector.

10. *On Fig. P9.1, indicate the 'acoustic' and 'optical' modes. Can you explain the characteristics of the modes observed experimentally?*

9.4: Thermodynamic Properties

Given dispersion relations $\omega(\mathbf{q})$ like the ones in Fig. P9.1, describing the phonon eigenmodes, we can determine the density of eigenmodes $g(\omega)$, i.e., the number $g(\omega)d\omega$ of oscillators with frequencies between ω and $\omega+d\omega$. Their effects on thermodynamic properties can thus be found in a completely general way.

11. *Show that the specific heat is given by*

$$ C = k_B \int_0^\infty \left[\frac{\beta \hbar \omega}{2 \sinh\left(\beta \hbar \omega/2\right)} \right]^2 g(\omega) d\omega \ . $$

12. *Deduce that, at high temperatures, the specific heat is given by the Dulong–Petit law, whatever the frequency distribution $g(\omega)$.*

In contrast, the low temperature behaviour of the specific heat depends on the form of the function $g(\omega)$ at low frequencies. The Debye model takes the phonon spectrum to be a straight line $\omega(\mathbf{q}) = c|\mathbf{q}|$, and bounds it by the Debye frequency ω_D such that the total number of modes is $3N$.

13. *Explain why this model is realistic for low temperature properties. Determine $g(\omega)$ and the Debye temperature $\theta_D = \hbar\omega_D/k_B$.*

14. *Show that this Debye model explains the observed behaviour $C \propto T^3$. What would be the T dependence in a d-dimensional space?*

Problem 9: Phonons in Solids

9.5: Resistivity

The aim here is to study the temperature dependence of the resistivity due to phonons in pure metals.

15. *Show that, at high temperatures, the number of phonons in a mode* **q** *is proportional to T. Why should the resistivity be proportional to this number?*

At low temperatures, the almost universal temperature dependence $\rho(T)$ of the resistivity shown in Fig. 4.8 goes as T^5 (Bloch–Grüneisen law).

16. *Use conservation of quasi-momentum to show that only phonons with energies close to $k_B T$ can be absorbed or emitted during an electron collision. What can you deduce about the number of phonons that can take part in a collision process?*

In order to justify the observed low temperature behaviour, we must also understand why the different collision processes suffered by the electrons do not equivalently limit the current in the direction of the applied electric field.

17. *Show that this leads to an extra factor $1 - \cos\theta$ in the scattering probability, where θ is the scattering angle. Does this produce a T^5 dependence in the resistivity? Would another temperature-dependent factor be conceivable?*

Solution

Einstein Model

1. With the quantisation of the levels

$$E_n = \left(n + \frac{1}{2}\right)\hbar\omega, \qquad n = 0, 1, 2, \ldots, \qquad (13.12)$$

the canonical partition function Z_N is given as a function of temperature by

$$Z = \left(\sum_{n=0}^{\infty} e^{-\beta E_n}\right)^{3N} = \left(\frac{e^{-\beta\hbar\omega/2}}{1 - e^{-\beta\hbar\omega}}\right)^{3N} = \left(\frac{e^{\beta\hbar\omega/2}}{e^{\beta\hbar\omega} - 1}\right)^{3N}. \qquad (13.13)$$

It follows that the energy is

$$\boxed{U = -\frac{\partial \log Z}{\partial \beta} = 3N\left(\frac{\hbar\omega}{2} + \frac{\hbar\omega}{e^{\beta\hbar\omega} - 1}\right).} \qquad (13.14)$$

2. The specific heat is therefore

$$\boxed{C = \frac{\partial U}{\partial T} = 3Nk_B \left[\frac{\beta\hbar\omega}{2\sinh(\beta\hbar\omega/2)}\right]^2.} \qquad (13.15)$$

3. In non-magnetic insulating solids, the valence and conduction bands are separated by a band gap E_g generally greater than 1 eV. In the ground state at $T = 0$, the conduction band is empty and the valence band is fully occupied. When T increases, a small number of electrons are excited into the conduction band. There is indeed an increase in internal energy of electronic origins, but it is very small because it is determined by the occupation factor, which goes as $\exp(-E_g/k_B T)$. When $T \to 0$, this same exponential factor will thus also dominate the behaviour of the specific heat. The latter is therefore low, even for semiconductors.
Likewise for the Einstein model, the temperature dependence of the specific heat per atom obtained in (13.15) is dominated at low T by the factor $\exp(-\hbar\omega/k_B T)$. Once again this arises because of the energy gap $\hbar\omega$ between the ground state and excited states. The Einstein model cannot therefore explain the dependence $C \propto T^3$ found experimentally at low T in non-magnetic insulating solids.

Problem 9: Phonons in Solids

Debye Model

a. Eigenmodes

4. The Hamiltonian is thus

$$\mathcal{H} = \sum_n \left[\frac{p_n^2}{2M} + \frac{1}{2}M\omega_0^2(u_{n+1} - u_n)^2 \right]. \tag{13.16}$$

The motions are first decomposed into 'modes':

$$u_n(t) = \frac{1}{\sqrt{N}} \sum_q \tilde{u}_q(t) e^{iqna}, \qquad p_n(t) = \frac{1}{\sqrt{N}} \sum_q \tilde{p}_q(t) e^{iqna}. \tag{13.17}$$

As for the electronic band structure, periodic boundary conditions lead to quantisation $q = 2\pi m/Na$ of the values of q, and the translational symmetry of the crystal lattice allows us to restrict the representation of the modes to values of q belonging to a primitive cell of the reciprocal lattice, here the first Brillouin zone $qa \in (-\pi, \pi]$. The number of values of q and hence the number of modes is precisely the number N of atoms.

To rewrite the Hamiltonian, note that

$$\sum_n p_n^2 = \frac{1}{N}\sum_{n,q,q'} \tilde{p}_q \tilde{p}_{q'} e^{ina(q+q')} = \sum_q |\tilde{p}_q|^2, \tag{13.18}$$

because, for the quantised values of q given above, we have

$$\sum_{n=1}^N e^{iqna} = 0, \quad \text{for } q \neq 0, \text{ and then } \sum_{n=1}^N e^{i(q+q')na} = N\delta_{q,-q'}.$$

Likewise,

$$u_{n+1} - u_n = \frac{1}{\sqrt{N}} \sum_q u_q e^{iqna}(e^{iqa} - 1), \tag{13.19}$$

$$\sum_n (u_{n+1} - u_n)^2 = \frac{1}{N} \sum_{n,q,q'} u_q u_{q'} (e^{iqa} - 1)(e^{iq'a} - 1) e^{ina(q+q')} \tag{13.20}$$

$$= \sum_q 4\sin^2\left(\frac{qa}{2}\right) |u_q|^2. \tag{13.21}$$

The Hamiltonian can then be rewritten as a sum of independent harmonic oscillator terms:

$$\mathcal{H} = \sum_q \left(\frac{|\tilde{p}_q|^2}{2M} + \frac{1}{2} M \omega_q^2 |\tilde{u}_q|^2 \right), \quad \text{with } \omega_q = 2\omega_0 \left| \sin\left(\frac{qa}{2}\right) \right|. \quad (13.22)$$

This dispersion relation is symmetrical with respect to $q = 0$. The modes for $q < 0$ and $q > 0$ represent identical modes propagating in opposite directions. The dispersion relation can thus be represented in the half-Brillouin zone $(0, \pi/a)$, as shown on the right of Fig. P9.2.

5. In the long wavelength limit, (13.22) is linear:

$$\omega_q = cq, \quad c = \omega_0 a = a\sqrt{K/M}. \quad (13.23)$$

Long wavelength vibrations correspond to sound waves in the solid, with c the speed of sound. According to (13.23), the speed of sound is determined by the interatomic distance a and the coupling constant K between atoms. Even though interatomic distances are greater in gases than in solids, the coupling constants K are much weaker in gases, so sound propagates much faster in solids than in gases, e.g., 1,300 m/s in lead, 3,000 m/s in copper, and 5,100 m/s in aluminium. In a gas like nitrogen, at room temperature and pressure, it is a mere 350 m/s.

b. Quantisation: Phonons

6. The Hamiltonian (13.22) thus describes an ensemble of quantised harmonic oscillators. It has eigenvalues

$$E = \sum_q \left(n_q + \frac{1}{2} \right) \hbar \omega_q, \quad (13.24)$$

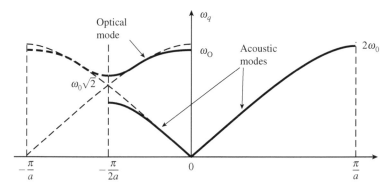

Fig. P9.2 Phonon dispersion relation for a monatomic linear chain (*right*) and for a diatomic linear chain (*left*)

where $n_q = 0, 1, 2, \ldots$. As we saw in (13.14), the energy of a harmonic oscillator at finite temperature is

$$U = \frac{\hbar\omega}{2} + \frac{\hbar\omega}{e^{\beta\hbar\omega} - 1}. \tag{13.25}$$

The energy of the chain at finite temperature is then

$$\langle E(T) \rangle = \sum_q \left(\langle n_q \rangle + \frac{1}{2} \right) \hbar\omega_q, \qquad \langle n_q \rangle = \frac{1}{e^{\beta\hbar\omega_q} - 1}. \tag{13.26}$$

We see therefore that the vibrational excitations of the chain involve N independent modes. The vibrational energy in each mode is quantised. One mode of quasi-momentum q involves n_q phonons of energy $\hbar\omega_q$. The number of phonons excited in a mode represents the energy accumulated in this mode, which is classically directly related to the amplitude of the vibration by

$$E_q = M\omega_q^2 \langle u_q^2 \rangle. \tag{13.27}$$

The analogy with photons is clear. We recover the duality between wave and particle: photons are particles corresponding to excited electromagnetic modes in the vacuum. In the same way, the energy quanta called phonons can be thought of as quasi-particles associated with the wavelike displacement modes excited in a solid. This analogy also extends to magnons, which are magnetic excitations associated with the wavelike modes of spin waves in materials exhibiting magnetic order at $T = 0$. The numbers of photons, phonons, or magnons thermally excited are unbounded and *obey Bose–Einstein statistics*. These excitations can thus legitimately be considered as bosons, the total number of which is not conserved, i.e., the chemical potential is zero.

The creation and annihilation operator formalism developed for photons can be generalised to phonons. Indeed, it is commonly used in textbooks on quantum mechanics or statistical physics to describe the harmonic oscillator, and was applied in Chap. 12 to the case of magnon modes. We shall see later that, during an interaction between an incident particle and a solid, the particle can be scattered by absorbing or emitting a phonon. In this process, energy and momentum, or rather quasi-momentum, are conserved, up to addition of the reciprocal lattice constant (a reciprocal lattice vector in the 3D case).

c. Diatomic Linear Chain

7. In this case, atoms of masses M_1 and M_2 are located at $2na$ and $(2n+1)a$, respectively. If u_{2n} and u_{2n+1} denote their respective displacements, the Hamiltonian is

$$\mathcal{H} = \sum_n \left[\frac{p_{2n}^2}{2M_1} + \frac{p_{2n+1}^2}{2M_2} + \frac{1}{2} K(u_{2n+1} - u_{2n})^2 \right]. \tag{13.28}$$

This leads to equations of motion that couple the motions of the atoms of different masses:

$$M_1 \frac{d^2 u_{2n}}{dt^2} = -K(2u_{2n} - u_{2n-1} - u_{2n+1}) \qquad (13.29)$$

$$M_2 \frac{d^2 u_{2n+1}}{dt^2} = -K(2u_{2n+1} - u_{2n} - u_{2n+2}). \qquad (13.30)$$

Here we look for solutions for harmonic modes of wave vector q, defining distinct vibrational amplitudes u_q and v_q for the atoms of masses M_1 and M_2, respectively:

$$u_{2n}(t) = u_q e^{i(2qna + \omega_q t)}, \qquad u_{2n+1}(t) = v_q e^{i[q(2n+1)a + \omega_q t]}. \qquad (13.31)$$

We obtain the two equations

$$-M_1 \omega_q^2 u_q = -K \left[2u_q - v_q (e^{iqa} + e^{-iqa}) \right], \qquad (13.32)$$

$$-M_2 \omega_q^2 v_q = -K \left[2v_q - u_q (e^{iqa} + e^{-iqa}) \right], \qquad (13.33)$$

or

$$u_q \left(M_1 \omega_q^2 - 2K \right) + v_q (2K \cos qa) = 0, \qquad (13.34)$$

$$u_q (2K \cos qa) + v_q \left(M_2 \omega_q^2 - 2K \right) = 0. \qquad (13.35)$$

There are only non-trivial solutions for u_q and v_q if the determinant of the system is zero, i.e., if

$$\omega_q^4 - 2K \left(\frac{1}{M_1} + \frac{1}{M_2} \right) \omega_q^2 + \frac{4K^2}{M_1 M_2} \sin^2 qa = 0. \qquad (13.36)$$

This second order equation in ω_q^2 has solutions

$$\boxed{\omega_q^2 = K \left(\frac{1}{M_1} + \frac{1}{M_2} \right) \pm K \left[\left(\frac{1}{M_1} + \frac{1}{M_2} \right)^2 - \frac{4}{M_1 M_2} \sin^2 qa \right]^{1/2}}. \qquad (13.37)$$

8. For $q \to 0$, the modes are thus given by

$$\boxed{\omega_O^2 = 2K \left(\frac{1}{M_1} + \frac{1}{M_2} \right), \qquad \omega_A^2 = \frac{2K}{M_1 + M_2} q^2 a^2}. \qquad (13.38)$$

9. There are two values of ω_q for each q, i.e., two branches on the dispersion curves. As for the monatomic chain, the values of ω_q are the same for $q > 0$ and $q < 0$, so

Problem 9: Phonons in Solids

the phonon dispersion relations can be represented in half of the Brillouin zone. Here the primitive cell of the chain has dimension $2a$, and the dispersion curves are represented in $(-\pi/2a, 0)$, on the left of Fig. P9.2.

One branch of the dispersion curves vanishes for $q \to 0$ and has the same q dependence as for the monatomic chain. Since this solution corresponds to $u_q/v_q \to 1$ for $q \to 0$, the two types of atom vibrate in phase. For $q \to 0$, there is therefore a long wavelength rigid displacement of the crystal structure. This corresponds to what is expected in mechanics for an acoustic deformation mode of the medium. It is therefore called the *acoustic branch*, as for the monatomic chain.

Regarding the other branch, we find that $u_q/v_q \to -M_2/M_1$ for $q \to 0$ and the two types of atom then vibrate completely out of phase. In ionic crystals like NaCl, where the two types of atom carry opposite charges, this vibrational mode can be excited by coupling with an electromagnetic wave, since its electric field will exert opposite forces on the different atoms. As this phonon mode can be excited optically, phonon branches for which the energy does not vanish when $q = 0$ are traditionally known as *optical branches*, even when the atoms carry charges of the same sign.

Note that, when $M_1 = M_2$, the chain that previously had a primitive cell of length $2a$ becomes once again a monatomic chain of lattice constant a. Its Brillouin zone is then $(-\pi/a, \pi/a)$, and the two branches

$$\omega_q^2 = \frac{2K}{M_1}(1 \pm \cos qa)$$

are degenerate for $q = \pi/2a$, where they correspond to solutions $u_q \to \pm v_q$. They do indeed correspond to the solution $\omega_q = (4K/M_1)^{1/2} \sin|qa/2|$ obtained for the monatomic chain, which is folded in the half-Brillouin zone.

Experimental Detection of Phonons

10. In graphite, composed purely of carbon, all the atoms have the same atomic mass. However, there are two carbon atoms per unit cell, with non-identical environments. Whether the atomic sites are distinguished by a difference in their mass, their position in the unit cell, or the force constant K, the result is in all cases analogous to the one obtained above for the diatomic chain. Experimentally, it is found that the phonon dispersion curves determined for graphite have acoustic and optical branches (see Fig. P9.3). Since we are considering 3D vibration modes, there are three non-degenerate accoustic modes and three non-degenerate optical modes in this case. The structure of graphite is highly anisotropic. The interatomic distance from plane to plane is 3.35 Å. This is determined by Van der Waals interactions. On the other hand, within a plane, C–C bonds correspond to a distance of only 1.42 Å. As a consequence, vibration

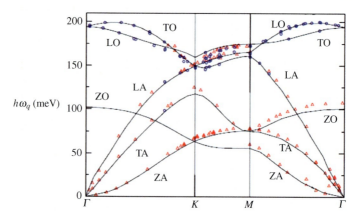

Fig. P9.3 Acoustic and optical phonon modes: longitudinal (LA, LO), transverse in the graphite plane (TA,TO), and transverse to the graphite plane (ZA, ZO). Note the lower energy of the latter

modes transverse to the graphite planes (ZA, ZO) are naturally of much lower energy than those within a plane.

Phonon mode calculations with the harmonic coupling model can be generalised to the case where the couplings are not limited to nearest neighbours. In Fig. P9.3, continuous curves are modes calculated taking into account up to fifth neighbours. The agreement with experiment is remarkable.

These observations are in fact very general. The total expected number of phonon modes is equal to the number of degrees of freedom of the crystal. For a crystal with p atoms per unit cell, *we expect 3p phonon branches*, each involving N modes for a crystal with N primitive cells. This is illustrated in Fig. P9.4, which shows an experimental determination of the phonon modes in the metallic compound Sr_2RuO_4. The Fermi surface of this compound is shown on the cover of this book. It has the same atomic structure as La_2CuO_4 (see Problem 17: *Electronic Properties of La_2CuO_4*). In this case, there are $p = 7$ atoms per unit cell, hence 21 phonon branches, as observed by inelastic neutron scattering. Note that the three acoustic modes split away from most of the optical modes. Once again, harmonic oscillator calculations up to fifteenth neighbours account exceptionally well for the experimental observations.

Thermodynamic Properties

11. The specific heat is

$$C = k_B \sum_q \left[\frac{\beta \hbar \omega_q}{2 \sinh \left(\beta \hbar \omega_q / 2 \right)} \right]^2 = k_B \int_0^\infty \left[\frac{\beta \hbar \omega}{2 \sinh \left(\beta \hbar \omega / 2 \right)} \right]^2 g(\omega) d\omega ,$$

(13.39)

where $g(\omega)$ is the mode density.

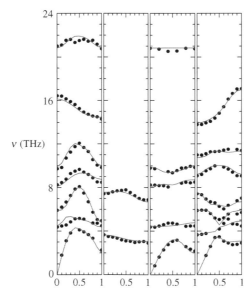

Fig. P9.4 Experimental phonon dispersion curves obtained by inelastic neutron scattering on Sr_2RuO_4. Only shown are the x dependences of the 21 branches in the direction $(x00)$ of the reciprocal lattice associated with the body-centered tetragonal structure of this compound. *Continuous curves*: Dispersion relations calculated using the harmonic oscillator approximation up to fifteenth neighbours. Figure courtesy of M. Braden, from Braden, M., Reichardt, W., Sidis, Y., Mao, Z., Maeno, Y.: Phys. Rev. B **76**, 014505 (2007)

12. At high temperatures, i.e., $k_B T$ greater than the energy $\hbar\omega_0$ of the eigenmode of highest frequency, the specific heat is given by the Dulong–Petit law, viz.,

$$C = 3Nk_B ,$$

whatever the frequency distribution $g(\omega)$. However, the low temperature behaviour depends on the low frequency behaviour of the function $g(\omega)$.

13. For $k_B T \ll \hbar\omega_0$, the exponential factor cuts off the integral at $k_B T$, so we only need to know $g(\omega)$ for low frequencies, where the spectrum is linear, i.e., $\omega = c|\mathbf{q}|$. The number of oscillators $g(\omega)d\omega$ with frequencies in the range from ω to $\omega + d\omega$ is

$$g(\omega)d\omega = 3\left(\frac{L}{2\pi}\right)^3 4\pi q^2 dq = \frac{3}{2}\frac{L^3}{\pi^2}\frac{\omega^2}{c^3}d\omega . \tag{13.40}$$

The Debye model shown schematically in Fig. P9.5 amounts to introducing a cutoff frequency such that the total number of modes remains equal to $3N$:

$$\int_0^{\omega_D} g(\omega)d\omega = 3N . \tag{13.41}$$

Fig. P9.5 Phonon density of states in the Debye model

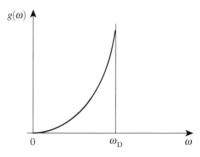

It follows that

$$\omega_D = \left(\frac{N}{V}\right)^{1/3} c \left(6\pi^2\right)^{1/3},$$

and the mode density becomes

$$g(\omega) = \begin{cases} 9N\omega^2/\omega_D^3, & \text{for } \omega < \omega_D, \\ 0 & \text{for } \omega > \omega_D. \end{cases} \quad (13.42)$$

The Debye temperature θ_D is defined by $k_B \theta_D = \hbar \omega_D$. It takes values between 72 K for mercury and 640 K for silicon.

14. To determine the low temperature behaviour of the specific heat in the Debye model, we write

$$C = k_B \int_0^{\omega_D} \left[\frac{\beta \hbar \omega}{2 \sinh(\beta \hbar \omega/2)}\right]^2 9N \frac{\omega^2}{\omega_D^3} d\omega = 72 N k_B \left(\frac{kT}{\hbar \omega_D}\right)^3 \int_0^{\beta \hbar \omega_D /2} \frac{x^4}{\sinh^2 x} dx. \quad (13.43)$$

As β tends to infinity, the upper bound of the right-hand integral tends to infinity and the integral is found to have the value $\pi^4/30$. The specific heat then becomes

$$\boxed{C = 12 \frac{\pi^4}{5} \left(\frac{T}{\theta_D}\right)^3 N k_B}. \quad (13.44)$$

Hence C goes as T^3 at low temperatures. In a d-dimensional space, it would go as T^d.

Resistivity

15. We saw earlier that the resistivity of a metal is due to electron scattering by defects perturbing the periodicity of the crystal structure. When there are no

Problem 9: Phonons in Solids

impurities, phonons are the main cause of this deviation from strict periodicity. Classically, the resistivity is therefore dominated by the scattering cross-section due to these vibrational modes, itself proportional to the square of the mean atomic displacement, viz., $\langle u_q^2 \rangle$ for the given mode. As we saw in question 6, the latter is proportional to the number of phonons in this mode.

When $k_B T > \theta_D$, the number of phonons in the mode ω_q is given by

$$\langle n_q \rangle = \frac{1}{\exp(\hbar\omega_q/k_B T) - 1} \sim \frac{k_B T}{\hbar\omega_q}. \tag{13.45}$$

It is proportional to T. Since this is true for all phonon modes, the resistivity due to phonons must indeed be linear in temperature, as observed experimentally, for example, for copper and its alloys (see Fig. 4.9).

In more detail, the electron scattering processes that dominate the resistivity involve the scattering of an electron with emission or absorption of a phonon of energy $\hbar\omega(\mathbf{q})$. If the initial electronic state has wave vector $\mathbf{k}$ and energy $\varepsilon_\mathbf{k}$, it can only be scattered into a final unoccupied electronic state of wave vector $\mathbf{k}' = \mathbf{k} \pm \mathbf{q}$. Furthermore, energy conservation implies

$$\hbar\omega(\mathbf{q}) = \pm(\varepsilon_{\mathbf{k}\pm\mathbf{q}} - \varepsilon_\mathbf{k}).$$

This restricts the wave vectors of phonons that can scatter the electron $\mathbf{k}$ to a 2D surface in the space of wave vectors $\mathbf{q}$. Since the maximal phonon energy, typically $\hbar\omega_D \simeq 0.02$ eV, is much less than the Fermi energy $E_F \simeq 1$ eV, these conditions restrict available scatterings to phonons of wave vector $\mathbf{q}$ connecting a state $\mathbf{k}$ to a state $\mathbf{k}'$ on the constant energy surface $\varepsilon_\mathbf{k}$, i.e., almost E_F, to satisfy the rule that state $\mathbf{k}$ is occupied and state $\mathbf{k}'$ is not. At high temperatures $T > \theta_D$, scattering probabilities are proportional to the occupancy $n_\mathbf{q}$ of these phonons which are all proportional to T. *So we find that the resistivity is indeed proportional to T.*

16. At low temperatures, the temperature dependence is radically different, in fact, going as T^5. We shall see that this results from a number of different factors restricting the processes in which phonons can scatter electrons.

Note that, extending the above argument, it might be thought that the total number of phonons excited at temperature T would govern the scattering process. In the Debye approximation, it is given by

$$\sum_q \langle n_\mathbf{q} \rangle = \int d\omega D(\omega) \frac{1}{\exp(\hbar\omega/k_B T) - 1} \sim T^3. \tag{13.46}$$

However, the above constraints restrict the number of phonons that can take part in scattering processes. Indeed, for $T \ll \theta_D$, only phonons with $\hbar\omega(\mathbf{q}) < k_B T$ can be absorbed or emitted by scattering by electrons at the Fermi level. For absorption, this is obvious, because these are the only modes to be significantly populated. For emission, it is due to the need to find an unoccupied final state into which the electron can be scattered. This state is lower than the initial state by an amount $\hbar\omega_q$. Since the unoccupied states of lowest energy are located at $k_B T$ below the Fermi energy, while the occupied states of highest energy are at $k_B T$ above it, only phonons of energy $k_B T$ can actually be emitted.

Since $\omega(\mathbf{q}) = c\mathbf{q}$ in this Debye limit, the surface described above containing the vectors $\mathbf{q}$ of phonons involved in scattering has linear dimension $\propto T$, and hence size T^2. The effective number of phonons available for scattering processes thus goes as T^2.

17. It is easy to see that not all collision processes affect the electron drift current in the same way. Indeed, a scattering process in which an electron in state $\mathbf{k}$ is scattered into a state $\mathbf{k}'$ that is almost collinear with $\mathbf{k}$ is much less efficient than one in which $\mathbf{k}$ goes to $\mathbf{k}' = -\mathbf{k}$. This introduces a geometrical factor

$$1 - \cos\theta \sim (\mathbf{k} - \mathbf{k}')^2 = q^2 \propto T^2 \qquad (13.47)$$

into the scattering cross-section for electron transport. This therefore leads to a temperature dependence going as T^4 for the resistivity.

The extra factor of T comes from the fact that, in the above argument, we assumed that the scattering cross-section for electrons by phonons is independent of $\mathbf{q} = \mathbf{k}' - \mathbf{k}$. In fact, for small $\mathbf{q}$, the electron–phonon coupling constant has the form [4, Chap. 26]

$$|g_\mathbf{q}|^2 \simeq \hbar\omega(\mathbf{q}) E_F \simeq cq \,.$$

At low temperatures, this introduces the missing factor of T in the scattering cross-section that needs to be taken into account when describing electron transport.

A more mathematical discussion of the scattering cross-section is possible, starting with the Fermi golden rule [2, Chap. 17]. An account can be found in [10, Chap. 9]. It leads to the result

$$\rho(T) \sim \left(\frac{T}{\omega_D}\right)^5 \int_\theta^{T_D/T} dx \frac{x^5}{(e^x - 1)(1 - e^{-x})} \,. \qquad (13.48)$$

At low temperatures, the integral is constant and a T^5 dependence is obtained. At high temperatures, the integral goes as T^{-4} and we recover the linear temperature dependence discussed above. Equation (13.48) is the *Bloch–Grüneisen law*.

Problem 10: Thermodynamics of a Thin Superconducting Cylinder: Little–Parks Experiment

The aim in this problem is to investigate the thermodynamics of a superconducting cylinder of outer radius R as a function of its thickness d (see Fig. P10.1). We shall be particularly interested in the temperature dependence of its critical field. The tube is cooled in zero field at a temperature $T < T_c$. A small external field $B = \mu_0 H$ is then applied parallel to the z axis of the tube.

1. We assume that d is much greater than the London penetration depth λ. Specify the spatial configuration of the fields and the currents in the superconductor.

The field is now applied in the normal state, and the superconductor cooled in an applied field H.

2. Does the field or current configuration change, and if so, in what way?

The conductivity of the metal is measured by placing two contacts for the current I and two for the voltage V along the tube, as shown in Fig. P10.1. The tube is cooled in zero field until $T < T_c$ and the variation of V is monitored as the applied field H is increased.

2. Explain how V varies with H.

3. For what type of superconductor and under what experimental conditions could one thereby determine the condensation energy in the bulk superconductor?

Experiments show that the thermodynamic critical field measured in a bulk sample obeys the relation

$$H_c^b(T) = H_c(0)\left[1 - \left(\frac{T}{T_c}\right)^2\right],$$

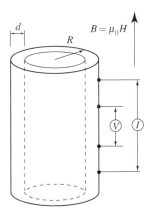

Fig. P10.1 Thin superconducting cylinder

where $\mu_0 H_c(0) = 0.03$ tesla and $T_c = 3.73$ K for tin. We now consider the case where d is not infinitely large compared with the penetration depth λ. The tube is cooled in a much higher applied field than the critical field $H_c^b(T)$ of the bulk material, down to $T < T_c$, and V is measured for a decreasing field H at fixed T. The transition fields $H_c^d(T)$ measured for a tin tube of thickness $d = 2500$ Å are plotted in Fig. P10.2.

5. Show that $H_c^d(T) > H_c^b(T)$, and give a qualitative explanation.

6. Assuming that the radial profile of the field is similar to the one observed for a thin film of thickness $d = 2a$, as studied earlier in Chap. 6, find the expected value of the critical field $H_c^d(T)$ in the limit $d \ll \lambda$ according to the London phenomenological model.

For the bulk material, the T dependence of λ is well modelled by

$$\lambda(T) = \lambda_0 \left[1 - \left(\frac{T}{T_c}\right)^4\right]^{-1/2},$$

with $\lambda_0 = 600$ Å for tin.

7. How is $H_c^d(T)$ expected to behave near $T = T_c$? Keep only the first term of the expansion in $t = 1 - T/T_c$.

8. How do the experimental results compare quantitatively with this result in the London model?

In fact, in the above argument, we have neglected flux quantisation, and also the microscopic physical origins of superconducting pair condensation. We begin by considering the first of these features.

9. Show that the superfluid current density J_s is quantised and given by

$$J_s = \frac{n_s \hbar q}{2m^* R}\left(N - \frac{\Phi}{\Phi_0}\right),$$

where $m^* = 2m_0$ is the mass of the superconducting carriers, $q = 2e$ is their charge, Φ is the magnetic flux, $\Phi_0 = h/2e$ is the flux quantum, and n_s is the density of superconducting electrons.

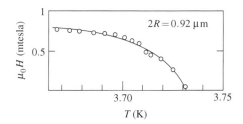

Fig. P10.2 Fields $H_c^d(T)$ measured for a tin tube of thickness $d = 2500$ Å

Fig. P10.3 Variation of V versus the applied field for different temperatures. From Groff, R.P., Parks, R.D.: Phys. Rev. **176**, 567 (1968), with the permission of the American Physical Society (© 1968 APS). http://link.aps.org/doi/10.1103/PhysRev.176.567

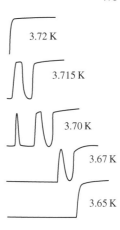

In fact, experimentally, one measures the behaviour of V, as shown in Fig. P10.3, for $2R = 1.35$ μm, showing that there are transitions at several field values for a given value of T. (Figure P10.2 concerns only the transition corresponding to the weakest field.) This suggests that, in the above arguments, we have neglected some contribution to the free energy that is sensitive to flux quantisation. We shall express it in the form $E^* = \gamma J_s^2$, where γ is a constant of proportionality.

10. *What could such a contribution correspond to physically? What coefficient γ would you use?*

Figure P10.4 shows all transitions obtained as a function of T, for experiments carried out on tin tubes of different radius R but the same thickness $d = 2,500$ Å. Curves joining the experimental points are guides to the eyes.

11. *How does this extra free energy term depend on the applied field? For what values of H does it vanish? What would you expect if this term was the only one responsible for the dependence of T_c on H?*

12. *Check the main features that can be explained this way against the experimental results.*

The London hypothesis fixes the condensation energy in the superconducting state as

$$G_s - G_n = -\frac{1}{2}\mu_0 \left[H_c^b(T)\right]^2,$$

but does not explain its physical origin. Under the London hypotheses, the density of superconducting carriers and the condensation energy are assumed to be independent of the dimension of the superconducting material. But we have seen that condensation into Cooper pairs implies that n_s can only vary in space over a distance greater than a characteristic length.

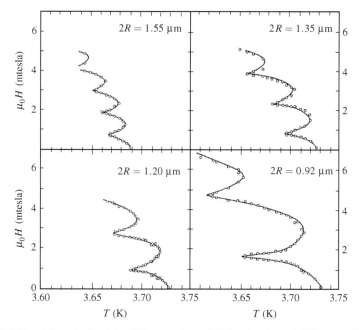

Fig. P10.4 Transitions obtained for different values of T for tin tubes of different radius R but the same thickness $d = 2,500$ Å. From Groff, R.P., Parks, R.D.: Phys. Rev. **176**, 567 (1968), with the permission of the American Physical Society (© 1968 APS). http://link.aps.org/doi/10.1103/PhysRev.176.567

13. *If the material considered here is type I and if $d < \lambda$, can n_s vary over the thickness of the tube?*

To account for the fact that n_s does not necessarily have the value expected for a bulk superconductor, Ginzburg and Landau suggested that, close to T_c, the zero-field free energy of condensation in the superconducting state could be written as an expansion in terms of n_s of the form

$$G_s - G_n = An_s + Bn_s^2/2,$$

where A and B may depend on $t = 1 - T/T_c$. Its value is determined by minimising with respect to n_s, which leads to conditions on A and B that can be obtained for the bulk material.

14. *To obtain $n_s = 0$ for $T > T_c$, show that A must change sign at T_c, and that B must be positive.*

The simplest form is thus $A = \alpha(T_c - T) = \alpha T_c t$, taking α and B constant near T_c.

15. *Show that in this case $n_s = \alpha(T - T_c)/B$ when $T < T_c$ for a bulk superconductor.*

Problem 10: Thermodynamics of a Thin Superconducting Cylinder

16. *Using the experimental forms of $H_c^b(T)$ and $\lambda(T)$ close to $T = T_c$, determine α and B as a function of $H_c(0)$, λ_0, and T_c.*

For the thin tube in an applied field, surface currents flow around the superconductor. We therefore add the free energy term E^* considered above and the energy difference between superconducting and normal states is given by

$$G_s - G_n = An_s + \frac{1}{2}Bn_s^2 + \gamma J_s^2.$$

17. *Show that the critical temperature depends on the applied magnetic flux and can be written in the form*

$$T_c(\Phi) = T_c\left[1 - \frac{\xi_0^2}{R^2}\left(N - \frac{\Phi}{\Phi_0}\right)^2\right],$$

where ξ_0 is the coherence length of the superconductor. Does this result explain the experimental observations?

18. *Combining this with the solution to question 7, do we obtain a quantitative explanation for all the experimental results?*

Solution

1. The tube is cooled in zero field at a temperature $T < T_c(0)$. The flux through the cylinder, which is initially zero, remains zero when the field $B = \mu_0 H$ outside is increased, provided that $H < H_c^b(T)$ for a type I superconductor and $H < H_{c_1}^b(T)$ for a type II superconductor. The magnetic induction remains zero inside the superconductor, because persistent currents are induced throughout a thin layer of thickness λ at the cylinder surface (see Fig. P10.5).

2. When the system is cooled in an applied field $H < H_c^b(T)$ [or $H_{c_1}^b(T)$], the flux within the cylinder remains unchanged (up to Φ_0). Currents flow on the outer face and on the inside of the cylinder in such a way as to cancel B in the bulk of the superconductor (see Fig. P10.5).

3. V is zero as long as $H < H_c^b(T)$ [or $H_{c_2}^b(T)$]. It acquires a value independent of the field, associated with the resistance of the tube in the normal state when $H > H_c^b(T)$ [or $H_{c_2}^b(T)$].

4. If the superconductor is type I (type II, respectively), this measurement gives $H_c^b(T)$ [respectively, $H_{c_2}^b(T)$]. For a type I superconductor, this determines the condensation energy in the superconducting state, which is given by

$$G_s - G_n = -\frac{1}{2}\mu_0 \left[H_c^b(T)\right]^2.$$

5. Near T_c, the experimental results are approximated by the straight line

$$H_c^b(T) = 2H_c(0)\left(1 - \frac{T}{T_c}\right),$$

the dashed line in Fig. P10.6, with $\mu_0 H_c(0) = 0.03$ tesla and $T_c = 3.73$ K for tin. We observe that this straight line lies below the experimental curve $H_c^d(T)$

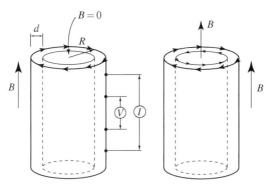

Fig. P10.5 Persistent currents induced by a field B in a superconducting cylinder after cooling in zero field (*left*) and upon cooling in the applied field (*right*)

Fig. P10.6 Linear approximation to $H_c^b(T)$ near T_c for a bulk sample (*dashed line*), and temperature dependence of the critical field expected for the small thickness ($d < \lambda$) of the cylindrical superconductor (*thick continuous curve*)

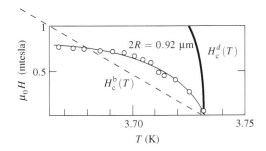

obtained for a cylinder of thickness $d = 2,500$ Å, for $T > 3.68$ K. This corresponds to the fact that, for a thin superconductor compared with the penetration depth, the magnetic induction will not be zero in the superconductor, so the field exclusion energy cost is reduced compared to the bulk superconductor. It is thus natural to find a higher critical field than the one observed for the bulk sample.

6. If the field profile is similar to the one for a thin film of thickness $d = 2a$ with $d \ll \lambda$ (which strictly applies for $R \gg d$), the increase in the critical field is then given as in Chap. 6 by

$$\boxed{H_c^d(T) = 2\sqrt{3}\frac{\lambda(T)}{d}H_c^b(T)}.$$

7. Near $T = T_c$, setting $t = 1 - T/T_c$, we have

$$\lambda(T) = \lambda_0\left[1 - (1-t)^4\right]^{-1/2} \simeq \lambda_0(4t)^{-1/2},$$

and then

$$\boxed{H_c^d(T) \simeq 2\sqrt{3}\frac{\lambda_0}{d}H_c(0)t^{1/2}}.$$

8. For $\lambda_0 = 600$ Å, $d = 2,500$ Å, and $\mu_0 H_c(0) = 0.03$ tesla, we have

$$\mu_0 H_c^d(T) = 2\sqrt{3}\frac{6}{25}0.03t^{1/2} = 0.025t^{1/2} \quad \text{(tesla)}.$$

The corresponding prediction, plotted as a continuous curve in Fig. P10.6, lies well above the curve observed experimentally, showing that there is another energy term affecting the critical field.

9. Consider a circuit (C) in the superconductor. Since this necessarily lies less than a distance λ from the edge of the tube, the quantised quantity is

$$\Phi^* = \Phi + \frac{m}{q^2\phi^2}\oint_{(C)}\mathbf{j}\cdot d\mathbf{l} = N\Phi_0,$$

with N an integer and the notation used in Chap. 5 ($\phi^2 = n_s/2$ and $m = m^*$). Since $\mathbf{j} = J_s$ varies little across the tube thickness, (C) can be taken on the outer diameter of the tube, whereupon

$$2\pi R J_s = \frac{q^2 n_s}{2m^*}(N\Phi_0 - \Phi),$$

and

$$\boxed{J_s = \frac{n_s \hbar q}{2m^* R}\left(N - \frac{\Phi}{\Phi_0}\right).}$$

10. The energy to be taken into account here is the kinetic energy of the supercurrents. These are given by $J_s = -n_s e v_s$, and their kinetic energy is

$$E^* = \frac{1}{2} n_s \frac{m^*}{2} v_s^2 = \frac{m^*}{n_s q^2} J_s^2,$$

so that

$$\boxed{\gamma = m^*/n_s q^2.}$$

11. Substituting in the quantised version of J_s, we obtain

$$\boxed{E^* = \frac{n_s \hbar^2}{4m^* R^2}\left(N - \frac{\Phi}{\Phi_0}\right)^2.}$$

The constant in front of the bracket is independent of H for fixed T, so this kinetic energy depends only on H through the flux Φ. Consequently, for fixed N, E^* is quadratic in Φ, i.e., quadratic in the applied field H. This kinetic energy term vanishes for $\Phi = N\Phi_0$. The energy cost becomes enormous if Φ deviates much from $N\Phi_0$. It is easy to show that this kinetic energy term is minimised if N changes by one unit each time $|\Phi - N\Phi_0| > \Phi_0/2$. Thermodynamic equilibrium is achieved by this adjustment of N which minimises the current. The latter grows linearly and changes sign when Φ/Φ_0 is a half-integer (see Fig. P10.7). The corresponding dependence of E^* on Φ is also shown in the figure. When the applied field has no other effect on the value of T_c, we therefore expect a *periodic dependence of T_c on the field*, where the latter resumes its maximal value each time $\Phi = N\Phi_0$.

12. This periodicity is indeed observed in the experimental curves (see Fig. P10.8). The maxima occur for values of the field H that are multiples of a fixed value H_p for each of the four tubes. The minima of T_c also occur for values equal to half this period H_p. In fact, H_p varies approximately linearly with R^{-2} (see Fig. P10.9). From the slope of the straight line, we obtain

Fig. P10.7 Dependence of J_s and E^* on Φ

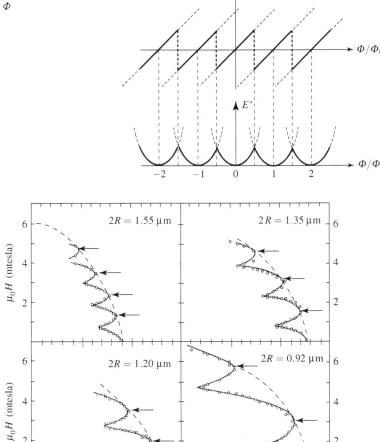

Fig. P10.8 Transitions obtained for different values of T for tin tubes of different radius R but the same thickness $d = 2,500$ Å

$$\Phi_0 = \pi R^2 \mu_0 H_p \simeq (\pi/4)(0.92)^2(3)10^{-15}$$
$$\simeq 1.99 \times 10^{-15} \text{ weber,}$$

which agrees reasonably well with the value $\Phi_0 = 2.07 \times 10^{-15}$ weber given that the experimental error in the tube diameter is at least 5%. Note also that the amplitude of the oscillations in $T_c(H)$ increases as R^{-2}, as predicted by the above equation (see Fig. P10.9). We conclude that this kinetic energy term does account for most of the experimental results.

Fig. P10.9 Linear dependence of H_p and the amplitude of the oscillations in $T_c(H)$ on R^{-2}

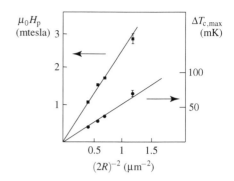

13. Since $d < \lambda$, and since $\lambda < \xi$ in a type I material, the fact that $d < \lambda$ implies $d < \xi$, which implies that n_s cannot vary significantly over the thickness of the tube.

14. Setting

$$G_s - G_n = An_s + Bn_s^2/2,$$

and minimising this free energy with respect to n_s yields

$$\boxed{n_s = -A/B},$$

with $B > 0$ for this condition to correspond to a minimum. Since $n_s \geqslant 0$, A and B have opposite signs for $T < T_c$ (hence $A < 0$), and the same sign for $T > T_c$ (to avoid there being a physical solution with n_s positive). Moreover, for $T = T_c$, we must have $n_s = 0$, and hence $A = 0$ for $T = T_c$. The parameter A must therefore grow steadily and change sign at T_c.

15. With $A = \alpha(T_c - T) = \alpha T_c t$, where $\alpha < 0$ and B is constant close to T_c, we do indeed obtain $n_s = \alpha(T - T_c)/B$ when $T < T_c$ for the bulk superconductor, and

$$\boxed{G_s - G_n = -\frac{A^2}{2B} = -\frac{1}{2}\frac{\alpha^2 T_c^2}{B}t^2}.$$

16. This value $n_s = \alpha(T - T_c)/B = -(\alpha T_c/B)t$ should be compared with the experimental expression for the temperature dependence of λ given in Chap. 5. Close to T_c, this takes the form

$$\lambda(T) = \left(\frac{m_0}{n_s e^2 \mu_0}\right)^{1/2} \simeq \lambda_0 (4t)^{-1/2},$$

or

$$\frac{n_s e^2 \mu_0 \lambda_0^2}{m_0} \simeq 4t.$$

We may thus conclude that

Problem 10: Thermodynamics of a Thin Superconducting Cylinder

$$\boxed{-\frac{\alpha T_c}{B} = \frac{4m_0}{e^2 \mu_0 \lambda_0^2}.}$$

Moreover, the experimental parabolic shape of $H_c^b(T)$ implies that, near T_c, we have $H_c^b(T) = 2H_c(0)t$, that is

$$G_s - G_n = -\frac{1}{2}\mu_0 \left[H_c^b(T)\right]^2 = -2\mu_0 H_c^2(0)t^2 = -\frac{1}{2}\frac{\alpha^2 T_c^2}{B}t^2.$$

We thus have

$$\boxed{-\alpha T_c m_0 = \left[e\mu_0 \lambda_0 H_c(0)\right]^2.}$$

17. We have

$$G_s - G_n = \alpha(T_c - T)n_s + \frac{1}{2}Bn_s^2 + \frac{m^* J_s^2}{n_s q^2}.$$

With the form of J_s obtained in question 9, we observe that the free energy E^* is linear in n_s, so

$$G_s - G_n = \alpha(T_c - T)n_s + \frac{1}{2}Bn_s^2 + \frac{m^*}{4}n_s v_s^2,$$

where $v_s = \hbar(N - \Phi/\Phi_0)/m^*R$ is independent of n_s. Minimising this free energy with respect to n_s, we obtain

$$Bn_s = -\alpha(T_c - T) - \frac{m^*}{4}v_s^2.$$

The superconducting state disappears when $n_s = 0$, i.e., when

$$T_c(\Phi) = T_c + \frac{m^*}{4\alpha}v_s^2.$$

We thus obtain

$$\boxed{T_c(\Phi) = T_c \left[1 - \frac{\xi_0^2}{R^2}\left(N - \frac{\Phi}{\Phi_0}\right)^2\right],}$$

where

$$\boxed{\xi_0^2 = -\frac{\hbar^2}{4\alpha m^* T_c}}$$

defines the coherence length of the superconductor. This is similar to the expression in question 11, but does not involve a value of n_s independent of the applied flux. It is in fact *the vanishing of n_s that defines the critical temperature under any experimental conditions*. The other experimental aspects are described in the same way as in question 11.

18. For $\Phi = N\Phi_0$, flux quantisation does not affect T_c. Its dependence on the applied field *is then entirely due to the free energy associated with flux penetration.* The value
$$\mu_0 H_c^d(T) = at^{1/2}, \qquad a = 0.025 \text{ tesla},$$
estimated in question 7 can therefore describe the temperature dependence of the fields corresponding to the maxima of T_c of the curves $T_c(H)$. Indeed, the dotted lines in Fig. P10.8 show quite clearly that these fields do vary as $(T_c - T)^{1/2}$. The numerical constant $a = 0.0416$ obtained experimentally is the same for all the tubes and is slightly greater than indicated above.

Note: Continuous curves in Fig. P10.8 are the results of a detailed calculation in which the magnetic energy is included in the Ginzburg–Landau formalism.

Problem 11: Direct and Alternating Josephson Effects in Zero Magnetic Field

The aim here is to study the characteristics of a Josephson junction subjected to a static or alternating excitation. The Josephson effect is a tunnelling of the macroscopic wave function through an insulating barrier between two superconductors (1 and 2). Recall first the characteristic equations relating the current I and the voltage V across the terminals of the junction to the phase difference $\gamma = \theta_2 - \theta_1$ of the wave function on either side of the junction:

$$I = I_c \sin \gamma, \qquad \hbar \frac{d\gamma}{dt} = 2eV. \qquad (13.49)$$

11.1: Model Josephson Junction

The junction is connected to a current source and we assume that initially $I = 0$ and $\gamma = 0$. The current $I(t)$ supplied by the current source is slowly increased up to a constant value I_0, where $|I_0| \leq I_c$.

1. What stationary state is obtained for the junction from (13.49)?

2. What energy is supplied by the current source in order to establish the current I_0 starting from $I = 0$?

We now connect the junction to a constant voltage $V = V_0$ at time $t = 0$, the earlier values of I and γ being zero.

3. What current passes through the junction?

4. Consider the case where $V_0 = 10\,\mu V$. Do you think this current could be detected directly by means of an ammeter?

11.2: Realistic Josephson Junction

In the model Josephson junction described above by (13.49), the current cannot exceed I_c and the direct voltage is zero. In a real Josephson junction, if we start with $I = 0$ and apply a voltage source V across the junction, the experimentally observed characteristic representing the steady-state current I measured for different values of V is as shown in Fig. P11.1 (left).

5. Explain the physical origin of the current arising for large V and the observed asymptotic behaviour.

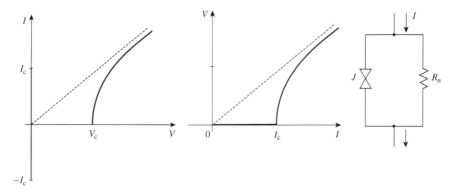

Fig. P11.1 Realistic Josephson junction: Experimentally observed characteristics $I(V)$ (*left*) and $V(I)$ (*center*) and equivalent circuit when supplied with current I (*right*)

Likewise, when a current is passed through the Josephson junction, the observed characteristic $V(I)$ is as shown in Fig. P11.1 (center).

6. *Explain the physical origin of the current for $I > I_c$ and the asymptotic behaviour observed for large I.*

We thus represent the realistic Josephson junction with current supply by the equivalent electric circuit shown in Fig. P11.1 (right), where the ideal Josephson junction J described by (13.49) is connected in parallel with a resistance R_n.

7. *Write down the equation for the behaviour of the phase γ for this equivalent circuit.*

8. *Show that, for $I(t) = I$ constant with $|I| < I_c$, we recover the same response as for the ideal junction.*

For $I > I_c$, the phase γ is no longer bounded. The equation found in question 7 remains unchanged if γ is increased by 2π. Therefore, the time dependence of γ can be decomposed into a linear variation on which is superposed a periodic variation with period τ. There corresponds a periodic variation of $V(t)$ about an average value $\overline{V}$.

9. *Determine τ and hence $\overline{V}$ as a function of I, using the fact that*

$$\int_0^{2\pi} \frac{dx}{1 - b\sin x} = \frac{2\pi}{\sqrt{1-b^2}}. \tag{13.50}$$

10. *Is the result obtained here compatible with the characteristic of Fig. P11.1 (center)? Examine the asymptotic behaviour. What is the physical meaning of R_n?*

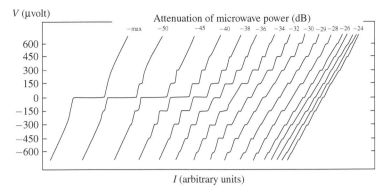

Fig. P11.2 Characteristics $V(I)$ observed for an Nb/NbO/Nb junction for different values of the applied microwave irradiation power (72 GHz). From Grimes, C.C., Shapiro, S.: Phys. Rev. **169**, 397 (1968). With the permission of the American Physical Society (© 1968 APS). http://link.aps.org/doi/10.1103/PhysRev.169.397

11.3: Josephson Junction in a Microwave Field

We consider a Josephson junction Nb/NbO/Nb (superconductor–insulator–superconductor), with a current supply, and study the characteristic $V(I)$ of this junction when it is irradiated by a microwave field of angular frequency ω. The measured characteristics $\overline{V}(I)$ are shown in Fig. P11.2 for different powers of the microwave source at frequency $\omega/2\pi = 72$ GHz. Note that, since the junction is symmetric by construction, the characteristics are symmetric with respect to the origin, i.e., $\overline{V}(-I) = -\overline{V}(I)$. The curves in the figure have thus been shifted arbitrarily along the horizontal axis, with increasing microwave power from left to right.

11. *Give a qualitative description of the observed phenomena, without attempting to interpret them for the moment.*

12. *Looking back at the results of questions 3 and 4, can you identify the main physical effect observed here?*

In order to give a quantitative explanation for the experimental results in Fig. P11.2, we shall now examine the case of a junction with a voltage supply. Analytic calculations are not feasible for a junction with a current supply. However, we assume that the observed effects will be similar.

It is assumed that the effect of the microwave field is to superpose upon the voltage V a microwave voltage $v_s \cos \omega t$ of strength v_s proportional to the amplitude of the incident microwave field.

13. *Write down the voltage–phase Josephson equation (P11.1) under these conditions and deduce an expression for the current. The relation*

$$e^{ib\sin x} = \sum_{n=-\infty}^{+\infty} J_n(b) e^{inx}$$

may be useful, where J_n is the order n Bessel function of the first kind, a real-valued function satisfying $J_{-n}(b) = (-1)^n J_n(b)$.

14. *Is the current expected to change at zero frequency? Can you explain certain experimental features noted in question 11 as observed in Fig. P11.2? Hint: For small x,*

$$J_0(x) \simeq 1 - \frac{x^2}{4}, \qquad J_n(x) \simeq \frac{1}{n!}\left(\frac{x}{2}\right)^n.$$

Solution

Model Josephson Junction

The starting point here is

$$I = I_c \sin\gamma, \qquad \hbar\frac{d\gamma}{dt} = 2eV.$$

1. When I is increased to I_0, γ increases as a function of time to reach the value $\gamma = \arcsin(I_0/I_c)$. During this variation $d\gamma/dt$ is nonzero and a voltage $V(t)$ appears, but $d\gamma/dt$ vanishes when $I = I_0$, and then $V = 0$. The stationary state thus corresponds to

$$\boxed{\gamma = \arcsin(I_0/I_c), \quad I = I_0, \quad V = 0}.$$

2. The energy supplied by the current source is

$$\Delta W = \int_0^\infty I(t)V(t)dt = \frac{\hbar I_c}{2e}\int_0^\infty \sin\gamma(t)\frac{d\gamma(t)}{dt}dt$$

$$= -\frac{\hbar I_c}{2e}\bigl[\cos\gamma(t)\bigr]_0^\infty = \frac{\hbar I_c}{2e}(1-\cos\gamma).$$

The energy of the junction therefore increases and is minimal for γ a multiple of 2π.

3. When a voltage V_0 is applied at $t = 0$, with $I = 0$ and $\gamma = 0$ initially, we obtain

$$\gamma(t) = \frac{2eV_0}{\hbar}t, \qquad \text{hence} \quad I(t) = I_c \sin\left(\frac{2eV_0}{\hbar}t\right).$$

The phase grows continuously and I is an alternating current at the Josephson frequency

$$\boxed{\nu_J = 2eV_0/h},$$

so $\nu_J/V_0 = 483.6 \times 10^{12}$ (Hz/volt). This assumes that $eV_0 < 2\Delta$, where Δ is the superconductor gap. Otherwise one also expects superconducting pairs to be broken, an effect neglected in this model.

4. For $V_0 = 10\ \mu\text{volt}$, $\nu_J \simeq 5$ GHz. An alternating current at such a high frequency could not of course be detected by a simple device like an ammeter. One might attempt to detect the electromagnetic power emitted from the junction at this frequency. This is possible but difficult, since the signal is very weak!

Realistic Josephson Junction

5. For the characteristic observed in Fig. P11.1 (left), a current flows when the applied voltage is such that eV exceeds $eV_c = 2\Delta$, where Δ is the superconductor gap. Indeed, this voltage would be required to break the Cooper pairs and give rise to 'normal' electrons. The latter can tunnel from 1 to 2. Such an electronic tunneling current is dissipative, unlike superconducting pair tunneling. When V is increased, the number of broken pairs is also increased, and for large enough V, there will be no more superconducting pairs. The tunneling current therefore tends asymptotically toward the linear characteristic for electron tunneling between normal metals, such as would be observed above T_c for this same junction. Its slope is $1/R_n$, where R_n is the *normal state junction resistance*.

6. In a setup with a current supply, when I exceeds I_c, *superconducting pairs are broken and 'normal' electrons appear*. It is the tunneling current of these electrons that adds to the current of superconducting pairs. When I is increased, the number of broken pairs also increases and for large enough I, tunneling by superconducting pairs disappears. The tunneling current therefore tends asymptotically to the linear characteristic $V = R_n I$ for tunneling by normal electrons.

7. In the circuit of Fig. P11.1 (right), the current I is the sum of the currents in the two branches, viz.,

$$I = I_c \sin\gamma + \frac{V}{R_n} = I_c \sin\gamma + \frac{d\gamma}{dt}\frac{\hbar}{2eR_n},$$

and then

$$\boxed{\frac{d\gamma}{I/I_c - \sin\gamma} = \frac{2eR_n I_c dt}{\hbar}}.$$

8. For $I < I_c$, we find that γ increases up to the value $I = I_c \sin\gamma$. As in question 1, this leads to $V = 0$.

9. For $I > I_c$, we find that γ is no longer bounded, increasing indefinitely, according to the above relation. γ increases by 2π over a period τ, and hence $2e\overline{V} = \hbar\overline{d\gamma/dt} = h/\tau$. We thus have

$$\tau = \frac{\hbar}{2eR_n I_c}\int_0^{2\pi}\frac{d\gamma}{I/I_c - \sin\gamma} = \frac{h}{2eR_n}\left(I^2 - I_c^2\right)^{-1/2},$$

using the given integral, and this yields

$$\boxed{\overline{V} = R_n\sqrt{I^2 - I_c^2}}.$$

Problem 11: Direct and Alternating Josephson Effects in Zero Magnetic Field   487

10. V is indeed zero up to $I = I_c$ and grows as $\sqrt{I - I_c}$ thereafter. The limiting behaviour does indeed correspond asymptotically to $\overline{V} = R_n I$. The above equation gives a good description of the behaviour in Fig. P11.1 (center).

Josephson Junction in a Microwave Field

11. We note the following points with regard to Fig. P11.3:
    (a) At zero microwave power, the observed characteristic does correspond to a realistic Josephson junction with current supply, as in Fig. P11.1 (center).
    (b) *Voltage plateaus* appear in the characteristic curves *for discrete values of the voltage*, which seem to be quantised and independent of the microwave power.
    (c) The width of the voltage plateau at $V = 0$, initially $2I_c$, seems to steadily diminish as the microwave power increases.
    (d) New plateaus have current widths which grow to a maximum, then decrease with increasing microwave power.
    (e) For strong irradiation, the characteristic is almost linear, and its slope is the large-current limit corresponding to the characteristic without microwave irradiation, i.e., the tunneling resistance R_n of the junction in the normal state.

12. As soon as $V \neq 0$ across the terminals of the junction, there is an alternating Josephson current, according to questions 3 and 4. The incident microwave field produces an alternating electric field in the junction. It seems likely that this field couples with the alternating Josephson field in the junction and that a resonance effect occurs when the frequencies coincide. Indeed, a Josephson frequency of 72 MHz corresponds to a voltage of about 150 μvolt. The observed plateaus, known as *Shapiro steps* after the physicist who first demonstrated this

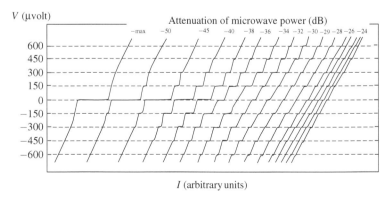

Fig. P11.3 Characteristics $V(I)$ observed for an Nb/NbO/Nb junction for different values of the applied microwave irradiation power (72 GHz)

phenomenon, occur at integer multiples of 150 μvolt. This implies that all multiples of the irradiation frequency are detected. These observations thus *reveal the existence of the alternating Josephson effect*.

13. With $V(t) = V_0 + v_s \cos \omega t$, solving the voltage–phase relation, we obtain

$$\gamma(t) = \gamma(0) + \frac{2eV_0}{\hbar}t + \frac{2ev_s}{\hbar\omega}\sin\omega t.$$

This leads to a current

$$I = I_c \sin\left[\gamma(0) + \frac{2eV_0}{\hbar}t + \frac{2ev_s}{\hbar\omega}\sin\omega t\right].$$

As the current–phase relation is non-linear, we see that the current response is not at the frequency of the microwave excitation. Indeed, it also contains the alternating Josephson frequency of questions 3 and 4 and the harmonics of this frequency.

This response can be Fourier analysed using the given expression:

$$e^{ib\sin x} = \sum_{n=-\infty}^{+\infty} J_n(b)e^{inx},$$

where J_n is the order n Bessel function of the first kind. We then have

$$e^{i(a+b\sin x)} = \sum_{n=-\infty}^{+\infty} J_n(b)e^{i(a+nx)} = \sum_{n=-\infty}^{+\infty} (-1)^n J_n(b)e^{i(a-nx)}.$$

In the second equality, we used $J_{-n}(b) = (-1)^n J_n(b)$. Taking the imaginary part of this expression with

$$a = \gamma(0) + \frac{2eV_0}{\hbar}t = \gamma(0) + 2\pi \nu_J t, \qquad b = \frac{2ev_s}{\hbar\omega},$$

we obtain

$$\boxed{I = I_c \sum_{n=-\infty}^{+\infty} (-1)^n J_n\left(\frac{2ev_s}{\hbar\omega}\right) \sin\left[\gamma(0) + (2\pi \nu_J - n\omega)t\right]}.$$

14. We check the points (b)–(e) noted for question 11:

(b) Note that the current I due to superconducting pairs has a direct component each time $2\pi \nu_J = n\omega$, i.e., for all Josephson frequencies that are *integer multiples* of the microwave irradiation frequency ω. We thus have a direct current of superconducting pairs for all discrete values V_n of V_0 such that $2eV_n = n\hbar\omega$, i.e., $V_n = n(\hbar\omega/2e)$.

Problem 11: Direct and Alternating Josephson Effects in Zero Magnetic Field   489

(c) The amplitude of this current depends on the value of $\gamma(0)$, and can take any value $|I| < I_c J_n(2ev_s/\hbar\omega)$ for the voltage V_n.

This dependence on $\gamma(0)$ is exactly analogous to what was obtained for the characteristic without irradiation at $V_0 = 0$. There the phase adjusted itself to fix the component of the supercurrent in the junction.

In the experiment where the junction is supplied with a current, the pair current adds to the one present when there is no irradiation, due to the resistance in parallel with the junction (that is, to the tunneling of 'normal' electrons). So in the presence of the microwave field, the supercurrent at zero voltage can take any amplitude up to the maximal value $|I| < I_c J_0(2ev_s/\hbar\omega)$, which itself depends on the irradiation power. Expanding J_0 to lowest order in its argument, we obtain

$$|I| < I_c \left[1 - 4 \left(\frac{ev_s}{\hbar\omega} \right)^2 \right].$$

This explains the reduction of the plateau at zero voltage as a function of irradiation power.

(d) Likewise, since the $J_n(x)$ go as x^n for small x, the current plateaus at higher voltages grow broader with the irradiation power according to

$$2I_c J_n(2ev_s/\hbar\omega) = \frac{2I_c}{n!} \left(\frac{ev_s}{\hbar\omega} \right)^n.$$

This shows that the first order plateau appears first, then the second order plateau, and so on. At increasing powers, $J_n(x)$ goes through a maximum and subsequently oscillates. The first plateau at 150 µvolt can be seen to increase and then fall off with the power.

(e) The condition $2eV_1 = \hbar\omega$ can be considered to correspond to pair tunneling helped by absorption of a microwave photon of energy $\hbar\omega$. Likewise, at the V_n plateau, n photons are absorbed simultaneously. For strong irradiation, the absorption of many photons breaks the superconducting pairs, and the normal electron current dominates the response. We then recover the linear characteristic which is the strong current limit when there is no microwave irradiation.

Problem 12: Josephson Junction in a Magnetic Field[7]

The aim here is to understand the effect of a magnetic field on certain properties of a Josephson junction: current distribution and analogies with type II superconductors (12.1), screening of the magnetic field inside a junction (12.2), and the observation of plasma resonance phenomena (12.3).

Note: 12.3 is independent of 12.2.

Consider a Josephson junction comprising an insulator placed between two identical superconductors, as in Fig. P12.1. The two superconductors occupy the regions $x < -w/2$ and $x > w/2$, labelled S_1 and S_2, respectively. The insulator is in the region J specified by

$$-w/2 < x < w/2, \qquad -L/2 < y < L/2.$$

The system is assumed infinite in the z direction. The local magnetic induction lies along the z axis, viz., $\mathbf{B} = B\mathbf{z}$, with $B \geq 0$. The field applied outside corresponds to the magnetic induction $B_a \mathbf{z} = \mu_0 H_a \mathbf{z}$.

Recall that an electric current flows between the two superconductors by tunneling of Cooper pairs. The local current density is

$$j_x(-w/2, y) = j_x(w/2, y) = j_0 \sin \delta(y), \qquad (13.51)$$

where j_0 is characteristic of the junction (it can be shown that $j_0 = -|j_0|$ is negative) and $\delta(y)$ is the local phase difference between the two superconductors. (In general

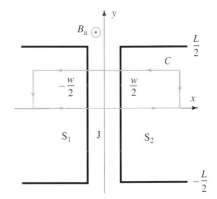

Fig. P12.1 Configuration of the Josephson junction and applied field. The superconducting (S_1, S_2) and insulating (J) regions are indicated

---

[7] This problem has been designed with P. Ledoussal.

$$\delta(y) = \theta(w/2, y) - \theta(-w/2, y) - \frac{q}{\hbar} \int_{-w/2}^{w/2} dx A_x(x,y),$$

with $q = -2|e|$, but here we can choose a gauge with $A_x = 0$.)

Let I be the total current through the junction, from S_1 to S_2, per unit length in the z direction, and λ the London penetration depth. The superconductors S_1 and S_2 are type II with an applied field $H_a \ll H_{c_1}$.

In 12.1 and 12.2, no voltage is applied and the problem is thus time-independent. In the geometry of Fig. P12.1, we write $\mathbf{B} = B(x,y)\mathbf{z}$ and

$$\mathbf{j} = j_x(x,y)\mathbf{x} + j_y(x,y)\mathbf{y}.$$

We have

$$\mathbf{j} = \mathbf{j}^M + \mathbf{j}^J, \tag{13.52}$$

where $\mathbf{j}^J$ is the Josephson current and $\mathbf{j}^M$ the Meissner screening current when we set $j_0 = 0$ (i.e., when each superconductor is considered alone). We thus have $\mathbf{j}^M = 0$ in the region J.

12.1: Current Distribution

Throughout this section we assume that $|j_0|$ is small enough to justify completely neglecting its influence on the field, which is therefore uniform and equal to the applied field $\mathbf{B} = \mathbf{B}_a$ outside the superconductors, in particular in region J. We begin by setting $j_0 = 0$ (and hence $I = 0$).

1a. Write down the two equations relating $\mathbf{j}$ and $\mathbf{B}$ in a superconductor. Deduce the equation determining $\mathbf{B}$ inside the superconductors S_1 and S_2.

1b. Without solving these equations, sketch the current lines $\mathbf{j}_M$ when $B > 0$. Over what thickness does $\mathbf{j}_M$ flow?

1c. Express $B = B(x)$ as a function of B_a in the three regions when $L/2 - |y| \gg \lambda$.

We now assume that $L \gg \lambda$ and hence neglect the edge regions ($L/2 - |y| \sim \lambda$) in all calculations. We consider $j_0 \neq 0$.

2a. Write down the equation relating $\mathbf{j}$, θ, and $\mathbf{A}$ in a superconductor.

2b. Calculate $\delta(y)$ in terms of $\delta(0)$, B_a, the flux quantum $\Phi_0 = h/2|e|$, λ, w, and j_0. Hint: Consider the closed path C of Fig. P12.1. Use the equation obtained in question 2a on the portion C' of C lying in the superconductors, and relate $\delta(y) - \delta(0)$ with the magnetic flux $\Phi(y)$ through the surface generated by C. Neglect the contribution from $\mathbf{j}^J$ in the superconductors.

In the following, Φ_{tot} denotes the total flux in the junction, defined here as the total magnetic flux in S_1, S_2, and J.

Problem 12: Josephson Junction in a Magnetic Field

3. *Calculate the total current I in the junction and express it in terms of Φ_{tot}.*

Now set $\delta(0) = 0$ (so that $I = 0$).

4. *Plot the current $j_x^J(y)$ as a function of y for B_a corresponding to the two cases (a) $\Phi_{\text{tot}} = 2\Phi_0$, (b) $\Phi_{\text{tot}} = 4\Phi_0$.*

5. *Sketch the current lines $\mathbf{j}^J$ in the three regions S_1, S_2, and J for the two cases. Explain the qualitative analogy with a vortex arrangement in the mixed phase of a superconductor.*

12.2: Screening of the Magnetic Field

We no longer assume **B** to be uniform in the region J, as the aim now is to estimate the spatial variation of **B** in the junction (screening). In the region J, we no longer neglect the effect of the current $\mathbf{j}^J$ on the field, which becomes a function of y. We still assume that $|j_0|$ is small, so that we may continue to neglect $|j^J| \ll |j^M|$ in the superconductors (but not in the junction) when calculating the field. We also assume that $j_y^J = 0$ throughout the region J and hence that $B = B(y)$ and $j_x^J(y)$ only depend on y.

1. *Show that, for $L/2 - |y| \gg \lambda$, the field profile has the form*

$$B(x,y) = \begin{cases} B(y)e^{-(x-w/2)/\lambda} & \text{(region } S_1\text{)}, \\ B(y)e^{(x+w/2)/\lambda} & \text{(region } S_2\text{)}. \end{cases} \qquad (13.53)$$

We now determine $B(y)$ and $\delta(y)$.

2. *Show that $\delta(y)$ satisfies the differential equation*

$$\frac{d^2\delta(y)}{dy^2} = \frac{1}{\lambda_J^2} \sin \delta(y), \qquad (13.54)$$

and express the characteristic length λ_J as a function of the flux quantum Φ_0, $|j_0|$, w, and λ. Hint: Use Maxwell's equation in the junction:

$$\frac{dB}{dy} = \mu_0 j_x^J, \qquad (13.55)$$

and the closed path C and C' defined in question 2b (Fig. P12.1).

We now determine the boundary conditions for solving (13.54) for fixed B_a and I. Define $B_\pm = B(0, \pm L/2)$ and $\delta'_\pm = (d\delta/dy)_{y=\pm L/2}$.

3. *Write $\delta'_\pm$ as a function of $B_\pm$. Show that $B_\pm = B_a \pm \mu_0 I/2$. Relate the total flux Φ_{tot} in the junction to $\delta_+ - \delta_-$.*

In order to analyse the solutions of (13.54), it is useful to note the analogy with the equation for the oscillations of a pendulum, if δ is the angle measured from the highest point (position of unstable equilibrium), and if we take y proportional to time.

4a. Write down the conserved first integral (analogous to the total mechanical energy of the pendulum), viz.,

$$E = \frac{1}{2}\delta'^2 + U(\delta).$$

Fit the constant of the potential energy $U(\delta)$ so that it vanishes at $\delta = 0$.

From now on we consider the case $\delta(0) = 0$, $I = 0$, and $B_a > 0$. By symmetry, we may choose a solution with $\delta(-y) = -\delta(y)$, and therefore

$$\delta'_+ = \delta'_-, \qquad \delta_+ = -\delta_-. \tag{13.56}$$

Note that $\delta(y)$ is a strictly decreasing function and $E \geq 0$.

4b. Express E as a function of B_a and Φ_{tot}.

We now investigate in turn the three following regimes: B_a very large (question 5), B_a very small (question 6), and B_a of intermediate value (questions 7, 8, and 9). We begin by examining the situation where E is large (kinetic energy much greater than the potential energy).

5. Show that we then recover the situation in which the field is uniform in the region J, $B = B_a$, with

$$\delta(y) = -\frac{2\pi}{\Phi_0}(2\lambda + w)B_a y. \tag{13.57}$$

Show that this limit corresponds to

$$B \gg B^*, \tag{13.58}$$

where B^* is a characteristic field, and give an expression for this characteristic field up to a numerical prefactor. Show that B^* vanishes when $|j_0| \to 0$ or $\lambda_J \to +\infty$.

6a. Show that, if B is very small, then (13.54) can be linearised. Deduce the equation satisfied by $B(y)$ in this limit. Discuss the analogy with Meissner screening in a superconductor. What is the penetration depth?

6b. Calculate $B(y)$ and $\delta(y)$. Show that, if the junction is long enough, there will be total screening of the field within it. What is the exact condition on B_a, Φ_0, λ, w, λ_J, and L for the linearised equation to provide a good approximation? Show that, for large L, we do indeed recover the condition $B \ll B^*$.

Problem 12: Josephson Junction in a Magnetic Field

In the following we shall consider an intermediate field B_a of order B^*. Consider the solution with $E \approx 0$ such that the pendulum has just enough kinetic energy to go above the point $\delta = 0$.

7a. Plot $U(\delta)$ and indicate the trajectory. Show that this situation corresponds to very long junctions $L \gg \lambda_J$. Show that δ is given at the edges $y = \pm L/2$ by

$$\cos\delta_\pm = 1 - 2\left(\frac{B_a}{B^*}\right)^2. \tag{13.59}$$

Give an exact expression for B^*.

7b. What value of the total flux Φ_{tot} does $B_a = B^*$ correspond to?

7c. In the analogy with a type II superconductor, what characteristic field H does $B_a = B^*$ correspond to? Explain. Is screening total for $B > B^*$?

8. What condition must λ_J satisfy for the superconductors S_1 and S_2 to be effectively in the Meissner phase?

9. Find the analytic solution for $\delta(y)$ and $B(y)$ when $E = 0$. Show that we recover the solution obtained for question 6b when $B_a \ll B^*$ and $L \gg \lambda_J$ and that it describes the profiles of **B** and **j** more precisely near the edges over the whole interval $0 < B_a < B^*$. Note the similarity with the equation describing the shape of a Bloch wall and the differences in the boundary conditions.

12.3: Josephson Plasma Resonance

When a voltage V is applied, the Josephson current is still determined by (13.51), but the phase difference acquires a time dependence:

$$\frac{\partial \delta}{\partial t} = -\frac{2|e|V}{\hbar}, \tag{13.60}$$

where

$$V = V_1 - V_2 = \int_{-w/2}^{w/2} \mathscr{E}_x dx$$

is the potential difference across the terminals of region J (using the gauge $A_x = 0$).

Assume that $B_a > 0$ and $\mathscr{E}_x$ are uniform and time-independent in region J and outside the superconductors. (As in 12.1, we neglect the effect of j_x^J on B in the junction.)

1. Show that the Josephson current is a plane wave

$$j_x^J(y,t) = j_0 \sin\left[\delta(0) + k_0 y + \omega_0 t\right]. \tag{13.61}$$

Determine k_0, ω_0, and the velocity v_0 of the wave as a function of the constant voltage V_0 applied across the terminals of the junction. In the analogy with the type II superconductor, what corresponds to the velocity v_0?

We now consider the problem for zero applied field $B_a = 0$, but when there is a nonzero total Josephson current I, still neglecting the effect of the latter on the field. We may thus assume that $B = 0$ everywhere and $\delta(y) = \delta$ is uniform. We study small oscillations of the electric field $\mathscr{E}_x(t)$, the current $j_x(t)$, and $\delta(t)$ in the junction as a function of time, assuming all these quantities to be spatially uniform.

2. *Use Maxwell's equation in the presence of a time-dependent electric field to show that*

$$\frac{d^2\delta}{dt^2} + \omega_J^2 \sin\delta = 0. \tag{13.62}$$

Find ω_J and the corresponding frequency ν_J (Josephson plasma mode).

In the following experiment, small voltage oscillations were measured while exciting the system at a fixed microwave frequency ν_p, as a function of the direct current I injected into the junction. A resonance is observed for a certain current $I = I_p$. Figure P12.2 shows the values of ν_p obtained as a function of the current injected into an Sn/SnO/Sn junction.

3. *Show that (13.62) explains the experimentally observed dependence of ν_p on I_p. What qualitative dependence of j_0 on the superconductor gap can be deduced from the temperature dependence? Estimate T_c.*

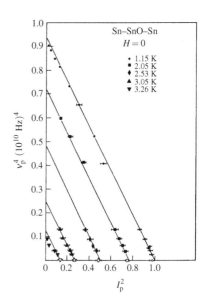

Fig. P12.2 Dependence of ν_p^4 on I_p^2 at different temperatures. Current values I_p have been normalised by the critical current I_0 of the junction measured at the lowest temperature. From Dahm, A.J., Denenstein, A., Finnegan, T.F., Langenberg, D.N., Scalapino, D.J.: Phys. Rev. Lett. **20**, 859 (1968). With the permission of the American Physical Society (© 1968 APS). http://link.aps.org/doi/10.1103/PhysRevLett.20.859

Problem 12: Josephson Junction in a Magnetic Field

Solution

Current Distribution

1a. The equations determining the current and magnetic field in a superconductor (in the Meissner phase) are

$$\mu_0 \mathbf{j} = \text{curl } \mathbf{B} \qquad \text{(Maxwell)}, \qquad (13.63)$$

$$\lambda^2 \mu_0 \text{ curl } \mathbf{j} = -\mathbf{B} \qquad \text{(London)}, \qquad (13.64)$$

where $\lambda^2 = m_0/n_s e^2 \mu_0 = m/4\phi^2 e^2 \mu_0$ is the London penetration depth, with m_0 the electron mass, n_s the density of superconducting electrons, $\phi^2 = n_s/2$ the density of Cooper pairs, m the mass of a pair, and $-|e|$ the electron charge. We also have $\text{div } \mathbf{j} = \text{div } \mathbf{B} = 0$. These imply

$$\nabla^2 \mathbf{B} = \frac{1}{\lambda^2} \mathbf{B}. \qquad (13.65)$$

With the geometry of Fig. P12.3, the field $\mathbf{B} = B(x,y)\mathbf{z}$ is determined by solving

$$\boxed{\left(\frac{\partial^2}{\partial x^2} + \frac{\partial^2}{\partial y^2}\right) B(x,y) = \frac{1}{\lambda^2} B(x,y)}, \qquad (13.66)$$

in S_1 and S_2, with $B(x,y) = B_a$ on the boundaries of S_1 and S_2.

1b. The current $\mathbf{j} = \mathbf{j}^M$ is obtained from the solution $B(x,y)$ of (13.66). Its components are then

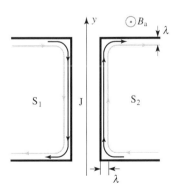

Fig. P12.3 Geometry of the junction and currents induced by the applied field

$$\mu_0 j_x(x,y) = \frac{\partial B(x,y)}{\partial y}, \qquad \mu_0 j_y(x,y) = -\frac{\partial B(x,y)}{\partial x}, \qquad (13.67)$$

with the boundary conditions $j_y(x,\pm L/2) = 0$ for $|x| > |w|/2$. The current flows through a layer of thickness roughly equal to λ near the surfaces. $\mathbf{j}^M$ is parallel to the edges very close to them (at distances $\gg \xi$ and $\ll \lambda$). It decreases roughly exponentially over distances of the order of λ as we move away from the surfaces of S_1 and S_2. The current lines are shown in Fig. P12.3.

1c. For $L/2 - |y| \gg \lambda$ (far from the horizontal edges), the current flows in the y direction. Therefore, $j_x^M = 0$ and $B(x,y) = B(x)$. The solution described in question 1a is thus

$$B(x) = \begin{cases} B_a, & |x| < w/2, \\ B_a e^{(x+w/2)/\lambda}, & x < -w/2, \\ B_a e^{-(x-w/2)/\lambda}, & x > w/2. \end{cases} \qquad (13.68)$$

2a. In a superconductor, (5.25) gives

$$\boxed{\mu_0 \lambda^2 \mathbf{j} = \frac{\hbar}{q} \nabla \theta - \mathbf{A}}, \qquad (13.69)$$

where $q = -2|e|$ is the charge of a Cooper pair.

2b. The closed path C is the rectangle shown in Fig. P12.4, with sides $y = 0$, y, and $x = \pm X$. We integrate (13.69) over the portion C' of C within the superconductor (where it is valid):

$$\mu_0 \lambda^2 \int_{C'} \mathbf{j} \cdot d\mathbf{l} = \frac{\hbar}{q} \int_{C'} \nabla \theta \cdot d\mathbf{l} - \int_C \mathbf{A} \cdot d\mathbf{l} + \int_{C \setminus C'} \mathbf{A} \cdot d\mathbf{l}. \qquad (13.70)$$

Choosing the gauge $A_x = 0$, the integral along the missing path is zero, i.e., $\int_{C \setminus C'} \mathbf{A} \cdot d\mathbf{l} = 0$. Moreover,

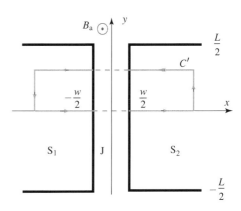

Fig. P12.4 Josephson junction configuration and illustration of the rectangular circuit C and its part circuit C' within the superconductor used to integrate (13.69)

Problem 12: Josephson Junction in a Magnetic Field

$$\int_{C'} \mathbf{j} \cdot d\mathbf{l} = \left(\int_{-\infty}^{-w/2} dx + \int_{w/2}^{\infty} dx \right) \left[j_x(x,0) - j_x(x,y) \right] + O\!\left(e^{-|X|/\lambda}\right) \approx 0, \tag{13.71}$$

since j_x^M is zero for $L/2 - |y| \gg \lambda$, far from the edges which we neglect here, and since we neglect j^J. Taking $|X| \to +\infty$, and with $\phi_0 = h/2|e|$, Eq. (13.70) reduces to

$$-\frac{\phi_0}{2\pi} \int_{C'} \nabla\theta \cdot d\mathbf{l} = \int_C \mathbf{A} \cdot d\mathbf{l} = \Phi(y),$$

or

$$-\frac{\phi_0}{2\pi} \int_{C'} \nabla\theta \cdot d\mathbf{l} = -\frac{\phi_0}{2\pi} \Big[\theta(w/2,y) - \theta(w/2,0) $$
$$+ \theta(-w/2,0) - \theta(-w/2,y) \Big]$$
$$= -\frac{\phi_0}{2\pi} \left[\delta(y) - \delta(0) \right]. \tag{13.72}$$

In addition,

$$\Phi(y) = \int_\Sigma B_z dx dy = \int_{-\infty}^{+\infty} \int_0^y B(x,y) dx dy = (2\lambda + w) B_a y, \tag{13.73}$$

where Σ is the surface generated by C. Hence,

$$\boxed{\delta(y) = \delta(0) - \frac{2\pi}{\phi_0}(2\lambda + w) B_a y}. \tag{13.74}$$

3. The total current in the junction is

$$I = \int_{-L/2}^{L/2} dy j_x(y) = j_0 \int_{-L/2}^{L/2} dy \sin\!\left[\delta(0) - \frac{2\pi}{\phi_0}(2\lambda + w) B_a y \right], \tag{13.75}$$

or

$$\boxed{I = j_0 L \sin\delta(0) \frac{\sin(\pi \Phi_{\mathrm{tot}}/\phi_0)}{\pi \Phi_{\mathrm{tot}}/\phi_0}}, \tag{13.76}$$

where $\Phi_{\mathrm{tot}} = (2\lambda + w) B_a L$ is the total flux in the junction (S_1, S_2, and J).

4. When $\delta(0) = 0$ and $\Phi_{\mathrm{tot}} = n\phi_0$, we write

$$\delta(y) = -n\pi \frac{y}{L/2}, \qquad j_x^J(y) = |j_0| \sin\!\left(n\pi \frac{y}{L/2} \right). \tag{13.77}$$

Figure P12.5 represents (13.77) for $n = 2$ and $n = 4$.

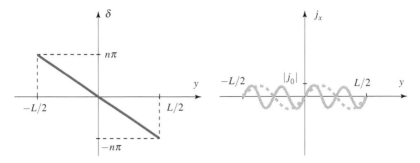

Fig. P12.5 Variation across the junction (*left*) of the phase of the wave function and (*right*) of the current for $n=2$ (*dashed line*) and $n=4$ (*full line*)

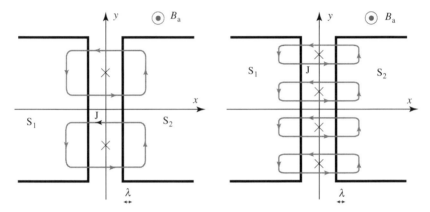

Fig. P12.6 Illustration of the periodic current configurations in the junction for $n=2$ and 4

5. We now plot the maximum current lines corresponding to question 4. For $n = 2$ and 4, they close naturally, and the configuration is periodic (see Fig. P12.6). This configuration of pair currents tunneling through the junction is reminiscent of the currents associated with a periodic row of n vortices along the y axis, as would appear in a thin type II superconductor above the critical field H_{c_1}. The average spacing between these vortices, called *Josephson vortices*, is $a = \Phi_0/[(2\lambda + w)B_a]$ here.

Screening of the Magnetic Field

1. We continue to neglect $|\mathbf{j}^J| \ll |\mathbf{j}^M|$ in the superconductors, and $\mathbf{j}^M$ still lies along the y direction. Since the London equation is

$$\frac{B}{\lambda^2} = \frac{\partial}{\partial y}(\mu_0 j_x) - \frac{\partial}{\partial x}(\mu_0 j_y), \tag{13.78}$$

Problem 12: Josephson Junction in a Magnetic Field

we still have
$$\frac{1}{\lambda^2}B = \left(\frac{\partial^2}{\partial x^2} + \frac{\partial^2}{\partial y^2}\right)B \approx \frac{\partial^2}{\partial y^2}B \qquad (13.79)$$

inside the superconductors, which amounts to assuming that $\lambda_J \gg \lambda$ (see below). The solution with $B = B(y)$ in region J is

$$B(x,y) = \begin{cases} B(y), & |x| < w/2, \\ B(y)e^{(x+w/2)/\lambda}, & x < -w/2, \\ B(y)e^{-(x-w/2)/\lambda}, & x > w/2. \end{cases} \qquad (13.80)$$

It remains to determine $B(y)$.

2. We have
$$\frac{\partial B}{\partial y} = \mu_0 j_x^J.$$

We still assume that $j_y^J = 0$, so $\partial B/\partial x = 0$ and $\partial j_x^J/\partial x = 0$, so j_x^J only depends on y. Then
$$\frac{\partial}{\partial y}B(y) = \mu_0 j_0 \sin\delta(y). \qquad (13.81)$$

Furthermore, we still assume that j_x^M and j_x^J can be neglected on the closed contour. Therefore, the relation (13.72), viz.,
$$\Phi(y) = -\frac{\Phi_0}{2\pi}[\delta(y) - \delta(0)],$$

is still valid. Differentiating, we have
$$\frac{d}{dy}\delta(y) = -\frac{2\pi}{\Phi_0}\frac{d\Phi}{dy} = -\frac{2\pi}{\Phi_0}(2\lambda + w)B(y), \qquad (13.82)$$

$$\frac{d^2\delta(y)}{dy^2} = -\mu_0 j_0 \frac{2\pi}{\Phi_0}(2\lambda + w)\sin\delta(y) = \frac{1}{\lambda_J^2}\sin\delta(y), \qquad (13.83)$$

with

$$\boxed{\lambda_J = \sqrt{\frac{\Phi_0}{2\pi\mu_0|j_0|(2\lambda+w)}}} \qquad (13.84)$$

The length λ_J diverges when $|j_0| \to 0$.

3. According to (13.82), the boundary conditions are
$$\delta'_+ = \delta'(L/2) = -\frac{2\pi}{\Phi_0}(2\lambda + w)B_+,$$
$$\delta'_- = \delta'(L/2) = -\frac{2\pi}{\Phi_0}(2\lambda + w)B_-. \qquad (13.85)$$

Integrating $\mu_0 j_x = \partial B/\partial y$ in the junction, B is found by superposition as the sum of the field at $I = 0$ and the field produced by the current I:

$$B_+ = B(L/2) = B_a + \mu_0 I/2,$$
$$B_- = B(-L/2) = B_a - \mu_0 I/2. \tag{13.86}$$

Integrating (13.82), we find the total flux Φ_{tot} in the junction:

$$\boxed{\delta_+ - \delta_- = -\frac{2\pi}{\Phi_0}\Phi_{\text{tot}}}. \tag{13.87}$$

Note that we also have

$$I = j_0 \int_{-L/2}^{L/2} \sin\delta(y) = j_0 \lambda_J^2 \int_{-L/2}^{L/2} \frac{d^2\delta(y)}{dy^2} = j_0 \lambda_J^2 (\delta'_+ - \delta'_-). \tag{13.88}$$

4a.   There is an analogy with a pendulum, where δ is the angle measured from the highest point:

$$m\ell^2 \frac{d^2\delta}{dt^2} - mg\ell \sin\delta = 0.$$

Comparing with (13.83), putting $t \to y$, we find that $\lambda_J^{-2} \to g/\ell = \omega_0^2$ corresponds to the natural frequency of the pendulum. The conserved total energy is

$$E = \frac{1}{2}\left[\frac{d\delta(y)}{dy}\right]^2 - \frac{1}{\lambda_J^2}[(1-\cos(\delta)] = K + U(\delta). \tag{13.89}$$

4b.   We now consider $\delta(0) = 0$ and $I = 0$, but $B_a > 0$. We can thus choose a solution with $\delta(-y) = -\delta(y)$ and

$$\delta'_+ = \delta'_- = -\frac{2\pi}{\Phi_0}(2\lambda + w)B_a, \qquad \delta_+ = -\delta_- < 0. \tag{13.90}$$

Since $\delta(0)$ and U vanish for $y = 0$, whereas the kinetic energy does not, we must have $E > 0$. This means that there are no trajectories that return along the same path, and then $\delta(y)$ is a monotonic decreasing function. δ_- can then be related to the total flux by

$$\delta_- = \pi \frac{\Phi_{\text{tot}}}{\Phi_0}, \tag{13.91}$$

and E can be identified with its value at $y = -L/2$, $\delta = \delta_-$:

$$E = \frac{1}{2}(\delta'_-)^2 - \frac{1}{\lambda_J^2}[1 - \cos(\delta_-)], \tag{13.92}$$

or

$$\boxed{E = E(B_a, \Phi_{\text{tot}}) = \frac{1}{2}\left[\frac{2\pi}{\Phi_0}(2\lambda + w)\right]^2 B_a^2 - \frac{1}{\lambda_J^2}\left[1 - \cos\left(\pi \frac{\Phi_{\text{tot}}}{\Phi_0}\right)\right].} \tag{13.93}$$

The problem is fully determined. Integrating the equation expressing energy conservation, using the fact that $\delta(y)$ is a monotonic decreasing function, we obtain a further (complicated) relation which can in principle be used to calculate Φ_{tot} in terms of B_a and L:

$$L = \int_{\delta_+}^{\delta_-} \frac{d\delta}{\sqrt{2E + 2(1-\cos\delta)/\lambda_J^2}}$$

$$= \int_{-\pi\Phi_{\text{tot}}/\Phi_0}^{+\pi\Phi_{\text{tot}}/\Phi_0} \frac{d\delta}{\sqrt{2E(B_a, \Phi_{\text{tot}}) + 2(1-\cos\delta)/\lambda_J^2}}. \tag{13.94}$$

5. We assume that the total energy E is large (strong field). In this case, the pendulum will oscillate very fast and we may neglect the potential energy U. We thus obtain

$$\delta(y) = -\frac{2\pi}{\Phi_0}(2\lambda + w)B_a y. \tag{13.95}$$

This solution is valid for

$$E \approx \frac{1}{2}(\delta'_+)^2 = \frac{1}{2}\left[\frac{2\pi}{\Phi_0}(2\lambda+w)\right]^2 B_a^2 \gg \frac{1}{\lambda_J^2}, \tag{13.96}$$

which is equivalent to

$$B \gg \lambda_J \mu_0 j_0 = \sqrt{\frac{\Phi_0 \mu_0 |j_0|}{2\pi(2\lambda+w)}} = \frac{\Phi_0}{2\pi \lambda_J(2\lambda+w)}. \tag{13.97}$$

We therefore define the characteristic field

$$\boxed{B^* = 2\frac{\Phi_0}{2\pi \lambda_J(2\lambda+w)}}, \qquad (13.98)$$

where the numerical constant will be justified in question 7a. It does indeed vanish for $\lambda_J \to +\infty$ and hence when $j_0 \to 0$.

6a. For very small B, we have δ_- very small and so can linearise to obtain

$$\partial_y^2 \delta(y) = \frac{1}{\lambda_J^2} \delta(y). \qquad (13.99)$$

Differentiating, this also leads to

$$\boxed{\partial_y^2 B(y) = \frac{1}{\lambda_J^2} B(y)}, \qquad (13.100)$$

which is just the Meissner equation (13.53), where λ_J plays the role of the London penetration depth. We conclude that the field $B(y)$ *is screened and cannot penetrate the junction beyond a depth* λ_J.

6b. We find the solutions

$$B(y) = B_a \frac{\cosh(y/\lambda_J)}{\cosh(L/2\lambda_J)}, \qquad (13.101)$$

$$\delta(y) = -\frac{2\pi}{\Phi_0}(2\lambda+w)\lambda_J B_a \frac{\sinh(y/\lambda_J)}{\cosh(L/2\lambda_J)}. \qquad (13.102)$$

The linearisation condition $\delta_+ \ll \pi$ thus becomes

$$B_a \ll \frac{\Phi_0}{2(2\lambda+w)\lambda_J} \coth(L/2\lambda_J). \qquad (13.103)$$

For long junctions, this condition is equivalent to $B \ll B^*$, but for short junctions $L \ll \lambda_J$, we may in fact exploit the results of question 5 and the condition becomes

$$\Phi_{\text{tot}} \ll \Phi_0. \qquad (13.104)$$

7a. We now assume that $E \approx 0$. This corresponds to a pendulum started at $\delta_- > 0$ with a speed $\delta'_- < 0$ that is just enough to get it past the highest point $\delta = 0$ with speed 0^+. The function $U(\delta)$ is shown in Fig. P12.7. The trajectory spends a long time near the highest point (an infinite time in the limit $E \to 0$ because its speed then vanishes there), so $E \approx 0$ corresponds to very large L. Equation (13.93) can be rewritten using (13.98) to obtain

Problem 12: Josephson Junction in a Magnetic Field 505

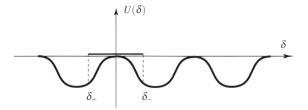

Fig. P12.7 The function $U(\delta)$

$$E(B_a, \Phi_{\text{tot}}) = \frac{1}{\lambda_J^2}\left[2\left(\frac{B_a}{B^*}\right)^2 - (1-\cos\delta_-)\right], \quad (13.105)$$

so $E = 0$ corresponds to

$$\boxed{\cos\delta_- = \cos\left(\pi\frac{\Phi_{\text{tot}}}{\Phi_0}\right) = 1 - 2\left(\frac{B_a}{B^*}\right)^2}. \quad (13.106)$$

7b. The above equation, relating δ_+ to B_a, only has solutions for $B_a < B^*$. The limiting case $B_a = B^*$ corresponds to $\delta_- = \pi$, hence $\Phi_{\text{tot}} = \Phi_0$, or exactly *one* flux quantum through the junction.

7c. We thus have the following analogy:

- $B_a < B^*$ corresponds *to the Meissner phase of a type II superconductor, with total screening inside*. For $B_a < B^*$, even for very large L, the total flux remains bounded by $\Phi_{\text{tot}} < \Phi_0$.
- $B_a = B^*$ corresponds to the field H_{c_1} for which vortices first appear in the superconductor. For $B_a > B^*$ (and L large), the pendulum goes round several times, and this corresponds to the presence of several vortices, hence the *mixed phase*. For L large, the number of vortices per unit length, or the reciprocal of their spacing, $n/L = 1/a = \Phi_{\text{tot}}/L\Phi_0$ is a function of B_a which vanishes for $B_a \leq B^*$, becomes nonzero for $B_a > B^*$, and converges to $n/L = 1/a = (2\lambda + w)B_a/\Phi_0$ for $B_a \gg B^*$.
- For $B_a > B^*$, the pendulum goes round more than once. For all y, we may write

$$E = \frac{1}{\lambda_J^2}\left\{2\left[\frac{B(y)}{B^*}\right]^2 - [1-\cos\delta(y)]\right\}, \quad (13.107)$$

so the field attains its maximal value $B_{\max}$ for $\delta = (2m+1)\pi$ and its minimal value $B_{\min}$ for $\delta = 2m\pi$, with

$$\frac{1}{2}E\lambda_J^2 = \left(\frac{B_{\min}}{B^*}\right)^2 = \left(\frac{B_{\max}}{B^*}\right)^2 - 1. \quad (13.108)$$

For $(B_{\max} \geq)B_a > B^*$, we have $B_{\min} > 0$ and *the field is no longer fully screened inside*.

8. To observe these phenomena, the applied external field $H^* = B^*/\mu_0$ must be very small compared with H_{c_1} for the superconductors S_1 and S_2. Since in general $\mu_0 H_{c_1} \sim \Phi_0/\lambda^2$ and here

$$B^* \sim \frac{\Phi_0}{\lambda_J(2\lambda + w)},$$

we must have, to within an order of magnitude,

$$\boxed{\lambda_J \gg \lambda + \frac{w}{2} \approx \lambda}, \qquad (13.109)$$

for w is usually very small (nanometers).

Note: Likewise, to justify neglecting the edge regions $L/2 - |y| \sim \lambda$ in the expressions obtained here, we must have $\min(L, \lambda_J) \gg \lambda$.

9. For $E = 0$, the equation integrates formally to

$$\frac{y}{\lambda_J} = \int^\delta \frac{d\delta}{\sqrt{2(1-\cos\delta)}} = \frac{1}{2}\int^\delta \frac{d\delta}{|\sin(\delta/2)|} = \ln|\tan(\delta/4)|, \qquad (13.110)$$

as for the Bloch wall. Since the integral diverges at $\delta = 0$, this solution corresponds to $L = +\infty$ (bearing in mind that, for $E = 0$, the time required for the pendulum to go past its highest point is strictly infinite). The equation can nevertheless be integrated, taking into account the boundary conditions (not the same as for the wall) in the following way:

$$\frac{L/2 - |y|}{\lambda_J} = \frac{1}{2}\int_{|\delta|}^{\delta_-} \frac{d\phi}{\sin(\phi/2)} = \ln\tan(\delta_-/4) - \ln\tan(|\delta|/4). \qquad (13.111)$$

Therefore,

$$\boxed{\delta(y) = -\operatorname{sgn}(y) 4 \arctan\left[e^{|y|-L/2}\tan\left(\frac{\delta_-}{4}\right)\right]}, \qquad (13.112)$$

with

$$\delta_- = \arccos\left[1 - 2\left(\frac{B}{B^*}\right)^2\right] = 2\arcsin(B/B^*), \qquad 0 \le \delta_- \le \pi.$$

Since L is in principle strictly infinite, this solution actually corresponds to the limit $L \to +\infty$ with $|y| - L/2$ fixed, so it describes the profile near the edges (over a layer of thickness roughly $\lambda_J \gg \lambda$). The field $B(y)$ is obtained by differentiating:

$$B(y) = B_a e^{(|y|-L/2)/\lambda_J} \frac{1+\tan^2\left[\frac{1}{2}\arcsin(B/B^*)\right]}{1+e^{2(|y|-L/2)/\lambda_J}\tan^2\left[\frac{1}{2}\arcsin(B/B^*)\right]} \ . \quad (13.113)$$

In the limit $B \ll B^*$ (δ_- small), we recover a simple exponential, which is indeed the large L limit, for fixed $L/2 - |y|$, of the relation (13.101). The above formula gives a valid description of the screening profile for arbitrary $0 < B \le B^*$. In particular, for $B = B^*$, we have the simple solution

$$B(y) = \frac{B_a}{\cosh\left[(|y|-L/2)/\lambda_J\right]} \ . \quad (13.114)$$

Josephson Plasma Resonance

1. Here $V = V_0$ and B_a are uniform throughout the region J and constant in time. In the superconductors, $B(x)$ is as in 12.1 and the argument with the closed path in question 2b applies identically to determine $\partial \delta/\partial y$. Since V determines $\partial \delta/\partial t$, we find

$$\delta(y) = \delta(0) + \omega_0 t + k_0 y \ , \quad (13.115)$$

$$\boxed{\omega_0 = -\frac{2|e|V_0}{\hbar}} \ , \quad (13.116)$$

$$\boxed{k_0 = -\frac{2\pi}{\Phi_0}(2\lambda + w)B_a} \ , \quad (13.117)$$

and the current $j_x^J(y)$ is therefore a plane wave with phase velocity

$$\boxed{v_0 = -\frac{\omega_0}{k_0} = -\frac{V_0}{(2\lambda+w)B_a}} \ , \quad (13.118)$$

in the y direction. Returning to the analogy between $j_x^J(y) = j(y - v_0 t)$ and the current distribution that would be produced by a periodic arrangement of $n = \Phi_{\text{tot}}/\Phi_0$ vortices along the y axis in a type II superconductor, we observe that, under an applied voltage, these vortices *flow with a constant velocity in the y direction.*

2. Maxwell's equation in vacuum is

$$\mathbf{curl\,B} = \mu_0 \mathbf{j} + \varepsilon_0 \mu_0 \partial_t \mathscr{E} \ . \quad (13.119)$$

In an insulator, we simply replace $\varepsilon_0 \to \varepsilon\varepsilon_0$, where ε is the dielectric constant. Since here $B = 0$ everywhere and $V = w\mathscr{E}_x$ ($\mathscr{E}$ uniform), we obtain in the x direction,

$$0 = \mu_0 j_x + \frac{1}{w}\varepsilon\varepsilon_0\mu_0\frac{\partial V}{\partial t}. \tag{13.120}$$

Combining with $j_x = j_0 \sin\delta$ and $\partial\delta/\partial t = -2|e|V/\hbar$, this yields

$$\boxed{\frac{d^2\delta}{dt^2} + \omega_J^2 \sin\delta = 0}, \tag{13.121}$$

with

$$\omega_J^2 = -\frac{2|e|wj_0}{\varepsilon\varepsilon_0\hbar} = \frac{2|e|w|j_0|}{\varepsilon\varepsilon_0\hbar}, \tag{13.122}$$

which is indeed positive. As can be seen by considering small oscillations $\delta \ll 1$ and linearising (13.115) around zero, ω_J is an angular frequency. The corresponding frequency, called the *Josephson plasma frequency*, is

$$\boxed{\nu_J = \frac{\omega_J}{2\pi} = \sqrt{\frac{|e|w|j_0|}{\pi\varepsilon\varepsilon_0 \hbar}}.} \tag{13.123}$$

3. The experiment measures the natural oscillation frequency ν_p of the Josephson junction. A resonance is observed when the imposed external frequency is equal to the eigenfrequency of the system. The experiment is carried out in the presence of a direct current I, so the average phase is $\delta = \delta_0$ determined by

$$I = I_0 \sin\delta_0, \tag{13.124}$$

with $I_0 = j_0 L$. In the presence of this current, the small natural oscillations of the system are obtained by setting $\delta(t) = \delta_0 + \phi(t)$ in (13.115) with $\phi(t)$ small. We may therefore linearise in ϕ to obtain

$$\frac{d^2\phi}{dt^2} + (\omega_J^2 \cos\delta_0)\phi = 0. \tag{13.125}$$

The new angular frequency is thus $\omega_p^2 = \omega_J^2 \cos\delta_0$, related to the current by

$$\boxed{\omega_p^4 = \omega_J^4\left[1 - \left(\frac{I}{I_0}\right)^2\right],} \tag{13.126}$$

exactly as observed in Fig. P12.2. The abscissa at the origin gives I_0^2 as a function of temperature. It can be seen from Fig. P12.8 that I_0^2 decreases when the

Fig. P12.8 Temperature dependence of I_0^2

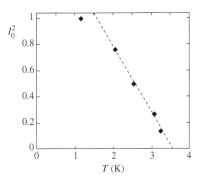

temperature is raised. It thus varies in the same way as the superconducting gap $\Delta(T)$. The latter vanishes at the normal–superconducting transition at $T = T_c$ so we may estimate that $T_c \approx 3.54$ K. The temperature dependence of I_0^2 is found to be almost linear near T_c with

$$I_0^2 = \alpha(T - T_c). \tag{13.127}$$

The experiment gives $\alpha = 0.50$ and $T_c = 3.54$ K according to Fig. P12.8. We also know that the superconductor gap $\Delta(T) \sim (T_c - T)^{1/2}$ near T_c. Hence I_0 varies approximately as $\Delta(T)$ and should saturate for $T \ll T_c$, as can be seen from Fig. P12.8.

Note on Gauge Invariance

We have

$$\mathscr{E} = -\nabla V - \frac{\partial \mathbf{A}}{\partial t}, \tag{13.128}$$

where V is the local electric potential and $\mathbf{A}$ is the vector potential. Combining this equation with the two following equations which hold in a superconductor, viz.,

$$\lambda^2 \mu_0 \mathbf{j} = \frac{\hbar}{q} \nabla \theta - \mathbf{A}, \tag{13.129}$$

$$\mathscr{E} = \lambda^2 \mu_0 \frac{\partial \mathbf{j}}{\partial t}, \tag{13.130}$$

we deduce that

$$\frac{\partial \theta}{\partial t} = -\frac{q}{\hbar} V = \frac{2|e|}{\hbar} V. \tag{13.131}$$

The time derivative of the phase, which is not gauge invariant, is proportional to the electric potential, itself not gauge invariant. On the other hand, the gauge invariant phase

$$\delta = \theta_2 - \theta_1 - \frac{q}{\hbar} \int_1^2 \mathbf{A}$$

satisfies

$$\frac{\partial \delta}{\partial t} = +\frac{q}{\hbar} \int_1^2 \mathcal{E} = -\frac{2|e|}{\hbar} \int_1^2 \mathcal{E}. \tag{13.132}$$

Being a physical observable, the electric field has to be gauge invariant. In the geometry we have here, we may use the gauge $A_x = 0$, and the two observables coincide, but this is not generally the case. In this problem, we had $V = V_1 - V_2$.

Problem 13: Magnetisation of a Type II Superconductor

The magnetisation curves of type II superconductors often exhibit marked hysteresis. In particular, in superconductors with high critical temperature, for a sample in the form of a platelet, and with a field applied parallel to the plane of the platelet, the measured cycle is totally asymmetric with respect to $M = 0$ (see Fig. P13.1).

1. *On the figure, indicate the region where a perfect Meissner effect prevails. How would you interpret the field for which the absolute value of the magnetisation suddenly decreases?*

To understand one of the mechanisms that opposes the transition to a state of thermodynamic equilibrium, we consider vortices in a superconductor in a non-equilibrium situation. We restrict here to type II superconductors for which $\lambda \gg \xi$. Then the second London equation, viz.,

$$\lambda^2 \mu_0 \mathbf{curl}\ \mathbf{j}(\mathbf{r}) + \mathbf{B} = 0, \qquad (13.133)$$

holds for a field $\mathbf{B} \parallel \mathbf{z}$ and the current $\mathbf{j}(\mathbf{r})$ in the superconducting region $r > \xi$. The core of the vortex parallel to the z direction, of radius ξ, is assumed normal.

Applying flux quantisation with a single quantum $\Phi_0 = h/2e$ per vortex, in the limit $\xi \ll \lambda$, we find

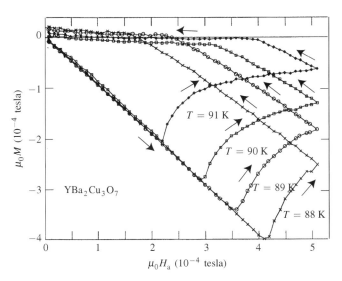

Fig. P13.1 Hysteresis cycles for $YBa_2Cu_3O_7$. Figure courtesy of M.M. Konczykowski (LSI, Ecole Polytechnique) from Konczykowski, M., Burlachkov, L.I., Yeshurun, Y., Holtzberg, F.: Phys. Rev. B **43**, 13707 (1991). http://link.aps.org/doi/10.1103/PhysRevB.43.13707

$$B_z = \begin{cases} \dfrac{\Phi_0}{2\pi\lambda^2} K_0\left(\dfrac{r}{\lambda}\right), & \text{for } r \geq \xi, \\ \dfrac{\Phi_0}{2\pi\lambda^2} K_0\left(\dfrac{\xi}{\lambda}\right), & \text{for } r \leq \xi. \end{cases} \quad (13.134)$$

In the above, $K_0(x)$ is the Bessel function such that $K_0(x) \simeq (\pi/2x)^{1/2} e^{-x}$ for $x \gg 1$ and $K_0(x) = \ln x$ for $x \ll 1$.

2. *Express the current* $\mathbf{j}(\mathbf{r})$ *in terms of* $K_1(x) = -\mathrm{d}K_0(x)/\mathrm{d}x$ *and represent it graphically.*

Such a vortex can exist in the superconductor out of thermodynamic equilibrium, and in particular when the external applied field is zero. Its energy per unit length is

$$W_{\mathrm{v}} = \frac{1}{2\mu_0} \int_S \left(\mathbf{B}^2 + \lambda^2 \mu_0^2 \mathbf{j}^2\right) \mathrm{d}S, \quad (13.135)$$

where integration is over a plane circular surface S perpendicular to the z axis, whose radius is allowed to tend to infinity.

3. *Specify the physical meaning of the two terms in the above integral.*

To handle energy questions concerning vortices, it can be shown that the flux quantisation condition can be directly taken into account by modifying the second London equation (13.133) to

$$\lambda^2 \mu_0 \mathbf{curl}\ \mathbf{j} + \mathbf{B} = \Phi_0 \delta(\mathbf{r})\mathbf{z} = \mathbf{V}(\mathbf{r}), \quad (13.136)$$

and reducing the normal core of the vortex to a point. $\delta(\mathbf{r})$ is a 2D Dirac delta function, for which $\int_S \delta(\mathbf{r})\mathrm{d}x\mathrm{d}y = 1$ for any integration surface S including $\mathbf{r} = 0$. In this case, the solution of (13.136) for B_z extends the first relation of (13.134) to $\mathbf{r} = 0$, but in order to correctly take into account the field in the vortex core, we shall take the value of B_z at $r = 0$ to be the one given by the second relation of (13.134).

By carrying out vector transformations of (13.135), it can be shown that the latter transforms to

$$W_{\mathrm{v}} = \frac{1}{2\mu_0}\left[\int_S \mathbf{B}\cdot\mathbf{V}(\mathbf{r})\mathrm{d}S + \lambda^2 \oint_C (\mathbf{B}\wedge\mathbf{curl}\ \mathbf{B})\cdot\mathrm{d}\boldsymbol{\ell}\right], \quad (13.137)$$

where $\mathrm{d}\boldsymbol{\ell}$ is a length element on the path (C) enclosing S.

4. *Show that the integral around the path (C) is zero in (13.137) when the radius R_{c} of this path tends to infinity.*

5. *Deduce the energy W_{v} of the vortex per unit length.*

Problem 13: Magnetisation of a Type II Superconductor

We now consider two vortices oriented along the z axis and located at $\mathbf{r}_1 = -a\mathbf{x}$ and $\mathbf{r}_2 = a\mathbf{x}$.

6. *Specify the spatial dependence of the magnetic field induced by these two vortices.*

The energy W_{2v} of the ensemble of two vortices is given by the same equation, namely (13.137), in which the vorticity is now

$$\mathbf{V}(\mathbf{r}) = \Phi_0 \big[\delta(\mathbf{r} - a\mathbf{x}) + \delta(\mathbf{r} + a\mathbf{x})\big] \mathbf{z} \,. \tag{13.138}$$

7. *Determine W_{2v} and deduce the interaction energy between the two vortices.*
8. *From this interaction energy, deduce the force induced per unit length of vortex 1 on vortex 2. Show that this force can be written in the form*

$$\mathbf{f}_{12} = \mathbf{j}_{12} \wedge \Phi_0 \mathbf{z} \,, \tag{13.139}$$

where $\mathbf{j}_{12}$ is the current created by vortex 1 at the core of vortex 2. (This force can be considered as a Lorentz force exerted by the magnetic field of one vortex on the current created by the other.)

We now consider two vortices with opposite vorticity, i.e., such that the fields at the core of the vortices at $a\mathbf{x}$ and $-a\mathbf{x}$ lie along the positive and negative z axes, respectively.

9. *Give the corresponding expressions for $\mathbf{B}$ and $\mathbf{j}$. What values do they take on the plane $x = 0$? In what direction is the force induced by one vortex on the other?*

Consider a semi-infinite superconductor occupying the half-space $x > 0$, and a vortex parallel to $\mathbf{z}$ located at $a\mathbf{x}$.

10. *What boundary conditions do we expect for $\mathbf{B}$ and $\mathbf{j}$ at $x = 0$ when no field is applied outside?*
11. *Show that these boundary conditions can be obtained if the fields and currents for $x > 0$ are those produced by superposing the field created by the vortex at $a\mathbf{x}$ and the one created by its image in the surface located at $-a\mathbf{x}$.*

A field $\mathbf{H}_a \parallel \mathbf{z}$ is now applied in the half-space $x < 0$.

12. *Determine the distributions of the field and the current $\mathbf{j}_S$ in the superconductor for $x > 0$ when there is no vortex.*

If a vortex is at $a\mathbf{x}$, with an external field $\mathbf{H}_a$, it is subject to two forces: $\mathbf{f}_I$ due to the interaction with its image, and $\mathbf{f}_S$ due to the interaction with the surface currents associated with penetration of the field $\mathbf{H}_a$. Assume that the latter also results from the Lorentz force, i.e.,

$$\mathbf{f}_S = \mathbf{j}_S \wedge \Phi_0 \mathbf{z} \,. \tag{13.140}$$

13. *Find the forces* $\mathbf{f}_I$ *and* $\mathbf{f}_S$.

14. *Assuming that the vortex forms at the surface and is located at a distance* $a = \xi$ *from the surface, what is the value* H_p *of* H_a *for which the vortex will be pushed inside the superconductor?*

15. *Recalling that*

$$H_{c_1} = \frac{\Phi_0}{4\pi \mu_0 \lambda^2} \ln \frac{\lambda}{\xi},$$

compare H_p *and* H_{c_1}.

The aim now is to analyse the various branches of the hysteresis cycles observed in Fig. P13.1.

16. *What remarkable behaviour is revealed by Fig. P13.1 for the branches corresponding to decreasing* H_a?

When $H > H_p$, the vortices penetrate into the superconductor. The measured magnetisation contains two contributions. One is $\mathbf{M}_S$, associated with surface currents, and the other is associated with vortices in the material. We assume that the distribution of vortices that have penetrated is spatially uniform, i.e., uniform average magnetic induction within the superconductor. In this case, only surface currents contribute to the measured magnetisation.

17. *What can we say about the variation of these surface currents and the variation of the vortex density along the branches of the cycle examined in question 16?*

18. *How may we explain this behaviour?*

Solution

1. The perfect Meissner effect is observed up to a point H_p that depends on the temperature T. (These points are clearly visible in Fig. P13.1. They are indicated on Fig. P13.7.) If the magnetisation were totally reversible, we would have $H_p(T) = H_{c_1}(T)$. The fact that it is irreversible shows that

$$\boxed{H_p(T) \geq H_{c_1}(T)}. \tag{13.141}$$

   $H_p(T)$ is the field for which vortices can penetrate the bulk of the sample.

2. We have $\mu_0 \mathbf{j} = \text{curl } \mathbf{B}$ and

$$\mathbf{j}_\perp = \begin{cases} -\dfrac{1}{\mu_0}\dfrac{\partial B_z}{\partial r}\mathbf{i}_\perp = \dfrac{\Phi_0}{2\pi\mu_0\lambda^3}K_1\left(\dfrac{r}{\lambda}\right)\mathbf{i}_\perp, & \text{for } r > \xi, \\ 0, & \text{for } r < \xi. \end{cases} \tag{13.142}$$

   $\mathbf{j}_\perp$ and its dependence on $\mathbf{r}$ are shown in Fig. P13.2a, b.

3. We have

$$W_v = \dfrac{1}{2\mu_0}\int_S (\mathbf{B}^2 + \lambda^2\mu_0^2\mathbf{j}^2)\mathrm{d}S.$$

   The first term here is the energy associated with the magnetic field produced by the vortex. Indeed, since we are considering a non-equilibrium thermodynamic state, the vortex is itself the source of a field. The second term is *the kinetic energy of the supercurrents*. Indeed, if the number of superconducting electrons per unit volume is n_s and their velocity at $\mathbf{r}$ is $\mathbf{v}(\mathbf{r})$, we have $\mathbf{j}(\mathbf{r}) = n_s e \mathbf{v}(\mathbf{r})$ and

$$\dfrac{1}{2}\lambda^2\mu_0\mathbf{j}^2(\mathbf{r}) = \dfrac{1}{2}\dfrac{m}{n_s e^2}[n_s^2 e^2 v^2(\mathbf{r})] = \left[\dfrac{1}{2}mv^2(\mathbf{r})\right]n_s.$$

   To get from (13.135) to (13.137), we observe that

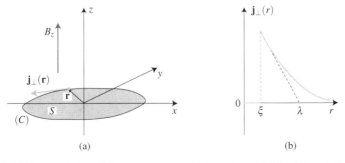

Fig. P13.2 (a) Representation of the current $\mathbf{j}_\perp$ around the vortex and (b) radial variation of $\mathbf{j}_\perp$

$$\mu_0 \mathbf{j} = \text{curl } \mathbf{B},$$

and hence that

$$\lambda^2 \mu_0^2 \mathbf{j}^2 = \lambda^2 \mu_0 \mathbf{j} \cdot \text{curl } \mathbf{B}.$$

Now,

$$\mathbf{j} \cdot \text{curl } \mathbf{B} = \mathbf{B} \cdot \text{curl } \mathbf{j} + \text{div}(\mathbf{B} \wedge \mathbf{j}).$$

We thereby obtain

$$W_v = \frac{1}{2\mu_0} \left[\int_S \mathbf{B} \cdot (\mathbf{B} + \lambda^2 \mu_0 \text{curl } \mathbf{j}) \mathrm{d}S + \oint_{(C)} \lambda^2 \mu_0 (\mathbf{B} \wedge \mathbf{j}) \cdot \mathrm{d}\boldsymbol{\ell} \right]$$

$$= \frac{1}{2\mu_0} \left[\int_S \mathbf{B} \cdot \mathbf{V}(\mathbf{r}) \mathrm{d}S + \lambda^2 \oint_{(C)} (\mathbf{B} \wedge \text{curl } \mathbf{B}) \cdot \mathrm{d}\boldsymbol{\ell} \right], \qquad (13.143)$$

which is Eq. (13.137) in the question (but this proof was not required).

4. $B_z(r)$ and **curl B** go as $e^{-r/\lambda}/r$ for large r, so

$$\oint_{(C)} (\mathbf{B} \wedge \text{curl } \mathbf{B}) \cdot \mathrm{d}\boldsymbol{\ell} \propto \int_0^{2\pi} \frac{1}{r} e^{-2r/\lambda} \mathrm{d}\theta \sim \frac{2\pi}{r} e^{-2r/\lambda} \longrightarrow 0.$$

5. The energy per unit length of the vortex is found by evaluating the first integral in (13.137), with the result

$$W_v = \frac{1}{2\mu_0} \int_S \mathbf{B} \cdot \mathbf{V}(\mathbf{r}) \mathrm{d}S = \frac{1}{2\mu_0} \int_S \mathbf{B}(\mathbf{r}) \cdot \Phi_0 \delta(\mathbf{r}) \mathbf{z} \mathrm{d}S = \frac{\Phi_0}{2\mu_0} B_z(0),$$

and hence

$$\boxed{W_v = \frac{\Phi_0^2}{4\pi \mu_0 \lambda^2} K_0(\xi/\lambda)}. \qquad (13.144)$$

6. The magnetic fields and currents satisfy (13.136), which is linear, so the fields and currents produced by two vortices are simply additive. With two vortices located at $\mathbf{r}_1 = -a\mathbf{x}$ and $\mathbf{r}_2 = a\mathbf{x}$, we thus have

$$\boxed{B_z(\mathbf{r}) = \frac{\Phi_0}{2\pi \lambda^2} \left[K_0 \left(\frac{|\mathbf{r} - a\mathbf{x}|}{\lambda} \right) + K_0 \left(\frac{|\mathbf{r} + a\mathbf{x}|}{\lambda} \right) \right]}. \qquad (13.145)$$

7. With $\mathbf{V}(\mathbf{r}) = \Phi_0 \big[\delta(\mathbf{r} - a\mathbf{x}) + \delta(\mathbf{r} + a\mathbf{x}) \big] \mathbf{z}$, choosing an integration surface normal to the two vortices and containing them, then letting it tend to infinity, the second term in (13.137) vanishes, while the first gives

Problem 13: Magnetisation of a Type II Superconductor

$$W_{2v} = \frac{\Phi_0^2}{4\pi\mu_0\lambda^2}\left[2K_0\left(\frac{\xi}{\lambda}\right) + 2K_0\left(\frac{2a}{\lambda}\right)\right]$$

$$= 2W_v + \frac{\Phi_0^2}{2\pi\mu_0\lambda^2}K_0\left(\frac{2a}{\lambda}\right), \qquad (13.146)$$

The interaction energy between the two vortices is thus

$$\boxed{W_{12} = \frac{\Phi_0^2}{2\pi\mu_0\lambda^2}K_0\left(\frac{2a}{\lambda}\right)}. \qquad (13.147)$$

8. The force exerted by the vortex at $-a$ on the one at $+a$ is given by

$$\mathbf{f}_{12} = -\frac{1}{2}\frac{\partial W_{12}}{\partial a}\mathbf{x} = +\frac{\Phi_0^2}{2\pi\mu_0\lambda^3}K_1\left(\frac{2a}{\lambda}\right)\mathbf{x}. \qquad (13.148)$$

This force is positive in the x direction. The two vortices repel one another. As can be seen from Fig. P13.3, the force can indeed be expressed in the form

$$\boxed{\mathbf{f}_{12} = \mathbf{j}_{12} \wedge \Phi_0 \mathbf{z}}, \qquad (13.149)$$

where $\mathbf{j}_{12}$ is the current produced by vortex 1 at the position of vortex 2, according to (13.142) of question 2.

9. For two vortices of opposite vorticity,

$$\boxed{B_z(\mathbf{r}) = \frac{\Phi_0}{2\pi\lambda^2}\left[K_0\left(\frac{|\mathbf{r}-a\mathbf{x}|}{\lambda}\right) - K_0\left(\frac{|\mathbf{r}+a\mathbf{x}|}{\lambda}\right)\right]}. \qquad (13.150)$$

Note that, in the plane $x = 0$, we have $|\mathbf{r} - a\mathbf{x}| = |\mathbf{r} + a\mathbf{x}|$ and hence $B_z(\mathbf{r}) = 0$. The symmetry of Fig. P13.4 is such that the total current in the plane $x = 0$ lies along the y axis and the contributions of the two vortices are additive in this plane. With $\mathbf{V}(\mathbf{r}) = \Phi_0[\delta(\mathbf{r} - a\mathbf{x}) - \delta(\mathbf{r} + a\mathbf{x})]$, we obtain

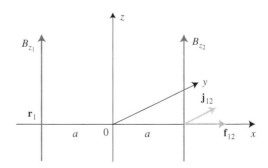

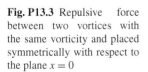

Fig. P13.3 Repulsive force between two vortices with the same vorticity and placed symmetrically with respect to the plane $x = 0$

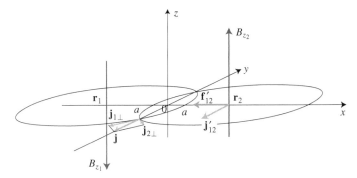

Fig. P13.4 Situation in the plane $x = 0$ for two vortices with opposite vorticity, placed symmetrically with respect to the plane $x = 0$

$$W'_{20} = \frac{\Phi_0^2}{4\pi\mu_0\lambda^2}\left[2W_v - 2K_0\left(\frac{2a}{\lambda}\right)\right]. \quad (13.151)$$

The interaction energy has changed sign compared with the previous case, and the two vortices attract one another with a force $\mathbf{f}'_{12}$ opposite to the one obtained in (13.149) and given by

$$\mathbf{f}'_{12} = \mathbf{j}'_{12} \wedge \Phi_0 \mathbf{z}. \quad (13.152)$$

10. For a vortex parallel to the surface of a semi-infinite superconductor, when no field is applied outside the superconductor, we expect $\mathbf{B} = 0$ at the surface. Surface currents flow in the plane of the surface and can only be oriented along the y axis if the vortex is parallel to the z axis.

11. We can therefore obtain the solution for the field and current by taking the vortex and its image at $-a\mathbf{x}$ with opposite vorticity. Indeed, we saw in questions 9 and 10 that we obtain $B_z = 0$ and currents in the plane $x = 0$ along the y axis. The solution for $B_z(\mathbf{r})$ is therefore precisely (13.150) for $x > 0$.

12. The London equation gives (see Fig. P13.5)

$$B = \mu_0 H_a \exp\left(-\frac{x}{\lambda}\right), \quad \mathbf{j}_S = \frac{H_a}{\lambda}\exp\left(-\frac{x}{\lambda}\right)\mathbf{y}. \quad (13.153)$$

13. The force acting on the vortex due to the interaction with its image was found in question 9 and given by (13.152). It pushes the vortex toward the surface. The force due to the applied field H_a is given by

$$\mathbf{f}_S = \mathbf{j}_S \wedge \Phi_0 \mathbf{z} = \Phi_0 \frac{H_a}{\lambda}\exp\left(-\frac{x}{\lambda}\right)\mathbf{x}. \quad (13.154)$$

Problem 13: Magnetisation of a Type II Superconductor

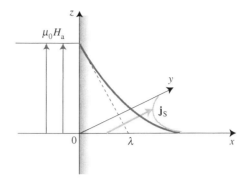

Fig. P13.5 Field and current at the surface of the superconductor in the absence of vortices

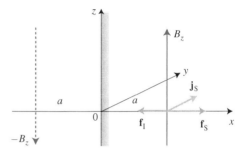

Fig. P13.6 Forces on a vortex lying parallel to the surface of the superconductor, in the presence of a field B

It pushes the vortex toward the interior of the superconductor. These two forces thus act in opposite directions (see Fig. P13.6).

14. When a vortex is located at the surface, it can only penetrate the bulk of the superconductor if the field H_a is large enough for the force $\mathbf{f}_S$ to exceed the attractive force of its own image. The penetration threshold H_p for vortices is thus specified by

$$\frac{\Phi_0 H_p}{\lambda} \exp\left(-\frac{\xi}{\lambda}\right) = \frac{\Phi_0^2}{2\pi \mu_0 \lambda^3} K_1\left(\frac{2\xi}{\lambda}\right). \qquad (13.155)$$

15. Since $\xi/\lambda \ll 1$, we have

$$K_1\left(\frac{2\xi}{\lambda}\right) \simeq \frac{\lambda}{2\xi}, \qquad \exp-\frac{\xi}{\lambda} \simeq 1,$$

which implies that

$$\boxed{H_p = \frac{\Phi_0}{4\pi \mu_0 \lambda \xi} = \frac{\lambda}{\xi} \frac{H_{c_1}}{\ln \lambda/\xi} \gg H_{c_1}}. \qquad (13.156)$$

H_p is thus of the same order of magnitude as the critical thermodynamic field $H_c = (\lambda/\xi) H_{c_1}$. Owing to their interaction with the surface, vortices forming at

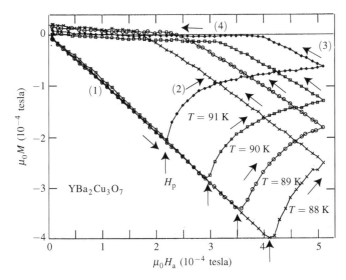

Fig. P13.7 Hysteresis cycles for YBa$_2$Cu$_3$O$_7$

H_{c_1} cannot therefore penetrate into the superconductor as long as $H_a < H_p \sim H_c$. We say that there is a *surface barrier*.

16. Beyond H_p, vortices can penetrate and surface currents gradually decrease on the second branch (2) of the hysteresis cycle in Fig. P13.7. When the field is reduced on branch (3) of the hysteresis cycle, the magnetisation varies with a slope almost equal to -1 [branch (3) is almost parallel to the straight line corresponding to the perfect Meissner effect]. This goes on until the magnetisation has almost vanished. Note that *branch (4) of the cycle corresponds to virtually zero magnetisation*.

17. If the vortices are uniformly distributed throughout the material, the average magnetic induction $\overline{B}_{int}$ over a volume with elementary dimension greater than the distance between vortices is almost uniform. The decrease in the surface currents can be understood from the fact that the magnetic induction varies from $\mu_0 H_a$ to $\overline{B}_{int}$ over the distance λ. It is this reduction in the surface currents that is detected by a drop in the measured magnetisation. When the field decreases on branch (3), the gradient -1 corresponds to *a reduction of the screening currents, but the vortices do not leave the sample*. At the end of branch (3), when $\mathbf{M} = 0$, we have $\mu_0 H_a = \overline{B}_{int}$ and the average magnetic induction is uniform, so there is then no more surface current. When H_a is reduced, M remains zero and $\mu_0 H_a = \overline{B}_{int}$ right along branch (4). There the *vortices gradually leave the sample*.

18. The existence of $\mathbf{f}_I$ and $\mathbf{f}_S$ corresponds to a dependence of the vortex energy on the distance from the surface which has a maximum. It thus constitutes a *surface barrier* for penetration or expulsion of vortices. When $H > H_p$, the surface barrier vanishes and the flux penetrates, but the surface currents fall off.

Problem 13: Magnetisation of a Type II Superconductor

The function $M(H_a)$ on branch (2) of Fig. P13.7 is such that the surface barrier is permanently zero. When H_a is reduced on branch (3), the surface barrier reappears and the flux can no longer be excluded along branch (3). It does indeed correspond to a decrease in the surface currents without modification of the vortex density. In contrast, on branch (4), when $\mathbf{M} = 0$, *the surface currents vanish*, $\mu_0 H_a = \overline{B}_{\text{int}}$ permanently, and the vortices are gradually excluded from the superconductor. We return at $H_a = 0$ to a situation in which almost no vortices are pinned.

Problem 14: Electronic Structure and Superconductivity of V_3Si[8]

In the 1960s, relatively high critical temperatures T_c were discovered in superconducting metal alloys based on vanadium and niobium. Whereas $T_c = 5$ K for vanadium and $T_c = 8$ K for niobium, T_c is 17 K in V_3Si and 23 K in Nb_3Ge. The best superconducting coils used today are made from Nb_3Sn ($T_c = 18$ K).

The aim here is to obtain a qualitative understanding of these high T_c values by investigating the electronic structure of these compounds. Their crystal structure is A-15, defined as follows for V_3Si. The Si atoms occupy sites in a body-centered cubic lattice with parameter a, while the V atoms are distributed along the faces of the cube at positions

$$\left(0, \frac{1}{4}, \frac{1}{2}\right), \quad \left(0, \frac{3}{4}, \frac{1}{2}\right), \quad \left(\frac{1}{2}, 0, \frac{1}{4}\right), \quad \left(\frac{1}{2}, 0, \frac{3}{4}\right), \quad \left(\frac{1}{4}, \frac{1}{2}, 0\right), \quad \left(\frac{3}{4}, \frac{1}{2}, 0\right).$$

1. *Sketch a primitive cell of V_3Si, indicating all the atoms of the basis?*

2. *Is it possible to predict a priori whether the compound V_3Si is metallic or insulating according to a band theory for independent electrons?*

It can be shown that the bands at the Fermi level are mainly built up from $3d$ orbitals of vanadium (bands resulting from the s levels are broad and contribute little at the Fermi level). To calculate the d band structure, we apply the LCAO method to these orbitals and assume that the orbitals of two neighbouring sites are orthogonal.

3. *Compare the distances between near-neighbour vanadium atoms located on the same face and on different faces. Deduce that the V atoms can be described as forming three series of mutually independent chains in the three space directions, and that the problem reduces to studying the band structure in a single direction.*

We consider a row of V atoms along the z axis.

4. *Indicate the corresponding first Brillouin zone.*

Assume that the $3d$ orbitals of each vanadium atom are degenerate with energy E_0.

5. *Considering only nearest neighbours, express the dispersion relation $E(k)$ in terms of the energy E_0 and hopping integrals t_0 and t_1, assuming that they are the same for all the $3d$ orbitals.*

6. *State and plot the corresponding density of states $D(E)$.*

In the following, we take $t_0 \simeq 0$.

7. *Determine the Fermi energy in terms of E_0 and t_1.*

---

[8] This problem has been designed with F. Rullier-Albenque.

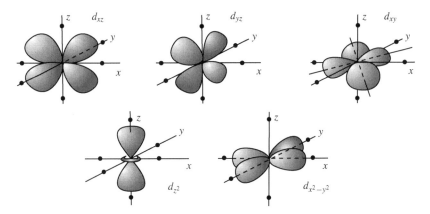

Fig. P14.1 Spatial representation of the five different 3d orbitals

In fact the hopping integrals t_1 are not the same for the different 3d orbitals. Their spatial representations are shown in Fig. P14.1.

8. *Show that, by symmetry, the 3d orbitals fall into three different groups with increasing hopping integrals aong the chains.*

Take the values of the hopping integrals to be $t_1/4$, $t_1/2$, and t_1. We still assume that the integrals t_0 are negligible.

9. *Describe the new band structure, indicating the atomic orbitals associated with each subband, and plot the resulting density of states $D(E)$.*

10. *Indicate the approximate position of the Fermi level expected in each case.*

Figure P14.2 shows the temperature dependence of the specific heat of a V_3Si single crystal.

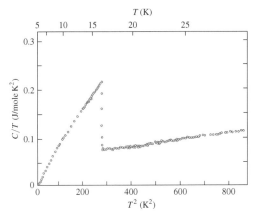

Fig. P14.2 Temperature dependence of the specific heat of a V_3Si single crystal. Adapted from Muto, Y., Toyota, N., Noto, K., Akustu, K., Isimo, M., Fukase, T.: J. Low Temp. Phys. **34**, 617 (1979)

Problem 14: Electronic Structure and Superconductivity of V_3Si

11. What characterises the superconducting transition? Determine the value of T_c in this sample.

12. How could you deduce the superconducting gap from these data? (Do not actually do this.) The value found is $\Delta = 2.4$ meV. Can V_3Si be described by the BCS theory?

13. From these data, find the electronic specific heat $C_v = \gamma T$ in the normal state.

For vanadium, $\gamma = 8.8$ mJ/mole/K^2.

14. Can you suggest an explanation for the high value of T_c in V_3Si based on the band structure model and these experimental data?

In contrast, Mo_3Ge has a very low $T_c = 1.5$ K.

15. Is this compatible with the above model?

Solution

1. The primitive cell is cubic. The basis contains two V_3Si molecules, as shown in Fig. P14.3.

2. There is an even number of V atoms and of Si atoms, and hence an even number of electrons in the primitive cell. It is not therefore possible to predict a priori whether this compound is metallic or insulating using a band theory for independent electrons.

3. The distance between near-neighbour vanadium atoms on a given face is $a/2$, but $(a/4)\sqrt{6} \approx 1.22 a/2$ for atoms belonging to orthogonal faces. The hopping integrals between vanadium atoms will be greater along the three directions x, y, and z. We may thus consider that the vanadium atoms form three series of orthogonal chains numbered 1, 2, and 3 in Fig. P14.3. These chains can be obtained from one another by symmetry. It suffices therefore to study the electronic structure along some given direction.

4. As the distance between V atoms is $a/2$, the Brillouin zone is a straight line segment $[-2\pi/a, 2\pi/a]$ along the z axis.

5. The dispersion relation (given in Chap. 1) is

$$E(k) = E_0 - t_0 - 2t_1 \cos \frac{ka}{2}.$$

This band is ten-fold degenerate: (2 spins) × (5 orbitals).

6. The density of states per V atom is, assuming $t_1 > 0$,

$$D(E) = \frac{1}{2\pi t_1} \frac{1}{\sqrt{1 - (E - E_0 + t_0)^2/4t_1^2}} \times 2 \times 5,$$

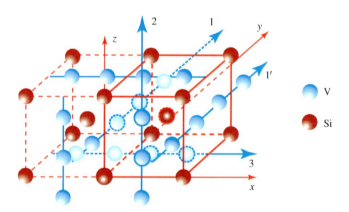

Fig. P14.3 Primitive cell and basis of V_3Si, showing the three linear V chains directed along the three axes

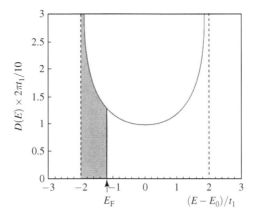

Fig. P14.4 Density of states $D(E)$ per V atom, showing the Fermi level

for $|E - E_0 + t_0| < 2t_1$. The two-fold spin degeneracy and five-fold orbital degeneracy of the band have been included. This density of states is shown for $t_0 = 0$ in Fig. P14.4.

7. The outermost electronic shells of vanadium are $3d^3\,4s^2$. There are thus 3 electrons per V atom in the d band. The Fermi energy E_F is thus given by

$$\int_{E_0-2t_1}^{E_F} D(E)\mathrm{d}E = 3\,.$$

This integral is more easily calculated in **k** space, since the density of states is constant there. It becomes $10(a/2\pi)\int_0^{k_F} \mathrm{d}k = 3$, or $k_F = 3\pi/5a$, and then

$$E_F = E_0 - 2t_1 \cos\frac{3\pi}{10} \approx E_0 - 1.18 t_1\,.$$

The position of the Fermi level is shown in Fig. P14.4, where the density of states is plotted as a function of the energy.

8. Consider for example a chain of V atoms along the z axis. The hopping integrals t_1 get bigger as the overlap of the atomic orbitals along z increases. Examining the d orbitals we find that the d_{z^2} orbitals, which yield a high probability along the z axis, correspond to the largest t_1 and will therefore form the broadest band. Then come the d_{xz} and d_{yz} orbitals, which yield significant probabilities in the (x,z) and (y,z) planes. Finally, the d_{xy} and $d_{x^2-y^2}$ orbitals, which are perpendicular to the z axis, correspond to the smallest t_1 and will thus form the narrowest bands.

9. The band structure now comprises three distinct bands arising respectively from:
   - the d_{z^2} orbitals: $E(k) = E_0 - 2t_1 \cos(ka/2)$, with two-fold degeneracy,
   - the d_{xz} and d_{yz} orbitals: $E(k) = E_0 - t_1 \cos(ka/2)$, with four-fold degeneracy,

- d_{xy} and $d_{x^2-y^2}$ orbitals: $E(k) = E_0 - (t_1/2)\cos(ka/2)$, with four-fold degeneracy.

The corresponding densities of states are:

$$D_1(E) = \frac{2}{2\pi t_1} \frac{1}{\sqrt{1-(E-E_0)^2/4t_1^2}},$$

$$D_2(E) = \frac{8}{2\pi t_1} \frac{1}{\sqrt{1-(E-E_0)^2/t_1^2}},$$

and

$$D_3(E) = \frac{16}{2\pi t_1} \frac{1}{\sqrt{1-4(E-E_0)^2/t_1^2}}.$$

These densities of states are plotted in Fig. P14.5. There are now *six singularities* located at $E = (E_0 \pm 2t_1)$, $E = (E_0 \pm t_1)$, and $E = (E_0 \pm t_1/2)$.

10. If there were only the bands 1 and 2, we would obtain $E_F = E_0$ (half-occupancy). In the present case, the Fermi level will lie between $E_0 - t_1/2$ and E_0 and closer to $E_0 - t_1/2$ than to E_0. A more careful calculation shows that, when the three d electrons of vanadium are placed, we begin by filling bands 1 and 2 up to the energy $E_0 - t_1/2$, which amounts to placing ~ 0.84 electrons in band 1 and 1.36 electrons in band 2. The Fermi level will thus be determined by placing the remaining ~ 0.8 electrons in the three bands starting from the energy $E_0 - t_1/2$. Owing to the divergence of the density of states of the third band for $E_0 - t_1/2$, we expect the Fermi level to be very close to this energy. Solving numerically, we do indeed find $E_F \approx E_0 - 0.41 t_1$. The corresponding Fermi level is shown in Fig. P14.5.

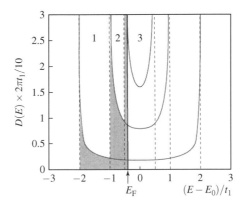

Fig. P14.5 Densities of states for the three distinct bands

Problem 14: Electronic Structure and Superconductivity of V_3Si

11. The superconducting transition is revealed by the discontinuity in the specific heat, implying that T_c is in this case equal to 16.7 K.

12. We expect the specific heat to go as $C \propto \exp(-\Delta/k_B T)$ in the superconducting state. Plotting the logarithm of C as a function of $1/T$, we should therefore obtain a straight line with slope Δ/k_B. In fact, in the present case this only applies for the very low temperature data, and we obtain $\Delta = 2.4$ meV. According to BCS theory, Δ is related to T_c by $2\Delta = 3.52 k_B T_c$. In the present case we find $k_B T_c = 1.39$ meV, so that $2\Delta/k_B T_c = 3.45$, which is close to 3.52. We may conclude that V_3Si is well described by the BCS theory.

13. In a metal, we know that the specific heat is the sum of electron and phonon contributions and takes the form $C = \gamma T + AT^3$. This is indeed what we observe in Fig. P14.2 since C/T varies linearly with T^2 in the normal state. Extending this straight line down to $T = 0$, we obtain the coefficient γ directly as 55 mJ/mole/K^2.

14. The coefficient γ of the electronic specific heat is related to the density of states at the Fermi level. The fact that γ is much smaller in vanadium than in V_3Si thus shows that $D(E_F)$ is higher in the latter. This agrees with the simplified calculation of the band structure, which shows that the Fermi level is located close to a singularity in the density of states. According to the BCS theory, T_c is given by

$$k_B T_c = 1.14 \hbar \omega_D \exp\left[-\frac{1}{V_0 D(E_F)}\right].$$

We thus expect T_c to increase as $D(E_F)$ increases. This explains the high values of T_c observed for V_3Si.

Note: It can be shown quite generally that T_c is high when the Fermi level lies near a van Hove singularity in the density of states. This is the case for the A-15 compounds, e.g., V_3Si, V_3Ga, Nb_3Sn. This is confirmed by the fact that T_c is highly sensitive to the order of the atoms in chains for this type of compound. If the chains are disordered, as occurs for instance if the compound is non-stoichiometric, the singularities in the density of states are broadened, whereupon $D(E_F)$ and hence T_c decrease very fast.

15. The molybdenum atom has one more electron than vanadium in its outer d shell and the simplified calculation of the band structure shows that the Fermi level in Mo_3Ge will be further from the van Hove singularity. We thus understand that $D(E_F)$ and hence T_c will be smaller.

Problem 15: Superconductivity of NbSe$_2$

NbSe$_2$ is a quasi-2D lamellar material that cleaves easily. The cleavage planes are almost perfect on the atomic scale and give rise to clean observations by scanning tunneling microscopy. The structure of the lamellas is shown in Fig. P15.1. The equilateral triangles with Se at the vertices have side $a = 3.45$ Å. The thickness of a lamella is $e = 2.75$ Å.

1. *What is the primitive cell of a lamella?*

In one crystalline form called 2H-NbSe$_2$, the lamellas are organised in such a way that the Se triangles have alternating orientations as shown in Fig. P15.2. The distance between lamellas is $d = 3.52$ Å.

2. *Specify a primitive cell of the 3D crystal and calculate its volume.*

From an electronic standpoint, the lamellas can be considered weakly coupled, and the band structure is then very similar to that of an isolated lamella. The atomic electronic structures of niobium and selenium are $4d^4\,5s$ and $4s^2 p^4$, respectively.

3. *Can we say whether a lamella is metallic or insulating?*

The scanning tunneling microscope can measure the electron current j between the metallic tip and the cleaved surface of the sample as a function of the tip–sample distance and the voltage V applied to the tip. When the tip is held at a fixed distance from the surface, we obtain the tunneling characteristic $G(V) = \mathrm{d}j/\mathrm{d}V$.

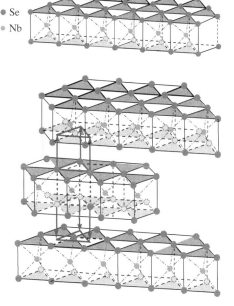

Fig. P15.1 Structure of one lamella of NbSe$_2$

Fig. P15.2 Three-dimensional structure of 2H-NbSe$_2$ showing the alternating stacking orientations of the Se triangles

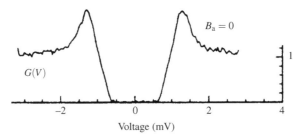

Fig. P15.3 Tunneling characteristic of NbSe$_2$ at $T = 0.3$ K in zero applied field. http://link.aps.org/doi/10.1103/PhysRevLett.62.214

4. Express the characteristic as a function of the densities of states of the tip and surface.

NbSe$_2$ is superconducting with critical temperature $T_c = 7.2$ K. The characteristic curve observed by tunneling at $T = 0.3$ K is shown in Fig. P15.3. In the figure, the conductance is normalised to unity for values of $|V|$ greater than 4 mV. The tungsten tip has energy-independent density of states for energies over a range of 0.1 eV on either side of its Fermi energy.

5. How does the density of states compare with the one expected according to BCS theory?
6. Estimate the width of the superconducting gap.
7. How does it compare with the BCS prediction for the given value of T_c?
8. Given that there is one electron per Nb in the conduction band, calculate the London penetration depth λ_L of NbSe$_2$ (consider the 3D structure).

A magnetic field H_a is applied perpendicular to the cleaved surface of the sample and the microscope tip moved parallel to the surface at height z. At each point of the surface, $G(V) = dj/dV$ is measured for two values of the voltage, viz., $V = 0$ mV and $V = 3$ mV. The values of $G(0)/G(3)$ for different x and y are used to produce a grey-scale image of the surface, where 0 = white and 1 = black. For an applied field $B_a = \mu_0 H_a$ less than 0.023 tesla, the image of the sample is completely white.

9. What can you deduce?

For a field B_a greater than 0.02 tesla, spots appear on the image. In particular, for a field of 1.041 tesla, the image is the one shown in Fig. P15.4.

10. Explain the physical origin of this observation. Would you expect to see the observed structure?
11. Do an accurate check to see whether the lattice parameter corresponds to the one expected for the given applied field.

Fig. P15.4 Grey-scale image of the cleaved surface for a field of 1.041 tesla. From Hess, H.F., Robinson, R.B., Dynes, R.C., Valles, J.M., Jr., Waszczak, V.: Phys. Rev. Lett. **62**, 214 (1989). With the permission of the American Physical Society (© 1989 APS). http://link.aps.org/doi/10.1103/PhysRevLett.62.214

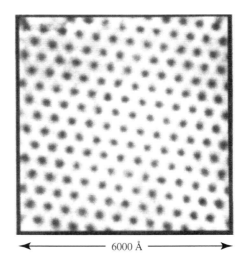

← 6000 Å →

In a simplified model, a vortex can be represented by a normal metal tube with radius equal to the coherence length ξ.

12. *What is the tunnelling characteristic $G(V)$ expected at the center of a vortex?*

13. *Can you use the image to estimate a value for ξ?*

When B_a exceeds 1.83 tesla, the observed image becomes completely black.

14. *Can this observation be used to estimate the penetration depth λ? How does it compare with the value found for λ_L in question 8?*

In a field of less than 0.05 tesla, the vortices are distant from one another and their electronic structure can be examined more carefully by plotting the characteristics $G(V)$ as a function of the distance d from the center of the vortex. The results of such a study are shown in Fig. P15.5, where the characteristics $G(V)$ for distances $d = 0$–585 Å are compared with the one obtained for zero field.

It can be shown that, when there is a supercurrent j_s, the gap in a BCS superconductor is lowered by a kinematic effect according to

$$\Delta \longrightarrow \Delta - \hbar k_F |v_s|,$$

where v_s is the average velocity of Cooper pairs. Recall that the field B around a vortex has the form $B_z(d) = B_0 K_0(d/\lambda)$, where z is the vortex axis, d the distance to the center of the vortex, and $K_0(x)$ a Bessel function.

15. *Give the expression for the resulting supercurrent $j_s(d)$. Note that $K_1(x) = -K_0'(x)$ and B_0 is determined by the flux quantisation condition.*

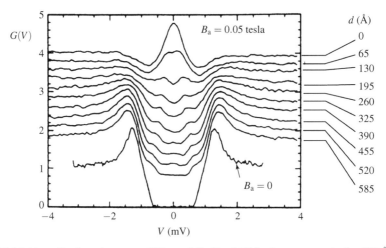

Fig. P15.5 Normalised conductances $G(V)$ are shifted by 0.75 for the curve acquired at 585 Å and a further 0.25 for each of the other curves. From Hess, H.F., Robinson, R.B., Waszczak, J.V.: Phys. Rev. Lett. **64**, 2711 (1990). With the permission of the American Physical Society (© 1990 APS). http://link.aps.org/doi/10.1103/PhysRevLett.64.2711

16. *Show that the variation of the gap can be written locally in the form*

$$\Delta(d) = \Delta \left[1 - \frac{\pi \xi}{2\lambda} K_1 \left(\frac{d}{\lambda} \right) \right].$$

Over what distance range does this formula apply?

17. *Show that, far from the vortex core, the curves of Fig. P15.5 vary in accordance with this kinematic effect.*

Using the curves in Fig. P15.5, we find the values of Δ as a function of d given in Table 13.3.

Table 13.3 Superconducting gap Δ as a function of the distance d from a vortex core

d	130	195	260	325	390	455	520	≫
Δ	–	0.33	0.41	0.47	0.52	0.57	0.61	0.7

18. *Given that $K_1(x) \sim 1/x$ for $x \ll 1$, find a value for ξ. Deduce a value for λ.*
19. *Is the curve observed at the vortex core the one expected from question 12?*

Problem 15: Superconductivity of NbSe$_2$

Solution

1. A primitive cell of the lamella contains one NbSe$_2$ unit. A *possible* primitive cell is shown in Fig. P15.6.

2. The alternating orientation of the lamellas is such that a primitive cell of the 3D crystal contains two NbSe$_2$ units, as shown in Fig. P15.7. The volume of the primitive cell is $v = a^2 c \sqrt{3}/2$, with $a = 3.45$ Å and $c = 2(2.75 + 3.52) = 12.54$ Å, therefore $v = 129.25$ Å^3.

3. The number of electrons per primitive cell of the lamella in the valence orbitals is $5 + (2 \times 6) = 17$, an odd number. According to band theory, this necessarily implies a metallic state, since at least one band will be partially occupied.

4. Recall that

$$j = A \mathscr{T} \int_{-\infty}^{+\infty} D_1(E) D_2(E + eV) [f(E) - f(E + eV)] dE .$$

For constant $D_1(E) = D_1(E_F)$,

$$\frac{dj}{dV} = A \mathscr{T} D_1(E_F) \int_{-\infty}^{+\infty} D_2(E') \frac{\partial f(E' - eV)}{\partial V} dE' ,$$

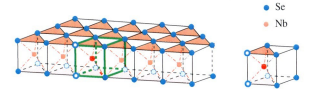

Fig. P15.6 Possible primitive cell for the NbSe$_2$ lamella

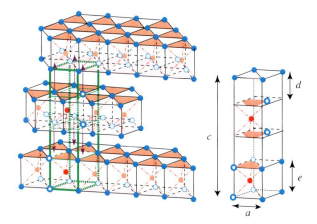

Fig. P15.7 Primitive cell of 3D crystal containing two NbSe$_2$ units

and for $T = 0$, this reduces to

$$\frac{dj}{dV} = -A\mathcal{T}eD_1(E_F)D_2(E_F+eV).$$

5. The density of states observed by electron tunneling is similar to the one shown in Chap. 7 for Sn (see Fig. 7.10), and this satisfies the BCS theory. There is indeed a range of values $|V| \lesssim 0.7$ meV for which $D_2(E_F+eV) = 0$, and this corresponds to the gap. However, the singularity at $eV = \Delta$ is not as sharp as would be expected from BCS theory at $T = 0$:

$$D_{\text{BCS}} = \begin{cases} D(E_F)\dfrac{|E-E_F|}{[(E-E_F)^2-\Delta^2]^{1/2}}, & \text{for } E-E_F > \Delta, \\ 0, & \text{for } E-E_F < \Delta. \end{cases}$$

This may be because the measurement was made at nonzero temperature ($T = 0.3$ K). However, in the case of Sn, for similar values of T and Δ, experimental and theoretical values correspond to a much smaller broadening of the transition, as shown by the comparison done in Fig. P15.8. This broadening of the density of states singularities thus has a distinct origin, such as a distribution of gap values in NbSe$_2$. According to Fig. 7.10, if we take into account a thermal broadening ~ 0.1 meV, we deduce that at zero temperature,

$$\boxed{\Delta_{\min}(0) = 0.7 \pm 0.05 \text{ meV}, \qquad \Delta_{\max}(0) = 1.1 \pm 0.05 \text{ meV}}.$$

Note: The structure of the NbSe$_2$ lamellas in the ab plane is not symmetrical. This anisotropy causes an anisotropy of the superconducting properties which may explain the anisotropy of the gap, why the vortex lattice is not hexagonal, and so on.

7. In BCS theory, we expect

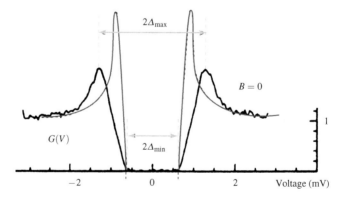

Fig. P15.8 Comparison of the data for NbSe$_2$ with expectation for a thermally broadened BCS density of states

Problem 15: Superconductivity of NbSe$_2$ 537

$$1.14\Delta = 2k_B T_c, \quad \text{hence} \quad \Delta = 1.75 k_B T_c.$$

For $T_c = 7.2$ K, this would correspond to $\Delta = 1.75 \times 25/300 = 1.05$ meV. There is therefore only a small discrepancy between the average gap 0.9 ± 0.05 meV and the value expected from the BCS theory.

8. We have $\lambda_L^2 = m_0/ne^2\mu_0$, with $n = (2/129.25) \times 10^{+30}/\text{m}^3$, which gives

$$\boxed{\lambda_L = 428 \text{ Å}}.$$

9. If the image is completely white, this means that the superconducting gap is present at all points of the sample. There is no normal region and we thus have a perfect Meissner effect for $\mu_0 H_a < 0.023$ tesla. Therefore,

$$\boxed{\mu_0 H_{c_1} = 0.023 \text{ tesla}}.$$

10. The appearance of black spots tells us that the superconducting state is modified in the black regions. The resulting lattice means there is a mixed state. The observed structure corresponds to a vortex lattice. This lattice appears to be hexagonal in Fig. P15.9. However, if we measure the lattice spacings in directions 1, 2, and 3, we find a triangular lattice, but not perfectly hexagonal. Indeed

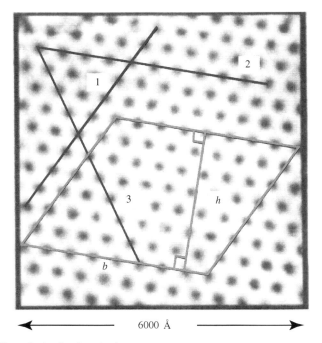

Fig. P15.9 Vortex lattice for the mixed state

the segments shown by dark lines, which correspond to 10 lattice constants are such that $10a_1 \to 81$ mm, $10a_2 \to 87$ mm, and $10a_3 \to 87$ mm.

11. The magnetic induction is such that $BS = \Phi_0$ if S is the area of the primitive cell of the vortex lattice. To determine S, we can consider the blue parallelogram in Fig. P15.9, for example, which corresponds to 56 cells of the lattice and has area

$$bh = \frac{69 \times 50.5}{(106)^2} 36 \times 10^6 \text{ Å}^2, \quad \text{and then } S = 0.1994 \times 10^6 \text{ Å}^2 \pm 0.5\%.$$

This corresponds to $B = 2.07/1.994 = 1.038 \pm 0.005$ tesla, which is perfectly compatible with the applied field of 1.041 tesla.

12. The characteristic for a normal region should be $G(V) = 1$.

13. Black regions correspond approximately to a region where $G(V) \simeq 1$. The average diameter of a black spot is 3 ± 0.3 mm, which corresponds to

$$\boxed{\xi = 85 \pm 9 \text{ Å}}.$$

14. When the sample gives a black image, it is completely in the normal state and

$$\boxed{\mu_0 H_{c_2} \simeq 1.83 \text{ tesla}}.$$

We thus have $H_{c_2}/H_{c_1} = (\lambda/\xi)^2$, or $(\lambda/\xi)^2 = 1.83/0.023 = 79.5$, which gives

$$\boxed{\lambda \sim 760 \pm 150 \text{ Å}},$$

slightly greater than λ_L. This is generally the case, as explained in Chap. 7.

15. The flux quantisation condition yields

$$B_0 = \frac{\Phi_0}{2\pi \lambda^2},$$

and $\mu_0 \mathbf{j} = \text{curl } \mathbf{B}$ yields $j_\perp = -\mu_0^{-1} \partial B_z/\partial r$, hence

$$\boxed{j_s(d) = \frac{\Phi_0}{2\pi \mu_0 \lambda^3} K_1\left(\frac{d}{\lambda}\right)}.$$

16. We have $j_s = n_s e v_s = n_{\text{pairs}} q v_s$. Hence

$$\Delta(d) = \Delta\left[1 - \frac{\hbar k_F}{\Delta n_s e} \frac{\Phi_0}{2\pi \mu_0 \lambda^3} K_1\left(\frac{d}{\lambda}\right)\right],$$

Problem 15: Superconductivity of NbSe₂

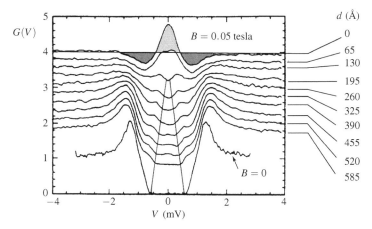

Fig. P15.10 Characteristic $G(V)$ near a vortex

with

$$\lambda^2 = \frac{m_0}{n_s e^2 \mu_0}, \qquad \Phi_0 = \frac{h}{2e}, \qquad \xi = \frac{h v_F}{\pi \Delta},$$

and therefore

$$\boxed{\Delta(d) = \Delta\left[1 - \frac{\pi \xi}{2\lambda} K_1\left(\frac{d}{\lambda}\right)\right]}.$$

This relation is only valid outside the vortex core, i.e., for $d > \xi$.

17. It is clear from Fig. P15.10 that *the gap decreases as we approach the vortex core*, for d decreasing from 585 to 195 Å. It even seems to disappear for $d \sim 130$ Å.

18. For $d/\lambda \ll 1$, we have

$$\Delta = \Delta_0 \left(1 - \frac{\pi \xi}{2} \frac{1}{d}\right).$$

Figure P15.11 shows Δ as a function of $1/d$. Taking into account the condition $d/\lambda \ll 1$, i.e., small values of d, the best agreement with the above equality is given by the straight line in Fig. P15.11. It has slope $p = 78$ mV Å, which permits the estimate

$$\xi = \frac{2}{\pi \Delta_0} p = \frac{2}{\pi} \frac{78}{0.7} \text{ Å} = 71 \text{ Å}.$$

From $(\lambda/\xi)^2 = 79.5$, we therefore deduce

$$\boxed{\lambda = 630 \text{ Å}}.$$

Fig. P15.11 Variation of the measured gap Δ with the distance d to the vortex core

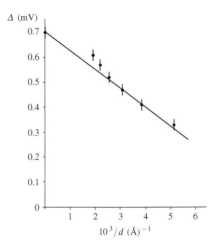

19. At the center of the vortex, we would expect $G(V) = 1$, so the fact that we observe a peak in the density of states at the Fermi level suggests that, in the vortex core, the electrons do not behave altogether as they would in a normal metal.

Note: The 'normal' electrons are confined within the vortex core and interact with the superconducting interface. This leads to a closer spacing of the levels near the Fermi level than in the normal metal. However, it can be checked that the integral of the density of states as a function of energy is equal to unity, as the light and dark regions have the same area in Fig. P15.10. This indicates that the number of electron states accessible in the vortex core remains the same as in the normal state.

Problem 16: Magnesium Diboride: A New Superconductor?

MgB_2 is an extremely simply binary compound, to be found on the shelf of almost any chemistry laboratory. It is quite astonishing that its superconductivity, with abnormally high $T_c = 39$ K for an apparently conventional material, was only discovered in 2001, while research groups have been carrying out systematic searches since the end of the 1970s to find the maximum possible value of T_c for electrotechnical applications. The aim here will be to present the characteristics of this material and its superconductivity, determined in the two years following its discovery.

16.1: Atomic and Electronic Structure of MgB_2

The lamellar atomic structure of this compound is depicted in Fig. P16.1. It comprises alternating honeycomb planes of B and planes of Mg. The latter are arranged vertically above the centers of the B hexagons.

1. *Specify the 3D Bravais lattice of MgB_2, together with a primitive cell and its basis.*

2. *What is the reciprocal lattice of this structure and its first Brillouin zone?*

3. *Can we deduce immediately whether this compound is insulating or conducting according to a band theory for independent electrons?*

4. *Experiment tells us that MgB_2 is a conductor. What can we conclude regarding its band structure, if such a band theory is applicable?*

In this structure, Mg is almost in the form Mg^{2+}. Its core levels and the $1s^2$ levels of B form narrow fully occupied bands.

5. *If the role of the Mg were simply to donate its two valence electrons to B, what well known band structure would the valence electrons of the boron plane adopt?*

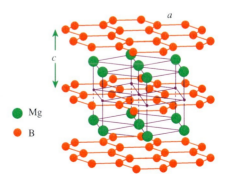

Fig. P16.1 Lamellar atomic structure of MgB_2

The calculated band structure for these valence electrons is shown for several directions of reciprocal space in Fig. P16.2. The directions can be identified in Fig. P16.3, which shows the Fermi surface in a unit cell of the reciprocal lattice.

6. *Plot the first Brillouin zone of question 2 on Fig. P16.3, taking Γ_1 as zone center.*

7. *The Fermi surface comprises four sheets corresponding to the last occupied states of different bands. Identify the curves $E(\mathbf{k})$ in Fig. P16.2 giving rise to the different sheets of the Fermi surface and associate them pairwise.*

8. *How could one find the velocity of an electron corresponding to a Bloch state represented by a point on the Fermi surface?*

9. *Identify the two sheets of the Fermi surface corresponding to electrons propagating in the B plane and the two sheets corresponding to bands with a more 3D character.*

The various band are built up from σ orbitals hybridising the boron $2s$ and $2p_x$, $2p_y$ orbitals on the one hand and the π orbitals built up from the boron $2p_z$ orbitals on the other.

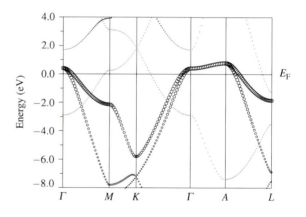

Fig. P16.2 Calculated band structure for the Mg valence electrons for several directions of reciprocal space. From Ann, J.M., Pickett, W.E.: Phys. Rev. Lett. **86**, 4366 (2001). With the permission of the American Physical Society (© 2001 APS). http://link.aps.org/doi/10.1103/PhysRevLett.86.4366

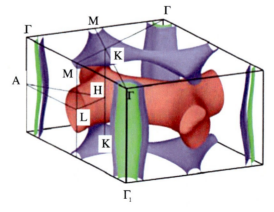

Fig. P16.3 Fermi surface in a unit cell of the reciprocal lattice. From Kortus, J., Mazin, I.I., Belachenko, K.D., Antropov, V.P., Boyer, L.L.: Phys. Rev. Lett. **86**, 4656 (2001). With the permission of the American Physical Society (© 2001 APS). http://link.aps.org/doi/10.1103/PhysRevLett.86.4656

Problem 16: Magnesium Diboride: A New Superconductor?

10. What are the dominant features of the different sheets of the Fermi surface?
11. What are the differences with the band structure of graphene?
12. Can you explain the effect of the Mg^{2+} ions on the band structure?

16.2: Superconductivity of MgB_2

a. Zero-Field Properties

Electron tunneling experiments have been carried out to investigate the superconducting properties of MgB_2. When a single crystal of MgB_2 is cleaved along a plane parallel to the B planes, the tunneling characteristics dI/dV parallel to c were observed at different temperatures and the results plotted in Fig. P16.4. For improved visibility, the curves are shifted vertically by 0.5 units for each temperature, while the conductivity is always equal to unity for $V = -15$ mV.

1. What electronic states are probed by this type of experiment? What quantity can be deduced?
2. Do we observe the value expected from BCS theory given that $T_c = 39$ K?

A tunneling experiment on a sample cleaved along the plane (a, c) gives a completely different result at low temperatures, as can be seen from Fig. P16.5.

3. Explain why certain electronic bands do note contribute to tunnelling when the tip is parallel to the c axis. What causes the modification observed in Fig. P16.5?
4. What is the physical origin of the two new peaks?

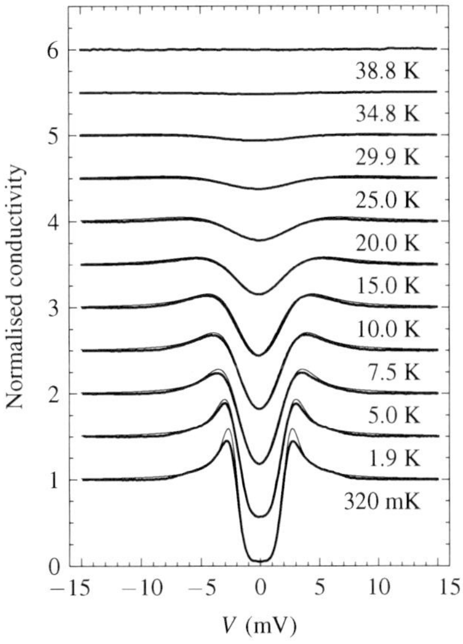

Fig. P16.4 Tunneling characteristics dI/dV parallel to c in an MgB_2 single crystal at different temperatures. From Eskildsen, M.R., Kugler, M., Lévy, G., Tanaka, S., Jun, J., Kazakov, S.M., Karpinski, J., Fisher, O.: Physica C **385**, 169 (2003). © Elsevier

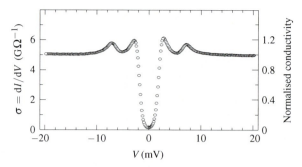

Fig. P16.5 Low temperature tunnelling characteristic dI/dV parallel to the plane (a,c) in an MgB$_2$ single crystal. From Eskildsen, M.R., Kugler, M., Lévy, G., Tanaka, S., Jun, J., Kazakov, S.M., Karpinski, J., Fisher, O.: Physica C **385**, 169 (2003). © Elsevier

b. Superconductivity in an Applied Field

When a magnetic field is applied along the c axis of the sample, the tunnelling characteristic dI/dV along c depends on the spatial position of the tip over the sample. For example, for an applied field of 0.05 tesla, we find that dI/dV does not depend on the voltage V at certain points of space. Moving the tip a distance r from such a point, we find that the characteristic dI/dV varies as shown in Fig. P16.6. In this figure, the curves dI/dV are shifted vertically by an amount proportional to the tip displacement and the total distance investigated is 250 nm.

5. *What produces this result? What can you conclude regarding the superconducting characteristics of MgB$_2$?*

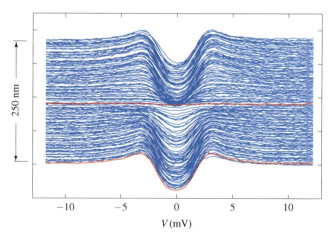

Fig. P16.6 Voltage dependence of dI/dV as the tip moves away from a point where the characteristic is independent of V. From Eskildsen, M.R., Kugler, M., Tanaka, S., Jun, J., Kazakov, S.M., Karpinski, J., Fisher, O.: Phys. Rev. Lett. **89**, 187003 (2002). With the permission of the American Physical Society (© 2002 APS). http://link.aps.org/doi/10.1103/PhysRevLett.89.187003

Problem 16: Magnesium Diboride: A New Superconductor? 545

6. *What characteristic length of the superconducting state can be deduced from this experiment? Estimate this length. Is it compatible with the BCS theory, given that the average velocity of the electrons at the Fermi level is $\langle |v_F| \rangle = 5 \times 10^5$ m/s?*

In a higher applied field, we find that many similar regions appear. Figure P16.7 maps the value of dI/dV taken at $V = 0$ across the sample surface. The value of dI/dV is represented by a scale from blue to red.

7. *Given that the applied field is $B = 0.5$ tesla, can you determine the size of the region in the figure?*

Another experimental method for studying the influence of a magnetic field on the properties of a superconductor is neutron diffraction. The field is applied perpendicularly to the surface in the direction of the c axis of a single-crystal sample. A monokinetic neutron beam produces diffraction peaks which are used to reconstruct the reciprocal lattice plane perpendicular to the c axis, as shown in Fig. P16.8 (left) for $B = 0.5$ tesla. The direct beam that would end up at the center of the figure is blocked beyond the sample and hence not detected.

8. *Analyse these experimental results. Are the observations compatible with the STM results? The neutrons used here have energy corresponding to a temperature of 9.5 K. What is the angle of diffraction for the observed spots?*

9. *What other characteristic length of the superconducting state can in principle be ascertained through this experiment?*

10. *How could it be determined?*

In an applied field of 0.9 tesla, the diffraction pattern changes to the one shown in Fig. P16.8 (right).

11. *What real space feature does this experimental result reflect?*

12. *Can you suggest the origin of the observed effect by considering the electronic band structure investigated in 16.1?*

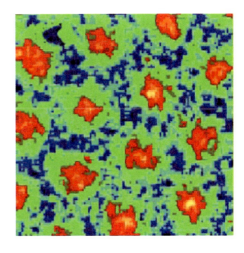

Fig. P16.7 Map of dI/dV taken at $V = 0$ across the sample surface. From Eskildsen, M.R., Kugler, M., Tanaka, S., Jun, J., Kazakov, S.M., Karpinski, J., Fisher, O.: Phys. Rev. Lett. **89**, 187003 (2002). With the permission of the American Physical Society (© 2002 APS). http://link.aps.org/doi/10.1103/PhysRevLett.89.187003

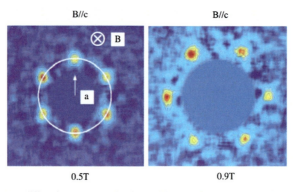

Fig. P16.8 Neutron diffraction patterns. Reciprocal lattice plane perpendicular to the c axis. *Left*: $B = 0.5$ tesla. *Right*: $B = 0.9$ tesla. From Cubitt, R., Eskildsen, M.R., Dewhurst, C.D., Jun, J., Kazakov, S.M., Karpinski, J.: Phys. Rev. Lett. **91**, 047002-1 (2003). With the permission of the American Physical Society (© 2003 APS). http://link.aps.org/doi/10.1103/PhysRevLett.91.047002

Solution

Atomic and Electronic Structure of MgB$_2$

1. The boron planes have exactly the same honeycomb structure as the graphene sheets making up the 3D structure of graphite. The 2D Bravais lattice of graphene can be specified by the two vectors **a** et **b** in Fig. P16.9a. In the orthonormal frame $(\mathbf{e}_x, \mathbf{e}_y, \mathbf{e}_z)$, they are given by

$$\mathbf{a} = \frac{3a}{2}\mathbf{e}_x - \frac{a\sqrt{3}}{2}\mathbf{e}_y, \qquad \mathbf{b} = \frac{3a}{2}\mathbf{e}_x + \frac{a\sqrt{3}}{2}\mathbf{e}_y.$$

The corresponding primitive cell of the honeycomb structure contains 2 boron atoms at the points $(0,0)$ and $(2/3, 2/3)$ relative to the frame $(\mathbf{a}, \mathbf{b})$. Graphite, in which there is no atomic plane between the graphene sheets, has a hexagonal 3D Bravais lattice obtained by adjoining the vector $\mathbf{c} = c\mathbf{e}_z$. This is also the Bravais lattice $(\mathbf{a}, \mathbf{b}, \mathbf{c})$ of MgB$_2$. In addition to the two boron atoms at $(0,0,0)$ and $(2/3, 2/3, 0)$, the primitive cell contains an Mg at $(1/3, 1/3, 1/2)$, as shown in Fig. P16.9b.

2. The reciprocal lattice of the 2D hexagonal lattice is also hexagonal with axes $\mathbf{a}^*, \mathbf{b}^*$ rotated by 90° with respect to $\mathbf{a}, \mathbf{b}$, where $|\mathbf{a}^*| = |\mathbf{b}^*| = 4\pi/a\sqrt{3}$. Its first Brillouin zone is a hexagon centered on the Γ point (see Fig. P16.10a). The reciprocal lattice of the 3D hexagonal lattice is obtained by adjoining the vector $\mathbf{c}^* = (2\pi/c)\mathbf{e}_z$. Its first Brillouin zone is a cylinder with hexagonal base centered on the Γ point (see Fig. P16.10b).

3. Magnesium is bivalent, so has an even number of electrons, like B$_2$. The primitive cell thus contains an even number of electrons. According to a band theory for independent electrons, it may be insulating or conducting, depending on the band width and occupancy.

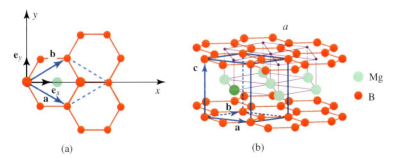

Fig. P16.9 (a) Two-dimensional Bravais lattice of graphene. (b) Three-dimensional hexagonal Bravais lattice of MgB$_2$

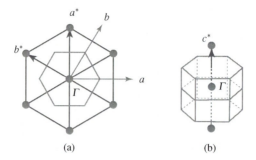

(a) (b)

Fig. P16.10 (a) Reciprocal lattice of the 2D hexagonal lattice showing axes $\mathbf{a}^*, \mathbf{b}^*$ rotated by 90° with respect to $\mathbf{a}, \mathbf{b}$ and the hexagonal first Brillouin zone centered on the Γ point. (b) Reciprocal lattice of the 3D hexagonal lattice showing the vector $\mathbf{c}^* = (2\pi/c)\mathbf{e}_z$ and cylindrical first Brillouin zone with base centered on the Γ point

4. The fact that MgB$_2$ is metallic necessarily means that at least two bands overlap, and hence that there are at least two partially occupied bands at the Fermi level. This gives rise to two independent sheets for the Fermi surface.

5. If the role of the Mg was simply to donate its 2 valence electrons to the B, there would be 4 valence electrons per boron atom, and we would recover precisely the same situation as in graphene. The two band structures would be identical.

6. Part of the first Brillouin zone is shown by the white lines in Fig. P16.11, where a second primitive cell, identical to the one in Fig. P16.3 has been represented below the point Γ_1. The part shown corresponds to only one third of the first Brillouin zone. The rest is obtained by two rotations, through 120° and −120°

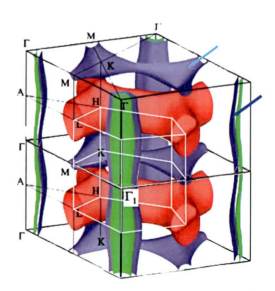

Fig. P16.11 Fermi surface in two unit cells of the reciprocal lattice. *White lines* represent part of the first Brillouin zone

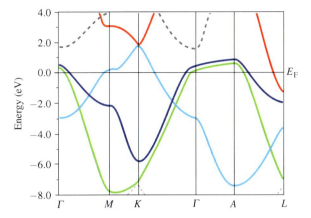

Fig. P16.12 Calculated band structure for the Mg valence electrons for several directions of reciprocal space. The bands of Fig. P16.2 are identified here by choosing colours which correspond to the sheets of the Fermi surface shown in Fig. P16.11

about the c^* axis. In fact, the representation here could have been restricted to one quarter of the volume shown, given the symmetries of the primitive cell.

7. Following the curves $E(\mathbf{k})$ in Fig. P16.12, it is easy to identify those that reach the value E_F at definite points of the line segments ΓK, ΓM, and AL. One can then identify the bands indicated by continuous green and dark blue curves in Fig. P16.12 which give rise to the green and blue Fermi surfaces lying parallel to the $\mathbf{c}^*$ axis in Fig. P16.11. A light blue band cuts the Fermi level near the point M on ΓM and near the point K on ΓK. It corresponds to the blue Fermi surface forming a ring around the point Γ_1 in Fig. P16.11. Finally, the fourth band (continuous red curve) cuts the Fermi surface near L on the axis AL. This corresponds to the red Fermi surface shown by a ring around the $\mathbf{c}^*$ axis. In the first Brillouin zone, it decomposes into two half-rings, symmetrical with respect to the horizontal plane passing through Γ_1.

8. The velocity of an electron corresponding to a Bloch state represented by a point $\mathbf{k}$ of the Fermi surface is given by

$$\boxed{\mathbf{v} = \hbar^{-1} \nabla_\mathbf{k} E}.$$

It is thus normal to the constant energy surface at the point corresponding to the given Bloch state.

9. The blue and green sheets with almost cylindrical shape parallel to the $\mathbf{c}^*$ axis are such that the velocity of electrons at the Fermi level in real space is perpendicular to the $\mathbf{c}$ axis and hence lies in the plane spanned by $\mathbf{a}$ and $\mathbf{b}$. Consequently, these bands have quasi-2D behaviour at the Fermi level. Note that the conductivity of these bands is hole conductivity. In contrast, the other bands

have electron velocities at the Fermi level which point in all space directions, and thus exhibit more 3D behaviour. One of these bands, the blue one, corresponds to holes, while the red one corresponds to electrons.

10. The bands giving rise to the quasi-cylindrical Fermi surfaces correspond to carriers propagating in the plane (**a**, **b**). They are therefore mainly formed by hybridisation of the sp_x et sp_y orbitals. The 3D bands contain a dominant contribution from the π orbitals built up from the p_z orbitals, which can be hybridised in the plane and also give rise to an interplane hopping integral, ultimately responsible for the 3D nature of the bands.

11. For graphene, as discussed in Sect. 3.3.4, the $sp\sigma$ bands have low energy and do not reach as far as the Fermi level. Likewise, it was noted that bonding and antibonding bands, π and π^*, respectively, meet at the Dirac points. Regarding MgB_2, it is clear from Fig. P16.2 that these two bands also meet at K. However, in MgB_2, the Dirac point is totally involved in excited states and plays no role at the Fermi level. Likewise, for graphite, the 3D interactions make the bands overlap, but the Fermi level remains close to the Dirac point. The electronic properties of graphite are thus close to those of hole-doped graphene (see the explanation below).

12. The Mg^{2+} ions produce an attractive Coulomb potential for the electrons. Since the π electrons are closer to the Mg than the $sp\sigma$ electrons, they are more strongly attracted by the Mg^{2+} ions, so their energy decreases. This explains the lower energy levels of the π bands as compared to the $sp\sigma$ bands when MgB_2 is compared with graphene. Likewise, the antibonding π^* bands have higher electron density close to the Mg^{2+} than the bonding π bands. Comparing with graphene, this explains why the antibonding π^* bands have lower energy than the bonding bands. Furthermore, the Coulomb potential of the Mg^{2+} ions attracts the π electrons toward the Mg, and this increases the hopping integrals between π orbitals along the **c** axis. This explains the more 3D nature of the π bands as compared with graphite, where they retain a 2D character.

Calculated Band Structures of MgB_2 and Graphite

The band structures of MgB_2 and graphite can be directly compared in Fig. P16.13. The bonding $sp\sigma$ bands, shown by white circles, are in fact very similar in the two electronic structures, but lower in energy in graphite. The bonding π and antibonding π^* bands of MgB_2, shown by dotted curves, are not so different from those of graphene either, and they do indeed meet at the K point. But in contrast to graphite, the bonding and antibonding bands of B overlap, a further reason for MgB_2 to be conducting. We thus have two π contributions at the Fermi level: the red and blue 3D bands. If the Fermi level of MgB_2 is raised by electron doping, for example by partially substituting Al^{3+} for Mg^{2+}, we may eliminate the holes in the $sp\sigma$ bands, but the π bands will remain metallic.

Problem 16: Magnesium Diboride: A New Superconductor?

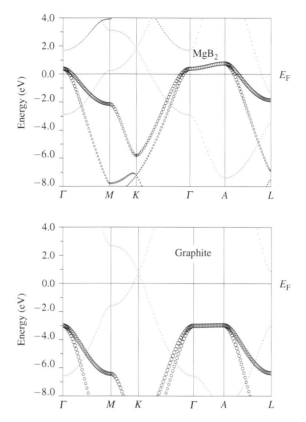

Fig. P16.13 Calculated band structures of MgB$_2$ (*upper*) and graphite (*lower*). From Ann, J.M., Pickett, W.E.: Phys. Rev. Lett. **86**, 4366 (2001). With the permission of the American Physical Society (© 2001 APS). http://link.aps.org/doi/10.1103/PhysRevLett.86.4366

Superconductivity of MgB$_2$

a. Zero-Field Properties

1. In an electron tunneling experiment, an electron is injected into a state accessible to the material, or an electron is extracted from an occupied state. Since electrons are paired in a superconductor, such an experiment thus probes excited electron states (or excited holes). In a superconductor, electrons can only be injected from a normal metal in states excited above the superconducting gap. It is therefore this gap that gets measured in such experiments.

2. In the experiment considered in Fig. P16.14, the band gap is clearly visible at low temperatures. It gradually closes as the temperature is raised. The value measured at zero temperature in Fig. P16.14 is $\Delta = 2.8$ meV. The energy depen-

Fig. P16.14 Tunneling characteristics dI/dV parallel to c in an MgB$_2$ single crystal at different temperatures

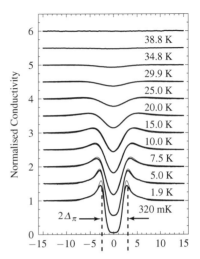

dence of the excited states does indeed match the prediction of BCS theory (as does the observed temperature dependence). However, in BCS theory, the value of Δ is directly related to T_c by $\Delta = 1.75 k_B T_c$. For the value $T_c = 39$ K obtained by resistivity and Meissner effect, one should obtain a gap $\Delta \simeq 6$ meV, twice the observed value. BCS theory would not therefore appear to be completely valid.

3. When electrons are injected with the tip parallel to the **c** axis, their velocities are mainly in the direction of the **c** axis. This will therefore be much less sensitive to electron states with velocities in the plane spanned by **a** and **b**, hence much less sensitive to electrons in the $p\sigma$ orbitals. It would therefore seem that the modification observed in the excited states when the tip is parallel to the plane (**a**, **b**) in Fig. P16.15 comes from the contribution of these $p\sigma$ electrons.

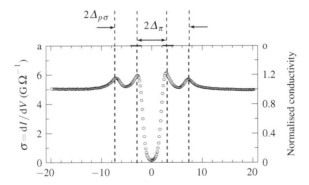

Fig. P16.15 Low temperature tunneling characteristic dI/dV parallel to the plane (a,c) in an MgB$_2$ single crystal

Note: This argument implicitly assumes that tunneling conserves $\mathbf{k}_\parallel$, the momentum component parallel to the sample surface. This is true for an elastic tunneling process, i.e., without energy loss, which have been the only ones considered in the lecture notes. However, inelastic processes can transfer electrons into Bloch states with **v** perpendicular to **c**.

4. The two peaks appearing in Fig. P16.15 are therefore excited states that we may try to associate with electronic states in the $p\sigma$ bands. The gap thus differs between the two types of electronic state! It therefore seems that the condensation energy of electrons in the superconducting state differs for $p\sigma$ and π electrons. The gap obtained for the former in Fig. P16.15 is $\Delta_{p\sigma} = 7.7$ meV, whereas an isotropic value of $\Delta_\pi = 2.8$ meV is obtained for the other gap. It is the average condensation energy of the two types of electron that gives rise to the overall superconductivity of the material. The average value of Δ is in better agreement with the one expected on the basis of BCS theory.

b. Superconductivity in an Applied Field

5. The constant characteristic dI/dV corresponds to the one observed in the normal material. There are therefore normal regions in the superconducting material. This is what happens in a vortex. But if there are vortices, this means that MgB$_2$ is a type II superconductor.

6. In Fig. P16.6, when the tip moves away from a vortex, the tunneling characteristic reaches a minimum at $V = 0$ which decreases gradually until dI/dV becomes identical to the characteristic observed in zero field. The amplitude of the variation at $V = 0$ corresponds to a gradual increase in the density n_s of superconducting pairs in the material as r increases. The dependence of the minimum on r, as plotted in Fig. P16.16, can be used to estimate the coherence length

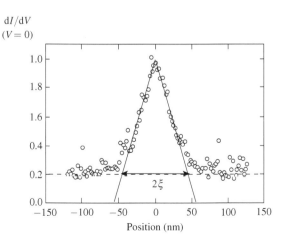

Fig. P16.16 Voltage dependence of dI/dV as the tip moves away from a point where the characteristic is independent of V. From Eskildsen, M.R., Kugler, M., Tanaka, S., Jun, J., Kazakov, S.M., Karpinski, J., Fisher, O.: Phys. Rev. Lett. **89**, 187003 (2002). With the permission of the American Physical Society (© 2002 APS)

$$\boxed{\xi \approx 500 \text{ Å}}.$$

BCS theory would imply

$$\xi = \frac{\hbar v_F}{\pi \Delta} = \frac{\hbar v_F}{1.75\pi k_B T_c}.$$

With $v_F = 5 \times 10^5$ m/s and $T_c = 39$ K, we obtain $\xi = 180$ Å. Since the measured value of ξ is greater by a factor of about 2, this implies that it is associated with a small value of Δ, close to the value of the gap Δ_π.

7. In a high field, we see that the vortices arrange themselves into an ordered hexagonal structure with parameter a'. Since the flux is quantised and has the value Φ_0 per primitive cell of the vortex lattice, the macroscopic magnetic induction, which is a spatial average, is $B = \Phi_0/S$, where S is the area of the primitive cell of the hexagonal lattice, i.e.,

$$a'^2 \frac{\sqrt{3}}{2} = \frac{\Phi_0}{B}.$$

It follows that

$$\boxed{a' = 69 \text{ nm}}.$$

By measuring $3a'$ to within 3% in Fig. P16.7, we can determine the dimensions of the figure, viz., 215 nm × 215 nm to within 3%.

8. The diffraction peaks observed in Fig. P16.8 (left) correspond to neutron diffraction by the magnetic field distribution. We thereby directly observe the 2D reciprocal lattice of the field distribution. Since the latter is a hexagonal lattice of side a', it has a reciprocal lattice of side

$$a'^* = 4\pi/a'\sqrt{3}.$$

From the figure we find $2a'^* = 0.0218$ Å^{-1} to within 3%, which yields

$$\boxed{a' = 665 \pm 20 \text{ Å}},$$

agreeing with the direct STM observation of the vortex lattice for the same field $B = 0.5$ tesla. The angle of diffraction is

$$2d \sin\theta = \lambda',$$

where $d = a'\sqrt{3}/2 = 576 \pm 20$ Å is the distance between lattice rows parallel to one of the sides of the lattice triangle, and λ' is the de Broglie wavelength of the neutrons, i.e., $\lambda' = h/p = h/(2m_n \varepsilon)^{1/2}$. This gives

$$\lambda'_{\text{Å}} = 0.286(\varepsilon_{\text{eV}})^{-1/2}.$$

Problem 16: Magnesium Diboride: A New Superconductor?

For energy ε corresponding to 9.5 K, $\varepsilon = 0.819$ meV and $\lambda' = 100$ Å. We then have

$$\boxed{\sin\theta \approx \theta = \frac{100}{1152} \simeq 5°}.$$

Note: To see this diffraction, the sample thus had to be tilted at an angle θ in order to satisfy the Bragg condition, and this for various directions in reciprocal space corresponding to nearest neighbours.

9. Neutrons have spin 1/2 and are deflected by the magnetic field they encounter along their trajectory. Diffraction is thus due to the magnetic field distribution between vortices. This clearly has the same periodicity as the superconducting pair distribution and corresponds to the same real space lattice. This is indeed borne out. On the other hand, the intensities of the diffraction peaks are determined by the Fourier transform of the magnetic field distribution, which is directly related to the penetration depth λ.

10. We may thus estimate λ by measuring the intensity of the Bragg peaks and hence the structure factor of the field distribution. When $B \gg \mu_0 H_{c_1}$, which is the case here, the lattice spacing is not very large compared with ξ, and B varies only slightly between vortices. In this case, the structure factor falls off rapidly with distance in reciprocal space. This explains why only those Bragg peaks close to the origin are actually observed. In the present case, λ was determined by studying the variation of the intensity of the Bragg peak for different values of the applied field and a value of λ of the order of 1,000 Å was obtained.

11. For $B = 0.9$ tesla, we still have a hexagonal reciprocal lattice in Fig. P16.8 (right), and its unit cell grows, which does indeed imply a smaller unit cell in the real space vortex lattice. This is to be expected, since the quantisation condition implies that $a' \propto B^{-1/2}$. We do indeed find that a' is multiplied by $\sqrt{9/5} = 1.34$. However, *the orientation of the reciprocal lattice has changed between 0.5 and 0.9 tesla.* The same therefore goes for the vortex lattice in real space.

12. In the neutron experiment considered here, the vortex lattice forms a large single crystal with fixed orientation relative to the crystal axes. There is therefore an anisotropic energy term favouring the observed choice of orientation. The rotation of the vortex lattice in high applied fields indicates that this anisotropy term changes with the applied field. This is related to the fact that the superconductivity gets contributions from both the $p\sigma$ and π bands. As the superconductivity of the π band corresponds to the small superconducting gap, it has lower condensation energy and is more easily destroyed for increasing magnetic field than the superconductivity due to $p\sigma$. The superconductivity is associated with both types of electron in zero field, but gradually becomes dominated by the $p\sigma$ contribution. The anisotropy term associated with the latter favours a different orientation of the vortices relative to the crystal axes.

Further Notes: In fact the experiment shows a gradual rotation of the axes of the vortex lattice and a gradual change in the values of λ and ξ between 0.5 and 0.9 tesla. The value of ξ corresponding to the $p\sigma$ bands is less than the one measured in a weak field for the π band. The value of H_{c_2} is also higher. The existence of *differing superconducting gaps for different bands* is a novel feature which has been revealed by the study of MgB_2 and is nowadays quite commonly found in *multiband* metallic systems.

> The fact that *the vortex lattice is linked to the crystal lattice* is in itself also a novel feature. In an isotropic superconducting material, the vortex lattice would in principle have no reason to depend on the crystal lattice. Vortices generally appear in a 'polycrystalline' form, because order nucleates at several independent points of the sample, and with random orientations. Since the many vortex crystals that form are distributed randomly, their Bragg spots have a single distance from the origin and a random orientation. They thus form a diffraction ring for each lattice distance.
>
> The existence of an anisotropic term can be understood qualitatively when we realise that the energy term associated with the field penetration can be minimal for certain configurations of the current lines relative to the crystal axes. In materials for which the coherence length ξ is small, these terms are of increasing importance close to H_{c_2} when the distance between vortices becomes of the same order of magnitude as ξ. In MgB_2, since the crystal lattice itself has hexagonal symmetry, only the orientation of the vortex lattice relative to the crystal lattice can be affected. In cubic materials, this can lead in some cases to the stabilisation of a square vortex lattice in strong fields, rather than a hexagonal one.

Problem 17: Electronic Properties of La_2CuO_4

The compound with chemical formula La_2CuO_4 has the 3D structure with body-centered tetragonal conventional cell shown in Fig. P17.1. Note that it comprises a stack of LaO and CuO_2 planes.

1. *On Fig. P17.1, indicate the primitive vectors and basis of a primitive cell in the 3D crystal.*

The electronic configurations of the neutral atoms making up this structure are

$\quad$ La: (Xe) $5d^1 6s^2$, $\quad$ Cu: (Ar) $3d^{10} 4s$, $\quad$ O: $1s^2 2s^2 2p^4$.

2. *According to an LCAO calculation, is this compound insulating or metallic?*

The LCAO calculation shows that the La and oxygen in the LaO planes are in the configurations La^{3+} and O^{2-} corresponding to fully occupied shells. Bands resulting from these orbitals are at energy levels well below those of the Cu(3d) and O(2p) bands of the oxygens in the CuO_2 planes. It is thus the latter that constitute the valence or conduction levels of the crystal. It therefore suffices to consider the electronic structure of the CuO_2 plane.

3. *Specify the primitive cell of the CuO_2 plane, along with its reciprocal lattice and first Brillouin zone.*

We shall assume that the last occupied band corresponds to one orbital per Cu site of atomic energy E_0. The band is calculated using the LCAO approximation, only taking into account nearest-neighbour hopping integrals t_0 and t_1.

4. *Find the eigenenergies $E(k_x, k_y)$ and plot the constant energy curves in the first Brillouin zone. What is the curve corresponding to an occupancy of one electron per unit cell?*

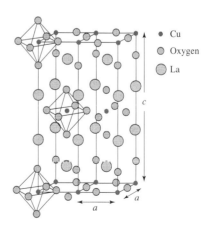

Fig. P17.1 Crystal structure of La_2CuO_4, showing body-centered tetragonal conventional cell

Fig. P17.2 Temperature dependence of the intensity of the neutron diffraction peak at $(\pi/a, \pi/a)$. Adapted from Yamada, K., Kudo, E., Endoh, Y., Hikada, Y., Oda, M., Suzuki, M., Murakami, T.: Solid State Commun. **64**, 753 (1987)

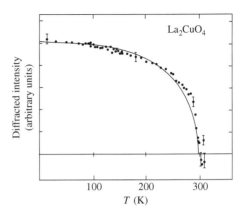

Conductivity measurements show that La_2CuO_4 is insulating. Moreover, neutron diffraction produces Bragg peaks at the four points $(\pm\pi/a, \pm\pi/a)$ of the plane below 300 K. The intensity of these peaks increases at low temperatures and saturates at $T = 0$ K, as shown in Fig. P17.2. X-ray diffraction only picks up the Bragg spots corresponding to the reciprocal lattice found in question 3.

5. *What can you deduce about the ground state of the electronic system of the CuO_2 plane? Specify the primitive cell of the real lattice of the CuO_2 plane for $T < 300$ K.*

A fraction of the La^{3+} is replaced by Sr, which has the ionic form Sr^{2+} in the crystal. In the compound $La_{1.85}Sr_{0.15}CuO_4$ it is observed that the spots at the four points $(\pm\pi/a, \pm\pi/a)$ of the plane disappear in a neutron diffraction experiment. Moreover, this compound is superconducting with a critical temperature of 40 K.

6. *What conclusions can be drawn from these observations? According to the LCAO approximation, what should be the Fermi surface of this metal?*

Problem 17: Electronic Properties of La$_2$CuO$_4$

Solution

1. Like the body-centered cubic cell, the body-centered tetragonal cell has two Bravais lattice points per conventional cell. The latter contains twice the chemical formula La$_2$CuO$_4$ (see Fig. P17.3a). A possible primitive cell containing the formula La$_2$CuO$_4$ only once is shown in Fig. P17.3b. It is constructed from the vectors $a(-1,0,0)$, $a(0,1,0)$, and $(-a/2, a/2, c/2)$.

2. 2La and 4O give an even number of electrons per unit cell. However, copper has an odd number of electrons. The primitive cell thus contains an odd number of electrons. There will therefore be at least one partially occupied band in an LCAO calculation and this leads to a metallic state.

3. The square primitive cell in the CuO$_2$ plane containing one Cu atom and two O atoms is shown in Fig. P17.4a. The square reciprocal lattice of side $2\pi/a$ and the first Brillouin zone are depicted in Fig. P17.4b.

4. The LCAO calculation gives $E(k_x, k_y) = E_0 - t_0 - 2t_1(\cos k_x a + \cos k_y a)$. The constant energy curves are

$$\cos k_x a + \cos k_y a = \frac{E_0 - t_0 - E}{2t_1}.$$

These are almost circular curves around $k = 0$ for $E \gtrsim E_0 - t_0 - 2t_1$ and around $k = (\pi/a, \pi/a)$ for $E \lesssim E_0 - t_0 + 2t_1$. For one electron per cell, the last filled level corresponds to a half-filled band, or

$$E = E_0 - t_0,$$

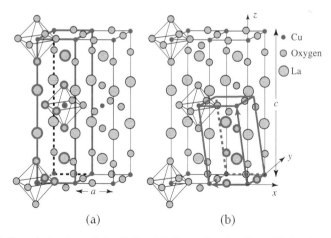

(a)   (b)

Fig. P17.3 Crystal structure of La$_2$CuO$_4$. (a) Conventional cell containing twice La$_2$CuO$_4$. (b) Possible primitive cell

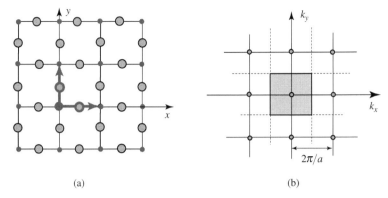

Fig. P17.4 (a) Square primitive cell of the CuO$_2$ plane. (b) Square reciprocal lattice and first Brillouin zone

Fig. P17.5 Constant energy curves and Fermi surface

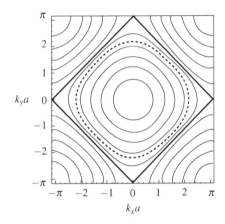

which corresponds to

$$\cos k_x a = -\cos k_y a.$$

This is represented by the square in Fig. P17.5, which would be the Fermi surface of the corresponding metal.

5. Since neutrons and X rays do not give the same Bragg spots below 300 K, this suggests that *magnetic order* occurs at this temperature. The further spots obtained with neutron diffraction at the four points $(\pm \pi/a, \pm \pi/a)$ can be used to determine the reciprocal lattice of the magnetic order. It corresponds to a square lattice of side $\pi\sqrt{2}/a$, with area equal to half that of the reciprocal lattice of the crystal structure (see Fig. P17.6a). The real magnetic lattice thus has a unit cell of area $2a^2$, twice that of the crystal lattice. The ground state is therefore antiferromagnetic with two sublattices.

Problem 17: Electronic Properties of La$_2$CuO$_4$

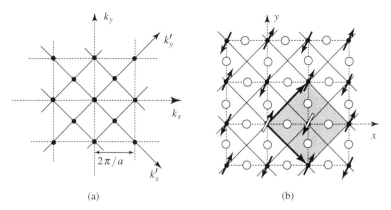

Fig. P17.6 (a) Reciprocal lattice of the magnetic structure. (b) Real space two-sublattice antiferromagnetic structure showing the primitive cell (*shaded*) and the two antiparallel Cu spin states of the magnetic basis

Since the La^{3+} and O^{2-} in the LaO planes donate one electron per primitive cell of the atomic structure to the CuO$_2$ plane, it seems likely that the electronegativity of the oxygen is satisfied and that the Cu is in a state Cu^{2+}, or $3d^9$. Therefore, the Cu^{2+} do carry the magnetic moment. The antiferromagnetic primitive cell is shown in Fig. P17.6b. This natural hypothesis regarding magnetism is borne out experimentally by measuring the intensities of the magnetic spots, which show that the corresponding structure factor is indeed that of Cu. The orientation of the moments in the unit cell cannot be obtained directly from this data, and are indeed more difficult to determine experimentally.

6. Replacing xLa^{3+} by xSr^{2+} amounts to transferring x fewer electrons per primitive cell to the CuO$_2$ planes. This seems to be enough to destroy the antiferromagnetic state and make the system metallic. Whereas one electron was lacking to saturate the $3d$ shell of Cu, $1+x$ are now lacking. Formally, there are therefore $1-x$ electrons or $1+x$ holes in the last occupied CuO$_2$ band. If we attempt to interpret this metallic state in a band model with $1-x$ electrons, this corresponds in the case $x = 0.15$ to a curve of constant E with area $0.85/2$ times the area of the first Brillouin zone. This would be roughly the area enclosed by the dotted curve in Fig. P17.5.

In reality, the compound La$_2$CuO$_4$ is a *Mott–Hubbard insulator*, and the holes created by substituting Sr are in fact holes on the oxygens in the planes. We are still a long way from understanding this type of metallic state, in which a magnetic moment associated with the hole on the Cu coexists with holes on the oxygen that play a dominant role in electron transport properties. The many conceptual difficulties sketched here explain in part the considerable effort required to understand the origins of high T_c superconductivity.

Problem 18: Properties of an Antiferromagnetic Solid[9]

The aim here is to determine the magnetic response of an antiferromagnetic solid in the mean field approximation and to evaluate the exchange interactions from these measurements.

18.1: Preliminaries: The Ferromagnetic Case

Consider an ensemble of N atoms each carrying a magnetic moment $\hat{\mu}_i$ associated with a spin 1/2, i.e., $\hat{\mu} = -2\mu_B \hat{S}$. The moments are non-interacting. They are subjected to a magnetic field $\mathbf{H}_0$ parallel to the z axis.

1. *Recall the expression*

$$M_z = N\langle\mu_z\rangle = f(H_0/T)$$

   for the temperature and field dependence of the magnetisation. Obtain the susceptibility

$$\chi = \mu_0^{-1} \left.\frac{\partial M}{\partial H_0}\right|_{H_0 \to 0}.$$

   What type of magnetic behaviour is characterised by the above laws?

The moments subjected to the field $\mathbf{H}_0$ also interact with their nearest neighbours by the exchange interaction, with a positive exchange constant $\mathscr{J}$.

2a. *Explain why in the molecular field approximation the magnetic state of the N magnetic moments is determined by*

$$M_z = f(H_{\text{eff}}/T), \qquad (13.157)$$

   where f is the function defined in question 1. Give the expression for H_{eff} in terms of $\mathscr{J}$, M, H_0, N, and the number z of nearest neighbours of an atom.

2b. *In zero field, (13.157) reduces to $M = g(M/T)$. In terms of $\mathscr{J}$, what is then the temperature T_c below which the system becomes ferromagnetic?*

2c. *For a weak external field, deduce the temperature dependence of the susceptibility χ for $T > T_c$.*

---

[9] This problem has been designed with J.C. Tolédano.

18.2: Antiferromagnetic Transition

Among the N magnetic moments considered in 18.1, $N/2$ occupy the corners of adjacent cubes forming the crystal lattice in a solid body, while $N/2$ occupy the centers of these cubes. The dipoles are thus located at the lattice points of two identical cubic sublattices A and B, shifted by one cube half-diagonal from one another.

Each moment interacts by the exchange interaction with its six nearest neighbours of the *same* sublattice, with the positive exchange constant $\mathcal{J}_1 > 0$, and with its eight nearest neighbours of the *other* sublattice, with the negative exchange constant $-\mathcal{J}_2$ (with $\mathcal{J}_2 > 0$). The Hamiltonian for this system is therefore

$$\mathcal{H} = -\mathcal{J}_1 \sum \mathbf{S}_\mathbf{R}^a \cdot \mathbf{S}_{\mathbf{R}'}^a - \mathcal{J}_1 \sum \mathbf{S}_\mathbf{R}^b \cdot \mathbf{S}_{\mathbf{R}'}^b + \mathcal{J}_2 \sum \mathbf{S}_\mathbf{R}^a \cdot \mathbf{S}_{\mathbf{R}'}^b$$
$$+ 2\mu_0 \mu_B \mathbf{H}_0 \cdot \left(\sum \mathbf{S}_\mathbf{R}^a + \sum \mathbf{S}_\mathbf{R}^b \right),$$

where the first three sums are over pairs of spins and the last two are over the $N/2$ spins of a sublattice. Indices a and b denote variables associated with sublattices A and B, respectively.

3. Adapting (8.37) of Chap. 8, write down $\mathcal{H}_\mathbf{R}^a$, the Hamiltonian containing all terms of $\mathcal{H}$ involving a spin $\mathbf{S}_\mathbf{R}^a$ at a given site of sublattice A. Similarly for $\mathcal{H}_\mathbf{R}^b$ for a given spin $\mathbf{S}_\mathbf{R}^b$.

4. Letting $\mathbf{M}_a$ and $\mathbf{M}_b$ be the magnetisations of the two sublattices A and B, deduce by analogy with 18.1 that the magnetic state of the system is determined in the molecular field approximation by

$$\mathbf{M}_a = \frac{1}{2} f(\mathbf{H}_{\text{eff}}^a/T), \qquad \mathbf{M}_b = \frac{1}{2} f(\mathbf{H}_{\text{eff}}^b/T), \tag{13.158}$$

where

$$\mathbf{H}_{\text{eff}}^a = \mathbf{H}_0 + W\mathbf{M}_a - W'\mathbf{M}_b, \qquad \mathbf{H}_{\text{eff}}^b = \mathbf{H}_0 + W\mathbf{M}_b - W'\mathbf{M}_a,$$

and f is the function defined in question 1. Express W and W' in terms of $\mathcal{J}_1$, $\mathcal{J}_2$, and N.

Equations (13.158) determine the appearance in zero external field of a nonzero magnetisation on each of sublattices A and B below a temperature T_N.

5. Explain qualitatively why the magnetisations of the sublattices should satisfy $\mathbf{M}_a = -\mathbf{M}_b$. Using this, calculate T_N as a function of W and W'.

When there is an external field $\mathbf{H}_0$, the total magnetisation $\mathbf{M} = \mathbf{M}_a + \mathbf{M}_b$ is no longer zero.

Problem 18: Properties of an Antiferromagnetic Solid

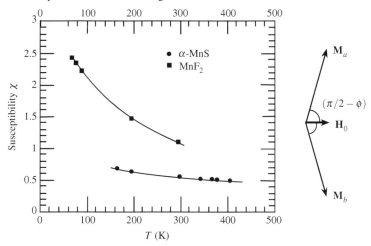

Fig. P18.1 Susceptibility measurements on MnF$_2$ and α-MnS (*left*). Tilting of the magnetisation in a small transverse external field (*right*). Adapted from Bizette, H., Tsai, B.: Comptes-Rendus de l'Académie des Sciences 1575 (1954)

6. From (13.158), calculate the temperature dependence of the susceptibility $\chi = \mu_0^{-1} (\partial M/\partial H_0)_{H_0 \to 0}$ for $T > T_N$ (M and H_0 small).

In the following, we set

$$\Theta = \frac{N\mu_0\mu_B^2(W' - W)}{2k_B},$$

Neutron diffraction experiments show that, in MnF$_2$ and α-MnS, the manganese moments are in an antiferromagnetic ground state with two sublattices, below the Néel temperatures $T_N = 68$ K and $T_N = 150$ K, for MnF$_2$ and α-MnS, respectively. Magnetic susceptibility measurements give the results shown in Fig. P18.1 (left).

7. How could you deduce $\mathcal{J}_1$ and $\mathcal{J}_2$ from these experimental results? Can the constant $\mathcal{J}_1$ be negative?

18.3: Susceptibility in the Antiferromagnetic State

We now consider $T < T_N$, and assume that the equilibrium directions of the magnetisations $\mathbf{M}_a$ and $\mathbf{M}_b$ in zero external field are parallel to the z axis. A small field $\mathbf{H}_0$ is then set up perpendicular to the initial magnetisation direction.

8a. Show using the Hamiltonian of question 3 and the expressions for W and W' obtained in question 4 that that the energy U of the magnetised lattices A and B is

$$\frac{U}{\mu_0} = -\frac{W}{2}(\mathbf{M}_a^2 + \mathbf{M}_b^2) + W'\mathbf{M}_a\mathbf{M}_b - \mathbf{H}_0(\mathbf{M}_a + \mathbf{M}_b). \qquad (13.159)$$

Assume that the 'transverse' $\mathbf{H}_0$ does not affect the magnitudes $|\mathbf{M}_a| = |\mathbf{M}_b| = M_0$ of the magnetisations, but tends to tilt them toward the field direction, as shown in Fig. P18.1 (right).

8b. *Find the tilt angle ϕ that minimises the energy. Deduce the transverse susceptibility $\chi_\perp$. How does it depend on $T < T_N$? How does it connect with the susceptibility found in question 4?*

8c. *What further contribution to the energy could have been considered here? What would be its effect?*

A small field $\mathbf{H}_0$ is now applied parallel to the z axis and in the same direction as $\mathbf{M}_a$. We assume that the changes in the magnitudes of $\mathbf{M}_a$ and $\mathbf{M}_b$ have opposite sign:

$$M_a = M_0 + \Delta(H_0), \qquad M_b = -M_0 + \Delta(H_0).$$

9. *Use (13.158) to find the temperature dependence of the longitudinal susceptibility $\chi_\parallel$ near T_N. Show that $\chi_\parallel(T = 0) = 0$ and that $\chi_\parallel(T_N) = \chi_\perp(T_N)$.*

10. *On Fig. P18.1 (left), sketch the behaviour expected for $\chi_\parallel$ and $\chi_\perp$ in the case of MnF_2. Are neutron diffraction measurements absolutely necessary to determine the characteristics of an antiferromagnetic solid?*

11. *Can there be domains and domain walls in an antiferromagnetic solid?*

Problem 18: Properties of an Antiferromagnetic Solid

Solution

Preliminaries: The Ferromagnetic Case

1. For an ensemble of non-interacting spin $1/2$ moments, the magnetisation is

    $$M_z = N\langle\mu_z\rangle = N\mu_B \tanh\frac{\mu_0\mu_B H_0}{k_B T} = f(H_0/T). \tag{13.160}$$

    Differentiating this,

    $$\boxed{\chi = \frac{N}{k_B}\frac{\mu_B^2}{T}}.$$

    These equations characterise the paramagnetic behaviour.

2a. The Hamiltonian for the system of N moments is

    $$H = -\mathcal{J}\sum_{\langle\mathbf{RR'}\rangle}\mathbf{S_R}\cdot\mathbf{S_{R'}} + 2\mu_0\mu_B\mathbf{H_0}\cdot\sum_{\mathbf{R}}\mathbf{S_R},$$

    where the first sum is over neighbouring pairs of spins and the second over all spins. The terms in the Hamiltonian involving a given spin $\mathbf{S_R}$ are

    $$H_\mathbf{R} = \mathbf{S_R}\cdot\left(-\mathcal{J}\sum_{\mathbf{R'}}\mathbf{S_{R'}} + 2\mu_0\mu_B\mathbf{H_0}\right),$$

    where the sum is over the z nearest neighbours of the spin $\mathbf{S_R}$. In the molecular field approximation, $\mathbf{S_{R'}}$ is replaced by $\langle\mathbf{S_{R'}}\rangle = -M_z/2N\mu_B$, its statistical average. Then,

    $$H_\mathbf{R} = \mathbf{S_R}\left(\frac{z\mathcal{J}M_z}{2N\mu_B} + 2\mu_0\mu_B H_0\right) = 2\mu_0\mu_B H_{\text{eff}}\mathbf{S_R},$$

    which defines the effective field

    $$H_{\text{eff}} = H_0 + \frac{z\mathcal{J}}{4N\mu_0\mu_B^2}M_z.$$

    For such a system, (13.160) is applicable with H_0 replaced by H_{eff}.

2b. In zero external field H_0, the equation $M_z = f(H_{\text{eff}}/T)$ reduces to

    $$\boxed{M_z = N\mu_B\tanh\left(\beta\frac{z\mathcal{J}}{4N}\frac{M_z}{\mu_B}\right)},$$

which only has nonzero solution if $\beta z \mathscr{J}/4 > 1$, that is, for temperatures $T < T_c = z\mathscr{J}/4k_B$. Below T_c, the system acquires a permanent magnetisation and becomes ferromagnetic.

2c. In a weak external field H_0 and above T_c, the magnetisation is also weak. It is given by the expansion

$$M_z = N\mu_B \tanh\left(\frac{\mu_0 \mu_B H_0}{k_B T} + \frac{T_c}{T}\frac{M_z}{N\mu_B}\right) \simeq \frac{N\mu_0 \mu_B^2 H_0}{k_B T} + \frac{T_c}{T}M_z.$$

It thus varies as $M_z = N\mu_0\mu_B^2 H_0/k_B(T - T_c)$, and the susceptibility χ diverges as

$$\boxed{\chi = \frac{N\mu_B^2}{k_B(T - T_c)}}.$$

Antiferromagnetic Transition

3. The Hamiltonians $\hat{\mathscr{H}}_\mathbf{R}^a$ and $\hat{\mathscr{H}}_\mathbf{R}^b$ which describe the coupling of the spins $\mathbf{S}_\mathbf{R}^a$ and $\mathbf{S}_\mathbf{R}^b$ with their surroundings are, respectively,

$$\hat{\mathscr{H}}_\mathbf{R}^a = \mathbf{S}_\mathbf{R}^a \cdot \left(-\mathscr{J}_1 \sum_{\mathbf{R}'} \mathbf{S}_{\mathbf{R}'}^a + \mathscr{J}_2 \sum_{\mathbf{R}'} \mathbf{S}_{\mathbf{R}'}^b + 2\mu_0\mu_B \mathbf{H}_0\right),$$

$$\hat{\mathscr{H}}_\mathbf{R}^b = \mathbf{S}_\mathbf{R}^b \cdot \left(-\mathscr{J}_1 \sum_{\mathbf{R}'} \mathbf{S}_{\mathbf{R}'}^b + \mathscr{J}_2 \sum_{\mathbf{R}'} \mathbf{S}_{\mathbf{R}'}^a + 2\mu_0\mu_B \mathbf{H}_0\right),$$

where sums are over nearest neighbours.

4. In the molecular field approximation, $\mathbf{S}_\mathbf{R}^a$ is replaced by its average value $\langle \mathbf{S}_\mathbf{R}^a \rangle = -\mathbf{M}_a/N\mu_B$. Likewise, $\langle \mathbf{S}_\mathbf{R}^b \rangle = -\mathbf{M}_b/N\mu_B$. (There are $N/2$ moments in each sublattice.) Consequently,

$$\hat{\mathscr{H}}_\mathbf{R}^a = \mathbf{S}_\mathbf{R}^a \cdot \left(\frac{6\mathscr{J}_1 \mathbf{M}_a}{N\mu_B} - \frac{8\mathscr{J}_2 \mathbf{M}_b}{N\mu_B} + 2\mu_0\mu_B \mathbf{H}_0\right) = 2\mu_0\mu_B \mathbf{H}_\text{eff}^a \cdot \mathbf{S}_\mathbf{R}^a.$$

The effective field $\mathbf{H}_\text{eff}^a$ is

$$\mathbf{H}_\text{eff}^a = \mathbf{H}_0 + W\mathbf{M}_a - W'\mathbf{M}_b, \qquad \boxed{W = \frac{3\mathscr{J}_1}{N\mu_0\mu_B^2}}, \qquad \boxed{W' = \frac{4\mathscr{J}_2}{N\mu_0\mu_B^2}}.$$

Likewise, $\mathbf{H}_\text{eff}^b = \mathbf{H}_0 + W\mathbf{M}_b - W'\mathbf{M}_a$. In the mean field approximation, sublattices A and B in effective fields $\mathbf{H}_\text{eff}^a$ and $\mathbf{H}_\text{eff}^b$, respectively, are statistically independent and we may apply the relation found in question 1 to each

Problem 18: Properties of an Antiferromagnetic Solid

sublattice. Then

$$\mathbf{M}_a = \frac{N\mu_B}{2} \tanh(\beta \mu_0 \mu_B \mathbf{H}_{\text{eff}}^a), \qquad \mathbf{M}_b = \frac{N\mu_B}{2} \tanh(\beta \mu_0 \mu_B \mathbf{H}_{\text{eff}}^b).$$

5. The negative exchange constant $-\mathscr{J}_2$ tends to orient the magnetisations $\mathbf{M}_a$ and $\mathbf{M}_b$ in opposite directions. Then since the two sublattices have identical structure, we can deduce that $\mathbf{M}_a = -\mathbf{M}_b$. Substituting this into the previous equations, we then have

$$\mathbf{M}_a = \frac{N\mu_B}{2} \tanh \beta \mu_0 \mu_B [\mathbf{H}_0 + (W + W')\mathbf{M}_a]. \tag{13.161}$$

In zero field, the equation

$$\mathbf{M}_a = \frac{N\mu_B}{2} \tanh \frac{\mu_0 \mu_B (W + W')\mathbf{M}_a}{k_B T}$$

has a nonzero solution if T is less than the temperature T_N determined by

$$\boxed{k_B T_N = \frac{N\mu_0 \mu_B^2 (W + W')}{2} = \frac{3\mathscr{J}_1 + 4\mathscr{J}_2}{2} > 0}. \tag{13.162}$$

6. In a weak external field, and above T_N, the magnetisation of a sublattice is small and (13.161) can be linearised:

$$\mathbf{M}_a = \frac{N\mu_0 \mu_B^2}{2k_B T}(\mathbf{H}_0 + W\mathbf{M}_a - W'\mathbf{M}_b),$$

$$\mathbf{M}_b = \frac{N\mu_0 \mu_B^2}{2k_B T}(\mathbf{H}_0 + W\mathbf{M}_b - W'\mathbf{M}_a).$$

Adding these, we obtain

$$\mathbf{M} = \mathbf{M}_a + \mathbf{M}_b = \frac{N\mu_0 \mu_B^2}{k_B T}\left(\mathbf{H}_0 + \frac{W - W'}{2}\mathbf{M}\right),$$

which yields

$$\boxed{\chi = \frac{N\mu_B^2}{k_B(T + \Theta)}}, \qquad \boxed{k_B \Theta = \frac{N\mu_0 \mu_B^2 (W' - W)}{2} = \frac{4\mathscr{J}_2 - 3\mathscr{J}_1}{2}}. \tag{13.163}$$

7. Neutron diffraction experiments are used to determine the temperature T_N below which antiferromagnetic order arises. (At the end of this problem, we shall see how T_N can be deduced from measurements of χ.) Furthermore, above T_N, the susceptibility varies as $1/(T + \Theta)$. We can thus find Θ and deduce $\mathscr{J}_1$ and $\mathscr{J}_2$ from these two characteristic temperatures:

Fig. P18.2 Obtaining Θ graphically from the plot of χ^{-1} vs. T

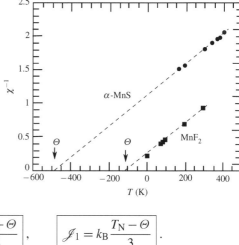

$$\mathscr{J}_2 = k_B \frac{T_N + \Theta}{4}, \qquad \mathscr{J}_1 = k_B \frac{T_N - \Theta}{3}.$$

Figure P18.1 (right) shows the temperature dependence of χ above T_N for the two compounds. We must first check that the variation does indeed have the form $1/(T + \Theta)$. This is generally done by plotting $1/\chi$ as a function of T. This should give a straight line that cuts the horizontal axis at $-\Theta$. This has been done in Fig. P18.2 for the experimental results of Fig. P18.1 (right).

We obtain $\Theta \simeq 108$ K for MnF$_2$ and $\Theta \simeq 482$ K for α-MnS. Using the T_N values obtained by neutron scattering, we find $\mathscr{J}_2 \simeq 44$ K and $\mathscr{J}_1 \simeq -13$ K for MnF$_2$ and $\mathscr{J}_2 \simeq 138$ K and $\mathscr{J}_1 \simeq -110$ K for MnS.

In both cases the coupling between sublattices is antiferromagnetic ($\mathscr{J}_2 > 0$), as required to obtain an antiferromagnetic state. We observe that, for both MnF$_2$ and α-MnS, the magnetic state occurs even with antiferromagnetic coupling within each sublattice ($\mathscr{J}_1 < 0$, see the definitions of $\mathscr{J}_1$ and $\mathscr{J}_2$ and their signs in the statement of the problem). In fact, from (13.162), the antiferromagnetic state described here is possible provided that $\mathscr{J}_2 > -3\mathscr{J}_1/4$, which is satisfied for the values found for MnF$_2$ and α-MnS. For $\mathscr{J}_1 > 0$, we would obtain $\Theta < T_N$.

Susceptibility in the Antiferromagnetic State

8a. Replacing the spins $\mathbf{S_R}$ by their thermal average, taking into account the fact that the first two sums in the Hamiltonian involve $3N/2$ terms, i.e.,

$$3N/2 = \frac{1}{2}(6 \text{ neighbours} \times N/2 \text{ spins}),$$

and that the third sum involves $4N$ terms ($N/2$ spins of one sublattice connected to 8 neighbours on the other sublattice), the average energy is

Problem 18: Properties of an Antiferromagnetic Solid

$$U = -\frac{3N}{2}\mathscr{J}_1\left(\frac{\mathbf{M}_a}{N\mu_\mathrm{B}}\right)^2 - \frac{3N}{2}\mathscr{J}_1\left(\frac{\mathbf{M}_b}{N\mu_\mathrm{B}}\right)^2 + 4N\mathscr{J}_2\frac{\mathbf{M}_a\mathbf{M}_b}{(N\mu_\mathrm{B})^2}$$
$$-\mu_0 H_0(\mathbf{M}_a + \mathbf{M}_b),$$

that is

$$\frac{U}{\mu_0} = -\frac{3\mathscr{J}_1}{2N\mu_0\mu_\mathrm{B}^2}(\mathbf{M}_a^2 + \mathbf{M}_b^2) + \frac{4\mathscr{J}_2}{N\mu_0\mu_\mathrm{B}^2}\mathbf{M}_a\mathbf{M}_b - H_0(\mathbf{M}_a + \mathbf{M}_b).$$

Given the expressions for W and W', we obtain (13.159) as required.

8b. We have

$$\frac{U}{\mu_0} = -W\mathbf{M}_0^2 - W'\mathbf{M}_0^2 \cos 2\phi - 2H_0 M_0 \sin\phi.$$

Minimising this with respect to ϕ, we obtain $\sin\phi = H_0/2W'M_0$. The magnetisation along the field H_0 is $2M_0 \sin\phi$, so the transverse susceptibility is $\chi_\perp = (\mu_0 W')^{-1}$. It is independent of temperature. It is easy to check that $\chi_\perp$ is equal to the value of $\chi(T = T_\mathrm{N})$ above the transition, which is given by

$$\boxed{\chi_\perp = \chi(T = T_\mathrm{N}) = \frac{N\mu_\mathrm{B}^2}{k_\mathrm{B}(T_\mathrm{N} + \Theta)}}. \qquad (13.164)$$

8c. We must take into account the magnetic anisotropy. This determines the magnetisation directions in space. It can be described by an energy of the form $K\sin^2\phi$. This changes the susceptibility to

$$\chi = \frac{1}{\mu_0 W' + K/2M_0^2}.$$

9. Setting $M_a = M_0 + \Delta$ and $M_b = -M_0 + \Delta$, the first equation of (13.158) becomes

$$M_0 + \Delta = \frac{N}{2}\mu_\mathrm{B}\tanh\beta\mu_0\mu_\mathrm{B}\left[M_0(W + W') + H_0 + \Delta(W - W')\right].$$

Since H_0 and Δ are assumed small here, the hyperbolic tangent can be expanded about

$$x = \beta\mu_0\mu_\mathrm{B} M_0(W + W').$$

Eliminating $M_0 = (N\mu_\mathrm{B}/2)\tanh x$, it follows that

$$\Delta = \frac{N}{2}\frac{\mu_0\mu_\mathrm{B}^2}{k_\mathrm{B}}\tanh' x \frac{H_0}{T + \Theta \tanh' x} \qquad \tanh' x = \frac{\mathrm{d}}{\mathrm{d}x}\tanh x = 1 - \tanh^2 x.$$

Fig. P18.3 Temperature dependence of longitudinal and transverse susceptibilities. Adapted from Bizette, H., Tsai, B.: Comptes-Rendus de l'Académie des Sciences 1575 (1954)

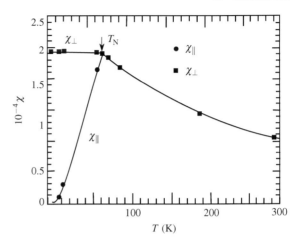

The susceptibility $\chi_\parallel = 2\Delta/\mu_0 H_0$ is

$$\chi_\parallel = \frac{N\mu_B^2}{k_B} \frac{\tanh' x}{T + \Theta \tanh' x}. \tag{13.165}$$

If $T = 0$, we have $x \to \infty$ and $\tanh' x = 0$, whereupon $\chi_\parallel(T = 0) = 0$.
If $T = T_N$, we have $M_0 \to 0$ and hence $x \to 0$ with $\tanh' x = 1$, so that

$$\chi_\parallel(T = T_N) = \frac{N\mu_B^2}{k_B(T_N + \Theta)}.$$

Note that, with an applied field H_0 parallel to $\mathbf{M}_a$, this solution corresponds to a high Zeeman energy for the two sublattices together. In reality, the Zeeman energy cost can reduce to zero if the magnetisations of the sublattices tilt to orient themselves in a direction perpendicular to H_0. Only some source of anisotropy could prevent the magnetisation tilting in this way. But the magnetisation will do this anyway if the applied field exceeds a critical value called the *spin flop field*. (This field is determined by the values of the anisotropy and exchange energies.)

10. The calculated values are plotted in Fig. P18.3 as continuous lines for $\chi_\parallel$ and $\chi_\perp$. The measured values for MnF$_2$ are also plotted there and found to be in perfect agreement with expectations. It should be pointed out that such susceptibility data do allow one to establish the existence of antiferromagnetic order at T_N.

11. There may be domains, associated with an interchange of the directions of $\mathbf{M}_a$ and $\mathbf{M}_b$ (antiphase domains).

Problem 19: Magnetism of Thin Films and Magneto-Optic Applications

Consider a thin ferrimagnetic single-crystal film of thickness $d = 4$ μm in the z direction and infinite in the x and y directions. We assume that the exchange energy is large enough for the magnetisation of the film to be uniform on a length scale greater than the interatomic distance. To begin with, consider a film A of a material with negligible magnetocrystalline anisotropy. The magnetisation is subject to demagnetising fields.

1. *Write down the demagnetisation energy of the film. In which direction does the magnetisation* **M** *point when there is no applied field?*

The film is subjected to a magnetic field H_a in a direction making an angle ψ with the z axis. The system is taken round a hysteresis cycle.

2. *Determine the cycle of the magnetisation in the direction of the applied field for $\psi = \pi/2$.*

3. *Repeat for $\psi = 0$.*

The composition of the ferrimagnetic material is modified in such a way as to significantly increase the uniaxial anisotropy. The new film B has the same thickness and its easy axis of magnetisation is perpendicular to the film. Setting $\theta = (O\mathbf{z}, \mathbf{M})$, the anisotropy energy per unit volume of the film is

$$E_a = K \sin^2 \theta \,.$$

4. *If we assume that the magnetisation of the film is still uniform, what condition must K satisfy for the zero-field magnetisation to differ from the one obtained for film A?*

We assume in the following that this condition is satisfied for film B. The aim now is to find out how these thin, insulating, transparent ferrimagnetic films A and B, can be used as indicators in magneto-optical imaging to carry out magnetic observations and measurements. An aluminium film is thus deposited on them to act as a mirror and they are observed by microscope under polarised light.

The polarisation of a polarised light wave will rotate through an angle $d\alpha$ given by

$$d\alpha = V\mathbf{M}_z dz \,,$$

when it passes through a slab of magnetic material of thickness dz in the direction of propagation z, where V is the Verdet constant and **M** the magnetisation of the medium. This is known as the Faraday effect in magneto-optics (see Fig. P19.1).

Figure P19.2 shows a suitable experimental arrangement. We analyse the polarisation of light crossing film A or B and reflected by the metal film. Since the angle of rotation α of the polarisation is small, the intensity of the light detected by the CCD camera is proportional to α. This intensity is grey-scaled.

Fig. P19.1 Faraday rotation of the polarisation of a light beam propagating through a magnetised ferrimagnetic sample

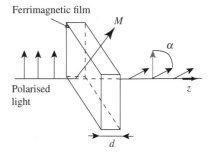

5. *What would you expect to observe for the above films A and B when there is no applied field?*

Figure P19.3 shows the image actually observed for film B.

6. *Can you explain this observation qualitatively? Which of our assumptions concerning film B is not borne out?*

7. *Without calculation, can you say which energy terms determine the width of the ribbons observed in Fig. P19.3?*

We wish to use films A and B as indicators for the magnetic structure inscribed on a plane magnetised object C. To do this, C is placed under film A or B, in contact

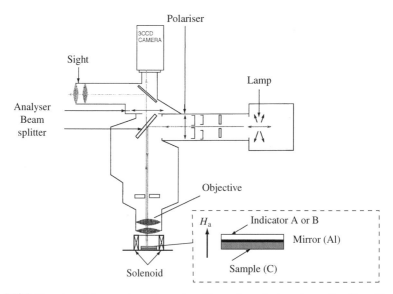

Fig. P19.2 Experimental magneto-optics setup. The *enlarged view* shows the disposition of indicator A or B on the sample C

Fig. P19.3 Magneto-optic image of film B in the absence of any applied field

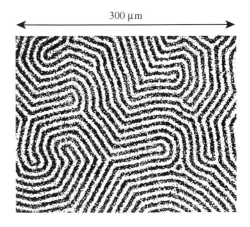

with it, and we observe any changes in the reflected light. Let $\mathbf{H}(x,y)$ be the magnetic field produced by C. We assume it uniform to begin with.

8. *Show that only the component of $\mathbf{H}(x,y)$ normal to the film will produce a visible effect for film A. But what about film B?*

Figure P19.4 (left) is the image observed with film B when C is a credit card with number in the form of a bar code on the magnetic strip. Note the difference of scale with Fig. P19.3.

9. *Can you explain the phenomena apparent in this image?*

Figure P19.4 (right) shows a different region of the same credit card observed with film A.

10. *Which of the two films A and B could be used to make an exact measurement of the magnetic field $H(x,y)$ induced by C? Explain why. What is the maximal value of H that could be measured in this way?*

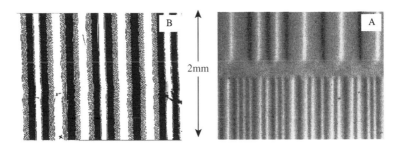

Fig. P19.4 Credit card. *Left*: Image of bar code using film B. *Right*: Image of another region of the card using film A. Images 19.1–4 courtesy of K. Van der Beek (LSI, Ecole Polytechnique)

Solution

1. The demagnetisation tensor is diagonal with $N_z = 1$, $N_x = N_y = 0$. The demagnetisation energy is

$$E_d = \frac{1}{2}\mu_0 \mathbf{M} \cdot \overline{\overline{\mathbf{N}}} \cdot \mathbf{M} = \frac{1}{2}\mu_0 M_z^2 .$$

For zero applied field, it is minimal when the magnetisation lies in the plane of the film. Under these conditions, the magnetic poles are repelled to infinity.

2. When a magnetic field H_a is applied in a direction making an angle ψ with the z axis, setting $\theta = (O\mathbf{z}, \mathbf{M})$, the total energy is

$$E_T = \frac{1}{2}\mu_0 M_0^2 \cos^2\theta - \mu_0 M_0 H_a \cos(\psi - \theta) .$$

The same calculation was carried out in Chap. 9 for the hysteresis cycle when there is a uniaxial anisotropy. We simply make the change of variables

$$\alpha = \frac{\pi}{2} - \theta, \qquad \phi = \frac{\pi}{2} - \psi, \qquad K = \frac{1}{2}\mu_0 M_0^2 ,$$

to recover the same equations. Indeed, the anisotropy due to the demagnetising fields has the same form as the uniaxial anisotropy. The difference here is just that the easy axis lies in the x direction of the plane. We obtain the same result, namely for $\psi = \pi/2$, i.e., $\phi = 0$, the field applied parallel to an easy axis leads to a square hysteresis cycle. The reversal field is simply $H_a = -M_0$.

3. On the other hand, for $\psi = 0$, i.e., $\phi = \pi/2$, the applied field is transverse with respect to the easy axis and the magnetisation perpendicular to the film, $M_z = M_0$ for $H_a > M_0$, rotates continuously when $H_a < M_0$, so that $M_z = H_a$.

4. The ferrimagnetic film B has two anisotropy terms with antagonistic easy magnetisation axes, the uniaxial anisotropy corresponding to an easy axis perpendicular to the film. The total anisotropy energy is

$$E_a = \frac{1}{2}\mu_0 M_0^2 \cos^2\theta + K\sin^2\theta = K + \left(\frac{1}{2}\mu_0 M_0^2 - K\right)\cos^2\theta .$$

For $K > \mu_0 M_0^2/2$, the sign of the anisotropy term changes and the zero field magnetisation points in a direction perpendicular to the plane.

5. For film A, the magnetisation $\mathbf{M}$ lies in the plane of the film. It is therefore perpendicular to the direction of propagation of the light. There is no Faraday effect and hence no rotation of the polarisation. For film B, the magnetisation is perpendicular to the film, and the polarisation is rotated through $\alpha = VMd$. In both cases we thus expect uniform polarisation of the reflected light.

6. The image of Fig. P19.3 shows that the polarisation of the light is rotated by different amounts at different points on film B. Therefore, the magnetisation of this film is not uniform. While a magnetisation direction perpendicular to the plane of the film minimises the uniaxial magnetocrystalline anisotropy, it does not minimise the demagnetisation energy. Quite the contrary, if the magnetisation were uniform, the latter would actually be maximal. In order to minimise the demagnetisation energy, magnetic domains appear that are alternately magnetised up and down. The Faraday rotation of the light polarisation is in opposite directions for domains magnetised up and down, and this is what leads to the lighter and darker strips in the image.

7. The configuration of these domains is the one minimising the sum of three energy terms, viz., the exchange energy which imposes a locally uniform magnetisation, the magnetocrystalline anisotropy energy which tends to align the magnetisation with the easy axis, and the demagnetisation energy which tends to minimise the field produced outside. In order to minimise the demagnetisation energy, the width of the domains must be reduced as far as possible to minimise the demagnetising fields of the magnetic domains and reduce the field produced outside. However, this cannot be done without increasing the number of Bloch walls. The energy involved in creating these interfaces is what prevents an infinite subdivision. So in practice it is the minimisation of the sum of the demagnetisation energy and the energy of the Bloch walls that determines the width of the domains.

8. For film A, the magnetisation lies in the plane of the film and can only acquire a transverse component under the effect of a perpendicular field. There is thus no Faraday effect if $\mathbf{H}(x,y)$ lies parallel to the film. For film B, a field parallel to the film will rotate the magnetisation of the domains and reduce the normal component of their magnetisation. With a sufficiently strong applied field, the domain structure can completely disappear so that the magnetisation lies in the plane. This disappearance of the domains is easy to observe magneto-optically.

9. The credit card has a magnetic bar code which encodes 0 and 1 by the direction of the magnetisation. The field produced by the magnetic zones has a component in the perpendicular direction which magnetises indicator B in such a way as to broaden domains corresponding to the field orientation. In Fig. P19.4 (left), the domains have completely disappeared in regions corresponding to the bar code. This shows that the indicator magnetisation was saturated there. In these zones, the magnetisation is $\mathbf{M} = \pm M_0 \mathbf{z}$, corresponding to the same magnetisation as in the domains of the map observed in zero field. In the regions of the map where there is no bar code, the field is zero and the indicator domains are visible. It can be checked that the Faraday rotation is the same in the domains of the film in zero field and in the magnetised zones (the image is made up solely of whites and blacks)

10. With indicator A, one can also observe the field distribution induced by the credit card in Fig. P19.4 (right). Recall that, in this case, only the field component perpendicular to the plane of the film is active. If $H_z(x,y) = H_z$ is

uniform, the magnetisation of the film is $M_z = H_z$, according to question 3. If $H_z(x,y) \ll M_0$, a spatial non-uniformity in $H_z(x,y)$ will be faithfully reflected by the same non-uniformity in $M_z(x,y) = H_z(x,y)$, provided that the spatial variation of $H_z(x,y)$ is only significant over large distances compared with the characteristic dimensions of the Bloch walls of the indicator. In this case, the image *provides a direct measurement of the spatial variation of $H_z(x,y)$*. The image in Fig. P19.4 (right) contains all shades of grey, and constitutes a faithful reproduction of $H_z(x,y)$. If $K \gg \mu_0 M_0^2/2$, the field that can be measured is limited only by the value of the anisotropy field $2K/\mu_0 M_0$. To measure strong fields, film A should have the greatest anisotropy possible.

With film B, the pattern of magnetic domains is influenced by the fields both parallel and perpendicular to the film, and these processes are not generally linear in the field. It cannot therefore be used to carry out accurate measurements of $H_z(x,y)$. This can be seen by comparing the left and right images of Fig. P19.4. In the first, no variation is visible in the magnetisation at the edges of the regions corresponding to the bar code. A weak field H_z suffices to saturate the magnetisation of the film.

The only possible linear response for film B would arise for a field $H_z(x,y)$ much weaker than the coercive field H_c of the film. In this case, the transverse field broadens (narrows) the domains depending on their direction of magnetisation. Integrating the light intensity over much greater spatial distances than the width of the ribbons, we obtain a quantity proportional to $H_z(x,y)$. However, in so doing, we lose a great deal of spatial resolution, and this approach can only be used to measure large-scale spatial variations.

Problem 20: Magnetism of a Thin Film[10]

The aim here is to investigate the magnetisation configuration in a thin magnetic film as a function of its thickness. The three parts of the problem can be tackled independently. However, it is important to obtain a general grasp of the issues here.

General Introduction

Near a surface, the magnetic anisotropy increases, mainly due to the lower symmetry of the environment in which the atoms find themselves. This has consequences for the magnetic properties of thin films, which may be relevant when they are used for the purposes of magnetic recording.

If we assume uniform magnetisation (constant magnitude M and parallel orientation of spins), the anisotropy energy per unit volume of a thin film contains three terms:

$$E_A = \left(K_V - \frac{1}{2}\mu_0 M^2 + \frac{2K_s}{t}\right)\sin^2\theta, \qquad \mu_0 = 4\pi \times 10^{-7}. \qquad (13.166)$$

Here θ is the angle between the magnetisation and the normal to the film, and K_V is the anisotropy energy per unit volume of the bulk material. The second term, the shape anisotropy, arises from the demagnetisation energy of the film, assumed of infinite extent (see Problem 19). The third term is due to the different environments of atoms located close to the surface, which have fewer nearest-neighbour atoms and therefore see an anisotropic exchange field. This anisotropy tends to impose a magnetisation perpendicular to the surface on these surface layers. It has less and less effect as the thickness t of the film increases, whence its form K_s/t, where K_s is the anisotropy energy per unit area.

In most cases, K_V is either zero (cubic symmetry) or very small. We shall take $K_V = 0$ throughout this problem. The image underlying (13.166) is that, despite the contradictory instructions given to the spins at the surface and in the bulk, the exchange interaction is strong enough to align all the spins in the same direction. We shall ascertain the conditions for this to provide a good description.

To do this, we consider a Heisenberg model on a simple cubic lattice. We assume that the film is infinite in the x and y directions and of thickness $t = La$ in the z direction, where L is the number of atomic layers and a is the lattice parameter [see Fig. P20.1 (left)]. We treat the spins as classical vectors of unit length. The magnitude of the spin will be absorbed by redefining the various constants. The spins make an angle θ with the z axis [see Fig. P20.1 (right)]. Under these conditions, the orientations of the spins only depend on z, the spins in plane i making angle θ_i with the z axis [see Fig. P20.1 (left)]. The energy per unit area is then

---

[10] This problem has been designed with H. Schulz.

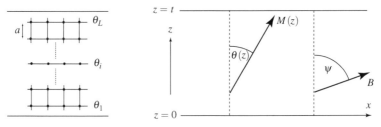

Fig. P20.1 *Left*: Simple cubic lattice. *Right*: Notation for spins and magnetic field

$$E = \frac{A}{a}\sum_{l=1}^{L-1}(\theta_l - \theta_{l+1})^2 - \frac{1}{2}\mu_0 M^2 a \sum_{l=1}^{L}\sin^2\theta_l$$

$$-BMa\sum_{l=1}^{L}\cos(\theta_l - \psi) + K_s(\sin^2\theta_1 + \sin^2\theta_L). \quad (13.167)$$

The first term, with $A > 0$, describes the exchange interaction, while the coupling with a magnetic field making angle ψ with the z axis [see Fig. P20.1 (right)] is described by the term in BM. The lattice parameter a is introduced to give the various constants the right physical dimensions.

20.1: Uniform Magnetisation

1. Assuming the θ_l independent of l, and when $B = 0$, show that (13.167) reduces to (13.166), up to a multiplicative constant t. For the relevant case $K_s > 0$, deduce that there is a critical thickness t_c for which the orientation changes suddenly from $\theta = \pi/2$ to $\theta = 0$ (or to $\theta = \pi$). Find t_c. Recall that $K_V = 0$ throughout.

2. Consider the case $B \neq 0$. Generalise (13.166) to this case. For $t < t_c$, show that there are metastable situations for a perpendicular field ($\psi = 0$), whereas there are not for a parallel field ($\psi = \pi/2$). Plot the hysteresis curves (magnetisation vs. applied field) for the two cases $\psi = 0$ and $\psi = \pi/2$.

When there is a uniaxial anisotropy, the equation of motion for the magnetisation is modified. We have

$$\frac{d\mathbf{M}}{dt} = \gamma \left[\mathbf{M} \times (\mathbf{B} + \lambda \mathbf{M}_z)\right], \quad (13.168)$$

$$\lambda = \frac{4K_s}{tM^2} - \mu_0, \quad (13.169)$$

with $\gamma = g\mu_B/\hbar$ and $\mathbf{M}_z = (0,0,M_z)$. Consider an applied field in the z direction for a thin film $t < t_c$.

3. Linearising (13.168), find the ferromagnetic resonance frequency when the field B is varied in the hysteresis diagram determined earlier.

Fig. P20.2 Ferromagnetic resonance field B_r vs. angle ψ between the field and the normal to the film. Figure courtesy of Hurdequint, H.: Laboratoire de Physique des Solides, Orsay

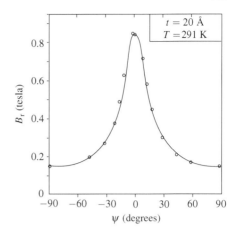

We now consider a thicker film, with $t > t_c$.

4. *What is the resonance frequency for a field applied in the x direction, i.e., with $\psi = \pi/2$?*

Even for $t > t_c$, a strong enough field parallel to the z axis can align the magnetisation with that axis.

5. *Find the minimal value B_0 of the field required to do this. Using the result of question 3, give the resonance frequency for $B > B_0$. (For weaker fields, the resonance frequency vanishes.)*

The ferromagnetic resonance of a cobalt film of thickness 2 nm is observed with a spectrometer at fixed frequency $\omega/2\pi = 9.23$ GHz, while varying the strength and orientation of the applied field B. In Fig. P20.2, the field B_r corresponding to the ferromagnetic resonance is plotted against the angle ψ between the field and the normal to the film. The values of the resonance field for $\psi = 0$ and $\psi = \pi/2$ should correspond to the situations investigated above.

6. *Given that $\mu_0 M = 1.7$ tesla for cobalt at room temperature, find the critical thickness t_c for cobalt. Could thin cobalt films be used to make perpendicular magnetic recordings? (Note that, for electron spins, $\omega/2\pi B$ is 27.98 GHz/tesla.)*

20.2: Non-uniform Situations

7. *Write down the equations minimising the energy in (13.167). Note that the surface variables θ_1 and θ_L satisfy different equations to the bulk variables $\theta_2, \ldots, \theta_{L-1}$. Here and in the following, take $B = 0$.*

Assume in the following that the variables θ_l vary slowly with l. We then have $\theta_l = \theta(z)$, where $z = la$ is treated as a continuous variable. Hence,

$$\theta_{l+1} - \theta_l = a\frac{d\theta}{dz}, \quad 2\theta_l - \theta_{l+1} - \theta_{l-1} = -a^2\frac{d^2\theta}{dz^2}. \tag{13.170}$$

8. Show that the equations for the bulk variables become the differential equation

$$4A\frac{d^2\theta}{dz^2} + \mu_0 M^2 \sin 2\theta = 0, \tag{13.171}$$

while the equations for θ_1 and θ_L become the boundary conditions

$$-2A\left.\frac{d\theta}{dz}\right|_{z=0} + \tilde{K}_s \sin 2\theta(0) = 0, \tag{13.172}$$

$$2A\left.\frac{d\theta}{dz}\right|_{z=t} + \tilde{K}_s \sin 2\theta(t) = 0, \tag{13.173}$$

where $\tilde{K}_s = K_s - \mu_0 M^2 a/2$.

In the following, $\tilde{K}_s > 0$.

9. For $B = 0$, show that there are three trivial solutions to (13.171). What physical situations do they correspond to?

Comments

In the following, still assuming $B = 0$, we shall consider the possibility of solutions in which θ is space dependent. We note first that the symmetry of the problem requires $\theta(t/2+z) = \theta(t/2-z)$. The function $\theta(z)$ thus takes its maximal value θ_m in the middle of the film: $\theta_m = \theta(t/2)$. We define the exchange length by

$$\Lambda = \sqrt{\frac{2A}{\mu_0 M^2}}. \tag{13.174}$$

After the transformation

$$\sin\theta = \sin\theta_m \sin\phi, \tag{13.175}$$

and setting $\phi_0 = \phi(z=0)$, the solution to (13.171) has the form

$$\frac{z}{\Lambda} = \int_{\phi_0}^{\phi(z)} \frac{d\phi}{\sqrt{1-\sin^2\theta_m \sin^2\phi}} = F\big(\phi(z), \sin^2\theta_m\big) - F\big(\phi_0, \sin^2\theta_m\big). \tag{13.176}$$

In particular, in the middle of the film, $\theta(t/2) = \theta_m$ and hence $\phi(t/2) = \pi/2$, so

$$\frac{t}{2\Lambda} = F(\pi/2, \sin^2\theta_m) - F(\phi_0, \sin^2\theta_m), \tag{13.177}$$

where $F(\phi, m)$ is a special function of the two variables ϕ and m called the incomplete elliptic integral of the first kind, defined by

Problem 20: Magnetism of a Thin Film

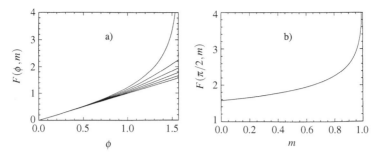

Fig. P20.3 (**a**) Dependence of $F(\phi,m)$ on ϕ for (*bottom to top*) $m = 0, 0.2, 0.4, 0.6, 0.8$, and 1. (**b**) Dependence of $F(\pi/2,m)$ on m

$$F(\phi,m) = \int_0^\phi \frac{dx}{\sqrt{1-m\sin^2 x}}. \tag{13.178}$$

Figure P20.3 plots some examples. When $m \to 1$, we have

$$F(\pi/2,m) \approx \frac{1}{2}\ln\left[16/(1-m)\right].$$

The solution (13.176) still contains two undetermined constants, θ_m and ϕ_0. Combining (13.176) with the boundary condition at $z = 0$, these are found to be related by

$$\sin^2\phi_0 = \frac{\rho + 1/\rho - \sqrt{(\rho+1/\rho)^2 - 4\sin^2\theta_m}}{2\rho\sin^2\theta_m}, \tag{13.179}$$

with

$$\rho = \frac{\tilde{K}_s}{\sqrt{\mu_0 M^2 A/2}}. \tag{13.180}$$

Note that $\sin^2\phi_0$ is an increasing function of θ_m. Equations (13.177) and (13.179) completely determine $\theta(z)$ for given thickness and constants specifying the material.

20.3: Detailed Investigation of Non-uniform Situations

For thin enough film (small t), the solution $\theta = 0$ is stable, minimising the surface anisotropy energy. We now consider a situation where the surface anisotropy is very large, i.e., $\rho \to \infty$. Equation (13.179) then gives $\rho_0 = 0$.

10. Sketch the function $\theta(z)$ approximately, choosing two or three values of the parameter θ_m.

11. Are these non-uniform situations observable? What experimental methods could you use?

12. Still for $\rho = \infty$, use (13.177) to show that there is a critical thickness t_c below which the solution (13.176) no longer exists, and we therefore have $\theta = 0$. Sketch the dependence of θ_m on t approximately. Sketch the function $\theta(z)$ for t just above t_c and for $t \gg t_c$. Comment on the result.

As in the previous case, for $\rho < \infty$, the critical thickness is given by the solution to (13.177) with $\theta_m = 0$.

13. Simplify this expression and obtain an explicit form for t_c as a function of ρ and Λ alone.

14. Show that, for $\rho > 1$, the solution given by (13.176) exists for any value of the thickness $t > t_c$. Use the fact that, in this case, $\phi_0 < \pi/2$ even when $\theta_m = \pi/2$. Plot $\theta(z)$ for $t \approx t_c$ and for $t \gg t_c$.

15. For small surface anisotropy $\rho < 1$, show that t remains finite when $\theta_m = \pi/2$ and that there is therefore an upper critical thickness t_{c_2}, and specify its dependence on ρ. Use the fact that, in this case for $\theta_m \to \pi/2$, we also have $\phi_0 \to \pi/2$.

If the two angles are close to $\pi/2$, (13.179) implies

$$(\pi/2 - \phi_0)^2 = \frac{\rho^2}{1 - \rho^2}(\pi/2 - \theta_m)^2 . \tag{13.181}$$

The integral in (13.177) thus simplifies. For $t > t_{c_2}$, we have a homogeneous solution $\theta = \pi/2$. The situation is completely dominated by the shape anisotropy.

16. With the numerical values found in 20.1, what are the corresponding orders of magnitude?

17. Plot a phase diagram in the (ρ, t) plane. Comment on the respective roles of the exchange interaction, surface anisotropy, and shape anisotropy in the three phases.

Solution

Uniform Magnetisation

1. We have

$$E = -\frac{1}{2}\mu_0 M^2 aL\sin^2\theta + 2K_s\sin^2\theta$$

$$= \left(-\frac{1}{2}\mu_0 M^2 + \frac{2K_s}{t}\right) t\sin^2\theta. \quad (13.182)$$

For small enough t, surface anisotropy dominates ($K_s > 0$). We thus minimise E by setting $\theta = 0$ or $\theta = \pi$. For greater thickness, shape anisotropy dominates, giving $\theta = \pi/2$. The critical value of t is found by setting the prefactor of $\sin^2\theta$ to zero, that is

$$\boxed{t_c = 4K_s/\mu_0 M^2}. \quad (13.183)$$

2. For $B \neq 0$, the calculation leading to (13.182) generalises to give

$$E = \left(-\frac{1}{2}\mu_0 M^2 + \frac{2K_s}{t}\right) t\sin^2\theta - BMt\cos(\theta - \psi). \quad (13.184)$$

For $t < t_c$, set

$$\boxed{B_0 = \frac{4K_s}{tM} - \mu_0 M}. \quad (13.185)$$

When $\psi = 0$ and $|B| < B_0$, the function $E(\theta)$ has two local minima at $\theta = 0$ and π. Metastable situations are thus possible for B perpendicular to the film. For $B > 0$, the solution $\theta = 0$ is the equilibrium case and $\theta = \pi$ is metastable, while the situation is reversed for $B < 0$. When $|B| > B_0$, there is only one minimum at $\theta = 0$ or $\theta = \pi$.

When $\psi = \pi/2$, there is only one local minimum for $E(\theta)$. When $|B| < B_0$, it corresponds to $\sin\theta = B/B_0$, and when $|B| > B_0$, it is at $\theta = \pm\pi/2$, with the same sign as B. The two cases are depicted in Fig. P20.4.

3. For $\mathbf{B} \| \hat{z}$, the equation of motion is

$$\frac{dM_x}{dt} = \gamma(B + \lambda M_z)M_y, \quad \frac{dM_y}{dt} = -\gamma(B + \lambda M_z)M_x. \quad (13.186)$$

In the linear response limit $M_z = \pm M$, we have precession with angular frequency

$$\omega^2 = \gamma^2(B + \lambda M)^2,$$

where $\lambda M = \pm B_0$, depending on the sign of M, and then

$$\boxed{\omega^2 = \gamma^2(B \pm B_0)^2}. \quad (13.187)$$

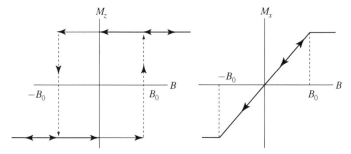

Fig. P20.4 Hysteresis curves for $\psi = 0$ (*left*) and $\psi = \pi/2$ (*right*). Thick lines are stable and *thin lines* are metastable

This frequency vanishes at the ends of the two metastable branches of Fig. P20.4.

4. In this case $\mathbf{M} \parallel \hat{\mathbf{x}}$, and we must write down the equations for $M_{y,z}$, viz.,

$$\frac{dM_y}{dt} = \gamma(B - \lambda M_x)M_z, \qquad \frac{dM_z}{dt} = -\gamma B M_y.$$

Linearising, i.e., putting $M_x \simeq M$, we have

$$\boxed{\omega^2 = \gamma^2 B(B - \lambda M)}. \tag{13.188}$$

5. We may return to (13.184) for the case $\psi = 0$ and $\lambda < 0$, with

$$\lambda = \frac{4K_s}{tM^2} - \mu_0. \tag{13.189}$$

It transpires that $\theta = 0$ is the absolute minimum of this function if

$$B \geq B_0 = |\lambda| M. \tag{13.190}$$

For $B \geq B_0$, the geometry is as in question 3 above. In this regime, we thus obtain the result (13.187). Since $\lambda < 0$ here, $\omega = 0$ for $B = B_0$. Subsequently, ω increases linearly with B for $B > B_0$.

6. It is the situations in question 4 for $\psi = \pi/2$ and question 5 for $\psi = 0$ that apply here. We write down the corresponding results, but slightly modifying the notation. For $\psi = \pi/2$, the resonance condition is

$$\left(\frac{\omega}{\gamma}\right)^2 = B_r(B_r + \mu_0 M - B_A), \tag{13.191}$$

whereas for $\psi = 0$, we have

$$\frac{\omega}{\gamma} = B_r - \mu_0 M + B_A, \tag{13.192}$$

Problem 20: Magnetism of a Thin Film

where $B_A = 4K_s/tM$. In the present case, the fixed spectrometer frequency gives

$$\frac{\omega}{\gamma} = \frac{9.23}{27.98} \text{T} = 0.33 \text{ T}. \tag{13.193}$$

For $\psi = 0$, we have $B_r = 0.85$ T, and with $\mu_0 M = 1.7$ T, we obtain $B_A = 1.18$ T. When $\psi = \pi/2$, we have $B_r = 0.15$ T, and from (13.191), we obtain $B_A = 1.12$ T. To determine the conditions under which there can be magnetisation perpendicular to the film, we write (13.182) in the form

$$E = \frac{1}{2}\left(\frac{4K_s}{tM} - \mu_0 M\right) Mt\sin^2\theta = \frac{1}{2}\left(1.18 \text{ T}\frac{2 \text{ nm}}{t} - 1.7 \text{ T}\right) Mt\sin^2\theta. \tag{13.194}$$

The coefficient of $\sin^2\theta$ becomes positive if $t < 1.39$ nm, so the magnetisation will be perpendicular to the film if

$$\boxed{t < t_c = 1.39 \text{ nm}}.$$

This effect has in fact been observed for films of thickness 1.13 nm [Chappert, C., et al.: Phys. Rev. B **34**, 3192 (1986)].

Non-uniform Situations

7. Differentiating E with respect to θ_l in (13.167), we find that

$$\frac{2A}{a}(2\theta_l - \theta_{l+1} - \theta_{l-1}) - \frac{1}{2}\mu_0 M^2 a\sin(2\theta_l) = 0, \quad 2 \leq l \leq L-1, \tag{13.195}$$

$$\frac{2A}{a}(\theta_1 - \theta_2) - \frac{1}{2}\mu_0 M^2 a\sin(2\theta_1) + K_s \sin(2\theta_1) = 0, \quad l = 1, \tag{13.196}$$

$$\frac{2A}{a}(\theta_L - \theta_{L-1}) - \frac{1}{2}\mu_0 M^2 a\sin(2\theta_L) + K_s \sin(2\theta_L) = 0, \quad l = L. \tag{13.197}$$

8. Using equations (13.170), (13.195)–(13.197) transform immediately into the required differential equation and boundary conditions.

9. $\theta = 0, \pi$ are the uniform solutions of (13.171). They correspond to magnetisation perpendicular to the film in the positive and negative z directions. These solutions minimise the surface anisotropy energy, but maximise the shape anisotropy energy. In contrast, for $\theta = \pi/2$, the magnetisation is parallel to the film, minimising the shape anisotropy energy, but maximising the surface anisotropy energy.

Fig. P20.5 Sketched curves for $\theta(z)$ for three different values of θ_m. *Lower curve*: $\theta_m \gtrsim 0$. *Upper curve*: $\theta_m \lesssim \pi/2$. *Center curve*: Intermediate value of θ_m

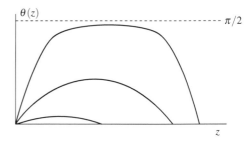

Detailed Investigation of Non-uniform Situations

10. For small θ_m, the function $F(\phi, \sin^2 \theta_m)$ can be approximated by a linear function of ϕ, as can be seen in Fig. P20.3, so that

$$\phi(z) \approx z/\Lambda. \qquad (13.198)$$

Furthermore, under these conditions, $\sin\theta(z) \approx \theta(z)$ for all z. From (13.175), we then obtain

$$\theta(z) = \sin\theta_m \sin(z/\Lambda). \qquad (13.199)$$

For $\theta_m \approx \pi/2$, the function $F(\phi, \sin^2 \theta_m)$ diverges almost everywhere when $\phi \to \pi/2$. $\phi(z)$ can be obtained graphically from Fig. P20.3. Under these conditions, we have $\theta(z) \approx \phi(z)$, and the same plot also gives $\theta(z)$, except in the immediate vicinity of $z = t/2$, where $\phi(z)$ approaches its maximum $\pi/2$ with a gentle but nonzero slope, while $\theta(z)$ reaches its maximum with zero slope. The curves are shown schematically in Fig. P20.5 for $\theta_m \approx 0$, $\theta_m \approx \pi/2$, and for an intermediate value. In particular, we observe that the parameter θ_m increases with the film thickness.

11. To identify such situations, we can try to measure the magnetisation transverse to the film. This might be feasible with SQUID sensitivity. However, there are some obvious problems, e.g., sample alignment, the need to orient the perpendicular component by applying a magnetic field, demagnetising field, etc. Clearly, θ_m must be large enough. The magnetisation profile can be ascertained by a technique in which polarised neutrons are reflected from the surface. This was developed by C. Fermon at the Léon Brillouin research institute, Saclay (France) in 1996.

12. When $\theta_m \to 0$, the above solution (13.198) becomes exact. This solution satisfies the boundary conditions for a thickness $t = \pi \Lambda$. Since θ_m increases with the film thickness, there is no non-constant solution satisfying the boundary conditions below the critical thickness given by

$$\boxed{t_c = \pi \Lambda}. \qquad (13.200)$$

Fig. P20.6 Dependence of θ_m on the film thickness t

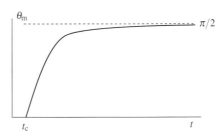

Below this thickness, the constant solution $\theta(z) = 0$ minimises the energy. From Fig. P20.3, we obtain the dependence of θ_m on the thickness, as shown in Fig. P20.6. For t close to t_c, we find that θ_m varies as $(t - t_c)^{1/2}$. Plots of $\theta(z)$ are the same as in Fig. P20.5. Small θ_m corresponds to $t \gtrsim t_c$, and $\theta_m \approx \pi/2$ to $t \gg t_c$.

Comments

- In the case investigated here, $\rho = \infty$, the surface anisotropy is large, and a magnetisation is imposed perpendicularly to the film at surfaces $z = 0, t$.
- For $t < t_c$, the shape anisotropy is *too small to 'pull' the magnetisation from the perpendicular orientation*.
- For $t \gtrsim t_c$, the shape anisotropy is just sufficient to shift θ slightly from $\theta = 0$, giving rise to a slightly arched curve $\theta(z)$.
- For $t \gg t_c$, the *shape anisotropy aligns the magnetisation parallel to the film almost everywhere*, except in a region of thickness approximately Λ close to the surface, where the surface anisotropy still dominates.

13. In contrast to the case $\rho = \infty$, for $\rho < \infty$, the surface magnetisation of the film is not strictly perpendicular. In particular, from (13.179) for $\theta_m \to 0$, we obtain

$$\sin^2 \phi_0 = \frac{1}{1+\rho^2} . \tag{13.201}$$

Equation (13.177) then yields

$$\boxed{t_c = 2\Lambda \left(\frac{\pi}{2} - \arcsin \frac{1}{\sqrt{1+\rho^2}} \right) = 2\Lambda \arctan \rho} . \tag{13.202}$$

14. For $\rho > 1$, when $\theta_m \to \pi/2$, (13.179) yields

$$\sin^2 \phi_0 \to \frac{1}{\rho^2} < 1 . \tag{13.203}$$

The second term on the right-hand side of (13.177) thus remains finite when $\theta_m \to \pi/2$, while the first diverges. Consequently, there is always a solution to

(13.177), for all values of $t > t_c$. Since $\phi_0 \neq 0$, we also have $\theta(0) = \theta(t) \neq 0$, and the magnetisation is not perpendicular, even at the surface. The curves $\theta(z)$ therefore have the shapes shown in Fig. P20.7.

15. Equation (13.177) can be rewritten in the form

$$\frac{t}{2\Lambda} = \int_{\phi_0}^{\pi/2} \frac{d\phi}{\sqrt{1 - \sin^2\theta_m \sin^2\phi}}. \tag{13.204}$$

Since $\phi_0 = \pi/2 - \delta$ here, with δ small, the $\sin^2$ under the square root sign can be replaced by a truncated expansion about $\phi = \pi/2$, namely

$$\sin^2\phi \approx 1 - (\phi - \pi/2)^2 = 1 - \delta^2.$$

Changing variables to $\phi' = \pi/2 - \phi$, we obtain

$$\frac{t}{2\Lambda} = \int_0^{\delta} \frac{d\phi}{\sqrt{\cos^2\theta_m + \sin^2\theta_m \phi^2}}$$

$$= \frac{1}{\sin\theta_m} \operatorname{arcsinh} \frac{\delta}{\cot\theta_m}. \tag{13.205}$$

Here $\theta_m \to \pi/2$, so $\sin\theta_m \to 1$ and $\cot\theta_m = \pi/2 - \theta_m$. Using (13.181), we obtain

$$\frac{t}{2\Lambda} = \operatorname{arcsinh} \frac{\rho}{\sqrt{1-\rho^2}} = \operatorname{arctanh}\rho. \tag{13.206}$$

The maximal allowed value of θ_m is $\pi/2$, corresponding to magnetisation parallel to the film. The thickness is then given by (13.206). For thicker films, the only solution is $\theta(z) = \pi/2$. As a consequence, there is an *upper critical thickness* t_{c_2}, above which the magnetisation is parallel to the film everywhere:

$$\boxed{t_{c_2} = 2\Lambda \operatorname{arctanh}\rho}. \tag{13.207}$$

This thickness increases for higher surface anisotropy and lower exchange energy.

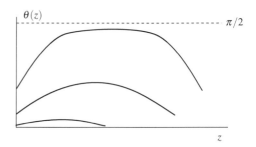

Fig. P20.7 *Lower curve:* $\theta(z)$ for $t \gtrsim t_c$. *Upper curve:* $\theta(z)$ for $t \gg t_c$. *Center curve:* $\theta(z)$ for an intermediate value of t

Fig. P20.8 Phase diagram showing the three possible configurations of the spatial variation of the magnetisation of a ferromagnetic film, depending on its thickness t and surface anisotropy parameter ρ

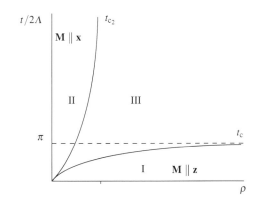

16. We have

$$\rho = \frac{t_c - 2a}{\Lambda}, \qquad \Lambda = \sqrt{\frac{2A}{\mu_0 m^2}}.$$

The experiment in 20.1 yields the value $\mu_0 M = 1.7$ tesla. A is determined by the exchange energy, which is not specified in the statement of the problem. The Curie temperature and crystallographic constants of Co give the estimate $\rho \sim 0.1$. The window for observing a configuration in which the magnetisation is not uniform is thus very narrow. In order to observe such a situation, it would be preferable to use a material with a low value of A, i.e., with a low Curie temperature.

17. There are 3 regions (see Fig. P20.8):

(I) $t < t_c$. *Surface anisotropy dominates* over shape anisotropy, and the magnetisation is perpendicular. This region grows in size when the exchange interaction increases, making the magnetisation more rigid.

(II) $t > t_{c_2}$. *Shape anisotropy dominates.* A strong exchange interaction also stabilises this phase, provided that the surface anisotropy is not too large.

(III) In the rest of the diagram, surface anisotropy dominates the film surfaces, but inside, shape anisotropy plays the dominant role. This region becomes significant when the exchange interaction is weak.

Appendix A
Physical Constants

Units

Angstrom	1 Å = 10^{-10} m ($\sim$ size of an atom)
Fermi	1 fm = 10^{-15} m ($\sim$ size of a nucleus)
Electron-volt	1 eV = 1.60218×10^{-19} J $\longleftrightarrow$ 11 600 K
Magnetic induction	tesla (1 gauss $\longleftrightarrow 10^{-4}$ tesla)

Fundamental Constants

Planck constant	$h = 6.6261 \times 10^{-34}$ J s $\hbar = h/2\pi = 1.05457 \times 10^{-34}$ J s
Speed of light	$c = 299\,792\,458$ m s^{-1}
Vacuum permeability	$\mu_0 = 4\pi \times 10^{-7}$ H m^{-1} ($\varepsilon_0 \mu_0 c^2 = 1$)
Boltzmann constant	$k_B = 1.38066 \times 10^{-23}$ J K^{-1} = 8.6174×10^{-5} eV K^{-1}
Avogadro number	$N_A = 6.0221 \times 10^{23}$
Electron charge	$e = 1.60218 \times 10^{-19}$ C
Electron mass	$m_0 = 9.1094 \times 10^{-31}$ kg
Proton mass	$m_p = 1.67262 \times 10^{-27}$ kg
Neutron mass	$m_n = 1.67493 \times 10^{-27}$ kg
Fine structure constant (dimensionless)	$\alpha = e^2/\hbar c = 1/137.036$
Classical electron radius	$r_e = e^2/m_0 c^2 = 2.818 \times 10^{-15}$ m
Compton wavelength of the electron	$\lambda_c = h/m_0 c = 2.426 \times 10^{-12}$ m
Bohr radius	$a_1 = \hbar^2/m_0 e^2 = 0.52918 \times 10^{-10}$ m
Hydrogen ionisation energy	$E_I = m_0 e^4/2\hbar^2 = \alpha^2 m_0 c^2/2 = 13.6057$ eV
Bohr magneton	$\mu_B = q_e \hbar/2m_0 = -9.2740 \times 10^{-24}$ J T^{-1} $= -5.7884 \times 10^{-5}$ eV T^{-1}
Nuclear magneton	$\mu_N = q\hbar/2m_p = 5.0508 \times 10^{-27}$ J T^{-1} $= 3.1525 \times 10^{-8}$ eV T^{-1}

Appendix B
Some Useful Functions and Relations

Euler Γ Function

$$\Gamma(t) = \int_0^\infty x^{t-1} e^{-x} dx = (t-1)\Gamma(t-1) = (t-1)!$$

$$\Gamma(1/2) = \sqrt{\pi}.$$

Riemann ζ Function

$$\zeta(t) = \sum_{n=1}^\infty \frac{1}{n^t}$$

with some specific values

t	2	3	4
ζ	$\pi^2/6$	1.202	$\pi^4/90$

Integrals of the Fermi–Dirac Function

$$f(x) = \frac{1}{e^x + 1} \quad \text{and} \quad -df/dx = \frac{1}{4\cosh^2(x/2)}$$

$$\int_0^\infty f(x) dx = \int_0^\infty -(df/dx) x \, dx = \ln 2.$$

For $t > 1$

$$\int_0^\infty f(x) x^{t-1} dx = \int_0^\infty \frac{x^{t-1}}{e^x + 1} dx = \frac{1}{4t}\int_0^\infty \frac{x^t}{\cosh^2(x/2)} dx = (1 - 2^{1-t})\Gamma(t)\zeta(t),$$

which yields some specific values

$$\int_0^\infty f(x)x\,dx = \frac{1}{8}\int_0^\infty \frac{x^2}{\cosh^2(x/2)}dx = \frac{1}{2}\zeta(2) = \frac{\pi^2}{12},$$

$$\int_0^\infty f(x)x^2\,dx = \frac{1}{12}\int_0^\infty \frac{x^3}{\cosh^2(x/2)}dx = \frac{3}{2}\zeta(3) \approx \frac{3}{2}1.202 \approx 1.8,$$

$$\int_0^\infty f(x)x^3\,dx = \frac{1}{16}\int_0^\infty \frac{x^4}{\cosh^2(x/2)}dx = \frac{21}{4}\zeta(4) = \frac{7}{120}\pi^4.$$

Bessel Functions

The Bessel functions of order α are solutions of the differential equation

$$x^2\frac{d^2y}{dx^2} + x\frac{dy}{dx} + (x^2 - \alpha^2)y = 0.$$

They are important to solve Laplace's equation in cylindrical (for integer α) or spherical (half-integer α) coordinates.

The Bessel functions *of the first kind* $J_n(x)$ with integer n are solutions which are constant for $x \to 0$ and are oscillating decreasing functions for $x \to \infty$. They have an integral form given by

$$J_n(x) = \frac{1}{\pi}\int_0^\pi \cos(nu - x\sin u)du$$

with

$$J_{-n}(x) = (-1)^n J_n(x).$$

For $n \geq 0$, they can be expanded for small x into

$$J_0(x) \approx 1 - \frac{x^2}{4} \quad \text{and} \quad J_n(x) \approx \frac{1}{n!}\left(\frac{x}{2}\right)^n.$$

The asymptotic form for $x \to \infty$ is

$$J_n(x) \approx \left(\frac{2}{\pi x}\right)^{\frac{1}{2}}\cos\left(x - \frac{n\pi}{2} - \frac{\pi}{4}\right).$$

Sinusoidal functions of sinusoidal functions can be developed into series of Bessel functions using

$$e^{ix\sin\phi} = \sum_{n=-\infty}^\infty J_n(x)e^{in\phi}.$$

The *modified Bessel functions of order* α are solutions of the differential equation

$$x^2\frac{d^2y}{dx^2} + x\frac{dy}{dx} - (x^2 + \alpha^2)y = 0.$$

Appendix B Some Useful Functions and Relations

The modified Bessel functions $K_n(x)$, with integer n, are solutions which diverge for $x \to 0$, and vanish exponentially for $x \to \infty$. Their integral representation is given by

$$K_n(x) = \frac{1}{2} \exp\left(in\frac{\pi}{2}\right) \int_{-\infty}^{\infty} \exp(-ix\, sht - nt)\, dt,$$

with

$$K_{-n}(x) = K_n(x).$$

Their expansion for small x is

$$K_0(x) \approx -\ln x \qquad \text{and} \qquad K_n(x) \approx \frac{1}{2n!}\left(\frac{2}{x}\right)^n.$$

Their asymptotic forms for $x \to \infty$ are

$$K_n(x) \approx \left(\frac{2}{\pi x}\right)^{\frac{1}{2}} \exp(-x).$$

These functions are related to one another through

$$\frac{d}{dx}\left[x^n K_n(x)\right] = -x^n K_{n-1}(x),$$

which gives as an example

$$dK_0/dx = -K_{-1} = -K_1.$$

Appendix C
Standard Notation

$A(T,\mu)$	Grand canonical potential
M	Atomic mass
I	Electric current
T	Temperature
U	Internal energy
V	Potential difference
Z	Atomic number
$Z_G(T,\mu)$	Grand partition function
e	Electron charge
h	Planck constant ($= 2\pi\hbar$)
k_B	Boltzmann constant
s	Electron spin ($s = 1/2$)
Φ	Flux of magnetic induction around a contour (C)
β	Reciprocal of $k_B T$
$\chi, \tilde{\chi}$	Magnetic susceptibility ($\tilde{\chi} = \mu_0 \chi$)
$\delta(x)$	Dirac delta
δ_{ij}	Kronecker delta
γ	Gyromagnetic ratio
μ	Chemical potential
μ_B	Bohr magneton
$\mathbf{A}$	Vector potential
$\mathbf{B}_a$	Applied magnetic induction ($= \mu_0 \mathbf{H}_a$)
$\mathbf{F}$	Force
$\mathbf{H}_a$	Applied magnetic field
$\mathbf{M}$	Magnetisation
$\mathbf{j}$	Electric current density
$\mathbf{k}$	Wave vector
$\mathbf{p}_i$	Momentum of ith electron
$\mathbf{v}_i$	Velocity of ith electron
$\boldsymbol{\mu}$	Atomic magnetic moment

Appendix D
Specific Notation

$a_{\mathbf{k}_0,\mathbf{k}_1}$	Atomic form factor $[=f(\mathbf{k}_1 - \mathbf{k}_0)]$
$B_J(x)$	Brillouin function with total angular momentum J as a function of $x = g_J\mu_0\mu_B H_a/k_B T$
$C_e(T)$	Specific heat • electronic
C_n	• in normal metal state
C_s	• in superconducting state
C_v	• at constant volume
D	Coefficient characterising magnetic anisotropy energy per ion (see K)
$D(E)$	Density of states • of energy E per atom and per spin direction
$D_d(E)$	• of a free electron gas in d dimensions
$D_s(E)$	• excited in the superconducting state
$D_\Omega(E)$	• of energy E per unit volume
D_χ	• at the Fermi level obtained by the Pauli susecptibility
D_c	• at the Fermi level obtained by the specific heat
$D_c(E - E_c)$	• of the conduction band of a semiconductor
$D_v(E - E_c)$	• of the valence band of a semiconductor
$E_1, E_2, \ldots E_n$	Atomic energy levels
E_v	Atomic energy level of a valence electron
$E_\mathbf{k}$	Electronic energy of state with quasi-momentum $\mathbf{k}$ (single band)
E_F	Fermi energy
E_Z	Energy of Zeeman coupling between magnetisation and magnetic field
E_d	Demagnetisation energy
E_{an}	Anisotropy energy as a function of orientation of $\mathbf{M}$ (see H_A)
E_{wall}	Energy of a Bloch wall per area of primitive cell
E_c	Energy at minimum of conduction band (semiconductor)
E_v	Energy at maximum of valence band (semiconductor)
E_c	Condensation energy in superconducting state
E_g	Semiconductor band gap $(= E_c - E_v)$
E_h	Energy of hole
$G_s(T, H_a)$	Gibbs free energy • of superconducting state
$G_n(T, H_a)$	• of normal metal
$G_{sf}(T, H_a)$	• of superconducting film
$G_{ms}(T, H_a)$	• of mixed state

$H_c, H_c(T)$	Critical thermodynamic field of a superconductor
H'_c	Critical field of thin superconducting film
H_{c_1}	Lower critical field for type II superconductor
H_{c_2}	Upper critical field for type II superconductor
$\hat{H}_Z^{rf}$	Hamiltonian for Zeeman coupling with a radiofrequency field
$\hat{H}_A$	Magnetic anisotropy Hamiltonain, e.g., $H_A = D\sum_{\mathbf{R}}(S_{\mathbf{R}}^z)^2$
H_K	Anisotropy field
H_d	Demagnetising field
I_c	Critical current of a Josephson junction
J	Total angular momentum of an atomic state ($J = L + S$)
J_{NN}	Current in tunnel junction • between two normal metals
J_{NS}	• between normal metal and superconductor
$\mathcal{J}$	Exchange interaction • between two spins $\mathbf{S}_1$ and $\mathbf{S}_2$
J_0	• interatomic
K	Magnetic anisotropy coefficient $= -NDS^2 = -DM_0^2/Ng^2\mu_0^2$
L	Spatial dimensions of a macroscopic crystal
L	Latent heat at normal–superconducting transition
$\mathcal{L}$	Lorenz number
$L(T,M(r))$	Ginzburg–Landau functional
$L(T,M_z)$	Landau function
$M_0(T)$	Spontaneous magnetisation of ferromagnet
M_s	Magnetisation at saturation
M_{alt}	Alternating magnetisation in antiferromagnet with two sublattices
N_e	Number of electrons in solid
N_n	Number of atomic nuclei in solid
$\overline{\overline{N}}$	Demagnetisation tensor
P_i	Momentum of ith atomic nucleus
R_W	Wilson ratio relating χ_P and C_e in metal
S_s	Entropy • of superconducting state
S_n	• of normal state
$S(\mathbf{k})$	Structure factor (primitive cell)
T_c	Critical temperature of superconductor
T_C	Curie temperature
T_N	Néel temperature
T_1	Relaxation time • spin lattice
T_2	• spin–spin
$\hat{T}_{\mathbf{R}_i}$	Operator effecting translation of crystal lattice by $\mathbf{R}_i$
$U(r)$	Magnetic potential
$V(\mathbf{r}_1 - \mathbf{r}_2)$	Interaction between two electrons at $\mathbf{r}_1$ and $\mathbf{r}_2$
V_{at}	Atomic potential
$V_\mathbf{K}$	Fourier component of periodic potential for vector $\mathbf{K}$ of reciprocal lattice
$V_c(\mathbf{r})$	Coulomb potential
$V_\mathbf{k}$	Fourier component of attractive potential $V(\mathbf{r}_1 - \mathbf{r}_2)$ between two electrons due to electron–phonon interaction

Appendix D Specific Notation

V_0	Interaction between electrons due to electron–phonon coupling in point interaction limit
a	Distance between nearest neighbour atoms
c	Specific heat capacity per particle
d	Width of tunnel barrier
$f(E)$	Fermi population factor
g_J	Landé factor of electron in ground state with total angular momentum **J**
j_en	Energy flux
k_F	Fermi wave vector for free electron gas
ℓ	Mean free path • of particle
ℓ_e	• of electron
$(m_\text{e})_{\alpha\beta}$	Effective electron mass tensor $[(m_\text{h})_{\alpha\beta}$ for hole]
m_e	Effective mass • of electron (isotropic case)
m^*	• of charge
$m(t)$	Magnetic response to pulse excitation at $t=0$
n	Band index (band structure of solid)
n or n_e	Number density • of conduction electrons
n_p	• of Bravais lattice points
n_s	• of electrons condensed in superconductor
n_h	• of holes in semiconductor
q	Electric charge of pair in superconductor ($q=-2e$)
t_1	Hopping integral • between nearest neighbours
$t_{n,\ell}$	• between sites n and ℓ
$u_{n,\mathbf{k}}(\mathbf{r})$	Coefficient of $e^{i\mathbf{k}\cdot\mathbf{r}}$ in Bloch function $\psi_{n,\mathbf{k}}(\mathbf{r})$
v	Volume of primitive cell in crystal
v_F	Electron velocity at Fermi level
z	Number of nearest neighbours of atom
Δk	Width • of wave packet in quasi-momentum space
Δr	• of wave packet in real space
Δ	Superconducting band gap
Ω	Volume of solid
Φ_0	Flux quantum ($=h/2e$)
Φ^*	Fluxoid (quantised quantity in superconductor)
α	Optical absorption coefficient
α	Exponent for isotopic effect
$\alpha = a/\lambda$	where a is thickness of superconducting film
$\alpha_{\ell-n}$	Overlap integral between atomic sites ℓ and n
$\chi'(\omega)$	Real part of alternating magnetic susceptibility (dispersion)
$\chi''(\omega)$	Imaginary part of alternating magnetic susceptibility (absorption)
χ_d	Diamagnetic susceptibility
$\chi_n(\mathbf{r})$	Atomic wave function • level n
$\chi_v(\mathbf{r})$	• valence state
χ_P	Pauli susceptibility of metal
δ	Binding energy of Cooper pair
ε	Dielectric constant of semiconductor

$\varepsilon_\mathbf{k}$	Energy of an electronic plane wave eigenstate ($\hbar^2 k^2 / 2m_0$)
$\varepsilon(\mathbf{k})$	Spin wave dispersion relation
$\|\phi(\mathbf{r})\|^2 = \phi^2$	Density of superconducting pairs ($= n_s/2$)
γ	Linear term in electronic specific heat of metal ($C_e = \gamma T$)
κ	Thermal conductivity
$\kappa = \lambda/\xi$	Ratio of penetration depth to coherence length
λ, $\lambda(T)$	Penetration depth
λ_0	• at zero temperature
λ_L	• London
λ	Spin–orbit coupling coefficient ($\lambda \mathbf{l} \cdot \mathbf{s}$)
λ^*	Molecular field coefficient
μ_e	Electron mobility
ω_c	Cyclotron angular frequency of charge in magnetic field
ω_D	Debye angular frequency
ω_L	Larmor angular frequency
$\psi_{n\mathbf{k}}(\mathbf{r})$	Bloch function with band index n and quasi-momentum $\mathbf{k}$
$\rho(\mathbf{r})$	Electronic density at $\mathbf{r}$
σ, σ_e, σ_h	Electron or hole conductivity
$\sigma(\omega)$	Alternating conductivity of a metal
$\sigma_s(\omega)$	Alternating conductivity of a superconductor
$\mathscr{T}$	Transmission coefficient of a tunneling barrier
τ	Relaxation time due to wall mobility in a ferromagnet
τ_e	Relaxation time of electron velocity distribution and electron collision rate
$\theta(\mathbf{r})$	Phase of wave function $\psi(\mathbf{r})$ (superconductor)
θ_D	Debye temperature characterising phonon spectrum in solid
ξ	Coherence length in the superconducting state
ζ	Thickness of Bloch wall
$\mathbf{B}_{\text{eff}}$	Effective magnetic induction at a site in a magnetic compound
$\mathbf{J}$	Total angular momentum for one ion
$\mathbf{H}_d$	Demagnetising field
$\mathbf{K}$	Vector in reciprocal lattice
$\mathbf{R}_l$	Vector in real lattice
$\|\mathbf{R}\rangle$	Excited state of Heisenberg model: single site at $\mathbf{R}$ has spin reversed with respect to ferromagnetic ground state
$\|\mathbf{R}_i\rangle$, $\chi_0(\mathbf{r} - \mathbf{R}_i)$	Atomic state and wave function for a nucleus at $\mathbf{R}_i$
$\mathbf{a}_1$, $\mathbf{a}_2$, $\mathbf{a}_3$	Vectors specifying unit cell • of real lattice
$\mathbf{a}_1^*$, $\mathbf{a}_2^*$, $\mathbf{a}_3^*$	• of reciprocal lattice
$\mathbf{b}_{\text{eff}}$	Effective field in rotating frame at magnetic resonance
$\mathscr{E}$	Electric field
$\mathbf{j}_c$	Critical current density of a superconductor
$\mathbf{j}_h$	Hole current density
$\hbar \mathbf{k}$	Quasi-momentum or crystal momentum
$\mathbf{k}_0$	Incident wave vector
$\mathbf{k}_1$	Scattered wave vector
$\mathbf{k}_h$	Quasi-momentum of hole

Appendix D Specific Notation

$\delta\mathbf{k}$	Displacement of Fermi surface in quasi-momentum space under effect of a force
$\lvert \mathbf{k}\rangle$	Eigenstate of Heisenberg model corresponding to a spin wave of wave vector $\mathbf{k}$
$\mathbf{m}$	Magnetisation in the rotating frame
$\langle v_{n,\mathbf{k}}\rangle$	Average velocity of electron in Bloch state $\psi_{n,\mathbf{k}}$
$\mathbf{v}_e$	Drift velocity of conduction electrons
$\mathbf{v}_h$	Hole velocity
$\mathbf{v}_g$	Group velocity of wave packet

References

Quantum Mechanics and Statistical Physics

1. Balian, R.: From Microphysics to Macrophysics, Vols. I and II. Springer-Verlag, Berlin, Heidelberg, New York, NY (1991)
2. Basdevant, J.L., Dalibard, J.: Quantum Mechanics. Springer-Verlag, Berlin, Heidelberg (2002)
3. Georges, A., Mézard, M.: Introduction á la théorie statistique des champs. Cours de l'Ecole Polytechnique, Palaiseau, France

Solid State Physics

4. Ashcroft, N.W., Mermin, N.D.: Solid State Physics. Saunders College Publishing, Philadelphia, PA (1976)
5. Dugdale, J.S.: The Electronic Properties of Metals and Alloys. Edward Arnold, London (1977)
6. Ibach, H., Luth, H.: Solid State Physics. Springer-Verlag, Berlin, Heidelberg, New York, NY (1995)
7. Kittel, C.: Introduction to Solid State Physics, 7th edn., Wiley, New York, NY (1996)
8. Olsen, J.L.: Electron Transport in Metals. Interscience, New York, NY (1962)
9. Voos, M., Drouhin, H.J., Drévillon, B.: Semi-conducteurs et composants. Cours de l'Ecole Polytechnique, Palaiseau, France
10. Ziman, J.M.: Electrons and Phonons. Oxford University Press, Oxford (1960)

Superconductivity and Magnetism

11. Abragam, A.: Principles of Nuclear Magnetism. Oxford University Press, Oxford (1994)
12. Becker, R., Döring, W.: Ferromagnetismus. Springer, Berlin (1939)
13. Chikamuzi, S.: Physics of Ferromagnetism. Clarendon Press, Oxford (1997)
14. Cullity, B.D.: Introduction to Magnetic Materials. Wesley, Reading, MA (1972)
15. Evetts, J.: Concise Encyclopedia of Magnetic and Superconducting Materials. Pergamon Press, Oxford (1992)
16. Lévy, L.P.: Magnetism and Superconductivity. Springer-Verlag, Berlin, Heidelberg, New York, NY (2000)
17. Orlando, T.P., Devlin, K.A.: Foundations of Applied Superconductivity. Addison Wesley, Reading, MA (1991)

18. Rose-Innes, A.C., Rhoderick, E.H.: Introduction to Superconductivity. International Series in Solid State Physics, Pergamon Press, Oxford (1978)
19. Tilley, D.R., Tilley, J.: Superfluidity and Superconductivity. Graduate Students Series in Physics, Institute of Physics Publishing, London (1990)
20. Tinkham, M.: Introduction to Superconductivity. McGraw-Hill Inc, New York, NY (1996)

Subatomic Physics

21. Rougé, A.: Introduction à la Physique Subatomique. Editions de l'Ecole Polytechnique, Palaiseau, France

Periodic Table of the Elements

Translated with the agreement of the author from the french edition of ref (16): "Magnétisme et Supraconductivité" by L. Lévy, Savoirs Actuels, InterEditions, CNRS Editions (1997).

1 H HEX — 1s 110 — Hydrogen																	
3 Li CC 1.76 [He]2s 400 55.1 Lithium	4 Be HEX 0.21 [He]2s² 1000 166 0.03 Beryllium																
11 Na CC 1.46 [Ar]3s 150 37.7 Sodium	12 Mg HEX 1.34 [Ar]3s² 318 82.3 Magnesium																
19 K CC 1.97 [Ar]4s 100 24.6 Potassium	20 Ca FCC 2.9 [Ar]4s² 230 — Calcium	21 Sc HEX 10.8 [Ar]3d4s² 359 — Scandium	22 Ti HEX 3.41 [Ar]3d²4s² 380 — 0.4 Titanium	23 V CC 9.04 [Ar]3d³4s² 390 — 5.35 Vanadium	24 Cr CC 1.46 [Ar]3d⁵4s 480 — Chromium	25 Mn CUB 16.6 [Ar]3d⁵4s² 400 127 Manganese	26 Fe CC 5.0 [Ar]3d⁶4s² 420 130 Iron	27 Co HEX — [Ar]3d⁷4s² 385 — Cobalt									
37 Rb CC 2.43 [Kr]5s 56 21.5 Rubidium	38 Sr FCC 3.64 [Kr]5s² 147 45.7 Strontium	39 Y HEX 10.1 [Kr]4d5s² 256 — Yttrium	40 Zr HEX 2.91 [Kr]4d²5s² 250 — 0.5 Zirconium	41 Nb CC 8.4 [Kr]4d⁴5s 275 61.6 9.25 Niobium	42 Mo CC 2.1 [Kr]4d⁵5s 380 — 0.92 Molybdenum	43 Tc HEX 4.06 [Kr]4d⁶5s 351 — 7.8 Technetium	44 Ru HEX 3.3 [Kr]4d⁷5s 382 — 0.5 Ruthenium	45 Rh FCC 4.6 [Kr]4d⁸5s 350 — Rhodium									
55 Cs CC 3.53 [Xe]6s 50 18.4 Cesium	56 Ba CC 2.72 [Xe]6s² 110 42.3 Barium	57 *La HEX 10.1 [Xe]5d6s² 132 4.9 Lanthanum	72 Hf HEX 2.4 [Xe]4f¹⁴5d²6s² — 225 0.13 Hafnium	73 Ta CC 5.84 [Xe]4f¹⁴5d³6s² — 225 4.4 Tantalum	74 W CC 1.22 [Xe]4f¹⁴5d⁴6s² 310 — 0.015 Tungsten	75 Re HEX 2.4 [Xe]4f¹⁴5d⁵6s² 416 — 1.7 Rhenium	76 Os HEX 2.35 [Xe]4f¹⁴5d⁶6s² 400 — 0.65 Osmium	77 Ir FCC 3.15 [Xe]4f¹⁴5d⁹ 430 — 0.14 Iridium									
87 Fr CC — [Xe]7s — — Francium	88 Ra — — [Xe]7s² — — Radium	89 †Ac FCC — [Xe]7s²6d — — Actinium															

*Lanthanides

58 Ce FCC — [Xe]4f²6s² 139 — Cerium	59 Pr HEX — [Xe]4f³6s² 152 — Praseodymium	60 Nd HEX — [Xe]4f⁴6s² 157 — Neodymium	61 Pm — — [Xe]4f⁵6s² — — Promethium	62 Sm ROM — [Xe]4f⁶6s² 166 — Samarium	63 Eu CC — [Xe]4f⁷6s² 107 — Europium

†Actinides

90 Th FCC 4.69 [Rn]6d²7s² 100 — 1.37 Thorium	91 Pa TET — [Rn]5f²6d7s² — — 1.3 Protactinium	92 U ORT 10.9 [Rn]5f³6d7s² 210 — 1.1 Uranium	93 Np ORT — [Rn]5f⁴6d7s² 188 — 0.08 Neptunium	94 Pu MCL — [Rn]5f⁶7s² 150 — Plutonium	95 Am — — [Rn]5f⁷7s² — — Americium

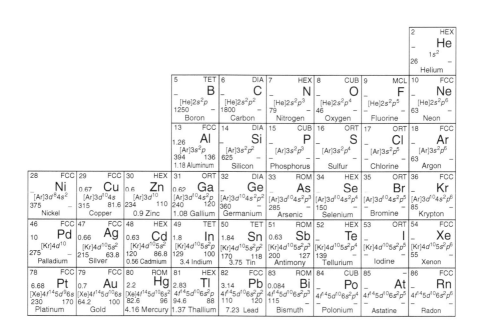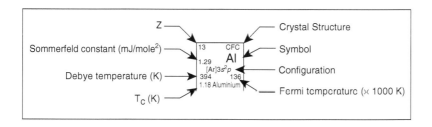

Index

A

Absorption, 209, 321, 328
Acceptor, 126
Alkaline earth metals, 204
Allowed energy band, 15
Alloy, 33
Alloys, electronic energy and stability of, 391
Alternating magnetisation, 252
Aluminium, Reflectance of, 373
Angle-resolved photoemission spectroscopy, 89
Antiferromagnetic Solid, 563
    Antiferromagnetic Transition, 564
    Preliminaries Case, 563
    Susceptibility, 565
Antiferromagnetism, 252, 341
Atomic force microscopy, 310
Atomic vibration, 118
Atom with partially filled shell, 237
Avogadro number, 593

B

Band gap, 62, 67, 76, 84, 215
Band structure, 15
    extended zone, 57
    restricted zone, 57
BCS theory, 215
Binding energy, 213
Bitter method, 288, 307, 308
Bloch function, 13, 14, 54, 111
Bloch's law, 350
Bloch theorem, 13, 53
Bloch wall, 283, 285, 322, 324
    energy of formation, 287
    thickness, 287
Body-centered cubic lattice, 29
Bohr magneton, 593
Bohr radius, 593
Boltzmann constant, 593
Bose condensation, 222
Boson, 222, 349
Bragg diffraction, 24, 37, 39
Bragg peak, 41, 43
Bragg plane, 39, 59, 66
Bravais lattice, 26, 28, 38
Brillouin function, 244
Brillouin zone, 66, 221

C

Causality, 321
Chemical shift, 340
Coercive field, 289
Coherence length, 185, 207, 222
Collision, 103
Collision time, 118
Compton wavelength, 593
Condensation energy, 210, 215
Conduction band, 77, 115
Constant energy surface, 69
Contact interaction, 340
Conventional cell, 28, 30, 31
Cooper pair, 161, 213
Critical current, 324
Critical exponents, 357
Critical field, 150, 178, 183, 186
    lower H_{c_1}, 191
    upper H_{c_2}, 192
Critical temperature, 147, 217, 259
Cryogenics, 204
Crystal lattice, 26, 28, 38
Crystallography, 205
Crystal momentum, 16, 56
Cubic lattice, 29, 32
Cuprate superconductor, 33, 65, 259

Curie paramagnetism, 235, 244
Curie temperature, 251, 353
Curie–Weiss law, 353
Cyclotron Resonance, 441
  Electron State, 441
  Metals, 444
  Silicon, 442

D
Debye frequency, 220
Debye temperature, 119
Debye–Waller Factor, 367
Defect, 116, 119
Demagnetisation energy, 282, 284
Demagnetising field, 279, 298, 303, 328
Density functional theory, 7
Density of states, 17, 71, 86, 235
Diamagnetism, 177, 234
Diffraction, 35
  Bragg, 37
Dipole interaction, 244, 277, 337
Disorder, 44
Dispersion, 321, 328
Dissipation, 320
Donor level, 125
Drude model, 102

E
Easy magnetisation axis, 272
Effective mass, 114
Elastic collision, 119
Electrical conductivity, 103, 108
Electron, 593
  nearly free, 56, 61, 66
Electron microscopy, 308
Electron mobility, 103
Electron–phonon interaction, 206, 220, 221
Energy bands, 20
Entropy, 179
EPR, 331
Ewald construction, 40
Exchange anisotropy, 273
Exchange constant, 246
Exchange energy, 285
Exchange Hamiltonian, 246
Exchange interaction, 245, 354
Excited state, 208, 216

F
Face-centered cubic lattice, 30, 34, 43, 59, 69
Faraday balance, 300
Faraday effect, 307
Fermi–Dirac statistics, 108
Fermi energy, 73, 108

Fermi sphere, 73, 108
Ferrimagnetism, 252, 255, 322
Ferrites, 304
Ferromagnetic resonance, 325, 329
Ferromagnetism, 236, 249, 340
  hard, 290
  soft, 290
Field effect (doping), 126, 130
Fine structure constant, 593
First Brillouin zone, 16, 57, 59, 65
Flux quantum, 160, 161, 192
Fluxoid, 159
Form factor, atomic, 43
Fourier transform, 336
Free energy, 178, 184
Free precession, 335
Fullerene, 33, 260

G
Gap, 220
Gauge, 159
Giant Magneto-Resistance, 264
Glass, 33
Graphene, 29, 78, 126, 130
Group velocity, 108, 111
Gyromagnetic ratio, 324

H
Hall effect, 126, 306
Hall voltage, 297
Hartree approximation, 7
Hartree–Fock approximation, 7
Heisenberg model, 247
Helium, 4
High Temperature Superconductors, 223, 259, 261, 306, 309
High Temperature superconductors, 32, 65, 89
Hole, 121
Hopping integral, 10
Hubbard model, 256
Hund rules, 240
Hydrogen atom, 236
Hyperfine interaction, 340
Hysteresimeter, 304
Hysteresis cycle, 276, 289, 304

I
Impurity, 116, 119
Insulator, 75
Insulator–Metal Transition, 419
  Alkali Elements and Hydrogen, 423
  Hydrogen-Like Orbitals, 419
  Insulator–Metal Transition in Si–P, 423
  Interactions Between Electrons, 420

Interface, 188
Interface energy, 188
Intermediate state, 303
Ising model, 359
Isotope effect, 206, 220

J
Josephson effect, 165, 217
Josephson Effects in Zero Magnetic Field, 481
    Josephson Junction in a Microwave Field, 483
    Model Josephson Junction, 481
    Realistic Josephson Junction, 481
Josephson Junction in a Magnetic Field, 491
    Current Distribution, 492
    Josephson Plasma Resonance, 495
    Screening of, 493
Joule effect, 118

K
Kerr effect, 307
Kramers–Kronig relations, 321, 334

L
La_2CuO_4, 557
Landau function, 355
Landau theory, 354
Landé factor, 240, 324, 333
Lanthanide, 204
Larmor frequency, 297, 324, 339
Larmor resonance, 150
Latent heat, 179
Lattice planes, 27
Levitation, 155
Linear atomic chain, 11
Linear response, 320
Liquid vortex phase, 306
Little–Parks Experiment, 469
London equations, 152, 156
Lorentz microscopy, 309
Lorenz number, 107, 111

M
Magnesium Diboride, 541
    Atomic and Electronic Structure, 541
    Superconductivity, 543
Magnetic anisotropy, 272, 285, 330
Magnetic domain, 283
    rotation of, 289
Magnetic force microscopy, 311
Magnetic hysteresis, 273, 275, 289
Magnetic pole, 278
Magnetic potential, 278
Magnetic susceptibility, 300
    complex, 320

Magnetisation, 299
    remanent, 289
    rotation of, 274
    of superconductors, 301
Magnetocrystalline anisotropy, 272
Magneto-optical effects, 308
Magnetoresistance, 297, 306
Magnon, 348, 351
Matthiessen's law, 118
Maxwell equations, 156
Mean field approximation, 251, 352, 359
Mean free path, 106
Meissner effect, 154, 177
Metal, 75, 341
Microstructure, 323
Miller indices, 38
Mixed state, 185, 188, 306
Molecular field, 249
Monovalent metal, 221
Monovalent Metals, Optical Response of, 399
Mössbauer effect, 341
Mott–Hubbard insulator, 256
Multielectron atoms, 237
Muon, 339

N
$NbSe_2$, 531
Near-field microscopy, 310
Néel state, 252
Néel temperature, 252
Neutron, 205, 593
NMR, 331, 339
    high resolution, 334
Noble metals, 204
Nuclear magneton, 593
Nuclear spin, 333

O
Ohm's law, 103, 156
One-dimensional system, 5
Optical absorption, 84
Order parameter, 356
Orientational disorder, 34
Overlap integral, 19

P
Paramagnetic insulator, 257
Pauli exclusion principle, 107, 213, 238, 245
Pauli paramagnetism, 178
Pauli spin paramagnetism, 235
Pauli susceptibility, 341
Penetration depth, 156, 182, 185, 192, 339
Periodic boundary conditions, 12
Perovskite structure, 32

Persistent current, 149, 156
Phase transition, 352, 354
Phonon, 118
Phonons in Solids, 453
   Debye Model, 453
   Detection of, 455
   Einstein Model, 453
   Resistivity, 457
   Thermodynamic Properties, 456
Photoemission, 88
Planck constant, 593
Plastic crystal, 34
Potential, atomic, 6
Pressure, 204, 205
Primitive cell, 26
Proton, 593
Pseudogap phase, 259
Pulse response, 321

Q
Quasi-momentum, 16

R
Rayleigh regime, 289
Reciprocal lattice, 35, 38, 43
Reflectivity, 209
Relaxation time, 103, 336
Renormalisation group, 359
Rotating frame, 326
Rydberg constant, 593

S
Scanning tunneling microscopy, 85, 87
Semi-classical approximation, 108
Semiconductor, 31, 76, 77, 204
   doped, 123
Semi-metals, 204
Singlet state, 238
Slater determinant, 8
Solid solution, 33
Specific heat, 81, 179
Spin echo, 337
Spin lattice relaxation time, 325
Spin–orbit interaction, 240
Spin singlet, 213
Spin wave, 348, 351
Spontaneous magnetisation, 235
SQUID, 166, 217, 299
Structural defect, 302
Structure factor, 42
Superconductivity, 147, 177
Superexchange interaction, 252

Superfluidity, 222
Susceptibility, Pauli, 82
Symmetry axes, n-fold, 28
Symmetry breaking, 248

T
Thermal conductivity, 105
Thermodynamic potential, 178
Thin Film, 579
   Non-uniform Situations, 581, 583
   Uniform Magnetisation, 580
Thin film, 181
Thin Films and Magneto-Optic Applications, 573
Tight-binding approximation, 63, 64, 68
Transfer integral, 10
Transition metals, 204, 221
Transverse relaxation, 338
Triplet state, 239
TTF-TCNQ Compounds, 405
   Dimerised Chain, 407, 413
   Isolated Chains, 405
   Observations, 406
   Peierls Transition, 409, 416
Tunnel effect, 85, 217
Two-fluid model, 157
Type I superconductor, 188
Type II Superconductor, 511
Type II superconductor, 188

V
V_3Si, 523
Vacancy, 120
Valence band, 77
Van der Waals interaction, 222
Vortex, 192
Vortex pinning, 198

W
Wave packet, 108
Wigner–Seitz cell, 27
Wilson ratio, 83

X
X ray, 35, 205

Y
$YBa_2Cu_3O_7$, Band Structure, 377
   Chain and Plane, 380
   Isolated Copper–Oxygen Chain, 378
   Isolated Copper–Oxygen Plane, 379
   Realistic Models, 381